Advanced Principles of Zoology

Advanced Principles of Zoology

Edited by Celeste Stewart

SYRAWOOD
PUBLISHING HOUSE

New York

Published by Syrawood Publishing House,
750 Third Avenue, 9th Floor,
New York, NY 10017, USA
www.syrawoodpublishinghouse.com

Advanced Principles of Zoology
Edited by Celeste Stewart

International Standard Book Number: 978-1-68286-725-9 (Hardback)

Cataloging-in-Publication Data

Advanced principles of zoology / edited by Celeste Stewart.
    p. cm.
Includes bibliographical references and index.
ISBN 978-1-68286-725-9
1. Zoology. 2. Biology. I. Stewart, Celeste.
QL47.2 .A38 2019
590--dc23

TABLE OF CONTENTS

PREFACE

Every book is initially just a concept; it takes months of research and hard work to give it the final shape in which the readers receive it. In its early stages, this book also went through rigorous reviewing. The notable contributions made by experts from across the globe were first molded into patterned chapters and then arranged in a sensibly sequential manner to bring out the best results.

Animals are multicellular eukaryotic organisms that fall under the kingdom of Animalia. The study of these diverse organisms is the focus of the discipline of zoology. It involves a detailed study of the biological characteristics of animals with respect to their morphology, embryology, reproduction, development, ecology and diversity. Animals are categorized into the domains of kingdom, phylum, class, order, family, genus and species according to their biological specifications. Depending on food habits, animals are also classified as herbivores, omnivores, carnivores, parasites and detritivores. The topics covered in this book will offer the readers new insights into the field of zoology. The aim of this book is to present researches that have transformed this discipline and aided its advancement. Students as well as researchers will benefit from an in-depth study of this book.

It has been my immense pleasure to be a part of this project and to contribute my years of learning in such a meaningful form. I would like to take this opportunity to thank all the people who have been associated with the completion of this book at any step.

Editor

Perception of emotional valence in horse whinnies

Elodie F. Briefer[1*], Roi Mandel[1,2†], Anne-Laure Maigrot[1,3†], Sabrina Briefer Freymond[4], Iris Bachmann[4] and Edna Hillmann[1]

Abstract

Background: Non-human animals often produce different types of vocalisations in negative and positive contexts (i.e. different valence), similar to humans, in which crying is associated with negative emotions and laughter is associated with positive ones. However, some types of vocalisations (e.g. contact calls, human speech) can be produced in both negative and positive contexts, and changes in valence are only accompanied by slight structural differences. Although such acoustically graded signals associated with opposite valence have been highlighted in some species, it is not known if conspecifics discriminate them, and if contagion of emotional valence occurs as a result. We tested whether domestic horses perceive, and are affected by, the emotional valence of whinnies produced by both familiar and unfamiliar conspecifics. We measured physiological and behavioural reactions to whinnies recorded during emotionally negative (social separation) and positive (social reunion) situations.

Results: We show that horses perceive acoustic cues to both valence and familiarity present in whinnies. They reacted differently (respiration rate, head movements, height of the head and latency to respond) to separation and reunion whinnies when produced by familiar, but not unfamiliar individuals. They were also more emotionally aroused (shorter inter-pulse intervals and higher locomotion) when hearing unfamiliar compared to familiar whinnies. In addition, the acoustic parameters of separation and reunion whinnies affected the physiology and behaviour of conspecifics in a continuous way. However, we did not find clear evidence for contagion of emotional valence.

Conclusions: Horses are thus able to perceive changes linked to emotional valence within a given vocalisation type, similar to perception of affective prosody in humans. Whinnies produced in either separation or reunion situations seem to constitute acoustically graded variants with distinct functions, enabling horses to increase their apparent vocal repertoire size.

Keywords: Emotional contagion, *Equus caballus*, Emotion expression, Familiarity, Playbacks, Vocalisations

Background

Emotions are intense, short-lived affective reactions to specific events or stimuli. They can be characterised using two main dimensions (dimensional approach): valence (negative/unpleasant or positive/pleasant) and arousal (bodily activation or excitation; e.g. calm versus excited) [1]. Emotional arousal can be considered as the intensity of bipolar valence, which comprises the defensive (negative valence) and the appetitive (positive valence) motivational systems described in humans and other species [2].

Emotions can be transmitted through olfactory signals (pheromones present in conspecifics' urine [3, 4]), visual signals (facial expressions [5]), acoustic signals [6, 7] or a combination of these [8, 9]. Perception of emotion expression can potentially induce the same emotion in the receiver as in the producer of the signal. This phenomenon is termed "state matching" or "emotional contagion", and is the basis of empathy [10, 11]. For example, a signal indicating a high-arousal state could increase the emotional arousal of receivers (i.e. contagion of emotional arousal). If this signal is positive, it could also trigger a change in emotional valence from negative or neutral to positive (and vice-versa for negative

* Correspondence: elodie.briefer@usys.ethz.ch
†Equal contributors
[1]Institute of Agricultural Sciences, ETH Zürich, Universitätstrasse 2, 8092 Zürich, Switzerland
Full list of author information is available at the end of the article

signals) in receivers (i.e. contagion of emotional valence). Unlike higher, cognitive forms of empathy (e.g. sympathetic concern), the transmission of emotions from one individual to another is widespread in the animal kingdom [12]. It is enhanced by social closeness, familiarity and similarity between partners [10, 13], improves information transfer through state sharing between individuals, and results in higher coordination among group members and stronger inter-individual bonds [10, 14]. Because vocalisations are a very effective communication system (e.g. they can be transmitted over long distances, around obstacles, and can be perceived in low visibility conditions [15]), they constitute a rapid means of transmitting information to conspecifics and are, as a result, a prominent channel for emotional contagion [16].

Variation in the structure of vocalisations associated with emotional valence and arousal (i.e. vocal expression of emotions) have been observed across species [17]. While both changes in call types (i.e. discrete calls, e.g. pig, *Sus scrofa*, grunts to squeals [18]) and modification in the acoustic structure of a given call type (i.e. graded calls, e.g. meerkat, *Suricata suricatta*, alarm calls [19]) have been observed with variation in the emotional arousal experienced by the producer, contexts of opposite valence are usually associated with different call types ([17, 20] e.g. change from horse, *Equus caballus*, whinnies to squeals, from dog, *Canis lupus familiaris*, bark to growl, or from human laughter to crying). However, acoustic variation within call types that are produced in both negative and positive situations (e.g. contact calls) can also occur (e.g. African elephant, *Loxodonta Africana*, rumbles [21]; bonobos, *Pan paniscus*, peeps [22]; goat, *Capra hircus*, bleats [23]; horse whinnies [24]).

Perception of the variation existing within specific call types as a function of the emotional arousal of the producer has been mainly studied in non-human animals in alarm contexts. These studies revealed that conspecifics respond more to alarm calls that have been artificially modified to mimic higher urgency levels (i.e. the parameters indicating urgency have been increased [25–28]). The ability to perceive indicators of arousal in other types of calls has also been shown (e.g. [29–31]). Additionally, clear evidence for vocal contagion of emotional arousal (i.e. matching between the emotion of the producer and the receiver) exists in zebra finches (*Taeniopygia guttata*); females show raised corticosterone levels when hearing distance calls emitted by their pair mate given orally administered exogenous corticosterone, compared to when hearing regular distance calls [32]. However, to our knowledge, it is not known if receivers are able to perceive variation occurring within a given type of vocalisation as a function of the emotional valence of the producer, and if emotional contagion occurs as a result. This ability could allow species with limited vocal repertoires to communicate different emotions using the same vocalisation type. Such acoustically graded variants could, as a result, be associated with different functions (e.g. trigger retreat or approach), in the same way as different call types, and might be as important as call-type differentiation for modulating social interactions [7, 33, 34].

We investigated if domestic horses can perceive indicators of emotional valence in whinnies of familiar and non-familiar conspecifics, independently of the context of reception (i.e. using only the acoustic features of whinnies), and if contagion of emotional valence occurs. As a highly social species [35], horses should benefit from acoustic perception of emotions, in order to regulate social interactions within harems (stallion, females and foals) or bachelor bands (young or old stallions without a harem) [35]. Eight call types have been described in this species: whinnies, nickers, squeals, blows, snores, snorts, roars, and groans [36, 37]. Whinnies provide information about sex, body size and individuality [38], reproductive success [39] and emotions (valence and arousal [24]), while squeals provide information about dominance status [40]. Conspecific receivers can decipher familiarity [38, 41] and stallion fertility [39] encoded in whinnies, as well as dominance status encoded in squeals [40]. Furthermore, horses are capable of cross-modal individual recognition of conspecifics, matching whinnies to visual/olfactory characteristics of the caller [42].

Whinnies are the most common call type produced by horses and can be emitted in both negative and positive contexts (e.g. separation and reunion with conspecifics, anticipation of both unpleasant and pleasant events, disturbances, frustration and curiosity [36]). Our previous study revealed that these calls are constituted by two fundamental frequencies ("F0" and "G0", suggesting biphonation), and that whinnies produced during social separation from either one or all group members (negative situations) are longer and have a higher G0 frequency than those produced during social reunion with one or all group members (positive situations) [24]. Separation and reunion whinnies thus constitute acoustically graded variants of the same call type. The negative and positive situations were also characterised by different behavioural responses in the producer; horses displayed less chewing motion (moving the lower jaw up and down without food [43]), and spent more time with the head high in the negative compared to the positive situation [24]. Here, we tested if information about emotional valence in whinnies can be deciphered by both familiar and unfamiliar conspecifics using playback experiments. We predicted that horses would show different physiological and behavioural responses to negative and positive whinnies, therefore validating emotion perception. If contagion of emotional valence occurs, we expected horses to display more behavioural indicators of negative emotions (head high) during playbacks of negative whinnies, and more behavioural indicators of positive emotions (chewing motion) during playbacks of positive whinnies (i.e. state matching between producer and receiver [12]). We also expected the acoustic features of whinnies to affect the responses of receivers in a graded way, with the time spent chewing decreasing and the time spent with the head high increasing with an increase in the duration and G0 of the calls played back, as predicted with a change from negative to positive emotions. As the acoustic channel is the main channel of communication in humans (speech), the study of vocal contagion of emotions in non-human animals is a promising way to understand the evolution of emotional contagion and empathy [10].

Results

We tested 18 horses of various breeds (Additional file 1) housed in five different farms with four playback treatments each: 1) separation (negative) whinnies from a familiar horse, 2) reunion (positive) whinnies from the same familiar horse, 3) separation (negative) whinnies from an unfamiliar horse, and 4) unfamiliar reunion (positive) whinnies from the same unfamiliar horse. Each playback consisted in three whinnies produced by the same horse. Subjects were tested with the four treatments over two consecutive days (two playbacks per day). The order of the treatments was counterbalanced within horses for valence and between horses for familiarity. Familiar whinnies were recorded from horses housed in the same farm as the subjects, while unfamiliar whinnies were recorded from horses housed in other farms. Separation whinnies were produced by horses during separation from either one or all the other horses from their farm ("group members"). Reunion whinnies were produced when these horses were reunited with one or all group members, following the separation situation. Horses are highly gregarious animals and separation from conspecifics is thus stressful for them (i.e. emotionally negative), while their motivation to reunite with conspecifics is high (i.e. emotionally positive) [35, 44]. Separation whinnies were thus assumed to be of negative valence, and reunion whinnies of positive valence [24]. In order to investigate if horses could perceive vocal indicators of valence independently of the context of reception (i.e. if valence cues are stimulus-independent), horses were tested in their home environment ("neutral" context). We measured both their physiological and behavioural responses to each whinny played back. We analysed three physiological and five behavioural parameters that were previously shown to be affected by emotional valence and/or arousal [24], in addition to the latency of the subjects to respond to the

playbacks (Table 1). We then included all these parameters in a principal component analysis to eliminate redundancy. We tested the effect of the valence and familiarity of the whinnies played back, and of the interaction between these two factors, on the scores of the resulting principal components (PC) with eigenvalue greater than 1 using linear-mixed effects models (LMMs). As responses to the playbacks are likely to be affected by the sex of the producer in respect to the sex of the subject, we also included a factor indicating whether the whinnies played back were produced by a horse of the same sex as the subject or not. Interactions between this factor and valence and familiarity were also fitted in the models.

Familiarity influenced PC1 scores (PC1: 27.24% of the variance, Table 2; LMM: $N = 18$ horses, $P = 0.022$; $R^2_{GLMM(m)} = 2.94\%$, $R^2_{GLMM(c)} = 59.68\%$); horses had shorter inter-pulse-intervals (RR; i.e. faster heart rates), moved more (Locomotion), moved their head more (HeadMov), had their head high for a longer duration (HeadHigh), vocalised more (VocRate) and responded faster (LatenceRes; i.e. any change in behaviour after the onset of the playback) when hearing unfamiliar whinnies (model estimates for PC1 score: mean [95% confidence interval] = -0.04 [-0.59, 0.55]) compared to familiar whinnies (-0.31 [-0.91, 0.35]). In addition, the interaction between valence and familiarity influenced PC2 scores (PC2: 20.56% of the variance, Table 2; LMM: $N = 18$ horses, $P = 0.044$; $R^2_{GLMM(m)} = 5.90\%$, $R^2_{GLMM(c)} = 39.90\%$). Post-hoc pairwise comparisons showed that PC2 scores differed between separation and reunion whinnies when these were familiar to the subject (Tukey post-hoc test: $Z = 2.87$, $P = 0.021$, $N = 18$ horses; $R^2_{GLMM(m)} = 11.86\%$, $R^2_{GLMM(c)} = 46.95\%$), but not when they were unfamiliar (Tukey post-hoc test: $Z = -0.74$, $P = 0.88$, $N = 18$ horses, $R^2_{GLMM(m)} = 0.70\%$, $R^2_{GLMM(c)} = 16.23\%$); horses had lower respiration rates (RespRate), moved their head more (HeadMov), had their head high for a longer duration (HeadHigh) and responded faster (LatenceRes) when hearing familiar reunion compared to familiar separation whinnies (Table 2; Fig. 1). The effects of valence, familiarity, sex of the horses or interactions between these factors on PC1 to PC3 not mentioned above were not significant (see Additional file 2 for statistical results of these factors and Additional file 3 for model estimates).

In order to test if separation and reunion whinnies also affected the responses of the horses in a continuous way, we tested the effect of the vocal parameters of the whinnies played back on the responses of the horses. To this aim, the calls played back were analysed by measuring vocal parameters previously shown to be affected by emotional valence and/or arousal ([24] Table 1). These parameters were included in a second principal component analysis to eliminate redundancy. The effect of the scores of the

Table 1 Abbreviations and descriptions of the physiological, behavioural and vocal parameters measured

	Abbreviation	Description	Arousal/Valence
Physiology	RR (ms)	Inter-heart-beat interval	**A**
	RespRate (breaths/s)	Respiration rate	**A**
	SkinT (°C)	Skin temperature	V + A
Behaviour	Locomotion	Proportion of time spent moving (walk, trot or canter)	**A**
	HeadMov (min-1)	Number of rapid head movements per minute	V + A
	HeadHigh	Proportion of time spent with the eye line above the tip of the shoulder	**V**
	Chewing	Proportion of time spent chewing (i.e. moving the lower jaw up and down in a chewing motion). This behaviour is performed without the presence of food in the mouth	**V** + A
	VocRate (min-1)	Number of vocalisations (whinnies or nickers) per minute	V (nickers)
	LatenceRes	Latency from the onset of the call played back to the first behavioural response (including all the above described behaviours)	-
Vocalisations	Dur (s)	Duration of the whinny	**V** + A
	G0Start (Hz)	Frequency value of G0 at the start of the whinny	**V** + A
	G0Max (Hz)	Maximum G0 frequency value across the whinny	V + A
	G0Mean (Hz)	Mean G0 frequency value across the whinny	**V** + A
	F0Start (Hz)	Frequency value of F0 at the start of the whinny	**A**
	F0Max (Hz)	Maximum F0 frequency value across the whinny	V + A
	F0Mean (Hz)	Mean F0 frequency value across the whinny	A
	AMVar (dB/s)	Cumulative variation in amplitude divided by the total whinny duration	V + A
	AMExtent (dB)	Mean peak-to-peak variation of each amplitude modulation	V + A
	Q25% (Hz)	Frequency value at the upper limit of the first quartiles of energy	V + A
	Q50% (Hz)	Frequency value at the upper limit of the second quartiles of energy	V + **A**
	Q75% (Hz)	Frequency value at the upper limit of the third quartiles of energy	V + **A**

Whether each parameter was significantly affected by emotional valence (V) or arousal (A) in our previous study [24] is indicated. Bold "V" indicates reliable cues to valence, i.e. parameters that were changing consistently with valence and were clearly more affected by valence than arousal. Bold "A" indicates reliable cues to arousal, i.e. parameters that were changing consistently with arousal and were clearly more affected by arousal than valence [24]

Table 2 Loadings of the physiological and behavioural parameters measured during the playbacks on the principal components with eigenvalue > 1 (PC1 to PC3 on a total of 9)

		Principal components		
	Parameters	PC1	PC2	PC3
Physiology	RR	**-0.74**	0.36	-0.08
	RespRate	0.37	**-0.58**	-0.15
	SkinT	0.38	-0.09	-0.39
Behaviour	Locomotion	**0.74**	-0.35	0.05
	HeadMov	**0.48**	**0.51**	0.20
	HeadHigh	**0.40**	**0.72**	0.02
	Chewing	0.03	0.15	**-0.89**
	VocRate	**0.60**	-0.23	0.11
	LatenceRes	**-0.58**	**-0.64**	0.05
Eigenvalue		**1.57**	**1.36**	**1.02**
% variance		**27.24**	**20.56**	**11.49**

Bold types indicate the heaviest factor loadings (|r| > 0.40). Eigenvalues and variances explained are given at the bottom of the table (see Table 1 for abbreviation of the parameters)

extracted principal components (PCv; eigenvalue > 1) on the PC scores corresponding to the physiological and behavioural responses of the subjects to the calls played back was then tested using LMMs. PC3v, which explained 15.27% of the variance in the vocal parameters of the calls played back (Table 3), influenced PC2 scores (LMM: $N = 18$ horses, $P = 0.022$; $R^2_{GLMM(m)} = 3.54\%$, $R^2_{GLMM(c)} = 39.33\%$); horses had slower respiration rates (RespRate), moved their head more (MovHead), had their head high for a longer duration (HeadHigh) and responded faster (LatencyRes; PC2, Table 2) when whinnies played back to them had lower fundamental frequencies (G0: G0Start, G0Max and G0Mean; and F0: F0Start and F0Max), were less modulated in amplitude (AMVar) and had a higher first quartile of energy (Q25%) (PC3v; Table 3; slope estimate ± SE: -0.16 ± 0.06). The effects of PC1v to PC4v on PC1 to PC3 not mentioned above were not significant (see Additional file 4 for statistical results of these factors, including slope estimates ± SE).

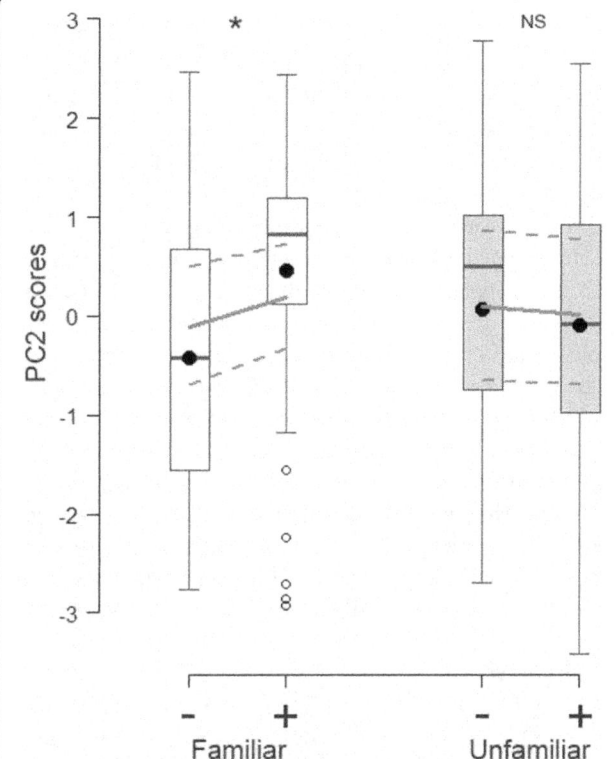

Fig. 1 Response of the horses to the playbacks. Scores of the second principal component (PC2) of the principal component analysis as a function of the four playback treatments (familiar (*white*)/unfamiliar (*grey*) * separation (-)/reunion (+); box plot: the horizontal line shows the median, the box extends from the lower to the upper quartile and the whiskers to 1.5 times the interquartile range above the upper quartile or below the lower quartile; open circles indicate outliers and black circles the mean; the grey lines show the model estimates (continuous line) and 95% confidence intervals (*dashed lines*)). More positive PC2 scores corresponded to horses that moved their head more, had their head high for a longer duration, responded faster and had a slower respiration rate (Table 2) (Tukey post-hoc test: * $p < 0.05$, NS = Non-Significant)

Table 3 Loadings of the vocal parameters extracted from the calls played back on the principal components with eigenvalue > 1 (PC1v to PC4v on a total of 12)

Parameters	Principal components			
	PC1v	PC2v	PC3v	PC4v
Dur	0.19	**0.64**	0.28	-0.14
G0Start	**-0.61**	0.20	**0.61**	-0.19
G0Max	**-0.81**	-0.11	**0.48**	0.02
G0Mean	**-0.81**	-0.13	**0.51**	0.01
F0Start	**0.56**	**0.62**	**0.41**	0.11
F0Max	**0.61**	**0.60**	**0.44**	0.04
F0Mean	0.30	**0.81**	-0.08	0.06
AMVar	**0.45**	-0.34	**0.44**	**0.54**
AMExtent	0.13	-0.36	0.15	**0.83**
Q25%	**-0.41**	**0.58**	**-0.50**	**0.41**
Q50%	**-0.73**	**0.52**	-0.20	0.31
Q75%	**-0.76**	**0.42**	-0.09	0.17
Eigenvalue	**1.99**	**1.70**	**1.35**	**1.17**
Cum % variance	**33.16**	**24.11**	**15.27**	**11.34**

Bold types indicate the heaviest factor loadings (|r| > 0.40). Eigenvalues and variances explained are given at the bottom of the table (see Table 1 for abbreviation of the parameters)

conspecifics. As the number of reliable indicators of emotions (revealed during our previous study) was limited (two indicators: chewing motion, time spent with head high [24]), we did not find evidence for negative emotions during playbacks of separation whinnies nor for positive emotions during playbacks of reunion whinnies. It is thus unclear whether contagion of emotional valence occurred. However, our study shows that horses are capable of perceiving variation in vocal parameters indicating emotional valence within whinnies. To our knowledge, this is the first demonstration of perception of changes linked to emotional valence within a given vocalisation type in a non-human species, and is similar to perception of affective prosody in humans (i.e. paralinguistic emotional information in speech, which differs from discrimination of laughter and crying). This ability might enable fine-tuned communication between horses within a given situation.

Valence perception
We previously found that whinnies produced in negative situations (i.e. separation from group member(s)) were longer (mean Dur = 2.23 s) and had a higher mean fundamental frequency (G0Mean = 1588.52 Hz) compared to those produced in positive situations (i.e. reunion with group member(s); mean Dur = 2.14 s, mean G0Mean = 1392.41 Hz [24]). Our playback experiment now confirmed that this variation in duration and frequency within whinnies produced in different contexts

Discussion
Using playback experiments, we tested if horses are able to discriminate whinnies produced during separation and reunion by both familiar and unfamiliar conspecifics, independently of the context (i.e. using only the acoustic features of the calls), as well as whether contagion of emotional valence occurs. Our results showed that horses reacted differently to separation and reunion whinnies when these calls were produced by familiar horses, but not when they were produced by unfamiliar individuals. In addition, some parameters of the whinnies played back, which had been previously shown to differ between separation and reunion situations ([24] F0, G0 and Q25%), affected the responses of the horses in a continuous way. This suggests that these two types of whinnies are graded into one another not only in their production [24], but also in the way they affect

can be perceived, at least in the whinnies of familiar horses. Previous studies of responses to different call types produced in negative and positive contexts showed that these call types can be discriminated and that the emotion they convey can be transmitted (e.g. rodents [34], marmosets [7]). Our results indicate that perception of emotional valence is also possible within a given vocalisations type, in the same way as what has been shown for emotional arousal (e.g. [32, 45]). Separation and reunion whinnies could constitute acoustically graded variants with distinct functions, thus increasing horse apparent vocal repertoire size and potential to transmit information. More generally, we suggest that within-call type variation could enable fine-tuned communication between individuals within a given situation, unlike between-call type variation, which is related to different contexts.

In addition to the difference in reaction to familiar separation and familiar reunion whinnies that we observed, we found that the acoustic parameters of whinnies affected horses' response in a continuous way, independently of the familiarity of the caller. Indeed, the same physiological and behavioural parameters (i.e. those loading on PC2) that differed in reaction to separation and reunion whinnies, were also affected by the parameters of the calls themselves (F0, G0 and Q25%). This suggests that the response of receivers to the negative and positive graded whinny variants is also graded. Similar results have been found by zebra finches; in addition to differences between the physiological and behavioural responses of females to the calls of their mates produced during corticosterone treatment and to regular contact calls, some of the acoustic parameters of the calls affected females' corticosterone concentrations in a continuous way [32]. Whether horses categorise separation and reunion whinnies as the same or different call types, and what is the minimum acoustic variation that they can perceive, could be tested using further playback experiments (e.g. habituation-dishabituation paradigm [46, 47]).

Interestingly, horses' behaviour and physiology significantly differed between playbacks of separation and reunion whinnies when these sounds were produced by familiar individuals (i.e. housed in the same farm), but not when produced by unfamiliar ones (i.e. housed in different farms), suggesting better perception of emotional valence in familiar compared to unfamiliar whinnies. Several phenomena could explain these results. First, horses could have perceived the emotional content of unfamiliar whinnies, but without reacting differently to separation and reunion whinnies, because the perceived difference might not have been meaningful to them [48]. Second, unfamiliar whinnies could be generally perceived as more negative than familiar whinnies,

independently of the valence that they convey, because of the potential aggressive interactions that accompanies encounters between two unfamiliar horses [49]. This hypothesis is supported by the fact that horses' reaction to unfamiliar whinnies suggests a negative state of high arousal; compared to playbacks of familiar whinnies, when hearing unfamiliar whinnies, horses moved more and had shorter inter-pulse-intervals (RR, i.e. higher heart rate), which indicates high arousal, and they had their head high for a longer duration, which indicates a negative emotion [24]. Third, emotional perception could be easier between individuals that are familiar with each other, notably as a result of past experiences [13]. In humans, although emotion recognition is cross-cultural, it is more accurate within cultures, due to cultural variations acquired through social learning [50, 51]. We could thus hypothesise that variation in the acoustic structure of whinnies between negative and positive situations can only be perceived by horses if they are familiar with the voice of the producer and have learned the range of changes that can occur in the producer's vocalisations.

An additional potential explanation for our results is that emotional perception could be stronger between familiar horses, because social affiliates are generally more empathic towards each other [13]. Enhanced emotion perception or emotional contagion between social affiliates seems widespread in the animal kingdom [10, 13]. For example, corticosterone resonance occurs between female zebra finches and their pair mate, while calls from unfamiliar males do not have such clear effect [32]. Micheletta et al. [52] found that crested macaques (*Macaca nigra*) attend more to playbacks of recruitment alarm calls if these are produced by close social affiliates. Rukstalis and French [53] revealed a decrease in stress (urinary cortisol levels) linked to isolation in marmosets when playing back contact calls of their pair mate, but not when playing back calls of an unfamiliar opposite sex individual. Similar enhanced reactions to the emotions experienced by familiar compared to unfamiliar individuals have also been highlighted in studies focussing on other sensory modalities than audition (e.g. [8, 54]; review [13]). At an ultimate level, enhanced emotion perception between social affiliates facilitates reciprocal altruism, which predicts a return of favour [55].

Emotional contagion

Emotional contagion occurs when the producer's emotion is transmitted to the receiver. The whinnies used in our study were recorded as part of a previous study aimed at finding indicators of emotions [24]. This study demonstrated that respiration rate and time spent moving were the best indicators of emotional arousal (indicated by heart rate), while time spent chewing (i.e.

moving the lower jaw in a chewing motion, without food) and time spent with the head high were the best indicators of emotional valence. If emotional valence matching had occurred during our playback experiment, we would have expected horses to have the head high for a longer duration (indicator of negative emotion) during playbacks of separation whinnies and to display more chewing motion (indicator of positive emotion) during playbacks of reunion whinnies. If emotional contagion was driven by the acoustic parameters of the whinnies, we would also have expected the time spent with the head high to increase, and the time spent chewing to decrease, with an increase in the duration and G0 of the calls played back, as predicted during a change from positive to negative emotions. Instead, chewing loaded highly on the third principal component of the PCA, which was neither affected by the valence nor by the familiarity of the calls, and the time spent with the head high was correlated positively with the scores of the first and second principal components (PC1 and PC2), which were higher during familiar reunion (i.e. positive) compared to familiar separation whinnies (i.e. negative). In addition, PC2 (indicating a higher proportion of time spent with the head high) was negatively affected by the third component of a PCA carried out on the parameters of the vocalisations. This effect indicated that whinnies that had lower fundamental frequencies (both F0 and G0), were less modulated in amplitude (AMVar) and had a higher first quartile of energy (Q25%) triggered a higher proportion of time spent with the head high. Although a low AMVar and a high Q25% suggested a negative emotion in our previous study (i.e. were significantly lower and higher, respectively, in the negative compared to the positive context), a low G0 and F0 indicated a positive emotion (particularly G0 [24]). Therefore, there is no clear evidence suggesting that contagion of emotional valence occurred during our playbacks.

One explanation for the higher proportion of time spent with the head high during familiar reunion compared to familiar separation whinnies is that horses could have been frustrated to hear reunion whinnies without seeing their group mate(s) arriving. However, frustration is a negative emotion that is likely to be of high arousal (e.g. [23]), and the responses of the horses to familiar reunion whinnies was also characterised by low respiration rates (negatively correlated with PC2), indicating low arousal [24]. Alternatively, the time spent with the head high could, in addition to indicate negative emotions in a normal situation, indicate a high level of attention to the playbacks in our experiment. This suggests that this behavioural parameter might not constitute a reliable indicator of valence in a playback situation. Further studies investigating emotional contagion

through vocalisations could include preference tests (e.g. [56]), in order to know if horses judge separation whinnies as negative, and reunion whinnies as positive.

One reason for the lack of evidence for emotional contagion in our study could be that the emotion elicited in the producers by the situations during our recordings (i.e. when valence indicators were established) was stronger than the emotion triggered in the receivers during the playbacks. This could be due to the fact that the receivers of the playbacks were in a different situation than the producers of the whinnies, resulting in incongruent or weaker emotional reactions. Alternatively, emotional contagion could have occurred, but be expressed through other parameters than the ones measured in our study (e.g. facial expressions [57], odours [3]).

Familiarity

Our results revealed increased emotional arousal (shorter inter-pulse intervals, i.e. higher heart rate; and more movements [24]) when hearing unfamiliar compared to familiar whinnies. Horses also had their head high for a longer period of time during playbacks of unfamiliar compared to familiar whinnies, which could indicate negative emotion (but see above). Encounters between unfamiliar horses can elicit aggressive interactions while the hierarchy is being established [49, 58]. The higher arousal elicited by unfamiliar compared to familiar whinnies might thus result from anticipation of such potential aggressive encounter. Similar differences in response to playbacks of unfamiliar and familiar whinnies have been observed previously in horses [38]. Lemasson et al. [38] showed that the angle of head rotation and level of postural alertness increased when the familiarity with the horse that produced the whinny decreased (group member < familiar non-group member < unfamiliar). The evidence thus suggests that horses can perceive the difference between familiar and unfamiliar whinnies, and even categorise conspecifics into several degrees of familiarity [38]. Individual vocal signatures present in whinnies might allow them to perform this sound categorisation [38]. This ability is widespread in the animal kingdom [30, 59], and can result from habituation. Within natural settings, it could enable horses to identify group members, with whom they establish long-term bonds, unlike members of other groups that are only met temporarily and that might represent a threat (e.g. competitor) [60].

Conclusions

Although we did not find clear evidence for contagion of emotional valence, our results show that horses have the ability to perceive information about emotional valence within familiar whinnies, similarly to perception of affective prosody in humans. In addition, we show that the acoustic parameters of separation and reunion whinnies affect the physiology and behaviour of conspecifics in a continuous way. These two graded whinny variants could constitute functionally distinct calls, increasing the horses' potential to transmit information and enable fine-tuned communication between individuals within a given situation.

Methods

Subjects and management conditions

Eighteen horses of various breeds, sex and age were tested in July and August 2013 (Additional file 1). All horses had been in their respective farms for at least 6 months (3–4 horses per farm). At night the horses were housed in single boxes ($N = 4$) or in boxes with paddocks, either individually ($N = 9$), or in groups of two to three horses ($N = 5$). During daytime, they were kept outdoors, either individually in adjacent fields allowing physical, visual and acoustic contact ($N = 5$), or in groups of two to four horses ($N = 13$). Horses from different farms had never encountered each other.

Playback treatments

The separation and reunion whinnies used to build the playback treatments had been recorded in May and June 2013 from the same horses, as part of an experiment on physiological, behavioural and vocal indicators of emotions [24]. The acoustic structure of these whinnies differed significantly. For details about these situations, how underlying emotions were validated using physiological and behavioural measures, and acoustic differences between separation and reunion whinnies, see [24].

Each playback comprised a sequence of three whinnies from the same horse, with 13.5 s of silence between each whinny (15 s on average for one call and the subsequent silence interval), in order to allow horses time to react to each whinny. Preparation of sequences involved selecting the three best quality whinnies (low level of background noise) from 13 horses that had vocalised the most in our previous study [24], scaled to a relative absolute peak amplitude of 0.99, and pasted successively using Praat 5.3.41 [61]. The number of horses used to prepare playbacks was maximised so that each horse was played to no more than four subjects, either as familiar or unfamiliar treatment (each horse was played to 2.92 ± 0.86 subjects; range = 2–4). Additionally, within a farm, the same familiar horse was played to no more than two subjects, and unfamiliar horses differed for each subject. In the few cases ($N = 5/26$ sequences) where it was not possible to obtain 3 different good quality whinnies to prepare a sequence, the same whinny was repeated three times. All sequences were then rescaled to the same maximum amplitude.

Playback procedure

The four treatments were played once to each subject individually over two consecutive days, with two treatments per day between 9 am and 5 pm. They were broadcast in an order that was counterbalanced within horses for valence (negative = "-", positive = "+") and between horses for familiarity (familiar = "F", unfamiliar = "U"; e.g. Horse 1, Day 1: F+, U-/Day 2: F-, U+; Horse 2, Day 1: U- F+/Day 2: U+ F-). Within a farm, for each of the 2 days, horses were tested one by one in the same order, with the first treatment followed by the second after about 1 h (interval between two treatments: 58 ± 12 min, range = 35 min to 1 h 30 min). To minimise behavioural reactions that would not be due to the playbacks (e.g. reaction to social separation), the subjects were tested in their home pen. The other horses from the farm (2–3 other horses for each farm) remained in their home pen also, but their view was totally occluded from the subject behind doors and fences before the playback started. This procedure did not seem to affect the normal behaviour of the horses.

The subject was equipped with the heart-rate monitor (see Physiological measures) and left for 5 min, undisturbed for habituation, which allowed the animals to return to normal activities. At the end of 5 min, the playback started. Sounds were broadcast with an Edifier S2000v loudspeaker (frequency response: 20Hz - 20 kHz), connected to a laptop where the sounds were stored in WAV format, at a sampling rate of 44.1 kHz and a bit rate of 705 kbps. Before the test started, the loudspeaker was placed behind a fence or door, at 5 m on average from the subject. To reduce habituation, the loudspeaker's location was randomly changed between conditions for each horse. Sounds were played at an intensity estimated to be normal for horses (85.19 ± 2.38 dB measured at 1 m using a sound level meter, C weighting (SoundTest-Master, Laserliner, UK)) [38, 42]. Playbacks stopped 30 s after the end of the last whinny, and the heart-rate monitor was removed from the subjects.

Physiological measures

Physiological measures were collected using a wireless non-invasive monitor (MLE120X BioHarness Telemetry System, Zephyr) [62, 63], fixed to a surcingle placed around the subject's heart girth. ECG gel was applied to the electrodes before each use. The data (continuous ECG trace, breathing wave, i.e. inhalation/exhalation cycle, and skin temperature) were transmitted and stored in real time to a laptop using LabChart software v.7.2 (ADInstrument) for later analyses. During tests, one experimenter entered comments in the software indicating when each of the three whinnies of the playback sequence was broadcast. This allowed us to measure the following physiological parameters precisely for up to

10 s (when possible, i.e. good quality signal, clearly visible heart beats on the ECG trace and respiration on the breathing wave) following the beginning of each call played back (selection duration = 8.05 ± 1.87 s): inter-heart-beat interval (RR), respiration rate and skin temperature (Table 1). Such short selections allowed us to identify short-term changes in physiology in reaction to the calls played back [64, 65]. For each selection, we ensured the software tracked the heart beats (ECG trace) correctly (as displayed by event markers on the screen) and the inspiration–exhalation cycles (breathing wave). Parts of the ECG trace when atrio-ventricular blocks could be observed (i.e. one heart beat missing every 3–4 beats) where excluded [66]. Then, the inter-heart-beat interval (RR), respiration rate (breaths/s, RespRate) and skin temperature (°C, SkinT) averages were then obtained automatically from the software. These three physiological parameters had been previously shown to be affected by emotional valence and/or arousal [24]. For 15 calls played back, the quality of the signal was not good enough to extract the physiological parameters. In addition, one group mate whinnied during one call of one playback, and we omitted the physiological response of the subject to this call. In total, we were thus able to obtain the physiological parameter values in response to 200 calls played back from a total of 216 (i.e. 18 subjects*4 playbacks*3 calls).

Behavioural measures

All tests were filmed using a Canon Legria FS2000 camcorder by an experimenter situated away from the loudspeaker. The behavioural parameters (Table 1) were scored from the videos of the tests using Interact software v. 9.0.7 (Mangold International GmbH, Arnstorf, Germany) for 15 s following the beginning of each played back call. They were scored either as occurrence (for discrete behaviours, indicated by "(min-1)" in Table 1) or as duration (for continuous behaviours). We then divided these values by the total scoring time for each call (15 s), hence obtaining frequency of occurrence for discrete behaviours (i.e. number of events per minute), and the proportion of time spent performing the behaviour for continuous behaviours. Analyses were carried out on these frequencies of occurrence or proportions. We considered for the analyses the five behaviours which were previously shown to be affected by emotional valence and/or arousal, and which had been observed during the playbacks (i.e. all except "Returns"; way and back movements along the fence or turns inside the stable [24]). In addition, we included in our analyses the latency to respond to the playbacks (see list, abbreviations and definitions of the parameters in Table 1). Because one group mate whinnied during one call of one playback, we omitted the behavioural response of the

subject to this call and thus obtained behavioural data in response to 215 calls from a total of 216 (i.e. 18 subjects*4 playbacks*3 calls).

Vocal parameters

In order to test the effect of the vocal parameters of the calls played back on horses' physiological and behavioural responses, we analysed all the whinnies used in the experiment ($N = 68$ different whinnies) following [24]. In the same way as for the physiological and behavioural parameters, we analysed the 12 vocal parameters that were significantly affected by emotional valence and/or arousal in our previous study ([24] see list, abbreviations and definitions in Table 1). These parameters were extracted using a custom built program in Praat, which batch processed the analyses and the exporting of output data [67]. In order to prevent biases linked to the settings used for the analyses, the same settings were used to analyse both negative and positive whinnies of each producer (for details about the setting used, see [24]). G0 could not be measured in three whinnies of one producer. All the other parameters could be extracted from the 68 whinnies.

Statistical analysis

We first tested the effect of the valence and familiarity of the calls played back on the physiological and behavioural responses of the horses (raw data are available in Additional file 5). We used a principal component analysis (PCA; prcomp function, library stats in R software 3.3.1.) to eliminate redundancy due to intercorrelation of the physiological and behavioural parameters [48]. To control for confounding factors that could have impacted on horses' responses, instead of including the original parameter values in the PCA, we included the residuals extracted from linear models (LMs, lm function in R) fitting the following control factor: 1) age of subjects (7–23 years old, Additional file 1); 2) because each subject was tested four times and could potentially hear and habituate to the calls played to the other horses, we also included the order of the playbacks for each farm (1–12 or 1–16 playbacks depending on the farm). The resulting residuals were independent of these factors and better approximated a normal distribution, after using a log transformation for LatenceResp and RespRate, and a logit transformation for Locomotion, HeadHigh, Chewing and VocRate (see Table 1 for abbreviations and description of the parameters). Because PCA does not handle missing data, responses to playback calls where the physiological response of the subjects ($N = 16/216$) or the behavioural response of the subject ($N = 1/216$) were missing (see above for reasons) were excluded (total included in the PCA = 200 data points from 18 horses).

The principal components with an eigenvalue greater than 1 (Kaiser's criterion) were extracted from the PCA (PC1 to PC3 of a total of 9). The effects of the valence and familiarity of the calls on PC1 to PC3 scores were then tested using LMMs (lmer function, lme4 library in R). These models (one for each PC as an outcome variable) included the valence of the calls (negative or positive), the familiarity of the calls (familiar or unfamiliar) and the interaction effect between familiarity and valence, as fixed factors. In addition, because the sex of the producer of the calls in respect to the sex of the subject might affect the responses, we added a fixed factor indicating whether the calls played back were produced by an individual of the same sex as the subject or not, as well as interaction terms between this factor and familiarity and valence. The inclusion of non-significant interaction terms in models makes the interpretation of main effects problematic [68]. On the other hand, model simplification, in which non-significant terms including interactions are dropped from the full model can lead to type 1 errors [69]. In order to be able to interpret main effects while leaving non-significant interactions in our models, we changed the contrasts of our factors (valence, familiarity and sex) from treatment contrasts (used by default by R) to sum contrasts [70]. In order to account for dependencies between data, our models included the following random effect; the playback number (each playback consisted of three calls), nested within the day of the playback (two playbacks per day), nested within the subject identity, nested within the farm where they were housed, crossed with the identity of horses whose whinnies were being played.

We then tested the effect of the vocal parameters of the calls played back on the physiological and behavioural responses of the horses (raw data available in Additional file 5). We first used a PCA in order to eliminate redundancy due to intercorrelation of the vocal parameters. To better approximate a normal distribution, we log-transformed beforehand all the vocal parameters except F0Mean and AMVar (see list in Table 1). Because the aim was to test the effect of the extracted PCs (hereafter "PCv") on the PCs corresponding to the physiological and behavioural responses of the horses, we excluded the acoustic data for which no response was available ($N = 16/216$ data points). In addition, because PCA does not handle missing data, two additional whinnies in which G0-parameters could not be measured were excluded (Total included in the PCA = 198 data points from 18 horses). The principal components with an eigenvalue greater than 1 (Kaiser's criterion) were extracted from the PCA (PC1 to PC4 of a total of 12; hereafter "PC1v to PC4v"). The effects of PC1v to PC4v on PC1 to PC3 scores corresponding to the physiological and behavioural responses of the horses to the playbacks

were then tested using LMMs (lmer function, lme4 library in R). These models (one for each PC as an outcome variable) included PC1v, PC2v, PC3v and PC4v as fixed factors, and the same random factors as listed above (playback number within day within subject within farm crossed with producer identity).

We checked the residuals of the models graphically for normal distribution and homoscedasticity [71]. P-values (PBmodcomp function, pbkrtest library in R), model estimates and confidence intervals (bootMer function, lme4 library in R), were calculated using parametric bootstrap methods (1000 bootstrap samples). To this aim, models were fitted with maximum likelihood. P-values calculated with parametric bootstrap tests give the fraction of simulated likelihood ratio test statistic values (LRT) that are larger or equal to the observed LRT value. This test is more adequate than the raw LRT test because it does not rely on large-sample asymptotic analysis and correctly takes the random-effects structure into account [72]. When an interaction effect was significant, we carried out Tukey post-hoc tests (glht function, multcomp library in R). The significance level was set at $\alpha = 0.05$. In addition, we calculated marginal ($R^2_{\text{GLMM(m)}}$) and conditional R^2 ($R^2_{\text{GLMM(c)}}$) of our models following [73]. $R^2_{\text{GLMM(m)}}$ corresponds to the proportion of variance explained by the fixed factors alone, while $R^2_{\text{GLMM(c)}}$ corresponds to the proportion of variance explained by both the fixed and random factors [73]. These two values were calculated for the full models, as well as for significant factors by including the significant factors and random effects only.

Additional files

Additional file 1: Characteristics of the horses used in the experiment. (DOCX 13 kb)

Additional file 2: Results of the models testing the effects of the valence and familiarity of the calls broadcast, as well as of the sex of the horses, on their responses to the playbacks. (DOCX 14 kb)

Additional file 3: Model estimations corresponding to the effects of valence, familiarity and sex on the responses of the horses to the playbacks. (DOCX 14 kb)

Additional file 4: Results of the models testing the effect of the vocal parameters of the calls broadcast on the horses' responses. (DOCX 14 kb)

Additional file 5: Raw data (physiological and behavioural responses of the horses to the playbacks, along with parameters of the vocalisations broadcast). (XLSX 52 kb)

Abbreviations

AMExtent (dB): Mean peak-to-peak variation of each amplitude modulation; AMVar (dB/s): Cumulative variation in amplitude divided by the total whinny duration; Chewing: Proportion of time spent chewing (i.e. moving the lower jaw up and down in a chewing motion). This behaviour is performed without the presence of food in the mouth; Dur (s): Duration of the whinny; F0Max (Hz): Maximum F0 frequency value across the whinny; F0Mean (Hz): Mean F0 frequency value across the whinny; F0Start (Hz): Frequency value of F0 at the start of the whinny; G0Max (Hz): Maximum G0 frequency value across the whinny; G0Mean (Hz): Mean G0 frequency value across the whinny; G0Start

(Hz): Frequency value of G0 at the start of the whinny; HeadHigh: Proportion of time spent with the eye line above the tip of the shoulder; HeadMov (min-1): Number of rapid head movements per minute; LatenceRes: Latency from the onset of the call played back to the first behavioural response (including all the above described behaviours); Locomotion: Proportion of time spent moving (walk, trot or canter); Q25% (Hz): Frequency value at the upper limit of the first quartiles of energy; Q50% (Hz): Frequency value at the upper limit of the second quartiles of energy; Q75% (Hz): Frequency value at the upper limit of the third quartiles of energy; RespRate (breaths/s): Respiration rate; RR (ms): Inter-heart-beat interval; SkinT (°C): Skin temperature; VocRate (min-1): Number of vocalisations (whinnies or nickers) per minute

Acknowledgements
We are very grateful to Amy E. Donnison, Christine J. Nicol and anonymous reviewers for helpful comments on this manuscript, and to Lorenz Gygax for statistical advice. We thank Solveig Pletscher, Anne-Sylvie and André Thévoz, Franziska and Beatrice Wohlfender, and Anja, Laurence and Alex Zollinger, for their help and access to the animals, and Sophie Masneuf for field assistance.

Funding
EFB and A-LM are funded by a Swiss National Science Foundation fellowship, and RM by a fellowship from the Universities Federation for Animal Welfare and by the Harry and Sylvia Hoffman Leadership and Responsibility Program at the Hebrew University, Israel.

Authors' contributions
All authors designed the study and contributed to the paper. EFB prepared the sounds for the playbacks, carried out the statistical analyses and drafted the manuscript. EFB, RM, A-LM and SBF collected the data. RM analysed the physiological responses and A-LM the behavioural responses. All authors gave final approval for publication.

Competing interests
The authors declare that they have no competing interests.

Author details
[1]Institute of Agricultural Sciences, ETH Zürich, Universitätstrasse 2, 8092 Zürich, Switzerland. [2]Koret School of Veterinary Medicine, Robert H. Smith Faculty of Agriculture, Food and Environment, the Hebrew University, Rehovot 76100, Israel. [3]Division of Animal Welfare, Veterinary Public Health Institute, Vetsuisse Faculty, University of Bern, Länggassstrasse 120, 3012 Bern, Switzerland. [4]Agroscope, Swiss National Stud Farm, Les Longs Prés, 1580 Avenches, Switzerland.

References
1. Russell J. A circumplex model of affect. J Pers Soc Psychol. 1980;39:1161–78.
2. Bradley M, Codispoti M, Cuthbert B, Lang P. Emotion and motivation I: defensive and appetitive reactions in picture processing. Emotion. 2001;1:276–98.
3. Boissy A, Terlouw C, Le Neindre P. Presence of cues from stressed conspecifics increases reactivity to aversive events in cattle: evidence for the existence of alarm substances in urine. Physiol Behav. 1998;63:489–95.
4. Amory JR, Pearce GP. Alarm pheromones in urine modify the behaviour of weaner pigs. Anim Welf. 2000;9:167–75.
5. Mineka S, Cook M. Mechanisms involved in the observational conditioning of fear. J Exp Psychol Gen. 1993;122:23–38.
6. Buchanan TW, Bagley SL, Stansfield RB, Preston SD. The empathic, physiological resonance of stress. Soc Neurosci. 2012;7:191–201.
7. Watson CFI, Caldwell CA. Neighbor effects in marmosets: social contagion of agonism and affiliation in captive Callithrix jacchus. Am J Primatol. 2010;72:549–58.
8. Langford DJ, Crager SE, Shehzad Z, Smith SB, Sotocinal SG, Levenstadt JS, et al. Social modulation of pain as evidence for empathy in mice. Science. 2006;312:1967–70.
9. Edgar JL, Lowe JC, Paul ES, Nicol CJ. Avian maternal response to chick distress. Proc R Soc B. 2011;278:3129–34.
10. de Waal FBM. Putting the altruism back into altruism: the evolution of empathy. Ann Rev Psychol. 2008;59:279–300.
11. Hatfield E, Cacioppo JT. Emotional Contagion. Cambridge: Cambridge University Press; 1994.
12. Edgar JL, Nicol CJ, Clark CCA, Paul ES. Measuring empathic responses in animals. Appl Anim Behav Sci. 2012;138:182–93.
13. Preston SD, de Waal FBM. Empathy: Its ultimate and proximate bases. Behav Brain Res. 2002;25:1–20.
14. Špinka M. Social dimension of emotions and its implication for animal welfare. Appl Anim Behav Sci. 2012;138:170–81.
15. Marler P. The structure of animal communication sounds. In: Bullock TH, editor. Recognition of complex acoustic signals. Berlin: Springer; 1977. p. 17–35.
16. Rendall D, Owren MJ. Vocalizations as tools for influencing the affect and behavior of others. In: Brudzynski SM, editor. Handbook of Mammalian Vocalization - An Integrative Neuroscience Approach. London: Academic; 2010. p. 177–85.
17. Briefer EF. Vocal expression of emotions in mammals: mechanisms of production and evidence. J Zool. 2012;288:1–20.
18. Marchant JN, Whittaker X, Broom DM. Vocalisations of the adult female domestic pig during a standard human approach test and their relationships with behavioural and heart rate measures. Appl Anim Behav Sci. 2001;72:23–39.
19. Manser MB. The acoustic structure of suricates' alarm calls varies with predator type and the level of response urgency. Proc Roy Soc Lond Ser B. 2001;268:2315–24.
20. Manser MB. The generation of functionally referential and motivational vocal signals in mammals. In: Brudzynski SM, editor. Handbook of Mammalian Vocalization - An Integrative Neuroscience Approach. London: Academic; 2010. p. 477–86.
21. Soltis J, Blowers TE, Savage A. Measuring positive and negative affect in the voiced sounds of African elephants (Loxodonta africana). J Acoust Soc Am. 2011;129:1059–66.
22. Clay Z, Archbold J, Zuberbühler K. Functional flexibility in wild bonobo vocal behaviour. PeerJ. 2015;3:e1124.
23. Briefer EF, Tettamanti F, McElligott AG. Emotions in goats: mapping physiological, behavioural and vocal profiles. Anim Behav. 2015;99:131–43.
24. Briefer EF, Maigrot A-L, Mandel R, Briefer Freymond S, Bachmann I, Hillmann E. Segregation of information about emotional arousal and valence in horse whinnies. Sci Rep. 2015;4:9989.
25. Fichtel C, Hammerschmidt K. Responses of redfronted lemurs to experimentally modified alarm calls: evidence for urgency-based changes in call structure. Ethology. 2002;108:763–78.
26. Fichtel C, Hammerschmidt K. Responses of squirrel monkeys to their experimentally modified mobbing calls. J Acoust Soc Am. 2003;113:2927–32.
27. Manser MB, Bell MB, Fletcher LB. The information that receivers extract from alarm calls in suricates. Proc R Soc B. 2001;268:2485–91.
28. Blumstein DT, Recapet C. The sound of arousal: The addition of novel non-linearities increases responsiveness in marmot alarm calls. Ethology. 2009; 115:1074–81.
29. Weary DM, Lawson GL, Thompson BK. Sows show stronger responses to isolation calls of piglets associated with greater levels of piglet need. Anim Behav. 1996;52:1247–53.
30. King LE, Soltis J, Douglas-Hamilton I, Savage A, Vollrath F. Bee threat elicits alarm call in african elephants. PLoS One. 2010;5:e10346.
31. Schehka S, Zimmermann E. Affect intensity in voice recognized by tree shrews (Tupaia belangeri). Emotion. 2012;12:632–9.
32. Perez EC, Elie JE, Boucaud ICA, Crouchet T, Soulage CO, Soula HA, et al. Physiological resonance between mates through calls as possible evidence of empathic processes in songbirds. Horm Behav. 2015;75:130–41.
33. Videan EN, Fritz J, Schwandt M, Howell S. Neighbor effect: evidence of affiliative and agonistic social contagion in captive chimpanzees (Pan troglodytes). Am J Primatol. 2005;66:131–44.

34. Seffer D, Schwarting RKW, Wöhr M. Pro-social ultrasonic communication in rats: Insights from playback studies. Meas Behav. 2014;234:73–81.

35. van Dierendonck MC. The importance of social relationships in horses. PhD thesis. Utrecht: Utrecht University; 2006.

36. Yeon SC. Acoustic communication in the domestic horse (*Equus caballus*). J Vet Behav. 2012;7:179–85.

37. Kiley M. The vocalizations of ungulates, their causation and function. Z Tierpsychol. 1972;31:171–222.

38. Lemasson A, Boutin A, Boivin S, Blois-Heulin C, Hausberger M. Horse (*Equus caballus*) whinnies: a source of social information. Anim Cogn. 2009;12:693–704.

39. Lemasson A, Remeuf K, Trabalon M, Cuir F, Hausberger M. Mares prefer the voices of highly fertile stallions. PLoS One. 2015;10:e0118468.

40. Rubenstein DI, Hack MA. Horse signals: The sounds and scents of fury. Evol Ecol. 1992;6:254–60.

41. Basile M, Boivin S, Boutin A, Blois-Heulin C, Hausberger M, Lemasson A. Socially dependent auditory laterality in domestic horses (*Equus caballus*). Anim Cogn. 2009;12:611–9.

42. Proops L, McComb K, Reby D. Cross-modal individual recognition in domestic horses (*Equus caballus*). Proc R Soc B. 2009;106:947–51.

43. Krueger K. Behaviour of horses in the "round pen technique". Appl Anim Behav Sci. 2007;104:162–70.

44. Lansade L, Bouissou M-F, Erhard HW. Reactivity to isolation and association with conspecifics: A temperament trait stable across time and situations. Appl Anim Behav Sci. 2008;109:355–73.

45. Kastein HB, Kumar VAK, Kandula S, Schmidt S. Auditory pre-experience modulates classification of affect intensity: evidence for the evaluation of call salience by a non-human mammal, the bat Megaderma lyra. Front Zool. 2013;10:75.

46. Fischer J. Barbary macaques categorize shrill barks into two call types. Anim Behav. 1998;55:799–807.

47. Fischer J, Metz M, Cheney DL, Seyfarth RM. Baboon responses to graded bark variants. Anim Behav. 2001;61:925–31.

48. McGregor P. Playback and studies of Animal Communication. New York: Plenum Press; 1992.

49. Hartmann E, Christensen JW, Keeling LJ. Social interactions of unfamiliar horses during paired encounters: Effect of pre-exposure on aggression level and so risk of injury. Appl Anim Behav Sci. 2009;121:214–21.

50. Elfenbein HA, Ambady N. On the universality and cultural specificity of emotion recognition: a meta-analysis. Psychol Bull. 2002;128:203–35.

51. Sauter DA, Eisner F, Ekman P, Scott SK. Cross-cultural recognition of basic emotions through nonverbal emotional vocalizations. Proc Natl Acad Sci U S A. 2010;107(6):2408–12.

52. Micheletta J, Waller BM, Panggur MR, Neumann C, Duboscq J, Agil M, et al. Social bonds affect anti-predator behaviour in a tolerant species of macaque, *Macaca nigra*. Proc R Soc B. 2012;279:4042–50.

53. Rukstalis M, French JA. Vocal buffering of the stress response: exposure to conspecific vocalizations moderates urinary cortisol excretion in isolated marmosets. Horm Behav. 2005;47:1–7.

54. Ben-Ami Bartal I, Rodgers DA, Bernardez Sarria MS, Decety J, Mason P. Pro-social behavior in rats is modulated by social experience. Elife. 2014;3:e01385.

55. Trivers RL. The evolution of reciprocal altruism. Q Rev Biol. 1971;46:35–57.

56. Westerath HS, Gygax L, Hillmann E. Are special feed and being brushed judged as positive by calves? Appl Anim Behav Sci. 2014;156:12–21.

57. Wathan J, Burrows AM, Waller BM, McComb K. EquiFACS: The equine facial action coding system. PLoS One. 2015;10:e0131738.

58. Briefer Freymond S, Briefer EF, Niederhäusern RV, Bachmann I. Pattern of social interactions after group integration: A possibility to keep stallions in group. PLoS One. 2013;8:e54688.

59. Bradbury JV, Vehrencamp SL. Principles of Animal Communication. Sunderland: Sinauer; 1998.

60. Waring GH. Horse behavior. Norwich: William Andrew Publishing; 2003.

61. Boersma P, Weenink D. Praat: doing phonetics by computer. 2009; Available from: http://www.praat.org/.

62. Johnstone JA, Ford PA, Hughes G, Watson T, Garrett AT. Bioharness(™) multivariable monitoring device: part. I: validity. J Sports Sci Med. 2012;11:400–8.

63. Johnstone JA, Ford PA, Hughes G, Watson T, Garrett AT. Bioharness(™) multivariable monitoring device: part. II: reliability. J Sports Sci Med. 2012;11:409–17.

64. von Borell E, Langbein J, Després G, Hansen S, Leterrier C, Marchant-Forde J, et al. Heart rate variability as a measure of autonomic regulation of cardiac activity for assessing stress and welfare in farm animals - A review. Physiol Behav. 2007;92:293–316.

65. Reefmann N, Wechsler B, Gygax L. Behavioural and physiological assessment of positive and negative emotion in sheep. Anim Behav. 2009;78:651–9.

66. Lilly L. Pathophysiology of Heart Disease. Philadelphia: Lippincott Williams and Wilkins; 2006.

67. Reby D, McComb K. Anatomical constraints generate honesty: acoustic cues to age and weight in the roars of red deer stags. Anim Behav. 2003;65:519–30.

68. Engqvist L. The mistreatment of covariate interaction terms in linear model analyses of behavioural and evolutionary ecology studies. Anim Behav. 2005;70:967–71.

69. Forstmeier W, Schielzeth H. Cryptic multiple hypotheses testing in linear models: overestimated effect sizes and the winner's curse. Behav Ecol Sociobiol. 2011;65:47–55.

70. Levy R. Using R formulae to test for main effects in the presence of higher-order interactions. 2014. Available from: arXiv preprint arXiv:1405.2094

71. Bates D. Fitting linear mixed models in R. R News. 2005;5:27–30.

72. Halekoh U, Højsgaard S. A Kenward-Roger approximation and parametric bootstrap methods for tests in linear mixed models: The R package pbkrtest. J Stat Softw. 2014;1:9.

73. Nakagawa S, Schielzeth H. A general and simple method for obtaining R2 from generalized linear mixed-effects models. Methods Ecol Evol. 2013;4:133–42.

2

The head morphology of *Pyrrhosoma nymphula* larvae (Odonata: Zygoptera) focusing on functional aspects of the mouthparts

Sebastian Büsse[1]* (iD), Thomas Hörnschemeyer[2] and Stanislav N. Gorb[1]

Abstract

Background: The understanding of concerted movements and its underlying biomechanics is often complex and elusive. Functional principles and hypothetical functions of these complex movements can provide a solid basis for biomechanical experiments and modelling. Here a description of the cephalic anatomy of *Pyrrhosoma nymphula* (Zygoptera, Coenagrionidae) focusing on functional aspects of the mouthparts using micro computed tomography (μCT) is presented.

Results: We compared six different instars of the damselfly *P. nymphula* as well as one instar of the dragonfly *Aeshna cyanea* and *Epiophlebia superstes* each. In total 42 head muscles were described with only minor differences of the attachment points between the examined species and the absence of antennal muscle M. scapopedicellaris medialis (0an7) in *Epiophlebia* as a probable apomorphy of this group. Furthermore, the ontogenetic differences between the six larval instars are minor; the only considerable finding is the change of M. submentopraementalis (0la8), which is dichotomous in the early instars (I1,I2 and I3) with a second point of origin at the postero-lateral base of the submentum. This dichotomy is not present in any of the older instars studied (I6, middle-late and pen-ultimate).

Conclusion: However, the main focus of the study herein, is to use these detailed morphological descriptions as basis for hypothetic functional models of the odonatan mouthparts. We present blueprint like description of the mouthparts and their musculature, highlighting the caused direction of motion for every single muscle. This data will help to elucidate the complex concerted movements of the mouthparts and will contribute to the understanding of its biomechanics not in Odonata only.

Keywords: Dragonfly (Anisoptera), Damselfly (Zygoptera), Functional morphology, Ontogenesis, Muscle equipment, Prehensile mask, Feeding apparatus, Micro computed tomography (μCT), Synchrotron radiation micro computed tomography (SRμCT)

Background

Insects evolved a staggering diversity of mouthparts and feeding modes [1, 2]. The feeding process usually requires a complex interaction of several specially shaped mouthpart elements – labrum, mandibles, maxillae and labium - moved in a concerted action by muscles through specialised joints supplemented by membranous regions. Such coordinated movements of the mouthparts were studied in exemplary insects using for example tomographic filming techniques [3], but remain poorly understood concerning muscle activation [4] and neuronal control [5].

Odonata larvae shows striking differences in their mouthpart organisation compared to adults and consequently differ in details of their feeding mode. Odonata larvae are aquatic predators, while the adults hunt their

* Correspondence: sbuesse@zoologie.uni-kiel.de
[1]Department of Functional Morphology and Biomechanics, Institute of Zoology, Christian-Albrechts-Universität zu Kiel, Am Botanischen Garten 9, 24118 Kiel, Germany
Full list of author information is available at the end of the article

prey on the fly [6]. The labium of Odonata larvae shows the most drastic differences compared to that of adults. The larval labium is modified into a prehensile labial mask that is used for capturing potentially fast moving organisms up to predator's own size. The mandibles and maxillae, however, show only minor differences compared to the adults. Some of these differences in the outer anatomy are reflected in the muscle configuration [7–9].

This transmutation of the labium into a prehensile mask is the most distinctive character of odonatan larvae and rather unique within the insects. Structural aspects of the larval mouthparts of Odonata and the functional mechanism of this catching apparatus have been investigated in various degrees of detail [7, 10–14]. Snodgrass [8], Pritchard [10], and Blanke et al. [9] provided detailed description of the head anatomy, focusing on the cuticular features and the muscle arrangement. Furthermore, Olesen [11, 12], Tanaka & Hisada [13] and Parry [14] focussed on the functionality of the labium, studying the mechanism that is responsible for extending the prehensile mask. Their results indicated that the main driving force contributing to the labium extension is an increase in haemolymph pressure generated by the respiratory system of Anisoptera using an internal rectal organ – the branchial chamber. Investigations on this topic in Zygoptera larvae are scarce and their overall morphology differs significantly from that of Anisoptera; in terms of respiration they use external gills – caudal lamellae – for respiration [15] instead of the internal organ mentioned for Anisoptera [6]. The internal anatomy of Zygoptera is controversial and differs from that of Anisoptera: i) extensively like described for example in Whedon [16] or ii) in some functional aspects as described in Miller [17]. However, more recent studies show that Zygoptera larvae are able to intake water into their hindgut and use this mechanism for respiration or supplementing respiration [15, 17–19].

This study was undertaken to better understand the functional morphology of the larval mouthpart system of Odonata. We describe the cephalic anatomy of *Pyrrhosoma nymphula* (Zygoptera, Coenagrionidae) with a focus on detailed 3D description of the mouthpart musculature and its potential function in the feeding process. The idea is to present hypothetical functions of different muscles in the complex movements of the odonatan larval feeding apparatus. This paper provides a solid basis for future biomechanical experiments and modelling. The results, obtained on the representative of Zygoptera, are compared with anisopteran mouthpart musculature and that of *Epiophlebia*, in order to allow more general conclusions for Odonata.

Results
Head capsule (Fig. 1a-d)
The strongly sclerotized and dorso-ventrally flattened head is almost three times as broad as long. It is prognathous and only sparsely covered with setae – some on the labrum and laterally at the posterior part of the occiput. The strongly convex and globular eyes protruding laterally are distinctly separated dorsally by almost three times their own width. The occipital ridge marks the anterior border of the occiput and is discernable but weakly developed. The postocciput is triangular and strongly developed. The frons is slightly rounded and declines towards the clypeus. The ocelli are barely discernable in SEM images, but they are present as the tomography data show. The clypeus is divided into a postclypeus and a small anteclypeus. The most important internal structure for muscle attachment is the tentorium. It consists of a rod-like corpotentorium, anterior, dorsal and posterior tentorial arms. The anterior tentorial arms are straight and connected via musculature (0te3) to the head capsule; tentorial pits are not discernable. The weakly developed posterior tentorial arms serve as attachment points of the maxillar muscles 0mx4 and 0mx4.

Musculature
M. tentoriofrontalis dorsalis (0te3) – O: apex of dorsal tentorial arms I: frons, posterior at antennal base.

Antennae (Fig. 1a,c)
The antennae reach out beyond the mouthparts and are composed of scapus, pedicellus and five flagellomeres. The pedicellus is twice as long as the scapus, and the flagellum is longer than scapus and pedicellus together. The tentorium is the attachment point of the antennal muscles. Antennal heart muscles are absent. Instead, the pharynx and a sac-like structure, situated in front of the brain, hardly discernable in the μCT-data, are connected with the antennal vessels, supporting haemolymph flow [20, 21].

Musculature (Figures cf. [8])
M. tentorioscapalis anterior (0an1) – O: mesal at the dorsal tentorial arm I: anterior at the base of the scapus. M. tentorioscapalis posterior (0an2) – O: mesal at the dorsal tentorial arm, dorsal to 0an1 I: posterior at the base of the scapus. M. frontopedicellarius (0an5) – O: dorsal tentorial arm, close to 0an2 I: ventral at the base of the pedicellus. M. scapopedicellaris lateralis (0an6) – O: antero-lateral at the base of the scapus I: antero-lateral at the base of the pedicellus. M. scapopedicellaris medialis (0an7) – O: meso-lateral

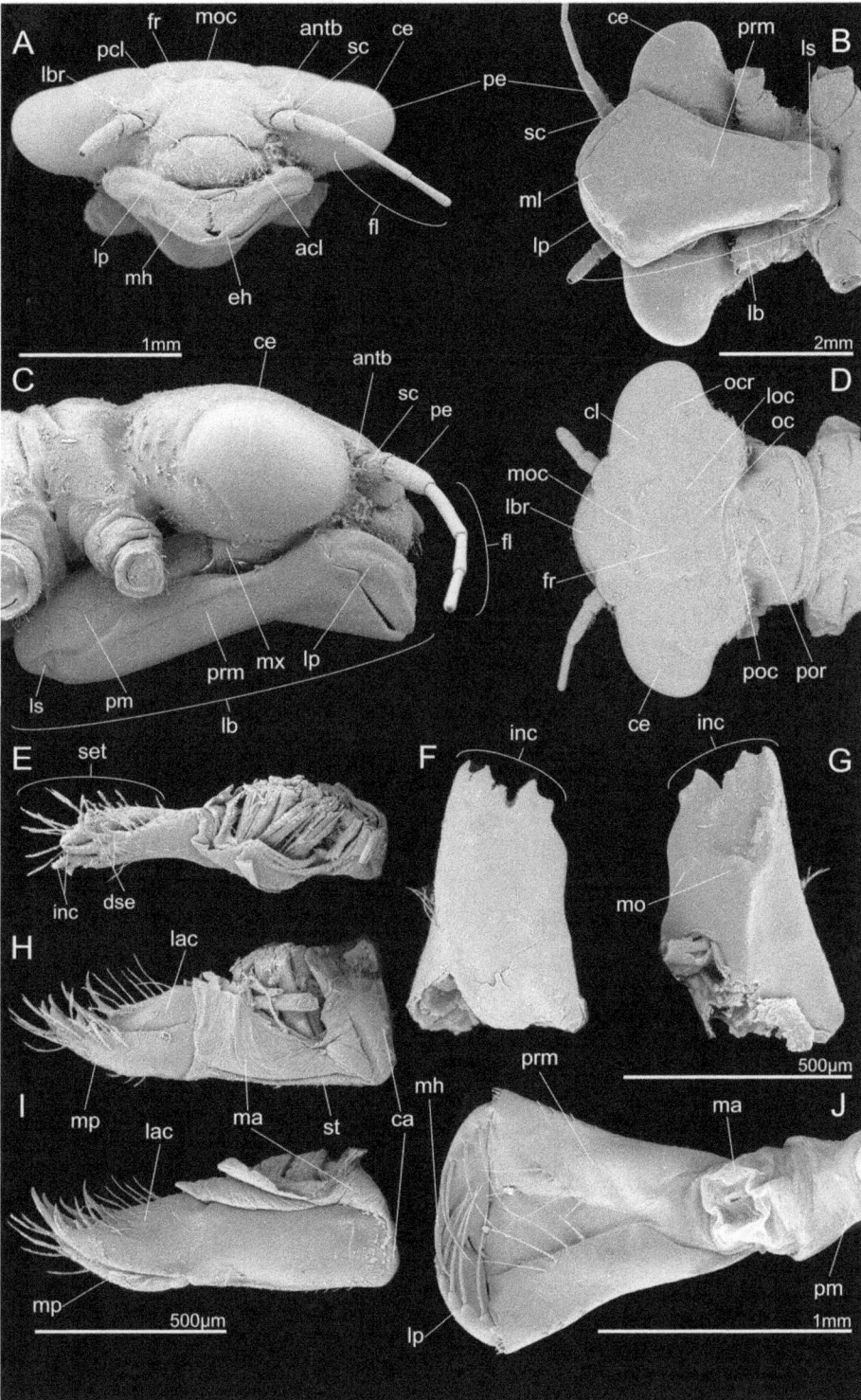

Fig. 1 SEM micrographs of *Pyrrhosoma nymphula* larva. **a-d** Head capsule **e-j**. Mouthparts **a**. Frontal view **b**. Ventral view **c**. Lateral view **d**. Dorsal view **e**. Maxilla, dorsal view **f**. Mandible, ventral view **g**. Mandible, dorso-lateral view **h**. Maxilla, median view **i**. Maxilla, lateral view **j**. Labium, dorsal view. Abbreviations: acl – anteclypeus, antb – antennal base, ce – compound eye, ca – cardo, cl – cleavage line, dse – dentisetae, eh – end hook, fl – flagellum, fr – frons, inc – incisivi, lac – lacinia, lb. – labium, lbr – labrum, lp – labial palp, ls – labial sutur, loc – lateral occellus, ma – membranous area, mh – movable hook, ml – median lobe, moc – median occellus, mp – maxillar palpus, oc – occiput, ocr – occipital ridge, pcl – postclypeus, pe – pedicellus, pm – postmentum, poc – post occiput, por – postoccipital region, prm – praementum, sc – scapus, set – setae, st – stipes

at the base of the scapus I: postero-lateral at the base of the pedicellus.

Labrum (Figs. 1a,d and 2)

The labrum is a roof-like structure at the anterior side of the head arching dorsally above the mandibles. In dorsal view it is semicircular and covered with setae. The tormae (small sclerites), which are almost Y-shaped, are the attachment points for *M. frontoepipharyngalis* (0 lb2).

Musculature

M. frontolabralis (0 lb1) – O: close to the interantennal ridge I: medial at the base of the labrum C: dichotomous over almost the entire length, unpaired at the very origin. *M. frontoepipharyngalis* (0 lb2) – O: antero-dorsal at the head capsule, lateral of 0 lb1 I: dorso-lateral at the tormae, postero-lateral of 0 lb1. *M. labroepipharyngalis* (0 lb5) – O: anterior at the inner labral wall, ventral of the insertion of 0 lb1 I: posterior at the inner epipharyngeal wall C: unpaired, median within the labrum.

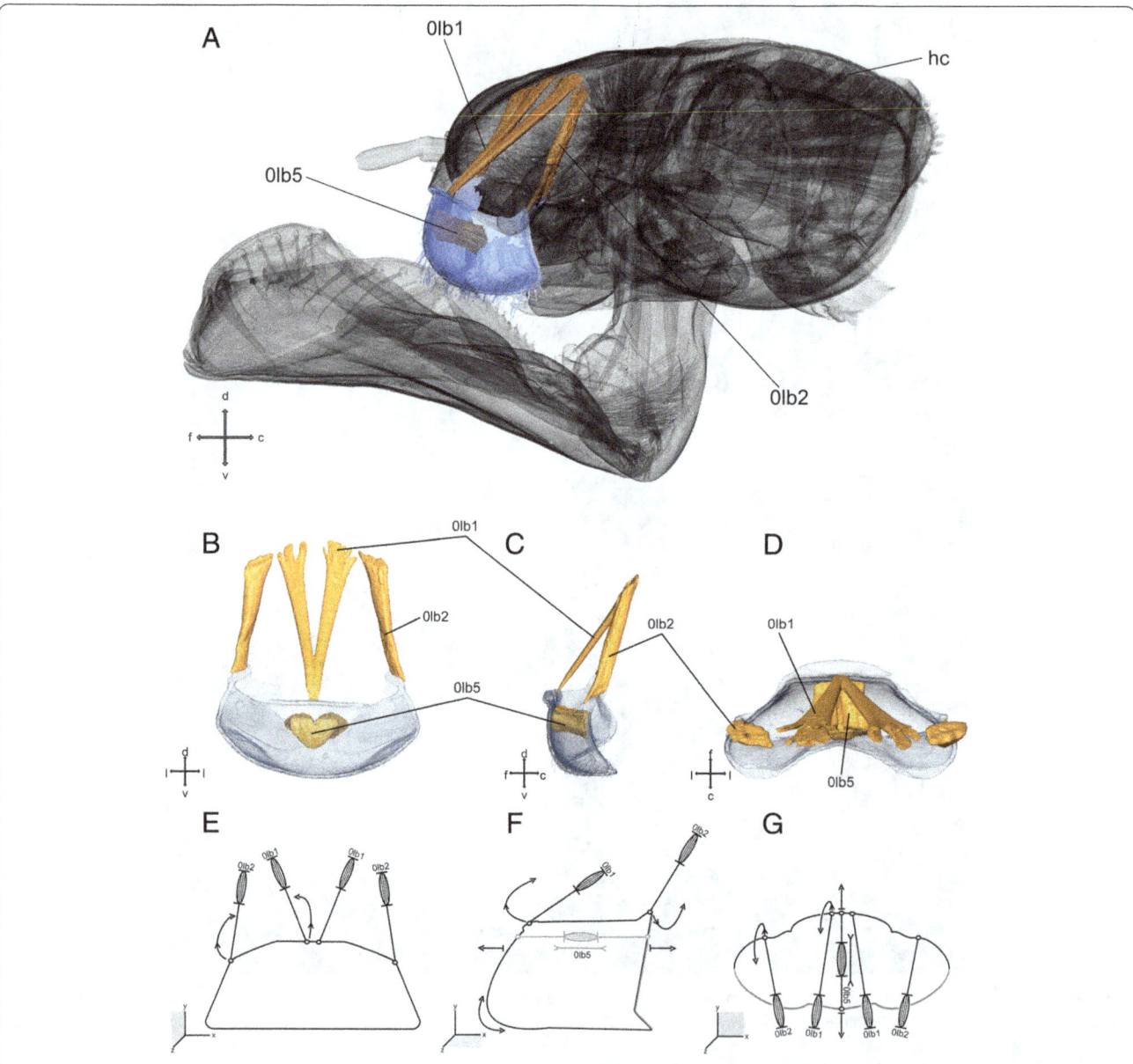

Fig. 2 Labrum and head capsule of *Pyrrhosoma nymphula*. **a-d** Three-dimensional visualisation from SRμCT data **e-g**. Sketch of the hypothetical function of the musculature – grey muscle-pictograms or parts of those indicates internal position within the respective mouthpart, grey additional outlines indicates membranous areas, lengths and broken lines are uninformative **a**. head capsule with labrum, fronto-lateral view **b**, **e**. Frontal view **c**, **f**. Lateral view **d**, **g**. Dorsal view. Abbreviations: c – caudal, d – dorsal, f – frontal, hc – head capsule, l – lateral, lb. - labrum, v – ventral

Mandibles (Figs. 1f,g and 3)

The mandibles are strongly sclerotized and show the typical dicondylic (two articulations points) ball-and-socket type. They are somewhat triangular – slightly elongated – from a dorsal view. The mandibles have four incisivi at the tip, which are broadly connected with each other at their bases, but divided into two groups by an incision. One lateral incisivus is located closer towards the mouth opening. The mola is separated into two areas and composed of one incisivus each. Laterally of the molar lobe, two reinforcing ridges run in cranial direction.

Musculature

M. craniomandibularis internus (0md1) – O: broad areas of the head capsule, postero-dorsal and postero-lateral I: postero-lateral at the mandible, abbuctor tendon C: the largest muscle within the head capsule. *M. craniomandibularis externus posterior* (0md3) – O: postero-lateral at the head capsule I: latero-ventral at the

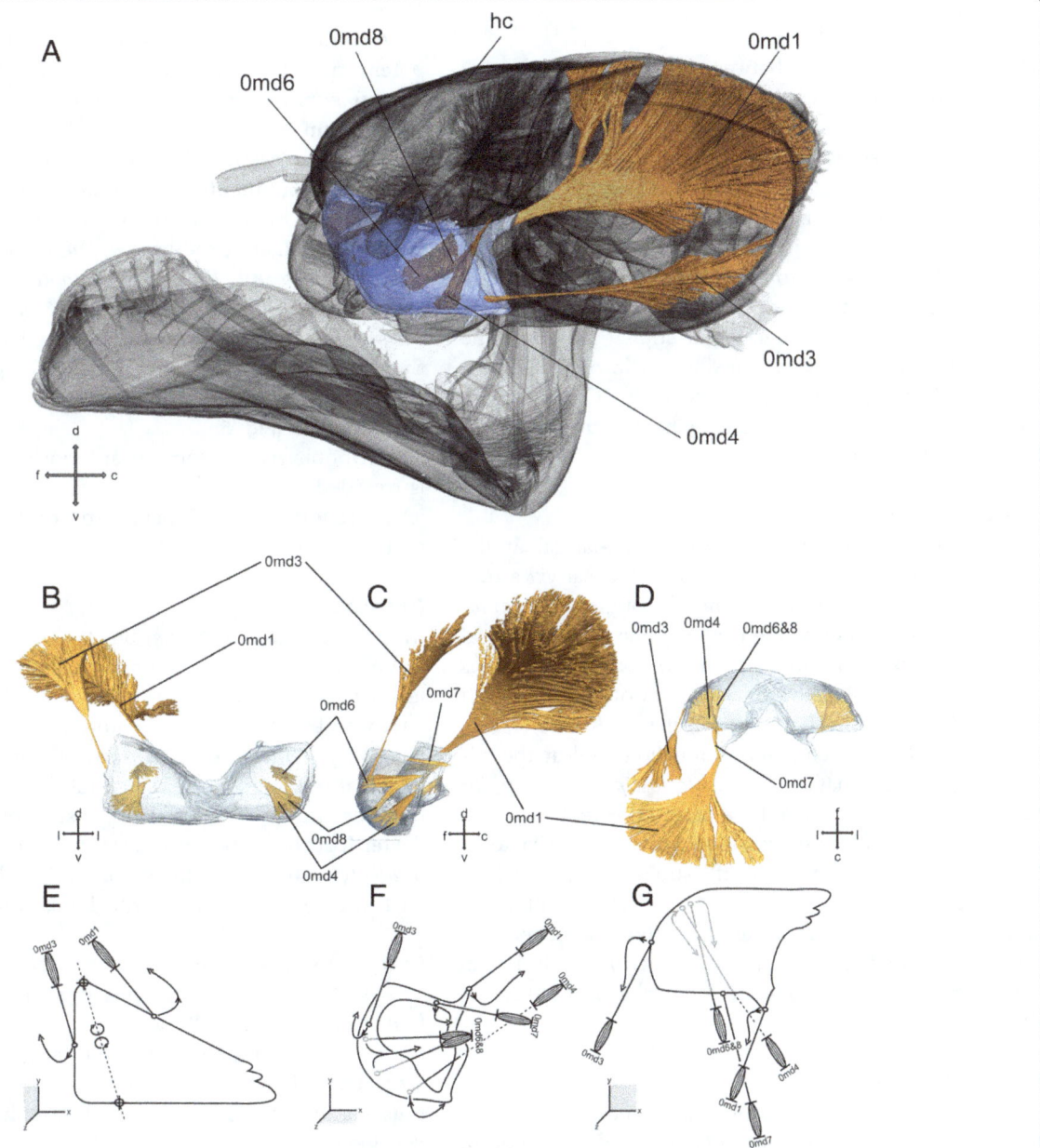

Fig. 3 Mandible and head capsule of *Pyrrhosoma nymphula*. **a-d** Three-dimensional visualisation from SRμCT data **e-g**. Sketch of the hypothetical function of the musculature – grey muscle-pictograms or parts of those indicates internal position within the respective mouthpart, grey additional outlines indicates membranous areas, lengths and broken lines are uninformative **a**. head capsule with labrum, fronto-lateral view **b**, **e**. Frontal view **c**, **f**. Lateral view **d**, **g**. Dorsal view. Abbreviations: c – caudal, d – dorsal, f – frontal, hc – head capsule, l – lateral, md – mandible, v – ventral

mandible, abductor tendon. *M. hypopharyngomandibularis* (0md4) – O: dorso-lateral at the suspensorial bar of the hypopharynx I: lateral within the mandible. *M. tentoriomandibularis lateralis inferior* (0md6) – O: anterior tentorial arm, via tendon I: lateral within the mandible C: same tendon as 0md8. *M. tentoriomandibularis medialis superior* (0md7) – O: dorsal tentorial arm I: postero-dorsal at the mandible C: weakly developed. *M. tentoriomandibularis medialis inferior* (0md8) – O: anterior tentorial arm, via tendon I: posterio-lateral within the mandible C: same tendon as 0md6.

Maxillae (Figs. 1e,h & i and 4)

The maxillae are located between the mandibles and the labium, dorso-lateral of the hypopharynx. They are developed to similar extent as in the adult [22]. The maxillae are composed of four parts: cardo, stipes, maxillar palp, and lacinia, a galea is absent. The undivided and triangular cardo is connected via the cardo-stipital membrane to the stipes (no joint is developed) – enabling its movement with respect to the stipes. The latter is almost rectangular in ventral view and divided into basistipes and mediostipes by a longitudinal stipital ridge. At the distal end of the stipes, the maxillary palp originates laterally, and the lacinia originates distally. The maxillary palp is connected via a socket, whereas the lacinia is fused with the stipes.

Musculature

M. craniocardinalis (0mx1) – O: ventro-lateral at the head capsule between 0md1 and 0md3 I: basal via a tendon at the cardo C: fan-shaped muscle (origin). *M. craniolacinialis* (0mx2) – O: at the head capsule, postero-dorsal to 0mx1 I: basal at the lacinia. *M. tentoriocardinalis* (0mx3) – O: lateral at the anterior tentorial arm I: within the cardo. *M. tentoriostipitalis anterior* (0mx4) – O: ventro-lateral at the corpotentorium I: ventral at the stipes, slightly fan-shaped. *M. tentoriostipitalis posterior* (0mx5) – O: ventro-lateral base of anterior tentorial arm, close to 0mx3 I: at the stipes. *M. stipitolacinialis* (0mx6) – O: ventro-lateral at the base of the stipes I: base of lacinia. *M. stipitopalpalis externus* (0mx8) – O: lateral within the stipes close to 0mx10 I: posterior on the base of the palpus. *M. stipitopalpalis internus* (0mx10) – O: lateral within the stipes close to 0mx8 I: anterior on the base of the palpus.

Labium (Figs. 1b–j and 5)

The labium is developed as prehensile mask responsible for prey capturing. It consists of post- and prementum, ligula, and palps with movable hooks. Glossa and paraglossa are not separated from the prementum by a particular ridge. The post- and prementum are connected via a cubital-like hinge joint, the so-called prementum-

postmentum joint (p-p joint), enabling movement of both parts relative to each other. The prehensile mask is connected to the head capsule ventrally via the postmentum. The postmentum shows the largest width at the apical end at the connection with the palps. It narrows towards the joint to almost half of its width. The labial palps are located at the apical tip of the prehensile mask. They are flexibly connected to the mask and show a blunt end hook and a pointed movable hook. Musculature inserting at the end hook is absent.

Musculature

M. tentoriopraementalis (0la5) – O: posterior at the corpotentorium I: lateral at the premental edge. *M. submentopraementalis* (0la8) – O: at the posterior part of the submentum I: dorsal at the prementum C: dichotomous in the early instars (I1,I2,I3) O2: at the postero-lateral base of the submentum. *M. praementopalpalis internus* (0la13) – O: median at the prementum, ventral of 0la14 I: antero-median at the base of the palpus. *M. praementopalpalis externus* (0la14) – O: median at the prementum, dorsal of 0la13 I: lateral at the base of the palpus. *M. praementomembranus* (0la15) – O: anterio-lateral at the postmentum I: postero-lateral at the prementum.

Hypopharynx (Fig. 6)

The hypopharynx is located anteriorly of the labium. It is rounded at the side facing towards the prementum. At its posterior base, a T-shaped rod originates and serves as attachment point for 0hy7.

Musculature

M. frontobuccalis lateralis (0hy2) – O: at the frons, ventral of the base of the antennae I: lateral at the suspensorial sclerites. *M. tentoriohypopharyngealis* (0hy3) – O: latero-ventral at the corpotentorium I: at the hypopharynx. *M. praementosalivaris anterior* (0hy7) – O: antero-lateral at the prementum; I: ventral at the T-rod. *M. oralis transversalis* (0hy9) – O: at the suspensorial sclerite, oral, dexter I: at the suspensorial sclerite, oral, sinister. *M. lororalis* (0hy10) – O: at the suspensorial sclerite, loral, dexter I: at the suspensorial sclerite, loral, sinister.

Pharynx and oesophagus

The wide lumen of the pharynx and oesophagus is folded dorsally, laterally and ventrally; these folds serve for muscle attachment and for an increase in the surface area. There is no contact with the corpotentorium. The musculature of the pharynx and oesophagus is strongly developed.

Musculature (Figures cf. [8])

M. clypeobuccalis (0bu1) – O: at the clypeus I: at the bucca. *M. frontobuccalis anterior* (0bu2) – O: close at

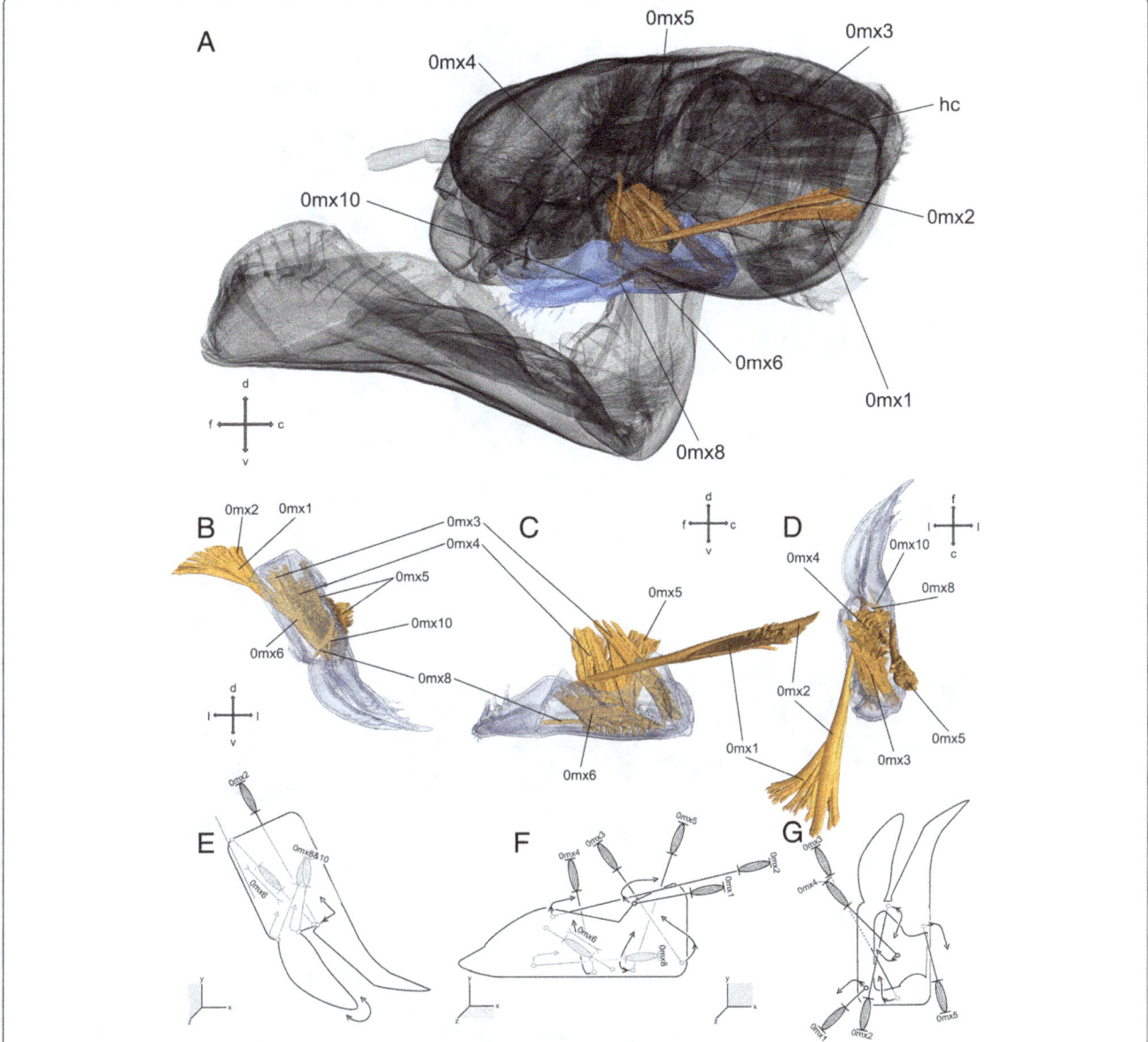

Fig. 4 Maxilla and head capsule of *Pyrrhosoma nymphula*. **a-d** Three-dimensional visualisation from SRµCT data **e-g**. Sketch of the hypothetical function of the musculature – grey muscle-pictograms or parts of those indicates internal position within the respective mouthpart, grey additional outlines indicates membranous areas, lengths and broken lines of are uninformative **a**. head capsule with labrum, fronto-lateral view **b**, **e**. Frontal view **c**, **f**. Lateral view **d**, **g**. Dorsal view. Abbreviations: c – caudal, d – dorsal, f – frontal, hc – head capsule, l – lateral, mx – maxilla, v – ventral

the interantenal ridge, posterior to 0 lb1 I: dorsal at the bucca. *M. frontobuccalis posterior* (0bu3) – O: posterior at the frons I: dorsal at the bucca. M. tentoriobuccalis lateralis (0bu4): O: base of dorsal tentorial arms; I: lateral at the bucca. *M. tentoriobuccalis anterior* (0bu5) – O: at the corpotentorium I: ventral at the bucca. *M. tentoriobuccalis posterior* (0bu6) – O: dorsal at the anterior tentorial arm I: ventral at the bucca. *M. verticopharyngalis* (0 ph 1) – O: at the occiput, medial of 0md1 I: dorsal at the pharynx. *M. tentoriopharyngalis* (0 ph 2) – O: at the corpotentorium I: ventral at the pharynx.

Discussion
Comparative morphology
Ninety-one head muscles are known so far for insects [23–25]. The most recent account for larval Odonata described 41 muscles for *Epiophlebia* [9]. Here we describe 42 head muscles for larval *Pyrrhosoma nymphula*.

The difference is due to the absence of antennal muscle 0an7 in *Epiophlebia* (*E. superstes*), which seems to be a peculiarity of *Epiophlebia*, since it is present in Anisoptera (*Aeshna cyanea*) as well as Zygoptera (*Pyrrhosoma nymphula*) – see also Blanke et al. [9]. Furthermore, there is a shift of the origin point of the muscle

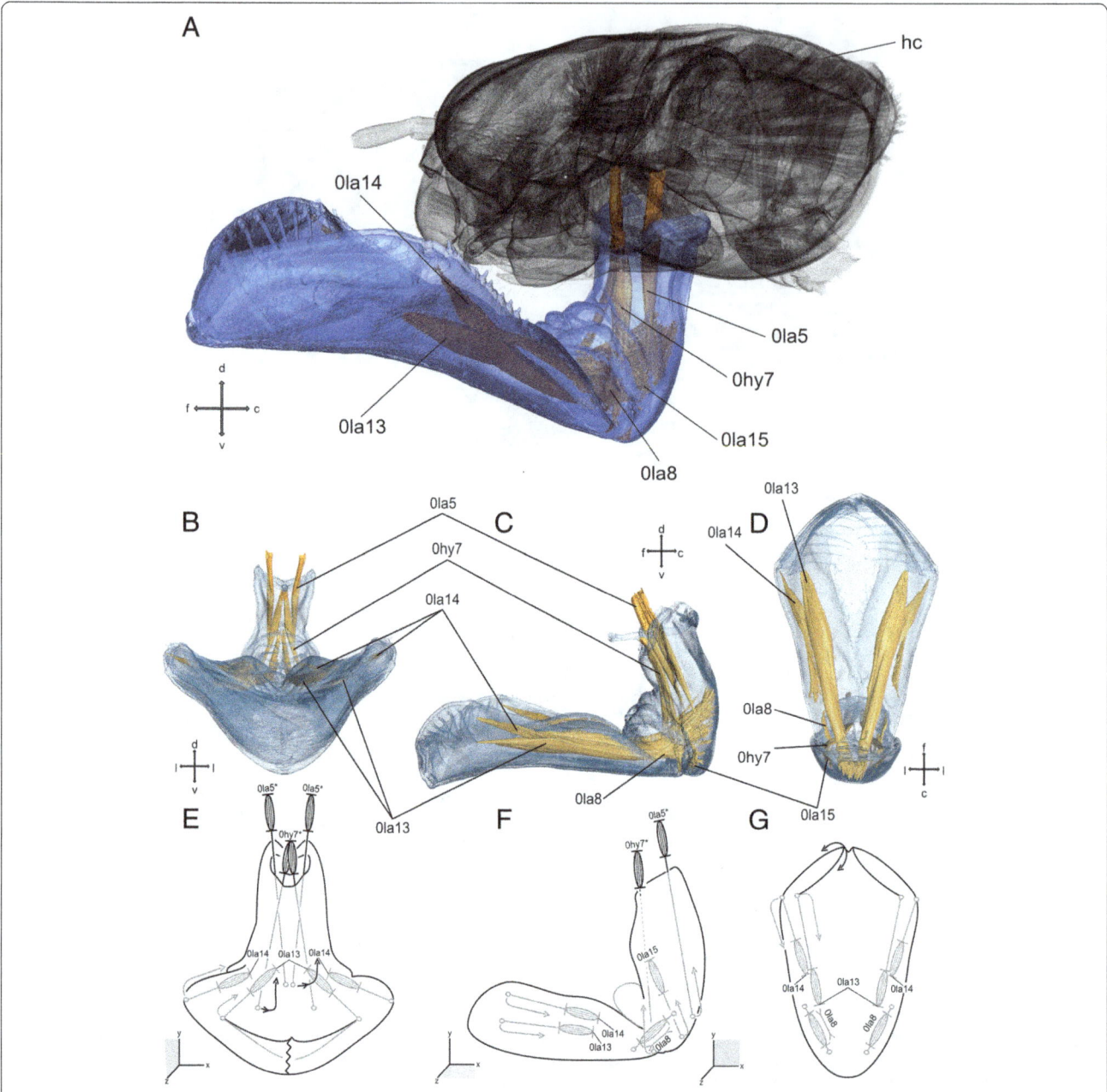

Fig. 5 Labium and head capsule of *Pyrrhosoma nymphula*. **a-d** Three-dimensional visualisation from SRµCT data E-G. Sketch of the hypothetical function of the musculature – grey muscle-pictograms or parts of those indicates internal position within the respective mouthpart, grey additional outlines indicates membranous areas, lengths and broken lines are uninformative **a**. head capsule with labrum, fronto-lateral view **b**, **e**. Frontal view **c**, **f**. Lateral view **d**, **g**. Dorsal view. Abbreviations: c – caudal, d – dorsal, f – frontal, hc – head capsule, hy – hypopharynx, l – lateral, la – labium, v – ventral

0an5 from the frons to the tentorium – 0an5 might take over the function of 0an7 – we follow the suggested homologization from Blanke et al. [9].

The peculiarities of the head musculature in Odonata larvae in comparison to adults mentioned by Blanke et al. [9] can be confirmed. More precisely, the absence of M. tentoriomandibularis lateralis superior (0md5), M. postoccipitopharyngealis (0 ph 3) and M. postoccipitalo-hypopharyngealis (0hy4) as well as the presence of the labial muscle M. tentoriopraementalis (0la5), antennal muscle M. tentoriofrontalis dorsalis (0te3), and the presence of the hypopharyngeal muscles (0hy2, 0hy3 0hy9, 0hy10), buccal muscles (0bu3, 0bu4, 0bu5), and pharyngeal muscle (0 ph 2) can be confirmed.

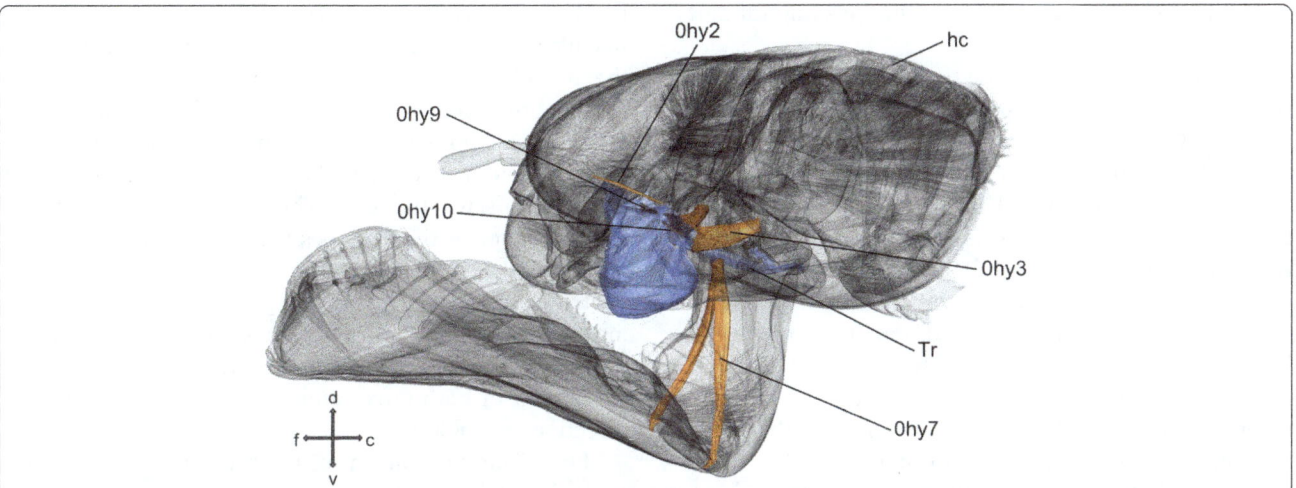

Fig. 6 Hypopharynx and head capsule of *Pyrrhosoma nymphula*. Three-dimensional visualisation from SRµCT data in fronto-lateral view. Abreviation: c – caudal, d – dorsal, f – frontal, l – lateral, v – ventral, Tr – T-rod

Muscle 0 lb1 of the labrum is a dichotomous muscle in *P. nymphula* while it is an unpaired muscle in *Epiophlebia*. The mandible muscle 0md7 is weakly developed compared to that in representatives of Epiprocta.

Muscle ontogenesis

Different instars of *P. nymphula* only show minor anatomical differences. Accompanied by the continuous increase in size, a slight shift of head proportions is noticeable during ontogenesis. Additionally, the prementum becomes elongated, if compared to the submentum, and the labium becomes elongated rather than widened. Small teeth-like sclerotized structures appear anteriorly at the lateral edge of the prementum [26] of instar 6. Generally, characters used in taxonomy [26], like the rows of setae on the head capsule especially at the dorso-caudal postocciput or on the labial palpus and ventro-medial of the prementum, are fully developed in later instars. Furthermore, the mode of crypsis changes towards instars 5 and 6 from a hyaline almost transparent body, which supposedly leads to near-invisibility in the body of water, to a piebald cuticle comprising different shades of beige and brown with spots of whitish and blackish to camouflage in the benthos [6].

The only muscular change within the larval stages is revealed concerning M. submentopraementalis (0la8), which is dichotomous in the early instars (I1,I2 and I3) with a second point of origin at the postero-lateral base of the submentum. This dichotomy is not present in any of the older instars studied (I6, middle-late and penultimate).

Compared to the adult, drastic differences in the overall morphology and muscle arrangement (cf. 'comparative morphology' section) occur not only because of the transmutation of the labium (from its insect ground pattern) into a prehensile mask. The re-orientation of the mouthparts from prognathous in the larva to orthognathous in the adult takes place. This modification leads to the adaptation to the different prey-capturing mode in the adult. The adults use their specialised legs as a basket for catching prey in flight [27] – the prey is brought to the mouth from ventral. Whereas, the larva uses its prehensile mask for prey-capturing under water – the prey is brought to the mouth from frontal. Furthermore, the presence of the hypopharyngeal muscles (0hy2, 0hy3 0hy9, 0hy10), buccal muscles (0bu3, 0bu4, 0bu5), and pharyngeal muscle (0 ph 2) and the strong development, within the larvae, of the latter two muscle groups might indicate further adaptations for ingestion within an aquatic habitat – ingestion via sucking the food "solved" in water. However, the most extensive change is the transmutation of the labium. Here, an elongation of the pre- and postmentum as well as the transformation of the labial palps into prey grasping organs occur. The abductor – M. praementopalpalis externus (0la14) – and adductor – M. praementopalpalis internus (0la13) – of the former palpi are greatly enlarged and the points of origin is translocated to the base of the postmentum to increase the applicable force. Two well-developed joints, prementum-head joint (P-H joint) ventral of the hypopharynx and the prementum-postmentum joint (P-P joint), lead to an increase of movability of the prehensile mask. Furthermore, the T-shaped rod or T-rod is characteristic for larval Odonata [8] (more precisely the hypopharyngeal apodem) serves as an important attachment structure for the redirected muscles (see also subsection 'labium' within the next section).

Functional significance of different groups of mouthparts muscles

The labrum (Fig. 2) limits the preoral cavity anteriorly. It is connected to the clypeal area by the membranous clypeolabral suture, which enables movement of the labrum. This movement is realised by the antagonistic action of muscles 0 lb1, 0 lb2. The muscle 0 lb5 compresses the labrum due to its attachment at its anterior and posterior (epipharyngeal) wall, the cuticle elasticity might function as muscle antagonist for returning of the labrum shape to its original condition. The muscle 0 lb1 is attached anterio-medially of the labrum and enables its dorso-lateral movement. The muscle 0 lb2 is the antagonist that is attached postero-laterally of the labrum and enables its dorso-medial movement.

The mandibles (Fig. 3) are strongly sclerotized jaws for crushing harder parts of the prey [6]. The movability of the mandibles is restricted to one axis due to the anterior and posterior ball-and-socket joints. This holds true for Odonata similar to other groups of insects [28]. These joints span a virtual axis of rotation and this rotation is produced by the antagonistic action of 0md1, the adductor muscle, and 0md3, the abductor muscle. Muscles 0md4, 6, 7 and 8 are often reduced in other winged insects [1]. Muscle 0md4 is attached antero-lateral within the mandible and at the hypopharynx, where it might support the movement of the hypopharyngeal sclerites. Muscles 0md6 and 0md8 are attached lateral and postero-lateral within the mandible, respectively. They enable a dorso-lateral and postero-dorsal movement of the mandible and support in this function the main abductor 0md3 [4].

The maxillae (Fig. 4) are specialised for manipulating, handling and sensing the food. The feeding movements of the maxillae are mainly protraction and retraction [8]. The muscles are able to move the maxillae in almost every direction, because flexible membranous regions and the cardo-cranial joint implement the suspension. The muscles 0mx1, 2, 3, 4 and 5 are used to move the maxilla very precisely: 0mx1 – is attached postero-lateral on the maxilla at the cardo (lateral to the head) and enables a dorso-median movement of the cranial part of the maxilla; 0mx2 – is attached antero-median on the maxilla at the lacinia and enables a dorsal (slightly dorso-lateral) movement of the apical part of the maxilla; 0mx3 – is attached postero-median on the maxilla within the cardo and enables a dorsal (slightly dorso-lateral) movement of the cranial part of the maxilla; 0mx4 – is attached median on the maxilla at the stipes and enables a dorsal movement of the maxilla; 0mx5 – is attached antero-lateral (medial to the head) and enables a dorso-lateral movement of the maxilla. During the protracting process the laciniae are used to grasp the prey, provided by the prehensile mask, by thrusting forward beyond the mandibles [8]. After retracting the maxillae, the food grasped by the laciniae, is delivered to the mandibles, where the digestive process starts. The muscle 0mx6 runs completely within the maxillae and is attached at the median base of the lacinia and might enable abduction. During or before this process the palps are used to sense the food. The maxillary palp is, due to its connection with the stipes via a ball and socket joint, very movable. The abductor muscle 0mx8 enables a ventro-lateral movement of the palp, whereas the opposing adductor muscle 0mx10 enables a dorso-medial movement of the palp. Combined with the dorso-ventral movement of both 0mx1 and 0mx2 the sensing in every direction is enabled.

The locking mechanism of the prehensile mask, as described by Olesen [12], where the maxillae are supposed to be used to prevent propelling of the prehensile mask could not be confirmed.

The labium (Fig. 5) performs the most interesting biomechanical movement in Odonata larvae (cf. Additional file 1: High-Speed Video). It represents a strong modification of mouthparts within the insects: the modification of the labium into a prehensile mask used for prey capturing. The process of prey capturing can be divided in to two different events: i) propelling of the prehensile mask towards the prey, and ii) grasping the prey using the movable pointed hooks.

The muscles 0la5, 0la8, 0la15, and 0hy7 are used for retracting the prehensile mask after the strike and to hold it in its resting position. They could presumably work as an antagonist against the pressure, while the jet propulsion in Anisoptera takes place, as suggested by Pritchard [10]. While muscles 0la5, 0la8 and 0hy7 might help to prevent propelling the prementum, muscle 0la15 might prevent the shearing of the postmentum against the head capsule. The movability of the prehensile mask is restricted by two joints, the prementum-head joint (P-H joint) and the prementum-postmentum joint (P-P joint). More precisely: 0la5 – is attached medially at the base of the postmentum and enables a dorsal movement towards the head, slightly restricted by the P-H joint for retraction/holding; 0la8 – is running from the ventral base of the prementum to the apical side of the postmentum; this muscle locks the movability within the p-p joint; 0la15 – is attached at the very base of the postmentum and might prevent the shearing of the postmentum against the head capsule and/or helps to lock the post- and prementum to each other; 0hy7 – is attached at the ventral base of the prementum directly at the labial articulation and enables a dorsal movement towards the head, strongly restricted by the P-H joint for retraction/holding. All these muscles might be included in the propelling movement of the prehensile mask, to steer during and/or positioning the mask. However, it becomes clear that the propelling of the prehensile labial mask cannot be powered by normal muscle contraction, since muscle contraction enables only the opposite movement.

However, both muscles 0la5 and 0hy7 become deflected in the completely drawn-in position of the prehensile labial mask. The muscle 0la5 is bent by the T-rod apodem (see above paragraph 'muscle ontogenesis), as described by Blanke and colleagues [9]: the bent ends of this apodem deflect the muscle 0la5 towards the thorax. The same is true for the muscle 0hy7: here the dorsal base of the prementum functions as 'deflection sheave' and redirects the muscle towards the tip of prementum.

So far the results of previous investigations indicate that the main force for the labium extension is produced by abdominal dorso-ventral muscles and transmitted to the labium as haemolymph pressure increases [11–14]. The same abdominal muscles are also involved in the mechanism of water intake into the digestive tract, to increase the body pressure, for respiration and swimming by jet propulsion [12, 29–31]; the latter is an escape mechanism in anisopteran larvae [6]. Epiprocta [Anisoptera + *Epiophebia*] use a dedicated internal respiration organ and are able to escape via jet propulsion (never observed in *Epiophlebia*) [32]. The ability to increase haemolymph pressure by closing their anal valve, in order to propel the prehensile mask, is therefore studied in anisopteran taxa only. The mechanism of propelling the prehensile mask in Zygoptera is not properly studied so far, due to missing experimental investigations and a differing larval anatomy as mentioned above. Caillère [33, 34] investigated the prey capturing strike of Zygoptera mentioning a convergent movement of the gills and the beginning of a forward movement of the digestive tract (abdomen) just before the prey capturing process. The extension of the prehensile mask and the forward movement of the abdomen stop simultaneously, and at almost the same time, the convergent movement of the gills ceases [33, 34]. This observation as well as the investigations by Eriksen [15], Miller [17, 18] and Sesterhenn and colleagues [19] confirm that also Zygoptera are able of an water intake into their digestive tract even though their external gills are responsible for a significant part of the gas exchange [15], as already Tillyard [7] suggested. However, the internal anatomy differs at least in the "lack of the diaphragm and the sub-intestinal muscle, which allows anisoptera larvae to suck water directly into the branchial chamber" ([17],p.386). Zygoptera are therefore restricted to a more gulping like ventilation, most likely due to the mentioned differences and a smaller volume of the related respiratory organs [17].

Since it is implausible – parsimony principle – that Zygoptera and Anisoptera use different mechanisms for such a highly complex biomechanical process, we safely can assume that the extension of the labium in Anisoptera as well as in Zygoptera is based on the same principles. Nevertheless, the mechanism of the mask propelling within the Odonata – considering Zygoptera and Anisoptera – should be reinvestigated because of the significant differences mentioned.

The second important process in prey capturing, is the prey grasping, which is realised by the strongest musculature found in the prehensile mask (Fig. 5). The adductor muscle 0la13 realizes the grasping and clutching of the prey at the tip of the labium by closing the labial palps with the movable hooks. The abductor muscle 0la14 realizes the opening of the labial palps, to release the prey remains and bring the palps in the initial position for new prey capturing process.

The hypopharynx (Fig. 5) is situated in the preoral cavity in front of the functional mouth [1]. The strongest muscle 0hy3 is the retractor of the hypopharynx, originating on the tentorium to allow a caudal movement and therefore a widening of the oral cavity. Muscle 0hy2 originating at the frons, most likely is able to protract the hypopharynx into the oral cavity, serving as antagonist to 0hy3. In Odonata, the unique T-rod or T-shaped rod is originating from the hypopharynx and 0hy7, which originates on the T-rod and inserts within the prehensile labial mask plays an important role in its movement (cf. paragraph on the labium). The small muscles 0hy9 and 0hy10 attaching at the left and the right side of the suspensorial sclerites might deform the hypopharynx laterally.

Methods

We studied six (L1, L2, L3, L6, middle-late, pen-ultimate) instars of *Pyrrhosoma nymphula* (Sulzer, 1776) (Zygoptera; Coenagrionidae), for comparison we also investigated larval specimens (late instars) of *Aeshna cyanea* (Müller, 1764) (Anisoptera; Aeshnidae), and *Epiophlebia superstes* (Sélys, 1889) (Epiprocta; Epiophlebiidae). All figures show the pen-ultimate instar. The specimens were fixed in alcoholic Bouin solution (= Duboscq-Brasil) and stored in 70–80% ethanol [35]. All applicable regulations concerning the protection of free-living species were followed. All necessary permits were obtained for collecting Odonata at the Billingshäuser Schlucht, Göttingen, Germany (permission granted by "Untere Naturschutzbehörde" file reference AZ.67.2.5 Wei). Prior to scanning (both CT and SEM), the samples were dehydrated in an ascending ethanol series and dried at the critical point (Balzers CPD030) or using Hexamethyldisilazan (HMDS) [35].

High resolution X-ray tomography (μCT) was carried out using a SkyScan 1172 desktop micro-CTscanner (Bruker micro-CT, Kontich, Belgium) at 40 kV and

250 μA with images taken every 0.25°. Additionally, we used the synchrotron radiation micro-computed tomography (SRµCT) setup at the Tomcat beamline of the Swiss Light Source, Villigen, Switzerland [36].

Segmentation and visualization of the data were done with Amira 5.4.3 (FEI SAS, France, www.vsg3d.com) and Photoshop CS3 (Adobe SystemInc.). Please refer to Betz et al. [37] for further information on the general setup for SRµCT and to Büsse et al. [38] for information on segmentation, labelling and visualization with Amira.

For SEM, the samples were dehydrated in an ascending ethanol series, critical point dried (Quorum E3000) and sputter-coated with gold-palladium (10 nm thickness; Leica Bal-TEC SCD500). Afterwards the samples were mounted on a rotatable sample holder [39] and examined in a Hitachi TM3000 scanning electron microscope at an accelerating voltage of 15 kV.

For the high-speed video recordings a Photron Fastcam SA1.1 (model 675 K–M1, www.vkt.de) equiped with a 105 mm/1:2.8 macro lens (Sigma, Japan, www.sigma-photo.co.jp) mounted on a Manfrotto055 tripod with Manfrotto410 geared head (Manfrotto, Italy, www.manfrotto.com) and two Dedocol COOLT3 light sources (Dedotech, Switzerland, https://dedotec.ch) was used. The labial strike was shot with 5400 frames per second (fps) (1/frame, Trigger Mode: End, Resolution 1024 × 1024). The video was saved as 16-bit TIFF image-stack and later reconstructed into a video format (AVI) using ImageJ 1.51e (National Institutes of Health, USA, https://imagej.nih.gov/ij/).

The juvenile stage in Odonata, Ephermeroptera and Plecoptera is suggested to be called "naiad "following the terminology proposed in Bybee et al. [40]. However, since for Odonata the more commonly used name is "larva", we decided to be consistent within the terminology, mainly used in the community, and used here the more general term [41, 42]. Anatomical structures are described using the nomenclature of Beutel et al. [43], muscle designations are made using the nomenclature of Wipfler et al. [25]. Muscles are described stating their origin (O) and their insertion (I) followed by (C) some special characteristics, if present.

Acknowledgements
We are grateful for the support by the members of the Functional Morphology and Biomechanics Group at Kiel University, especially to Esther Appel and Dr. Lars Heepe. We also want to thank Sebastian Boge for preparatory work. Furthermore, our special thanks goes to Dr. Alexander Blanke for proofreading and many elucidating discussions.

The support of the Swiss-Light-Source at the Paul-Scherrer Institut for granting beamtime at the TOMCAT beamline (Proposal No.: 20120025, granted to Dr. Sonja Wedmann, Senckenberg Gesellschaft für Naturforschung, Frankfurt a. M., Germany, as head of a combined fossil and extent entomology research group) is gratefully acknowledged.
Last but not least, we want to thank the unknown Reviewers for their constructive and helpful comments.

Funding
TH was supported through DFG grant HO2306/6–1, 2. The project was financed and SB directly supported through the DFG grant BU3169/1–1.

Authors' contributions
SB, TH and SNG designed the study. SB and SNG conducted the µCT and SEM analysis. SB and TH carried out the analysis of the raw data. SB wrote the manuscript. All authors read, revised and approved the manuscript.

Competing interests
The authors declare that they have no competing interests.

Author details
[1]Department of Functional Morphology and Biomechanics, Institute of Zoology, Christian-Albrechts-Universität zu Kiel, Am Botanischen Garten 9, 24118 Kiel, Germany. [2]Senckenberg Gesellschaft für Naturforschung, Senckenberganlage 25, 60325 Frankfurt, Germany.

References
1. Snodgrass RE. Principles of insect morphology. New York: Mc Graw-Hill Book Company; 1935.
2. Grimaldi D, Engel MS. Evolution of the insects. Cambridge: University Press; 2005.
3. Schmitt C, Rack A, Betz O. Analyses of the mouthpart kinematics in *Periplaneta americana* (Blattodea, Blattidae) by using synchrotron-based X-ray cineradiography. J Exp Biol. 2014;217:3095–107.
4. David S, Funken J, Potthast W, Blanke A. Musculoskeletal modelling of the dragonfly mandible system as an aid to understanding the role of single muscles in an evolutionary context. J Exp Biol. 2016;219:1041–9.
5. Gronenberg SJW. The control of mandible movements in the ant *Odontomachus*. J Insect Physiol. 1999;45:231–40.
6. Corbet PS. Dragonflies: behavior and ecology of Odonata. New York: Cornell Univ Press; 1999.
7. Tillyard RJ. The biology of dragonflies (Odonata or Paraneuroptera). Cambridge: University Press; 1917.
8. Snodgrass RE. The dragonfly larva. Smith Misc Coll. 1954;12:38.
9. Blanke A, Büsse S, Machida R. Coding characters from different life stages for phylogenetic reconstruction: a case study on dragonfly adults and larvae, including a description of the larval head anatomy of *Epiophlebia superstes* (Odonata: Epiophlebiidae). Zool J Linnean Soc. 2015;174:718–32.
10. Pritchard G. Prey capture by dragonfly larvae (Odonata; Anispotera). Canad J Zool. 1965;43:271–89.
11. Olesen J. The hydraulic mechanism of labial extension and jet propulsion in dragonfly nymphs. J Comp Physiol. 1972;81:53–5.
12. Olesen J. Prey-capture in dragonfly nymphs (Odonata; Insecta): labial protraction by means of a multi purpose ab- dominal pump. Vid Medd Dan Naturalist Foren. 1979;141:81–96.
13. Tanaka Y, Hisada M. The hydraulic mechanism of the preda- tory strike in dragonfly larvae. J Exp Biol. 1980;88:1–19.
14. Parry DA. Labial extension in the dragonfly larva Anax Imperator. J Exp Biol. 1983;107:495–9.

15. Eriksen CH. Respiratory roles of caudal lamellae (gills) in a Lestid damselfly (Odonata:Zygoptera). J North Am Benthol Soc. 1986;5:16–27.

16. Whedon AD. The comparative morphology and possible Adaptions of the abdomen in the Odonata. Trans Am Entomol Soc. 1918;44:373–437.

17. Miller PL. Responses of rectal pumping to oxygen lack by larval *Calopteryx splendens* (Zygoptera: Odonata). Physiol Entomol. 1993;18:379–88.

18. Miller PL. The responses of rectal pumping in some zygopteran larvae (Odonata) to oxygen and ion availability. J Insect Physiol. 1994;40:333–9.

19. Sesterhenn TM, Reardon EE, Chapman LJ. Hypoxia and lost gills: respiratory ecology of temperate larval damselfly. J Insect Physiol. 2013;59:19–25.

20. Pass G. Accessory pulsatile organs: evolutionary innovations in insects. Ann Rev Entomol. 2000;45:495–518.

21. Pass G. Phylogenetic relationships of the orders of hexapoda: contributions from the circulatory organs for a morphological data matrix. Arthropod Syst Phylo. 2006;64:165–203.

22. Blanke A, Beckmann F, Misof B. The head anatomy of Epiophlebia superstes (Odonata: Epiophlebiidae). Org Div Evol. 2013;13:55–66.

23. Kéler S v. Entomologisches Wörterbuch. Berlin: Akademieverlag Berlin; 1963.

24. Matsuda R. Morphology and evolution of the insect head. Mem Am Inst Ent. 1965;4:1–334.

25. Wipfler B, Machida AR, Müller RB, Beutel RG. On the head morphology of Grylloblattodea (Insecta) and the systematic position of the order, with a new nomenclature for the head muscles of Dicondylia. Syst Entomol. 2011;36:241–66.

26. Heidemann H, Seidenbusch R. Die Libellenlarven Deutschlands – Handbuch für Exuviensammler. Keltern: Goecke & Evers; 2002.

27. Leipelt KG, Suhling F, Gorb SN. Ontogenetic shifts in functional morphology of dragonfly legs (Odonata: Anisoptera). Zoology. 2010;113:317–25.

28. Gorb SN, Beutel RG. Head-capsule design and mandible control in beetle larvae: a three-dimensional approach. J Morph. 2000;244:1–14.

29. Pickard RS, Mill PJ. Ventilatory muscle activity in intact preparations of Aeschnid dragonfly larvae. J Exp Biol. 1972;6:527–36.

30. Mill PJ, Pickard RS. Anal valve movement and normal ventilation in *Aeshnid* dragonfly larvae. J Exp Biol. 1972;56:537–43.

31. Mill PJ, Pickard RS. Jet-propulsion in Anisopteran dragonfly larvae. J Comp Physiol. 1975;97:320–38.

32. Tabaru N. Larval development of Epiophlebia superstes in Kyushu. Tombo. 1984;27:27–31.

33. Caillère L. Long Term Learning in *Agrion* (Syn. *Calopteryx*) *splendens* Harris 1782 Larvae (Insecta, Odonata). Z Vergl Physiol. 1970;69:284–95.

34. Caillère L. Dynamics of the strike in a *Agrion* (syn. *Calopteryx*) *splendens* Harris. larvae (Odonata: Calopterygidae). Odonatologica. 1972;1:11-19.

35. Romeis B. Mikroskopische Technik. Urban und Schwarzenberg: München; 1987.

36. Stampanoni M, Mokso R, Marone F, Vila-Comamala J, Gorelick S, Trtik P, Jefimovs K, David C. Phase-contrast tomography at the nanoscale using hard X-rays. Physic Rev B. 2010;81:140105.

37. Betz O, Wegst U, Weide D, Heethoff M, Helfen L, Lee W-K, Cloetens P. Imaging applications of synchrotron X-ray phase-contrast microtomography in biological morphology and biomaterials science. I. General aspects of the technique and its advantages in the analysis of millimetre-sized arthropod structure. J Microsc. 2007;227:51–71.

38. Büsse S, Helmker B, Hörnschemeyer T. The thorax morphology of *Epiophlebia* (Insecta: Odonata) nymphs - including remarks on ontogenesis and evolution. Sci Rep. 2015;5:12835.

39. Pohl HA. Scanning electron microscopy specimen holder for viewing different angles of a single specimen. Microsc Res Tech. 2010;73:1073–6.

40. Bybee SM, Hansen Q, Büsse S, Wightman H, Branham MA. For consistency's sake: the precise use of nymph, larva and naiad within insecta. Syst Entomol. 2015;40:667–70.

41. Sahlén G, Suhling F, Martens A, Gorb SN, Fincke OM. For consistency's sake? A reply to Bybee *et al.* Syst Entomol. 2016;41:307–8.

42. Büsse S, Bybee SM. Larva, nymph and naiad – a response to the replies to Bybee et al. 2015 and the results of a survey within the entomological community. Syst Entomol. 2017;42:11–4.

43. Beutel RG, Friedrich F, Yang X-K, Ge S-Q. Insect morphology and phylogeny: a textbook for students of entomology. Berlin: De Gruyter; 2013.

3

Worker reproduction of the invasive yellow crazy ant *Anoplolepis gracilipes*

Ching-Chen Lee[1,2], Hirotaka Nakao[3], Shu-Ping Tseng[4,5], Hung-Wei Hsu[4], Gwo-Li Lin[4], Jia-Wei Tay[6], Johan Billen[7], Fuminori Ito[3], Chow-Yang Lee[8†], Chung-Chi Lin[1†] and Chin-Cheng (Scotty) Yang[5*†] ⓘ

Abstract

Background: Reproductive division of labor is one of the key features of social insects. Queens are adapted for reproduction while workers are adapted for foraging and colony maintenance. In many species, however, workers retain functional ovaries and can lay unfertilized male eggs or trophic eggs. Here we report for the first time on the occurrence of physogastric workers and apparent worker reproduction in the invasive yellow crazy ant *Anoplolepis gracilipes* (Fr. Smith). We further examined the reproductive potential and nutritional role of physogastric workers through multidisciplinary approaches including morphological characterization, laboratory manipulation, genetic analysis and behavioral observation.

Results: Egg production with two types of eggs, namely reproductive and trophic eggs, by physogastric workers was found. The reproductive egg was confirmed to be haploid and male-destined, suggesting that the workers produced males via arrhenotokous parthenogenesis as no spermatheca was discovered. Detailed observations suggested that larvae were mainly fed with trophic eggs. Along with consumption of trophic eggs by queens and other castes as part of their diet, the vital role of physogastric workers as "trophic specialist" is confirmed.

Conclusion: We propose that adaptive advantages derived from worker reproduction for *A. gracilipes* may include 1) trophic eggs provisioned by physogastric workers likely assist colonies of *A. gracilipes* in overcoming unfavorable conditions such as paucity of food during critical founding stage; 2) worker-produced males are fertile and thus might offer an inclusive fitness advantage for the doomed orphaned colony.

Keywords: *Anoplolepis gracilipes*, arrhenotokous parthenogenesis, physogastric workers, trophic eggs

Background

One of the hallmarks of higher social Hymenopterans (social bees, wasps, and ants) is the reproduction division of labor among nest members [1]. Queens are the reproductive caste that is morphologically adapted for dispersal and reproduction while workers are the non-reproductive caste specialized in foraging, nest maintenance and brood tending. A haplodiploid sex determination system is common to all hymenopterans, in which males arise parthenogenetically from unfertilized eggs (arrhenotoky) and are haploid, whereas females arise from fertilized eggs and are diploid [2, 3]. Such unique system results in an asymmetrical genetic relatedness among the colony members where workers

are more genetically related to the queen's daughters (their sisters) ($r = 0.75$) compared to their own daughters and sons ($r = 0.50$) in a monogynous colony headed by a singly mated queen [4]. According to Hamilton's kin-selection theory, this unusual asymmetry in relatedness appears to favor evolution of a sterile worker caste as workers gain indirect fitness (i.e., propagation of their own genes) by behaving altruistically and assisting in raising the queen's instead of their own offspring.

Reproductive constraints impair the worker reproduction either through behavioral mechanisms (e.g., worker policing) or by suppressing the development of the reproductive organs in workers [5]. However, the workers in most ant species retain functional ovaries, and are capable of producing viable male eggs and/or non-viable trophic eggs [6]. Bourke [7] reported that workers produce males in approximately 50 species from 24 genera. Trophic eggs are nutritional packets, and act as an important mechanism

* Correspondence: ccyang@rish.kyoto-u.ac.jp
†Equal contributors
[5]Research Institute for Sustainable Humanosphere, Kyoto University, Gokasho, Uji, Kyoto 611-0011, Japan
Full list of author information is available at the end of the article

for transferring nutrients or protein to the colony members, especially queens and larvae (reviewed in Wheeler [8]). Nevertheless, workers that have completely lost their reproductive organs only occur in a few genera (9 out of 283). These are *Solenopsis, Monomorium, Tetramorium, Hypoponera, Anochetus, Leptogenys, Pheidole* and *Carebara* [1, 9, 10]. It is interesting to note that in primitive ant species (e.g., Ponerinae), workers possess a spermatheca, and are capable of mating and produce fertilized eggs (i.e., gamergates) [11].

The yellow crazy ant *Anoplolepis gracilipes* has been listed as one of the world's top 100 invasive species due to their severe impacts on biological diversity and ecosystem sustainability [12]. This species is polygynous and forms supercolonies with individuals in physically separated colonies exhibiting limited aggression behavior towards each other [13]. *A. gracilipes* decimated over one-third of the entire population of endemic red crabs (*Gecarcoidea natalis*) in Christmas Island [14]. The displacement of native "keystone" species by this invasive ant indirectly impedes the litter breakdown process and causes the growth of sooty molds in canopy trees, which ultimately alters the island rainforest ecosystem. The numerical dominance of *A. gracilipes* negatively impacts the diversity and abundance of native invertebrate communities in introduced areas [15]. In addition, this species also attacks and kills populations of smaller vertebrates such as birds or new-born domestic animals, e.g. on the Seychelles [16–18].

So far, most of the well-studied invasive ants are known to possess a sterile worker caste [7], except for one previous study in which the presence of underdeveloped ovaries (i.e., absence of mature oocytes) was reported in a minority of *A. gracilipes* workers inspected [19]. While this study found little support for worker reproduction of *A. gracilipes*, our preliminary observation, in contrast, suggested that egg production often occurred in queenless *A. gracilipes* laboratory colonies, and that artificially-orphaned colonies are invariably found with the presence of "corpulent" workers, whose gaster sizes were conspicuously greater than those of "normal" foraging workers and appeared brown-whitish in color (hereafter referred to as "physogastric workers"). Such morphological difference leads to a possible link between the egg production and presence of physogastric workers, and merits further investigation. In this study, we therefore conducted a series of experiments addressing the following questions: 1) are physogastric workers present in queenright field colonies? 2) what is the anatomy of the reproductive organs of physogastric workers? 3) can *A. gracilipes* workers produce viable and/or trophic eggs under queenless condition? 4) if viable eggs are produced, what is the sex and ploidy level of such worker-produced offspring? In addition to understand the fundamental aspects of worker reproduction by

A. gracilipes, the origin, trophic function and evolution of worker reproduction in this invasive ant species also are discussed.

Results

Occurrence of physogastric workers and ovarian morphology of workers

In all three field-collected colonies, 7.23–11.74% of the workers were physogastric. Gaster widths of normal workers (GW: 1.09 ± 0.03 mm, Fig. 1a) were significantly smaller than those of physogastric workers (GW: 1.53 ± 0.02 mm, Fig. 1b; $Z = -5.475$, $P < 0.01$; Table 1). The clearly distinct external morphology of the queen is also illustrated in Fig. 1c.

We found normal workers possess ovaries, most of which, however, are underdeveloped and lacking of yolky oocytes (92%) (Fig. 1d). Physogastric workers tend to possess more well-developed ovaries (Fig. 1e) as the number of ovarioles/individual is higher than in normal workers (2.51 ± 0.09 vs. 1.62 ± 0.12; $Z = -5.652$, $P < 0.01$; Table 1), the number of yolky oocytes per ovariole (4.46 ± 0.10 vs. 1.73 ± 0.12; $Z = -10.416$, $P < 0.01$) and the total number of yolky oocytes were significantly higher in physogastric workers than in normal workers (11.21 ± 0.48 vs. 2.81 ± 0.31; $Z = -8.290$, $P < 0.01$). Note that numbers presented here were based on those ovarioles with at least a visible oocyte only. While no spermatheca was found in both types of the workers, yellow bodies that are characteristic of reproduction were visible in some physogastric workers (13%) (Fig. 1e). On average, queens of *A. gracilipes* had 44–52 ovarioles/individual and had a higher number of yolky oocytes (94.50 ± 6.63) than both types of workers. Yellow bodies were present in the ovaries of queens, along with a conspicuous spermatheca (Fig. 1f).

External and internal morphology of workers

Scanning electron microscopy revealed a noticeable difference in abdominal morphology between normal and physogastric workers (Fig. 2a and b). The abdomen of physogastric workers was greatly distended with exposed intersegmental membranes. Histological sections indicated that the fat body in the abdomen is far more abundant in physogastric than in normal workers (Fig. 2c and d). The absence of a spermatheca in physogastric workers was further confirmed by longitudinal histological sections (Fig. 2e), suggesting that sexual reproduction by workers of *A. gracilipes* is impossible.

Production of eggs by workers, sex, ploidy level and morphology of worker-produced offspring

After 4 months, we discovered that three out of nine artificially-orphaned colony fragments produced eggs and larvae, these three colonies fragments were designated as AGQLF01, AGQLF02 and AGQLF03. AGQLF01 was

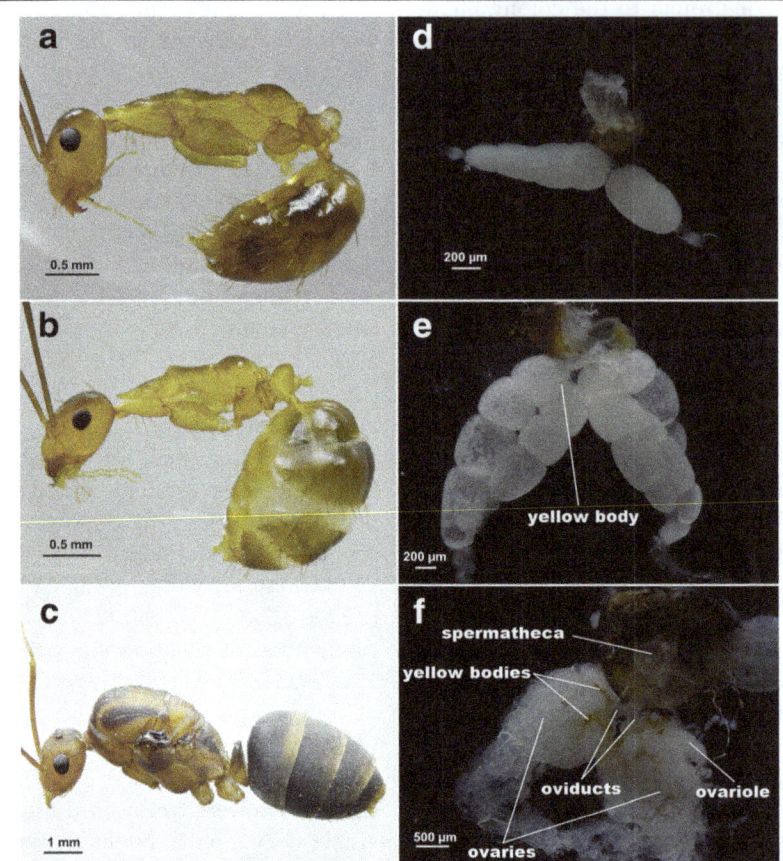

Fig. 1 Morphology and reproductive systems in worker and queen of *A. gracilipes*. Shown are the external morphology of normal worker (**a**), physogastric worker (**b**) and queen (**c**). Gaster dissection presenting ovarian morphology of normal worker (**d**), physogastric worker (**e**) and queen (**f**). Note difference in length of ovarioles and number of mature oocytes

isolated from the colony in Nantou County, while both AGQLF02 and AGQLF03 were isolated from the same source colony in Changhua County (Additional file 1: Figure S1). Morphological observations indicated the existence of two types of eggs produced by the workers, characteristically elongated oval shaped eggs and sub-spherically shaped eggs (Fig. 3). The former was confirmed viable with an obvious embryo, whereas the latter was embryoless and never hatched. Coupled with the fact that these non-viable eggs are consumed by larvae (Additional file 2: Video S1) and other castes (see "Fate of worker-laid trophic eggs" for more details), it is most likely that the eggs with sub-spherical shape serve as trophic eggs. Reproductive eggs in one (AGQLF03) of the three egg-producing colonies successfully developed into pupae that emerged as adult males ($n = 18$) 6 months after the

Table 1 Differences in gaster size and ovary development across three castes of *A. gracilipes*

Female castes	Ants dissected	Gaster width (mm)	Number of ovarioles/individual	Number of yolky oocytes per ovariole	Total number of yolky oocytes
Normal workers	$n = 90$	1.09 ± 0.03 [0.90–1.30]	1.62 ± 0.12 [1–4]	1.73 ± 0.12 [1–4]	2.81 ± 0.31 [1–8]
Physogastric workers	$n = 90$	1.53 ± 0.02 [1.40–1.70]	2.51 ± 0.09 [2–5]	4.46 ± 0.10 [1–9]	11.21 ± 0.48 [6–27]
Z		-5.475	−5.652	−10.416	-8.290
P		* < 0.01	* < 0.01	* < 0.01	* < 0.01
Queens	$n = 9$	2.77 ± 0.06 [2.60–3.00]	47.33 ± 1.23 [44–52]	2.00 ± 0.04 [1–3]	94.50 ± 6.63 [69–116]

Notes: Data are presented as mean ± standard deviation [range]; *p, statistically significant using Mann-Whitney U-test; Queens were not subjected to analysis due to its small sample size

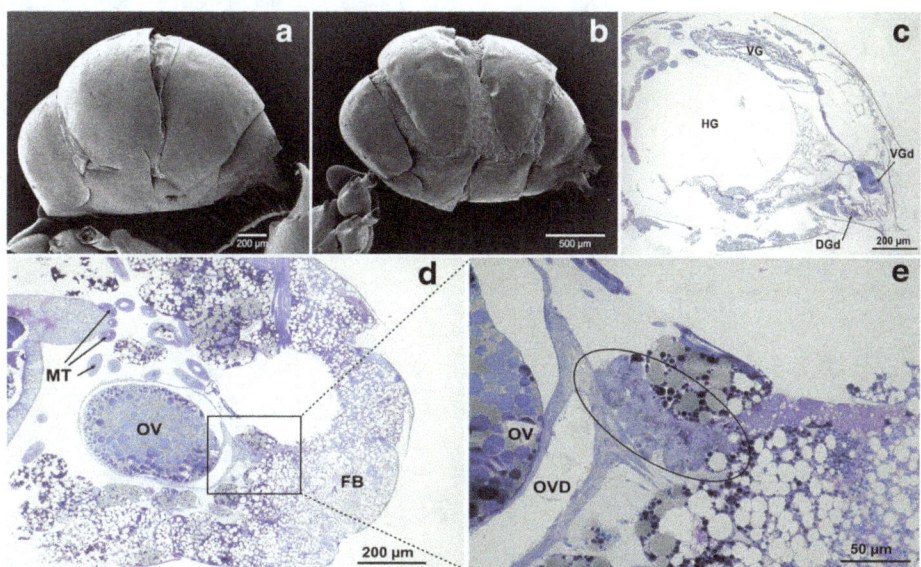

Fig. 2 Scanning electron micrographs and histological sections of two types of workers in *A. gracilipes*. SEMs of abdomen of normal worker (**a**) and physogastric worker (**b**), and longitudinal sections through posterior abdomen part of normal worker (**c**) and physogastric worker (**d**). Note large accumulation of fat body and absence of spermatheca in physogastric worker. The location where spermatheca is supposed to be found if it exists is highlighted with circled area in figure (**e**). DGd: Dufour gland duct, FB: fat body, HG: hindgut, MT: Malpighian tubules, OV: ovaries, OVD: oviduct, VG: venom gland, VGd: venom gland duct

start of the experiment, confirming the viability of the elongated oval shaped eggs. While larvae were present in AGQLF01 and AGQLF02 during the first 4 months, we failed to recover any adult male upon the end of the observation most likely due to cannibalism by nestmate (see Discussion).

Results of microsatellite genotyping revealed that all workers from AGQLF03 are heterozygotes across all loci with the presence of three major representing multi-locus genotypes (Table 2). Unlike the previously reported high frequency of heterozygous males [19, 20], we found that all

males from AGQLF03 possess homozygous multi-locus genotypes, harboring one of the maternal alleles at all loci, which suggests that the worker-produced males are invariably haploid. In contrast, virtually all males (90%) in the queenright colony (AGQR01) are diploid (heterozygous at least at one locus), a pattern consistent to previous studies that diploid males are common in the introduced ranges [19, 20].

Both head width and total body length of worker-produced male pupae (HW: 0.81 ± 0.01 mm; TL: 4.06 ± 0.07 mm) were significantly greater than those of queen-produced male pupae (HW: 0.70 ± 0.01 mm; TL: 3.69 ± 0.05 mm; Fig. 4a; HW: Z = −3.888, P < 0.01; TL: Z = −3.060, P < 0.01). Similarly, the two measurements of worker-produced males (HW: 0.80 ± 0.02 mm; TL: 4.58 ± 0.10 mm) were also greater than those of queen-produced males, respectively (HW: 0.71 ± 0.01 mm; TL: 4.00 ± 0.05 mm; Fig. 4b; HW: Z = −2.985, P < 0.01; TL: Z = −2.863, P < 0.01). Worker-produced males, however, shared similar genital structures (Fig. 4c) and had similar internal reproductive organs (Fig. 4d) as adult males in a queenright colony. Rupturing of seminal vesicles in worker-produced males further showed the presence of viable sperm (i.e., sperm bundle with apparent swimming ability, Additional file 3: Video S2).

Fate of worker-laid trophic eggs

We tracked the fate of 62 and 51 trophic eggs produced by physogastric workers in colonies AGTE01 and AGTE02, respectively (Table 3). Visual observations

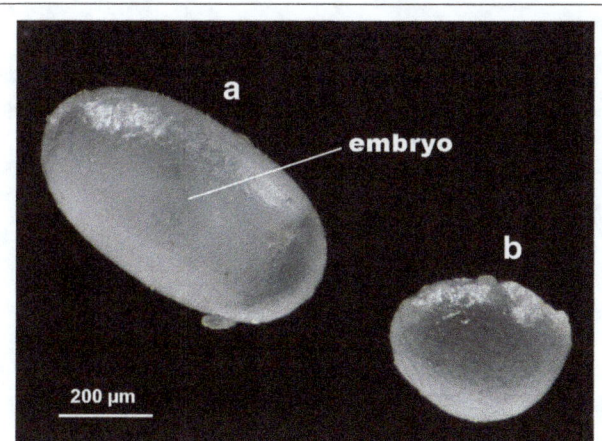

Fig. 3 Morphology of eggs produced by *A. gracilipes* workers. Light micrograph of a worker-laid reproductive egg (**a**) and a worker-laid trophic egg (**b**)

Table 2 Genotypic distribution for individuals of various castes (queen, workers and males) from a queenright and queenless colony

Caste	Sample size	Ano1		Ano3		Ano4		Ano5		Ano6		Ano8		Ano10	
AGQR01 (queenright colony)															
Queen	1	118		168		171		139		133		232		306	
Worker	1	112	118	152	168	173	177	121	139	133	145	210	232	262	306
Worker	2	112	118	152	178	171	177	121	135	133	145	210	232	262	306
Worker	2	112	118	152	168	171	177	121	135	133	145	210	232	262	290
Worker	3	112	118	152	168	171	177	121	135	133	145	212	234	262	308
Worker	1	112	118	152	168	171	177	121	135	133	145	212	234	262	292
Worker	3	112	118	152	178	173	177	121	135	133	145	210	232	262	306
Worker	1	112	118	152	178	173	177	121	139	133	145	210	232	262	306
Worker	1	112	118	152	178	173	177	121	135	133	145	212	234	262	308
Worker	1	112	118	152	168	173	177	121	135	133	145	212	234	262	308
Male	2	112	118	152	178	171	177	121	135	133	145	210	232	262	306
Male	1	112	118	152	168	171	177	121	135	133	145	210	232	−1	−1
Male	1	112	118	152	178	173	177	121	135	133	145	210	232	262	290
Male	1	112	118	152	168	171	177	121	135	133	145	210	232	262	306
Male	2	112	118	152	168	171	177	121	135	133	145	212	234	262	308
Male	1	112	118	152	178	173	177	121	135	133	145	212	234	262	308
Male	1	112	118	152	168	173	177	121	135	133	145	212	234	262	308
Male	1	112	118	152	178	173	173	121	135	133	145	210	232	262	306
Male	1	112	118	152	178	171	177	121	135	133	145	212	234	−1	−1
Male	1	112	118	152	178	171	177	121	135	−1	−1	212	234	262	308
Male	1	112	118	152	168	171	177	121	135	−1	−1	212	234	262	308
Male	1	112	118	152	178	177	177	121	135	133	145	210	232	262	262
Male	1	112	118	152	168	177	177	121	135	133	145	210	232	−1	−1
Male	1	112	118	152	178	177	177	121	135	133	145	212	234	262	308
Male	1	112	118	152	168	177	177	121	135	133	145	212	240	262	292
Male	1	112	118	152	168	177	177	−1	−1	145	145	210	210	−1	−1
Male	1	112		152		177		121		145		−1		262	
Male	1	112		152		177		121		145		210		-1	
AGQLF03 (queenless colony fragment)															
Worker	4	112	118	152	168	171	177	121	139	133	145	212	212	262	290
Worker	9	112	118	152	168	171	177	121	139	133	145	210	210	262	288
Worker	1	112	118	152	168	171	177	121	139	133	145	210	210	262	290
Male	1	118		152		171		139		133		−1		290	
Male	2	112		168		171		121		145		212		290	
Male	1	118		152		177		139		145		−1		292	
Male	1	118		168		171		121		145		−1		262	
Male	1	118		168		171		139		145		210		288	
Male	1	112		168		177		139		145		210		262	
Male	1	112		168		177		121		145		210		290	
Male	1	118		168		171		121		133		−1		262	
Male	1	118		152		177		139		133		−1		288	
Male	1	112		168		177		139		145		−1		288	

Table 2 Genotypic distribution for individuals of various castes (queen, workers and males) from a queenright and queenless colony *(Continued)*

Male	1	112	152	171	121	145	−1	290
Male	1	118	168	177	121	133	−1	262
Male	1	112	168	171	121	133	210	262

−1, amplification failure

indicated that most of the trophic eggs (≥ 63%) were offered to the larvae. Coupled with the fact that larvae received occasional trophallaxis from workers and never directly fed on solid prey items during the entire observation period, trophic eggs appear to be the main food source for larvae in *A. gracilipes*. Queens received both liquid food via oral trophallaxis and trophic eggs from workers, and the former seems to be their main diet (≥ 89%; Table 4). We also discovered that trophic eggs were occasionally offered to other castes such as workers and males.

Discussion

We performed both field survey and laboratory manipulation to study worker reproduction in the invasive yellow crazy ant, *A. gracilipes*. The results of our survey confirm the existence of physogastric workers in the field colonies, and subsequent gaster dissection reveals that the level of ovarian development is significantly higher in physogastric than in normal workers. Workers in artificially orphaned colonies produced both trophic and viable (reproductive) eggs. The viable eggs from one of the queenless colonies successfully developed into

Fig. 4 Morphological comparison between worker-produced and queen-produced offspring in *A. gracilipes*. **a** Male pupa from orphaned colony (*left*), worker pupa (*middle*) and male pupa from queenright colony (*right*); (**b**) external morphology of worker-produced male (*left*) and normal queen-produced male (*right*); (**c**) close up of external genital structure of worker-produced male (*left*) and queen-produced male (*right*); (**d**) internal organs of the male reproductive system of worker-produced male (*left*) and queen-produced male (*right*)

Table 3 The fate of worker-laid trophic eggs expressed by which caste/stage was a given trophic egg offered to after being laid

	Colony code	
	AGTE01	AGTE02
Number of queens	1	2
Number of workers	≈ 900	≈ 250
Total hours of observation	10.5	10
Total number of trophic eggs that had been followed	62	51
Number of trophic eggs given to		
dealate queens	3 (5%)	4 (8%)
males	-	8 (15%)
workers	3 (5%)	1 (2%)
larvae	56 (90%)	25 (49%)
queen larvae	-	7 (14%)
Number of trophic eggs not given to any of particular caste mentioned above	-	6 (12%)

Notes: Values in parentheses refer to proportion of trophic eggs given to respective individuals or castes relative to the total number of trophic eggs that we tracked; Neither males nor queen larvae were found in colony AGTE01 during the observation

males that were slightly larger than those produced by queens. All worker-produced males were haploid and possess a normal, functional reproductive system as their diploid counterparts do. Furthermore, our data suggest that the production of trophic eggs plays a crucial role in regulating colony nutrition, especially for larvae. Below we discuss how these findings, combined with additional evidence obtained from histological, SEM and behavioral observation, provide new insights into the role of physogastric workers in *A. gracilipes*.

Arrhenotokous parthenogenesis by physogastric workers
Our results clearly demonstrate that *A. gracilipes* workers are not functionally sterile, yet able to produce both trophic and reproductive eggs. In several ant species, workers are known for their ability to produce trophic eggs in queenright colonies and switch to produce reproductive eggs which develop into males once the queens die or disappear [21–23]. For instance, *Aphaenogaster senilis* workers produce unviable trophic eggs under queenright condition and begin to lay reproductive eggs that develop into males 4 months after being separated from the queens [24]. Similar to previous studies, our work showed that approximately 6 months

after queen removal, some viable reproductive eggs successfully developed into adult males in one of the experimental *A. gracilipes* colony fragments. To the best of our knowledge, this is the first study showing that worker reproduction occurs in *A. gracilipes*.

While we lack direct evidence on whether worker-produced males can copulate with queens or female alates, their seemingly functional genitalia, intact reproductive organs and presence of viable sperm lead us to speculate that worker-produced males may have equal reproductive capacities as queen-produced males. It is thus plausible that in field conditions the last cohort of worker-produced males might be able to copulate with female alates from other colonies in the proximity, and subsequently offer fitness advantage to the doomed orphaned colony.

Previous studies have suggested that thelytokous parthenogenesis (i.e., diploid daughter females are produced from unfertilized eggs) may have occurred in *A. gracilipes* based on the finding of high intracolonial relatedness among workers [20, 25, 26]. In contrast, our microsatellite analyses suggested that worker-produced reproductive eggs are invariably haploid, instead of diploid as expected when thelytokous parthenogenesis

Table 4 Dietary composition (trophallaxis vs. consumption of trophic egg) of queens of *A. gracilipes*

	Colony code & individual queen		
	AGTE01 Q1	AGTE02 Q1	AGTE02 Q2
Total hours of observation	10.5	10.5	10.5
Feeding on			
liquid food via oral trophallaxis from workers	324 times (89%)	74 times (97%)	60 times (92%)
trophic eggs	40 times (11%)	2 times (3%)	5 times (8%)

Notes: Values in parentheses refer to frequency of consumption relative to the total number of feedings by queens; one whole trophic egg was consumed per feeding operates. We also showed that virtually all males in the queenright colony are diploid, which is consistent with previous studies in which a high prevalence of diploid males in *A. gracilipes* colonies was discovered [19, 20]. The high proportion of diploid males in the field colonies collected from this study and elsewhere may imply that in field conditions the haploid males are either rarely produced by workers when colonies remain queenright or are mostly consumed by nestmates as food resources. In general, mechanisms (e.g., queen repression, worker self-restraint and/or worker policing [27–30]) contributing to low or absence of worker reproduction in a queenright colony are predictable with sex ratio optimization, relatedness asymmetry and/or kin structure [31]. For example, Chapuisat et al. [32] found that male larvae of *Formica exsecta* are preferentially cannibalized by nestmate workers at their late developmental stage not only to regulate the sex ratio of colony but also to feed the females as additional food. We, however, note that the presence of an unusual reproduction mode (e.g., asexual production of the queen) and a high frequency of diploid males of *A. gracilipes* [19, 20] may not satisfy the prerequisites of such prediction (e.g., classic haplodiploidy), thus leading the interpretation of a favorable scenario extremely difficult. Yellow bodies were found in some of the physogastric workers from queenright colonies in this study, however, it remains questionable whether yellow bodies can be an appropriate indicator for oviposition of viable eggs as they are also visible in trophic egg layers for some species [33–35].

Furthermore, the body size of haploid males produced by workers is, on average, greater than that of the diploid counterparts produced by queens, and such finding is opposite to what has been reported for other ant species (e.g., *Atta sexdens*, *Lasius sakagamii*, *Solenopsis invicta*) whose diploid males tend to be bigger than haploid ones due to diploidization and feminization [36–38]. Such inconsistency, however, might be explained by factors other than ploidy. Larger size and functional aspermy seem to be common feminized characteristics in diploid males of numerous hymenopterans [39]. The reproductive tracts in all diploid males of *A. gracilipes* we dissected, however, are fully functional with the presence of viable sperm, suggesting a negligible effect of ploidy level. We therefore regard the excess of food supply (to ensure worker survival since orphaned [40]) to the orphaned colony fragments or other factors such as social environment as an alternative contributing factor for the larger size of haploid males.

Physogastric workers as trophic specialist

Our observations have suggested that trophic eggs constitute a major dietary regime for larvae and approximately 11% of dietary regime in queens, suggesting physogastric workers that account for production of trophic eggs function as a trophic specialist in *A. gracilipes* colonies. At least three additional lines of evidence support such a nutritional role of the physogastric workers. First, while consumption of trophic eggs as main diet has been widely reported in ant species lacking the ability to share resources via trophallaxis [41], trophic eggs may hold as equal nutritional value in other trophallaxis-performing ant species [40, 42]. One plausible reason among is that trophic eggs serve as an essential food source for a specific caste and/or developmental stage in the colony [8, 43]. Our data are in perfect agreement with such prediction as larvae of *A. gracilipes* appear to mainly consume trophic eggs during our entire observation period. Moreover, trophic eggs also are occasionally fed to queens, males and nestmate workers of *A. gracilipes* despite the presence of trophallaxis, further confirming that trophic eggs may serve as additional nutritional sources under some circumstances. Secondly, physogastric workers were found to occur together with younger brood and queens in the royal chamber (Additional file 2: Video S1, Additional file 4: Figure S2), and never engaged in foraging or other tasks outside the royal chamber. One may expect that trophic eggs, once produced, could be fed to the queen and larvae right away as they all stay within close proximity. This interpretation is further supported by our video showing that a trophic egg was fed to the adjacent brood pile immediately after it was laid by a physogastric worker (Additional file 2: Video S1).

The third line of evidence linked to the trophic function of physogastric workers is that the proportion of physogastric workers in the colony appears to be higher during fall and winter based on a preliminary field observation (CCLee et al., unpublished data). *A. gracilipes* is well-known for its broad diet as they prey on a variety of invertebrates as protein-rich food source (e.g., insects, small isopods and arachnids [16]). However, prey items of such kind cannot be stored easily as they are perishable, and its availability also fluctuates on a seasonal basis [43]. The high proportion of physogastric workers in the colony likely results in an increasing production of trophic eggs and thus represents an innate response of the colony to the declining availability of arthropod prey during such seasons [44]. We therefore suggest that the production of trophic eggs can be further regarded as an adaptive strategy for *A. gracilipes*, allowing colonies to sustain during unfavorable climatic conditions or periods of food shortage (e.g., winter) as trophic eggs can be stored for a longer period [45] and easily redistributed within the colony when needed.

Evolution of worker production in *A. gracilipes*

The presence of males in only one out of nine well-fed artificially-orphaned fragments suggests that male production by *A. gracilipes* workers after dequeening under field conditions appear to be uncommon. Such pattern might be explained by three mutually non-exclusive mechanisms: 1) If worker-laid trophic eggs are essential in terms of nutrient provision in the colony of *A. gracilipes*, selection may favor increased reproductive potential of worker castes (i.e., physogastric workers). Thus, the occasional emergence of worker-produced males could simply represent a by-product of the reproductive workers possessing highly-developed ovaries [40, 41]. 2) Theoretically, each focal reproductive worker is expected to be more closely related to her own son than to the average worker-produced sons (nephews) [4]. Under this condition, kin selection theory predicts that potential conflict will arise among physogastric workers over male parentage as all physogastric workers are able to lay male eggs and will selectively remove work-laid brood (i.e., worker policing) to which they are less related [30, 46]. Nevertheless, extraordinarily high intracolony relatedness despite polygyny nature of *A. gracilipes* and unusual reproductive system [19, 20] indicate that worker-policing or competition for male parentage, if any, in this species possibly could not be explained by relatedness alone. 3) Aside from the relatedness hypothesis, low frequency of male production by workers in *A. gracilipes* may be attributed to selection for higher worker efficiency and colony-level productivity. An increasing number of studies have proposed that the cost of worker reproduction appears to underlie the regulation of worker policing and self-restraint in social insects [47, 48]. For instance, workers showed aggression behavior toward reproductive workers in the asexually reproducing ant, *Platythyrea punctata* where genetic conflicts are not expected as colony members are identical to each other due to clonality [49]. This is because reproductive workers invest less in non-reproductive tasks and hence may reduce the entire colony efficiency by disrupting the foraging activity or reducing life span in workers [27, 50, 51]. Similarly, it is highly possible that the male brood derived from workers in *A. gracilipes* is prevented from development by worker policing for optimizing priority task of physogastric workers, that is, the provisioning of nestmates with trophic eggs or other nursery-related tasks. Our data partially support this interpretation that male brood were found in all three viable egg producing colony fragments, but only one of which was observed with the presence of adult males.

Conclusion

Our study demonstrates for the first time that *A. gracilipes* workers possess functional ovaries and are able to produce both reproductive and trophic eggs. The former can be further developed into haploid males that may have equal reproductive fitness as their diploid counterparts, whereas the latter may have served as a critical

regulator for protein-rich food (especially for larvae), thus allowing the colonies of *A. gracilipes* to survive through periods of food shortage. Furthermore, the current study offers an excellent chance to study if production of trophic eggs functions as an adaptive strategy for *A. gracilipes* when encountering food shortage, and how such behavior contributes to the success and ecological dominance of this ant as invasive species. We are currently generating the necessary baseline data to elucidate the ecological role of physogastric workers, factors that trigger ovary development of workers, the reproductive value of worker-produced males and how the combination of these mechanisms contributes to the invasiveness of this ant species.

Methods
Existence of physogastric workers under natural conditions and reproductive organs of *A. gracilipes* workers

Between December 2015 and February 2016, three queen-right colonies of *A. gracilipes* were collected from Nantou (AGQR01), Changhua (AGQR02) and Miaoli (AGQR03) counties, Taiwan (Additional file 1: Figure S1), and brought to the lab for further inspection and experimental manipulations. Firstly, the presence of physogastric workers was visually inspected, and the percentage of physogastric workers in each colony was assessed. We define physogastric workers as workers whose gaster size is distinctly greater than that of normal foraging workers, and that appear brown-whitish in color. Thirty physogastric and thirty normal workers were randomly selected from each colony and dissected shortly after collection in the field (between 1 and 2 weeks). Prior to dissection, we measured gaster width (GW), maximum transverse distance across the gaster in dorsal view. Workers were anaesthetized with carbon dioxide followed by pulling of the last gastral tergite by forceps in PBS solution. Fat and tissue were removed to ease subsequent observation. To determine the ovarian development of workers (both physogastric and normal ones), the number of ovarioles/individual, number of mature or yolky oocytes per ovariole, and total number of yolky oocytes were counted for each worker inspected. As immature ovarioles are threadlike and difficult to visualize during dissection, we only focused on those ovarioles with at least one visible oocyte. The presence of yellow bodies and a spermatheca was also visually inspected in both types of workers. In addition, three queens per colony (a total of nine) were dissected to characterize the anatomical differences between queen and worker.

SEM analysis and histology

Physogastric and normal workers for scanning microscopy were critical point dried in a Balzers CPD 030 instrument, mounted on SEM-stubs, coated with gold, and examined in a JEOL JSM-6360 scanning microscope. To further confirm the presence of a spermatheca in both worker types, five physogastric and five normal workers were randomly selected from each colony (30 in total) mentioned above for histological sections. The posterior part of the gaster was cut off using microscissors and was fixed in cold 2% glutaraldehyde in a 50 mM Na-cacodylate buffer at pH 7.3 with 150 mM saccharose. After postfixation in 2% osmium tetroxide in the same buffer and dehydration in a graded acetone series, tissues were embedded in Araldite. Serial longitudinal sections with a thickness of 2 μm were made with a Leica EM UC6 ultramicrotome, stained with methylene blue and thionin and viewed in an Olympus BX-51 microscope. Voucher specimens were deposited in the Research Institute for Sustainable Humanosphere, Kyoto University, and are available upon request.

Production of eggs in artificially-orphaned colonies

A total of three colony fragments constituted of 100 randomly-selected normal workers were separated from each of the three original nests (n = 9). Caution was taken to avoid transfer of eggs and brood from the original colonies to ensure the presence of eggs in the colony fragment after isolation is the result of worker reproduction. Each colony fragment was cultured in a polyethylene container (39 × 31 × 10 cm) with its edges and inner surfaces coated with a thin layer of fluon to prevent escape of ants. Sugar water (10%), crickets, and honeybee larvae were provided *ad libitum*. The experimental colony fragments were maintained under constant environmental conditions of 26 ± 1 °C, 60 ± 5% relative humidity and a 12-h photoperiod. The egg and brood production were monitored (4 months) on a weekly basis, starting 4 weeks after colony fragmentation. If eggs were found, the morphology of the egg and eventual presence of an embryo were examined under a microscope. Some of the eggs were left uncollected and allowed to develop into pupa and adult stage if possible.

Sex, ploidy level and morphology of worker-produced offspring

If any worker-produced offspring was found at the end of the experimental period, both worker-produced offspring and several nestmate workers (randomly selected from the same colony fragment) were subjected to microsatellite genotyping. We genotyped a total of 14 worker-produced males and nestmate workers each from a queenless colony fragment (AGQLF03). To compare the genotypic distribution of individuals between queenright and queenless colony fragments, individuals of different castes including queen, workers and males from a queenright colony (AGQR01) were also genotyped. We genotyped a total of 15 workers and 20 males each from the queenright colony. Genomic DNA was extracted

from tissue of each individual ant using the Gentra Puregene cell and tissue kit (Qiagen, USA) according to the manufacturer's instructions. Individual genotypes were assessed at seven nuclear microsatellite loci, including *Ano1*, *Ano3*, *Ano4*, *Ano5*, *Ano6*, *Ano8* and *Ano10*, previously developed by Feldhaar et al. 2006 [52]. Microsatellite loci were amplified using the multiplex PCR method described by Blacket et al. 2012 [53]. The seven loci were amplified in two separate 15 μL multiplex-PCRs, each containing three to four pairs of primers (0.2 μM), 0.2 unit of SuperTherm Hot-start Taq DNA Polymerase (JMR Holdings, UK), 0.25 mM of each dNTP, 1X SuperTherm Gold PCR buffer (JMR Holdings, UK), and 10–20 ng of template DNA. Thermal cycling profiles were as follows: one cycle of 95 °C (10 min), followed by 35 cycles of 94 °C (30 s), primer-specific annealing temperature 55 °C (30s), and 72 °C (30 s), followed by a single final extension of 72 °C (30 min). The resulting PCR products were analysed on an ABI-3730 Genetic Analyzer (Applied Biosystems) by Genomics BioSci & Tech Co., Ltd. (Taipei, Taiwan). GeneMarker program (version 2.4.0, Softgenetics LLC) was employed to visualize and score alleles. Samples harboring homozygous multi-locus genotypes were considered haploid individuals, while those with heterozygote at one or more loci were considered diploid.

If any pupa or adult male successfully emerged in a worker-only colony fragment, both life stages were subjected to morphometric measurement. Pupa and male sizes were measured as head width (HW), maximum width of the head between the compound eyes and total body length (TL), and the total outstretched length from the mandibular apex to the gastral apex. The above-mentioned measuring procedures were repeated on the individuals collected from queenright colony as reference.

Fate of worker-laid trophic eggs

A colony fragment composed by individuals from different castes was separated from each of the two original nests (AGQR01 & AGQR02; n = 2). Each colony fragment was maintained in a polypropylene container in which several transparent plastic boxes were inversely placed for housing ants as nest chambers [54]. The bottom of the container was filled with moistened plaster of Paris. These two colony fragments were designated as AGTE01 and AGTE02. A nest chamber was randomly selected for observation. Egg-laying workers or workers carrying trophic eggs were identified and observed for a total of 10.5 h (30 min observation period; n = 21) and 10 h (30 min per observation period; n = 20) in AGTE01 and AGTE02, respectively. More specifically, after a trophic egg was laid, we observed the fate of a given trophic egg as expressed by which caste a given trophic egg was offered to. As queens generally consumed trophic eggs much faster than other castes (Ito et al., un-published data), we conducted a separate observation in which number of trophic eggs consumed by a given queen was recorded. Duration of observation was 10.5 h for each queen (30 min per observation period; $n = 21$).

Statistical analysis

The gaster size and reproductive parameters (i.e., number of ovarioles/individual, number of yolky oocytes per ovarioles, and total number of yolky oocytes) between physogastric and normal workers were compared and analysed with Mann-Whitney U-test using SPSS version 16.0 (SPSS, Chicago, IL, USA) at 95% confidence interval. The same test was also applied to examine the morphometric differences between worker- and queen-produced offspring.

Additional files

Additional file 1: Figure S1. Map of Taiwan showing the collection sites of three queenright colonies (AGQR01–03) used in the current study. (TIFF 63 kb)

Additional file 2: Video S1. Fate of trophic eggs. A physogastric worker (2nd worker in the upper left-hand corner) bends its gaster forward, seizes the freshly-laid egg with mandible and immediately offers the egg to an adjacent larvae pile. The video can be accessed through the URL https://www.youtube.com/watch?v=SvyrSZ-4n-s&feature=youtu.be. (MOV 2973 kb)

Additional file 3: Video S2. Sperm bundles. Motile sperm bundles in the seminal vesicle of worker-produced males. The video can be accessed through the URL https://www.youtube.com/watch?v=-AfHtSnak6A. (MOV 4352 kb)

Additional file 4: Figure S2. Physogastric workers in royal chamber. Physogastric workers were found tending younger brood (a) and form a dense retine around the queen (b). (JPEG 14314 kb)

Acknowledgements
The authors would like to thank Po-Cheng Hsu for technical assistance. We also are very grateful to An Vandoren and Alex Vrijdaghs for their assistance in section preparation and scanning microscopy.

Funding
This study was financially supported through a Prospection Visit Grant of Leuven University (JB), the Ministry of Science and Technology, Taiwan (CCY), NTU Career Development Aid Grant (CCY) and the Future Development Funding Program of the Kyoto University Research Coordination Alliance (CCY).

Authors' contributions
CCL and CCY carried out the analyses and drafted the manuscript. CCLin, CCY and CYL planned and coordinated the study. CCL, HWH and JWT carried out the dissections and field collection of ants. Histology sections and SEM examination were done by JB. Observations on trophic eggs and queens' feeding behavior were performed by HN and FI. Microsatellite analysis was done by SPT. The video fragments were filmed and prepared by GLL. All authors read, edited and approved the final version of the manuscript.

Competing interests
The authors declare that they have no competing interests.

Author details
[1]Department of Biology, National Changhua University of Education, No. 1, Jin-De Rd., Changhua 50007, Taiwan. [2]Master Program for Plant Medicine, National Taiwan University, No.1, Sec. 4, Roosevelt Rd., Taipei, Taiwan106. [3]Faculty of Agriculture, Kagawa University, Ikenobe, Miki 761–0795, Japan. [4]Department of Entomology, National Taiwan University, No.1, Sec. 4, Roosevelt Rd., Taipei, Taiwan106. [5]Research Institute for Sustainable Humanosphere, Kyoto University, Gokasho, Uji, Kyoto 611-0011, Japan. [6]Department of Entomology, University of California, Riverside, CA 92521, USA. [7]K.U. Leuven, Zoological Institute, Naamsestraat 59, box 2466, B-3000 Leuven, Belgium. [8]Urban Entomology Laboratory, Vector Control Research Unit, School of Biological Sciences, Universiti Sains Malaysia, 11800 Penang, Malaysia.

References
1. Wilson EO. The insect societies. Cambridge: Belknap Press of Harvard University Press; 1971.
2. Hamilton WD. The genetical evolution of social behaviour, I,II. J Theor Biol. 1964;7:1–52.
3. Hamilton WD. Altruism and related phenomena, mainly in social insects. Annu Rev Ecol Syst. 1972;3:193–232.
4. Bourke AFG, Franks NR. Social evolution in ants. Princeton: Princeton University Press; 1995.
5. Khila A, Abouheif E. Evaluating the role of reproductive constraints in ant social evolution. Phil Trans R Soc B. 2010;365:617–30.
6. Hölldobler B, Wilson EO. The ants. Cambridge: Harvard University Press; 1990.
7. Bourke AFG. Worker reproduction in the higher eusocial Hymenoptera. Q Rev Biol. 1988;63:291–311.
8. Wheeler DE. Nourishment in ants: patterns in individuals and societies. In: Hunt JH, Nalepa CA, editors. Nourishment and evolution in insect societies. Boulder: Westview; 1994. p. 245–78.
9. Oster G, Wilson EO. Caste and ecology in the social insects. Princeton: Princeton University Press; 1978.
10. Villet MH, Crewe RM, Duncan FD. Evolutionary trends in the reproductive biology of ponerine ants (Hymenoptera: Formicidae). J Nat Hist. 1991;25:1603–10.
11. Peeters C, Keller RA, Johnson RA. Selection against aerial dispersal in ants: two non-flying queen phenotypes in Pogonomyrmex laticeps. PLoS One. 2012;7:e47727.
12. Lowe S, Browne M, Boudjelas S, De Poorter M. 100 of the world's most invasive alien species: A selection from the global invasive species database. The Invasive Species Specialist group. http://rewilding.org/rewildit/images/IUCN-GISP.pdf. Accessed Dec 2000.
13. Abbott KL. Supercolonies of the invasive yellow crazy ant, Anoplolepis gracilipes, on an oceanic island: Forager activity patterns, density and biomass. Insect Soc. 2005;52:266–73.
14. O'Dowd DJ, Green PT, Lake PS. Invasional 'meltdown' on an oceanic island. Ecol Lett. 2003;6:812–7.
15. Holway DA, Lach L, Suarez AV, Tsutsui ND, Case TJ. The causes and consequences of ant invasions. Annu Rev Ecol Syst. 2002;33:181–233.
16. Haines IH, Haines JB, Cherrett JM. The impact and control of the crazy ant, Anoplolepis gracilipes (Jerd.), in the Seychelles. In: Williams DF, editor. Exotic ants. Biology, impact and control of introduced species. Boulder: Westview Press; 1994. p. 206–19.
17. Hill M, Holm K, Vel T, Shah NJ, Matyot P. Impact of the introduced yellow crazy ant Anoplolepis gracilipes on Bird Island, Seychelles. Biodivers Conserv. 2003;12:1969–84.
18. Matsui S, Kikuchi T, Akatani K, Horie S, Takag M. Harmful effects of invasive yellow crazy ant Anoplolepis gracilipes on three land bird species of Minamidaito Island. Ornithological Sci. 2009;8:81–6.
19. Gruber MAM, Hoffmann BD, Ritchie PA, Lester PJ. The conundrum of the yellow crazy ant (Anoplolepis gracilipes) reproductive mode: no evidence for dependent lineage genetic caste determination. Insect Soc. 2013;60:135–45.
20. Drescher J, Blüthgen N, Feldhaar H. Population structure and intraspecific aggression in the invasive ant species Anoplolepis gracilipes in Malaysian Borneo. Mol Ecol. 2007;16:1453–65.
21. Dietemann V, Peeters C. Queen influence on the shift from trophic to reproductive eggs laid by workers of the ponerine ant Pachycondyla apicalis. Insect Soc. 2000;47:223–8.

22. Gobin B, Peeters C, Billen J. Production of trophic eggs by virgin workers in the ponerine ant *Gnamptogenys menadensis*. Physiol Entomol. 1998;23:329–36.

23. Grangier J, Avril A, Lester PJ. Male production by workers in the polygynous ant *Prolasius advenus*. Insect Soc. 2013;60:303–8.

24. Ichinose K, Lenoir A. Reproductive conflict between laying workers in the ant *Aphaenogaster senilis*. J Ethol. 2009;27:475–81.

25. Gruber M, Hoffmann B, Ritchie P, Lester P. Crazy ant sex: genetic caste determination, clonality, and inbreeding in a population of invasive yellow crazy ants. In: Nash DR, den SPA B, Fine Licht HH, Boomsma JJ, editors. Copenhagen: XVI Congress of the International Union for the Study of Social Insects; 2010. p. 93.

26. Wenseleers T, Van Oystaeyen A. Unusual modes of reproduction in social insects: Shedding light on the evolutionary paradox of sex. BioEssays. 2011;33:927–37.

27. Cole BJ. The social behavior of *Leptothorax allardycei* (Hymenoptera: Formicidae): time budgets and the evolution of worker reproduction. Behav Ecol Sociobiol. 1986;18:165–73.

28. Endler A, Liebig J, Schmitt T, Parker JE, Jones GR, Schreier P, Hölldobler B. Surface hydrocarbons of queen eggs regulate worker reproduction in a social insect. Proc Natl Acad Sci U S A. 2004;101:2945–50.

29. Fletcher DJC, Ross KG. Regulation of reproduction in eusocial Hymenoptera. Annu Rev Entomol. 1985;30:319–43.

30. Ratnieks FLW. Reproductive harmony via mutual policing by workers in eusocial Hymenoptera. Am Nat. 1988;132:217–36.

31. Wenseleers T, Ratnieks FLW. Enforced altruism in insect societies. Nature. 2006;444:50.

32. Chapuisat M, Sundström L, Keller L. Sex-ratio regulation: the economics of fratricide in ants. Proc R Soc Lond B. 1997;264:1255–60.

33. Billen JPJ. Ultrastructure of the workers ovarioles in *Formica* ants (Hymenoptera: Formicidae). Int J Insect Morphol Embryol. 1985;14:21–32.

34. Peeters C, Liebig J, Hölldobler B. Sexual reproduction by both queens and workers in the ponerine ant *Harpegnathos saltator*. Insect Soc. 2000;47:325–32.

35. Dietemann V, Hölldobler B, Peeters C. Caste specialization and differentiation in reproductive potential in the phylogenetically primitive ant *Myrmecia gulosa*. Insect Soc. 2002;49:289–98.

36. Armitage S, Boomsma JJ, Baer B. Diploid male production in a leaf-cutting ant. Ecol Entomol. 2010;35:175–82.

37. Ross KG, Fletcher DJC. Genetic origin of male diploidy in the fire ant *Solenopsis invicta* (Hymenoptera, Formicidae) and its evolutionary significance. Evolution. 1985;39:888–903.

38. Yamauchi K, Yoshida T, Ogawa T, Itoh S, Ogawa Y, Jimbo S, Imai HT. Spermatogenesis of diploid males in the formicine ant, *Lasius sakagamii*. Insect Soc. 2001;48:28–32.

39. Zayed A, Packer L. Complementary sex determination substantially increases extinction proneness of haplodiploid populations. Proc Natl Acad Sci U S A. 2005;102:10742–6.

40. Dijkstra MB, Boomsma JJ. Are workers of *Atta* leafcutter ants capable of reproduction? Insect Soc. 2006;53:136–40.

41. Smith CR, Schoenick C, Anderson KE, Gadau J, Suarez AV. Potential and realized reproduction by different worker castes in queen-less and queen-right colonies of *Pogonomyrmex badius*. Insect Soc. 2007;54:260–7.

42. Heinze J, Cover SP, Hölldobler B. Neither worker, nor queen: an ant caste specialized in the production of unfertilized eggs. Psyche. 1995;102:173–85.

43. Crespi BJ. Cannibalism and trophic eggs in subsocial and eusocial insects. In: Elgar M, Crespi BJ, editors. Cannibalism: ecology and evolution among diverse taxa. Oxford: Oxford University Press; 1992. p. 176–213.

44. Bolger DT, Suarez AV, Crooks KR, Morrison SA, Case TJ. Arthropod in urban habitat fragments in southern California: area, age, and edge effects. Ecol Appl. 2000;10:1230–48.

45. Voss SH, Blum MS. Trophic and embryonated egg production in founding colonies of the fire ant *Solenopsis invicta* (Hymenoptera: Formicidae). Sociobiology. 1988;13:271–8.

46. Wenseleers T, Ratnieks FLW. Comparative analysis of worker reproduction and policing in eusocial Hymenoptera supports relatedness theory. Am Nat. 2006;168:E163–79.

47. Hammond RL, Bruford MW, Bourke AFG. Male parentage does not vary with colony kin structure in a multiple-queen ant. J Evol Biol. 2003;16:446–55.

48. Dijkstra MB, Boomsma JJ. The economy of worker reproduction in *Acromyrmex* leafcutter ants. Anim Behav. 2007;74:519–29.

49. Hartmann A, Wantia J, Torres JA, Heinze J. Worker policing without genetic conflicts in a clonal ant. Proc Natl Acad Sci U S A. 2003;100:12836–40.

50. Heinze J, Puchinger W, Hölldobler B. Worker reproduction and social hierarchies in *Leptothorax* ants. Anim Behav. 1997;54:849–64.

51. Tsuji K, Kikuta N, Kikuchi T. Determination of the cost of worker reproduction via diminished life span in the ant *Diacamma* sp. Evolution. 2012;66:1322–31.

52. Feldhaar H, Drescher J, Blüthgen N. Characterization of microsatellite markers for the invasive ant species *Anoplolepis gracilipes*. Mol Ecol Notes. 2006;6:912–4.

53. Blacket MJ, Robin C, Good RT, Lee SF, Milner AD. Universal primers for fluorescent labelling of PCR fragments–an efficient and cost-effective approach to genotyping by fluorescence. Mol Ecol Resour. 2012;12:456–63.

54. Ito F, Asfiya W, Kojima J. Discovery of independent-founding solitary queens in the yellow crazy ant *Anoplolepis gracilipes* in East Java, Indonesia (Hymenoptera: Formicidae). Entomol Sci. 2016;19:312–4.

4

Getting fat or getting help? How female mammals cope with energetic constraints on reproduction

Sandra A. Heldstab[*], Carel P. van Schaik and Karin Isler

Abstract

Background: Fat deposits enable a female mammal to bear the energy costs of offspring production and thus greatly influence her reproductive success. However, increasing locomotor costs and reduced agility counterbalance the fitness benefits of storing body fat. In species where costs of reproduction are distributed over other individuals such as fathers or non-breeding group members, reproductive females might therefore benefit from storing less energy in the form of body fat.

Results: Using a phylogenetic comparative approach on a sample of 87 mammalian species, and controlling for possible confounding variables, we found that reproductive females of species with allomaternal care exhibit reduced annual variation in body mass (estimated as CV body mass), which is a good proxy for the tendency to store body fat. Differential analyses of care behaviours such as allonursing or provisioning corroborated an energetic interpretation of this finding. The presumably most energy-intensive form of allomaternal care, provisioning of the young, had the strongest effect on CV body mass. In contrast, allonursing, which involves no additional influx of energy but distributes maternal help across different mothers, was not correlated with CV body mass.

Conclusions: Our results suggest that reproducing females in species with allomaternal care can afford to reduce reliance on fat reserves because of the helpers' energetic contribution towards offspring rearing.

Keywords: Allomaternal care, Cooperative breeding, Body fat, Paternal care, Helping behaviours, Reproduction, Allonursing, Provisioning

Background

Reproduction is energetically very expensive [1, 2] and several studies show that the amount of food available and hence the total amount of energy invested by the mother influences reproductive success in female mammals. Provisioning by humans generally leads to higher reproductive rates, shorter lactation periods, and shorter inter-birth intervals [3–5]. In natural animal populations, higher food abundance leads to higher birth rates [6–11]. In contrast, food restriction may delay sexual maturation and among adults may inhibit mating behaviour [12–14] or even produce acyclicity or anoestrus [15, 16].

In mammals that evolved in seasonal environments and thus face periods of food scarcity, a female's ability to bear the energy costs of pregnancy and lactation, and thus her reproductive success, may be affected by the amount of body fat she can deposit. That stored body fat plays an essential role in female reproduction has been proposed previously within the capital-income-continuum concept (for a review see [17]) and empirical evidence for this idea is abundant. For instance, in rhesus macaques (*Macaca mulatta*) and moose (*Alces alces*), the size of maternal fat stores positively affects pregnancy and birth rates [18, 19]. Furthermore, numerous studies show that heavier and fatter mothers produce heavier offspring that grow faster and are more likely to survive, suggesting that females in better body condition are able to allocate more stored resources to reproduction [20–25]. Finally, several studies in seals show that body fat is essential for lactation as seal

---

* Correspondence: sandra.heldstab@uzh.ch
Department of Anthropology, University of Zurich, Winterthurerstrasse 190, 8057 Zurich, Switzerland

mothers lose more than 50% of their stored body fat until the end of lactation ([26] and references therein). Significant seasonal fattening in females may also be found if they do not reproduce, e.g. to buffer environmental food fluctuations [27, 28]. However, because reproductive seasons and experienced seasonality in food intake are generally interrelated, it is usually impossible to disentangle these two reasons for body fat storage [29–33]. Female polar bears (*Ursus maritimus*) offer an extreme example of this. They store body fat to hibernate due to adverse environmental conditions for up to 8 months while simultaneously meeting the nutritional demands of gestation and lactation during this fasting period [23].

But the positive effect of fat stores on fitness is counterbalanced by their costs. Large fat reserves increase the energy costs of locomotion due to higher body weight [34–37], and also reduce agility and speed and so may compromise fitness by increasing predation risk or decreasing hunting success [38–42]. Furthermore, in arboreal species, body fat may also impede terminal branch feeding [43]. Indeed, arboreal species are less prone to store fat than terrestrial ones [44]. Therefore, we hypothesize that female mammals should minimize the amount of fat stores if they have an alternative to fuel their reproductive success.

All other things being equal, the energetic burden of reproduction on reproductive females is reduced when the costs of reproduction are distributed over several individuals. Thus, in species where other individuals provide energetic costly allomaternal care behaviours, breeding females might need to store less energy in the form of body fat themselves and could avoid the locomotion and predation costs resulting from high amounts of body fat. Allomaternal inputs are found in many mammals, comprising behaviours such as provisioning, carrying, huddling or communal nesting, babysitting, and protection from predators or defence of resources against conspecifics. The effects of such allomaternal care on offspring survival or fertility have been demonstrated within and between species [45–51]. One likely mechanism underlying this effect is load-lightening of pregnant or lactating females by helpers ('load-lightening' hypothesis [52]) which has been demonstrated in meerkats [53], callitrichids [54, 55] and siamangs [56]. This load-lightening effect has also been demonstrated in some species with facultative helping, where females can rear their pups solitarily, but under certain conditions share care for the young with one or more additional individuals. For instance, female prairie voles (*Microtus ochrogaster*) and pine voles (*Microtus pinetorum*) had shorter interlitter intervals in family groups consisting of the breeding pair and former offspring compared to families without previous offspring [57, 58]. In striped mice (*Rhabdomys pumilio*) living in the succulent karoo, offspring grew faster when the father was present, which may indirectly benefit females when young are weaned earlier [59]. In females of a facultatively cooperative breeding bird species, the splendid fairy-wren (*Malurus splendens*), the presence of helpers has been shown to increase survival of the breeding females and reduce the time for these females to renest after a brood [60]. Lastly, in another facultative cooperative breeder, the western bluebird (*Sialia mexicana*), the presence of helpers allowed the breeding female to lower her feeding rate, while nestlings still received more feeds at nests with helpers compared to nests without helpers present [60]. In sum, there is ample empirical evidence that distributing the costs of reproduction over two or more individuals yields an energetic benefit for mothers or offspring. We do not distinguish between the two, as a net fitness effect can be obtained by either.

Allonursing, the nursing of non-filial offspring, is another form of care that has been observed in every major mammalian lineage [61, 62]. However, allonursing events within a species are generally rare. For instance, in tufted capuchin monkeys (*Sapajus nigritus*) allosuckling accounted for 13% of all suckling events [63], in South American fur seals (*Arctocephalus australis*) for around 3% [64], and in red deer calves (*Cervus elaphus*) allosuckling was even less common [65]. Furthermore, the rejection rates of suckling of non-filial offspring are high. In guanacos (*Lama guanicoe*), for example, the rejection rate to non-filial offspring nursing attempts was three times higher than the rejection rate to filial nursing attempts [66]. Although allonursing may confer social benefits to the allonursed young [63, 67], the energetic benefits for offspring or mother are unclear. First, allonursing is more likely to occur when several females breed concurrently [62] and hence all females simultaneously bear the costs of reproduction. Therefore, the idea that allonursing functions as load-lightening mechanism for lactating females cannot apply [68], and instead allonursing may serve to more evenly divide maternal energy investment across different mothers [69]. Second, several studies show no apparent energetic benefits of allonursing for recipient offspring and/or mothers. For instance, red deer calves sucking only from maternal hinds increased faster in body weight than calves sucking maternal and non-maternal hinds [65]. Another study found no evidence that allonursing provides benefits to meerkat pups (*Suricata suricatta*) or mothers [70]: pups that received allonursing were not heavier at emergence and did not have a higher survival rate than pups that did not receive allonursing. Mothers whose litters were allonursed were not in better physical condition, did not reconceive faster and did not reduce their own nursing investment compared to mothers who nursed their litters alone. To sum up, allonursing does not necessarily provide energetic benefits for the mother or offspring.

With the exception of allonursing, all other allomaternal care behaviours can be performed by all sorts of helpers in cooperatively breeding species, including fathers or non-breeding group members. Whereas the help provided by adult males (potential fathers) might be unaffected by their body condition [71] or food abundance [72], other non-breeding group members generally adjust their helping efforts in relation to their body condition. Furthermore, subordinates can also start to breed themselves, in which case their help to the dominant female could end abruptly or be minimal to begin with [73, 74]. These results suggest that paternal care is more reliable and thus more important for females than the help of others. On the other hand, in cooperative breeders more helpers than just the father might be around to take over the energetic costs of female reproduction. The optimum amount of body fat stored by a female may therefore vary depending on whether they receive no care, paternal care or additional help from several non-breeding group members.

The aim of this study is to test whether energetic contributions towards offspring rearing through costly care allow reproductive females to reduce the amount of energy (stored as body fat) they themselves need to invest. As a proxy for the seasonal tendency to store body fat, we use data on seasonal body mass variation within a year, the coefficient of variation (CV) in body mass, which has been shown to correlate with the amount of body fat within [44] and across species (PGLS: $P = 0.03$, $N = 8$, $\lambda = 0$, $R^2 = 0.56$, $\beta = 0.19$, S.E. = 0.07, $t = 2.74$, calculated from data in [44]). Compared to single body fat values obtained from cadavers, CV body mass captures seasonal fluctuations, allows for a larger sample size for each species and can also be collected for wild animals [75]. In total, both reliable information on the nature and extent of allomaternal help and sufficient data on annual variation in body mass was available for 87 species from 9 mammalian orders.

We expect that an increased energetic contribution in the form of allomaternal care provided by the male or non-breeding group members is negatively correlated with annual variation in body mass in females, because storing fat and allomaternal subsidies independently stabilize the energetic costs for female reproduction. To test this prediction, we explore the effect of different types of allomaternal help on annual body mass variation in females. On the other hand, we do not expect a correlation between allonursing behaviour and annual variation in body mass in females.

Methods

CV body mass as a proxy for the tendency to store body fat

In mammals, body fat explained between 41 and 92% of the intraspecific variation in body mass, the amount of body fat was highly correlated with carcass weight for each age and sex; hence body weight was a good predictor of total body fat (for a summary, see references in [44]). We therefore used seasonal changes of body mass over a year as a proxy for the tendency to store body fat. For a given species, we calculated the coefficient of variation (CV = standard deviation/mean) over monthly means of adult female body mass, yielding a total sample of 87 mammalian species from 9 orders (Additional files 1 and 2). In a previous study we validated the use of CV body mass as a proxy for variation in body fat by showing that the monthly body mass correlated with percentage body fat in several studies that measured both in the same specimens [44].

We compiled monthly body mass data from the literature, including only those studies that reported monthly mean body mass for at least 4 months per year. If body mass data were given for four seasons, pooled across several months (e.g., spring, summer, autumn and winter), we set the number of months sampled to four (16 studies). In most species, monthly mean body mass data was distributed evenly across the year, except for *Antechinus stuartii*, *Lycaon pictus*, *Spermophilus franklinii* and *Zapus hudsonicus*. If several sources were available for one species, preference was given to the study with the largest sample size conducted in the wild.

Allomaternal care behaviours

In quantifying allomaternal care behaviour, we followed Isler and van Schaik [76] to obtain continuous data on the frequency of occurrence of the following care behaviours: provisioning, carrying, protection and a variable that comprises other energetically influential care behaviours such as huddling, communal nesting and pup retrieval (see Additional file 3 for a detailed description of the classification protocol). As the sample in [76] was restricted to species with known brain size, we expanded it by an additional 30 species for which data on both CV body mass and allomaternal care behaviour was available in the literature (Additional files 1 and 2). In total, CV body mass and data on allomaternal care behaviour were available for 87 species. We did not compile data for bats and cetaceans because reliable data on allomaternal care of both cetaceans and bats are notoriously difficult to obtain. Moreover, the amount of body fat and hence CV body mass as a proxy for the tendency to store body fat in these two groups may underlie different constraints than in other mammals [44, 77–79], precluding predictions for a combined sample.

In addition, to distinguish the effects of allomaternal care provided by males (paternal care) from that provided by other group members (care by others) we summed up the frequency of occurrence of all allomaternal care behaviours separately for the father and other group members. To investigate whether the results reported in this study are robust with respect to different coding schemes of allomaternal care, we additionally conducted all analyses by using a binary classification of all allomaternal care behaviours, with 1 indicating the presence and 0 the absence of the helping behaviour. Finally, we also conducted additional analyses with a binary classification of allomaternal care provided by males (paternal care) and that provided by other group members (care by others) (data from [76, 80, 81]).

Covariates

As captivity might affect body mass variation (for instance, under good husbandry conditions, most animals gain weight in captivity [82]), we added provenance (wild = 1/captivity = 0) as an additional factor in all analyses. Furthermore, we analysed the subsample of studies including only wild-caught females separately.

In a previous study we found that substrate use (arboreal versus terrestrial) influenced the amount of body fat of a species [44]. We therefore added substrate use as an additional factor in all analyses. Data from published sources were used to assign each species to one of two substrate use categories, terrestrial (0) or arboreal (1), based on their main habit. Species were classified as terrestrial when they spent more than 50% of observation time on the ground ([83–86], see Additional file 1).

We also controlled for several other potential methodological confounds. First, some studies include body mass data from pregnant and lactating females in the population mean, which may artificially increase annual body mass variation in seasonal breeders. Pregnancy affects a female's weight due to the added weight of the offspring and the associated tissues and fluids. To control for this effect, we added the variable "inclusion of reproductive females in the study" as a covariate. Second, we added the number of months sampled as covariate. Ideally, we would have preferred to use only those studies from the wild that reported the mean body mass for 12 consecutive months. However, in contrast to studies in captivity, most body mass data of wild living mammals have been recorded less frequently. Third, to control for allometric effects of size, we performed all analyses including log-transformed mean body mass as a covariate, taking the overall mean from the same specimens for which CV body mass was determined. Finally, as variation in female body mass may be influenced by life history traits such as litter size, neonatal mass, and the duration of gestation and lactation, we also included those as potential covariates.

Statistical analyses

Statistical analyses were done in JMP™ 12.0 [87] and in R3.1.3 [88]. In most species that exhibit allomaternal care, various kinds of care behaviours are observed,

potentially resulting in collinearity problems in the statistical analyses. We checked this by generating variance inflation factors (VIF) to assess potential multicollinearity in the full set of allomaternal care behaviours [89, 90] using non-phylogenetic generalized linear models and the function "vif" ("car" package: [91]) in R. VIFs quantify how much the variance of an estimated model parameter is increased because of multicollinearity between predictors. The VIF for carry by the male, carry by others, provisioning by the male and provisioning by others was higher than 5, which indicates a problematic amount of covariance among predictors [92]. To solve this, we summed up the frequency of occurrence of carrying by the male and by others to one single variable "carrying" and similarly provisioning by the male and provisioning by others to "provisioning". After this, the VIF of all allomaternal care behaviours in all models were less than 4, which indicates an acceptable amount of covariance among predictors (Additional file 4: Tables S1 and S2). Two life history traits (duration of gestation and neonatal mass) also showed VIFs consistently larger than 5 in all models (Additional file 4: Tables S1 and S2). To reduce the problematic multicollinearity in these models, we followed the method described in [93]: we first removed the life history variable with the highest VIF value from the models, the duration of gestation, and recalculated VIFs for the reduced models. Then, we removed neonatal mass, as it still had a VIF larger than 5. All remaining variables had VIFs lower than 5. We then repeated the analyses with the same specifications as the main analysis with these "reduced models" and assessed the relative contribution of each independent variable as described below.

We built phylogenetic generalized least-squares regressions (PGLS) models [94, 95] using the "caper" package [96] in R. Caper estimates PGLS model parameters in maximum likelihood [96] and the parameter lambda (λ), which quantifies the magnitude of the phylogenetic signal in the model residuals [94]. The value of λ can vary between 0, indicating no phylogenetic signal, and 1, indicating that the observed pattern fits a Brownian motion model of trait evolution along the branches of the phylogeny such that similarity between species is directly proportional to relatedness [94]. The phylogeny was based on a composite supertree from [97] (Additional file 5: Figure S1). CV body mass (used as a proxy for body fat) was the dependent variable, while measures of allomaternal care and all possible confounding variables (substrate use, provenance [wild / captivity], number of months sampled, inclusion of reproductive females, mean body mass and several life history variables) were independent variables in the PGLS models. We did not log-transform CV body mass values prior to the analysis as this would not have improved the skew of its distribution. Although the predictor CV body mass was skewed towards smaller values, the distribution of the residuals of the PGLS models were normally distributed and did not comprise any outliers.

We used a model selection approach based on the AICc (Aikaike Information Criterion with correction for finite sample size, [98]) to determine the most important allomaternal care behaviours for female CV body mass. We ran the model selection across all possible models built with the explanatory variables mentioned above. We accounted for uncertainty in the models by performing model averaging [99] in the candidate model set including models with ΔAICc <2 [100]. ΔAICc is the difference in AICc between the focal model and the AICc of the best-fitting model in the candidate model set. Estimates of each parameter were averaged across the candidate models (means were weighted by the Akaike weight of a given model). The relative importance of a predictor was obtained by summing the Akaike's weights of the models in the candidate model set including the focal predictor, following the method described by Symonds and Moussalli [101]. The method to perform model averaging with the PGLS function in the package "caper" [96] is described in [102] and the corresponding material is available at http://www.mpcm-evolution.org.

Results

The results confirmed our two main predictions. Model selection and averaging showed that the most important

effect among allomaternal care behaviours on female CV body mass was provisioning of the young by the male and other group-members (Relative importance = 1) (Table 1, Fig. 1a). This form of allomaternal care was negatively correlated with CV body mass in reproductive females, suggesting that an energetic contribution towards offspring rearing allows females to reduce the amount of stored body fat. In contrast, allonursing, which involves no additional influx of energy but distributes maternal help across different mothers, did not correlate with CV body mass (Relative importance = 0.06) (Table 1). Results using a binary coding scheme of allomaternal care behaviours are strikingly similar (Additional file 4: Table S6 and S8, Fig. 1b).

Using a continuous coding scheme of paternal care and the amount of allomaternal care provided by other group members, we found that only paternal care showed a negative relationship with CV body mass (Relative importance = 1) (Table 2, Fig. 2a and b). In contrast, using a binary coding scheme, both paternal care and the amount of allomaternal care provided by other group members had a negative effect on CV body mass, although the negative effect of paternal care was stronger than that of allomaternal care by other group members (Additional file 4: Table S7 and S9, Figure S2a and b).

Results for the subset of studies including only wild-caught females (N = 49 species) were largely similar to those obtained from the whole sample, although the effects were a bit weaker (Additional file 4: Tables S10-S15).

Table 1 Continuous classification of allomaternal care behaviours: Averaged parameter estimates and their relative explanatory importance for female CV body mass (N = 87). Gestation length and neonatal mass were excluded to reduce multicollinearity between predictors. Numbers in bold indicate predictors whose confidence intervals of their effect exclude zero

Predictors		Relative importance of predictors	Model averaging estimates[a]	95% CI
Intercept			0.126	**(0.100, 0.153)**
Provisioning		1.00	−0.040	**(−0.043, −0.036)**
Protecting		0.06	−0.001	(−0.002, 0.001)
Carrying		0.07	0.003	(−0.004, 0.010)
Communal nesting		0.06	0.001	(−0.002, 0.004)
Allonursing		0.06	0.005	(−0.010, 0.021)
Log mean body mass		0.44	−0.006	**(−0.010, −0.002)**
Provenance	captive	0.80	na	na
	wild		0.025	**(0.017, 0.032)**
Substrate use	terrestrial	1.00	na	na
	arboreal		−0.045	**(−0.050, −0.041)**
Number of months		0.53	−0.001	(−0.001, 0.001)
Inclusion of reproductive females		0.69	−0.019	**(−0.030, −0.008)**
Log litter size		0.56	0.027	**(0.013, 0.041)**
Log weaning age		na	0	0

[a]averaged model estimates based on 12 models with ΔAICc (AICc $_{focal\ model}$ − AICc $_{best\ model}$) < 2 since the best AICc model is not strongly weighted (weight = 0.15) [104]. A full list of models is given in Additional file 4: Table S4. Reference levels of categorical variables have an estimate of 0; na – not applicable; 95% CI - 95% confidence interval

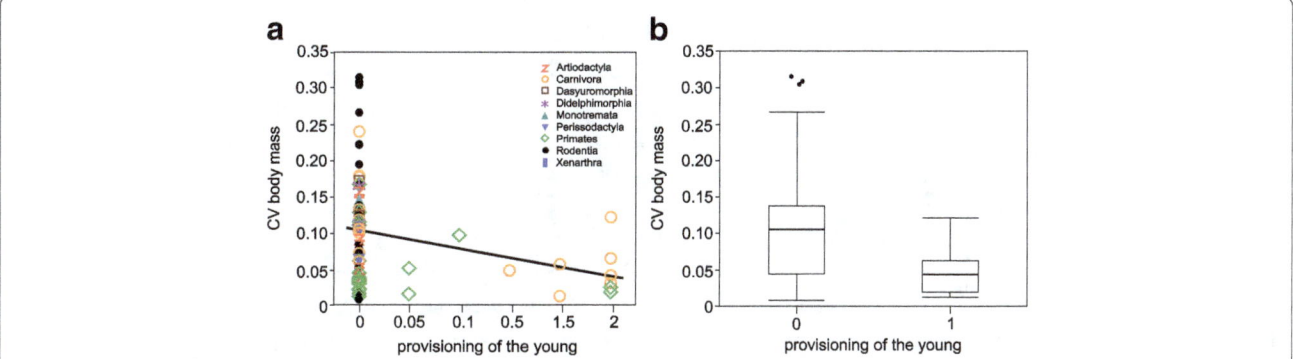

Fig. 1 a Female CV body mass as a function of provisioning of the young by the male and other group members, using the continuous coding scheme. **b** Female CV body mass is lower in species with provisioning of the young by the male and other group members (coded as 1) than in species without it (coded as 0). Details of phylogenetic models are shown in Table 1 and Additional file 4: Table S6. Species values are listed in the Additional file 1

In all analyses substrate use and provenance were correlated with CV body mass. Arboreal species had less body fat than terrestrial and semiaquatic species, as indicated by the negative correlation between CV body mass and substrate use. Furthermore, CV body mass was higher in wild-caught specimens compared to captive ones, suggesting that wild-caught individuals experience more variation in energy intake than provisioned specimens living in captivity. Controlling for further possible confounding variables (number of months sampled, inclusion of reproductive females, mean body mass, and several life history variables) did not change the effects of the main explanatory variables. In some models, both a lower species body mass and the inclusion of reproductive females in the study were related to a lower CV body mass, while species with a relatively high reproductive rate, as indicated by larger litters, exhibited a higher CV body mass. In some models, species for which fewer months were sampled showed a larger CV body mass (Tables 1 and 2 and Additional file 4: Tables S6, S7, S12 and S13).

Discussion

Using annual variation in body mass, we found that this CV body mass and the amount of allomaternal care show a pattern of correlated evolution among female mammals: females of those species with more contributions of non-mothers to offspring care exhibit reduced annual variation in body mass. From this, we conclude that allomaternal energy subsidies and fat storage are compensatory strategies to stabilise the energetic costs involved in female reproduction.

Table 2 Continuous classification of paternal care and care provided by other group members: Averaged parameter estimates and their relative explanatory importance for female CV body mass ($N = 87$). Gestation length and neonatal mass were excluded to reduce multicollinearity between predictors. Numbers in bold indicate predictors whose confidence intervals of their effect exclude zero

Predictors		Relative importance of predictors	Model averaging estimates[a]	95% CI
Intercept			0.148	**(0.127, 0.169)**
Care by others		na	0	0
Paternal care		1.00	−0.028	**(−0.029, −0.027)**
Log mean body mass		0.67	−0.008	**(−0.011, −0.004)**
Provenance	captive	0.80	na	na
	wild		0.024	**(0.017, 0.032)**
Substrate use	terrestrial	1.00	Na	na
	arboreal		−0.047	**(−0.050, −0.043)**
Number of months		0.38	−0.001	**(−0.002, −0.001)**
Inclusion of reproductive females		0.37	−0.011	**(−0.020, −0.002)**
Log litter size		0.24	0.007	(−0.001, 0.016)
Log weaning age		na	0	0

[a]averaged model estimates based on 11 models with ΔAICc (AICc $_{focal\ model}$ − AICc $_{best\ model}$) < 2 since the best AICc model is not strongly weighted (weight = 0.15) [104]. A full list of models is given in Additional file 4: Table S5. Reference levels of categorical variables have an estimate of 0; na – not applicable; 95% CI - 95% confidence interval

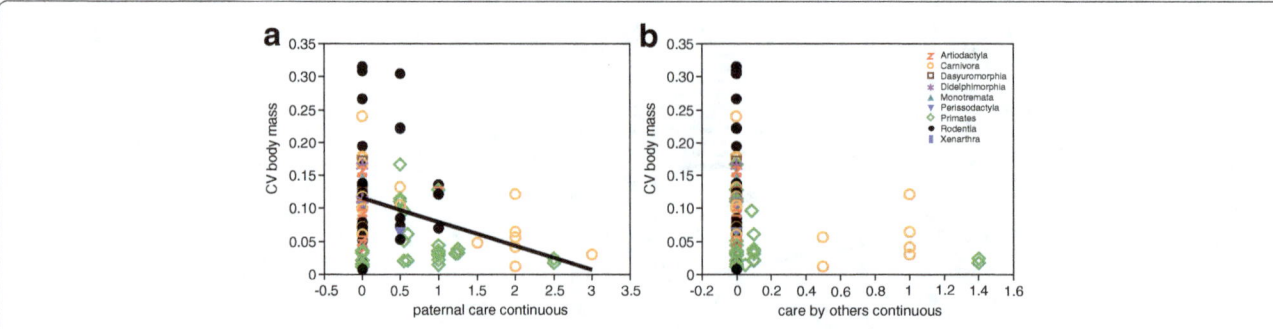

Fig. 2 Female CV body mass is lower in species with paternal care **a** but not with care provided by other group members (**b**), using the continuous coding scheme. Details of phylogenetic models are shown in Table 2. Species values are listed in the Additional file 1

First, we predicted that only an additional influx of energy in the form of costly allomaternal care behaviours by the male and other non-breeding group members towards the offspring and the mother would allow reproductive females to reduce the storage of body fat, whereas a mere redistribution of energy between mothers as in allonursing behaviour would not. As predicted, we only found a negative correlation between seasonal variation in body mass and the amount of allomaternal care in the form of provisioning of the young by the male and other group members, but not with allonursing. This suggests that if other conspecifics take over some of the maternal costs the need for these females to store extra body fat to fuel reproduction is relaxed.

This pattern across species is consistent with numerous intraspecific studies showing that extra energy delivered by costly care behaviours of helpers allows breeding females to reduce their maternal investment. For instance, in meerkats and cooperatively breeding bird species, an increased number of helpers enabled breeding females to maintain better condition and higher body mass and achieve a higher fitness [103–107]. In Campbell's dwarf hamsters (*Phodopus campbelli*) the presence of males protects females against extreme heat production in response to the exogenous heat requirements of the pups. As this acute increase in maternal temperature is thought to be a substantial cost to females, paternal presence likely allows females to decrease the energetic demands of reproduction [108]. Another study of the same species found that removal of the male not only decreased pup survival, growth, and readiness for dispersal by 18 days of age but also resulted in an additional 20% body weight loss in the female [109]. Lastly, a comparative study across mammals reveals that male care is associated with larger litters in some species or shorter lactation time in others, resulting in increased female fecundity [51].

Second, we investigated the effect of different types of allomaternal help (help of the male or other conspecifics)

on female fat stores. Both the help provided by the breeding male and the help provided by other group members showed a negative correlation with female CV body mass. However, the relative importance of allomaternal care provided by the breeding male was greater than the relative importance of help of other caretakers. This fits well with the often-reported finding that males care unconditionally, whereas care by helpers may be more conditional [71, 72, 110].

A broad comparative study as presented here can only provide an overview over potential patterns of correlated evolution and is limited by methodological issues. Ideally, we would have preferred to use individual variation in body fat over the year instead of the annual variation in body mass averaged over several females as used in this study. Although the published literature contains a variety of measures of adipose depots in living subjects such as palpation, skinfold thickness, perirenal adiposity, the number of adipocytes in bone marrow, and adipocyte volumes from tissue samples [111], these measures have not yet been compared to each other and each measure has only been applied to very few different species making broad phylogenetic comparisons impossible. Similarly, taking body fat values obtained from cadavers is problematic because they assess body fat at a single point in time, while the individual body fat fluctuations remain unknown [73].

It may be argued that, rather than taking annual variation in body mass, the costs of reproduction should be estimated by subtracting the maternal body weight at conception from the body weight at offspring weaning. However, such detailed data are rarely available, and may raise other issues, such as postpartum oestrus in lagomorphs, Callitrichid primates and several otariids, which means females suckle newborns while simultaneously being pregnant [112–115]. Even more importantly, in most mammals such as carnivores, rodents and primates allomaternal care and its beneficial effect for mothers continues post-weaning. Thus, offspring provisioning until independence allows females to invest more time in foraging, regain body condition more quickly and mate sooner [116], which we would not capture with the body weight difference of mothers between conception and offspring weaning.

In our study, some part of the variation in female body mass may result from the increasing weight of the foetus or litter during gestation. However, without dissection this cannot be disentangled from storing energy reserves during gestation for the subsequent lactation period, which is even more energetically demanding [33]. As a rough control for such effects, we included neonatal mass, litter size, gestation length and lactation time as potential correlates in the analyses, but this did not alter our findings. Moreover, because cooperative breeders tend to have higher reproductive efforts than independent breeder [117], this possibility cannot explain the reduced CV in body mass among species receiving allomaternal care.

In our data, we found a surprisingly weak phylogenetic signal of CV body mass and thus low values of λ for the model residuals, indicating that the phylogenetic disposition for fat disposition is partially masked by habitat-caused variation [118–120]. The fact that we still found significant relationships between CV body mass and allomaternal care would then make our case even stronger, because it implies that the underlying effect must be very strong.

Another unsolved question concerns the relationship between reproductive effort, seasonal fluctuations in climate or food abundance, and social factors such as allomaternal care. Reproductive seasons and experienced seasonality in food intake are generally interrelated in mammals [30]. There is evidence that species inhabiting more seasonal and less predictable habitats more often breed cooperatively [121, 122], and we also expect that they would benefit more from a higher ability to store body fat. However, because we found a negative, rather than the expected positive correlation between allomaternal care and the tendency to store body fat, this confirms that there is indeed a trade-off due to energetic costs of fat storage, and thus that social and physiological buffers are compensatory strategies to maintain fitness in a harsh environment. To further investigate these strategies, we would not only need data on environmental factors such as annual rainfall, vegetation indices or actual food abundance, but also of the seasonality experienced by the animals themselves, as expressed in dietary habits throughout the year, analogous to our studies of brain size and seasonality in primates [123–125].

Conclusions

In conclusion, several lines of evidence suggest that any allomaternal care, be it aimed at the mother or the offspring, and be it by the father or other conspecifics, allows females to reduce the amount of stored body fat.

In combination with intraspecific studies, our results further support the idea that the main reason for this negative correlation between the amount of allomaternal care and female CV body mass is the energetic contribution towards offspring rearing through costly care by males or helpers, which stabilises the energetic costs for female reproduction. Although our comparative approach has some limitations, our analyses indicate that female mammals have two different strategies of coping with energetic constraints on reproduction: either getting fat or getting help.

Additional files

Additional file 1: List of species and data used for this study. (XLSX 22 kb)

Additional file 2: References for the CV body mass data used for this study. (DOCX 60 kb)

Additional file 3: Compilation and quantification of allomaternal care behaviours. (DOCX 37 kb)

Additional file 4: Supplementary results: **Tables S1 and S2.** Results testing for collinearity among predictors. Variation inflation factors (VIF) for all the full models and all the reduced models after multicollinearity is considered. **Table S3.** Estimated phylogenetic signal (λ) in the individual variables. **Tables S4 and S5.** Model sets obtained after model selection based on ΔAICc <2 including best-supported models and multiple-model parameter estimates. **Tables S6-S9 and Figure S2.** Results of a binary coding scheme of allomaternal care behaviours as well as binary coded care provided by males (paternal care) or other group members (care by others). **Tables S10-S15.** Results for the subset of studies including only wild-caught females ($N = 49$). These remained largely identical to those obtained with the whole dataset (see also Tables 1 and 2 in the main text). (DOCX 163 kb)

Additional file 5: Figure S1. Phylogenetic tree of 87 mammal species used in this study visualised using Mesquite v. 3.11 [126]. (PDF 83 kb)

Acknowledgements
Many people and institutions have contributed data to our compilation, which we gratefully acknowledge. In particular, we thank Patricia Anne Fleming, Benedikt Gehr, Didier Julien-Laferrière, Sofia Silva and the Tierpark Hellabrunn München. We would also like to thank Redouan Bshary and Marcus Clauss for fruitful discussions and Dirk Ullrich from the Alpenzoo Innsbruck and Caroline Pond for sharing their data on mammalian body fat and body mass that we used for preliminary analyses.

Funding
Financial support was provided by the Swiss National Science Foundation grant no. 31003A-144,210, the A.H. Schultz Foundation and the University of Zurich.

Authors' contributions
SAH collected the data, performed the statistical analyses and wrote the paper. CvS and KI co-wrote the manuscript. All authors contributed to the design of the study, discussed the results and gave final approval for publication.

Competing interests
The authors declare that they have no competing interests.

References

1. McNab BK. The energetics of reproduction in endotherms and its implication for their conservation. Integr Comp Biol. 2006;46:1159–68.
2. Speakman JR. The physiological costs of reproduction in small mammals. Phil Trans R Soc. B. 2008;363:375–98.
3. Mori A. Analysis of population changes by measurement of body weight in the Koshima troop of Japanese monkeys. Primates. 1979;20:371–97.
4. Küster J, Paul A. Female reproductive characteristics in semifree-ranging Barbary macaques (Macaca sylvanus L. 1758). Folia Primatol. 1984;43:69–83.
5. Borries C, Koenig A, Winkler P. Variation of life history traits and mating patterns in female langur monkeys (Semnopithecus entellus). Behav Ecol Sociobiol. 2001;50:391–402.
6. Tyler NJC. Natural limitation of the abundance of the high arctic Svalbard reindeer. Cambridge: University of Cambridge; 1987.
7. Wauters LA, Lens L. Effects of food availability and density on red squirrel (Sciurus vulgaris) reproduction. Ecology. 1995;76:2460–9.
8. Heesen M, Rogahn S, Ostner J, Schülke O. Food abundance affects energy intake and reproduction in frugivorous female Assamese macaques. Behav Ecol Sociobiol. 2013;67:1053–66.
9. Takahashi H. Female reproductive parameters and fruit availability: factors determining onset of estrus in Japanese macaques. Am J Primatol. 2002;57:141–53.
10. Arlet ME, Isbell LA, Kaasik A, Molleman F, Chancellor RL, Chapman CA, et al. Determinants of reproductive performance among female gray-cheeked mangabeys (Lophocebus albigena) in Kibale National Park, Uganda. Int J Primatol. 2015;36:55–73.
11. van Noordwijk MA, van Schaik CP. The effects of dominance rank and group size on female lifetime reproductive success in wild long-tailed macaques, Macaca fascicularis. Primates. 1999;40:105–30.
12. Gill CJ, Rissman EF. Female sexual behavior is inhibited by short-and long-term food restriction. Physiol Behav. 1997;61:387–94.
13. Temple JL, Schneider JE, Scott DK, Korutz A, Rissman EF. Mating behavior is controlled by acute changes in metabolic fuels. Am J Physiol Regul Integr Comp Physiol. 2002;282:R782–90.
14. Cowlishaw G, Dunbar RI. Population biology. In: Cowlishaw G, Dunbar RI, editors. Primate Conservation biology. Chicago: University of Chicago Press; 2000. p. 119–57.
15. Wade GN, Schneider JE. Metabolic fuels and reproduction in female mammals. Neurosci Biobehav Rev. 1992;16:235–72.
16. Kauffman AS, Bojkowska K, Rissman EF. Critical periods of susceptibility to short-term energy challenge during pregnancy: impact on fertility and offspring development. Physiol Behav. 2010;99:100–8.
17. Jönsson KI. Capital and income breeding as alternative tactics of resource use in reproduction. Oikos. 1997;78:57–66.
18. Campbell BC, Gerald MS. Body composition, age and fertility among free-ranging female rhesus macaques (Macaca mulatta). J Med Primatol. 2004;33:70–7.
19. Testa JW, Adams GP. Body condition and adjustments to reproductive effort in female moose (Alces alces). J Mammal. 1998;79:1345–54.
20. Rödel HG, Valencak TG, Handrek A, Monclús R. Paying the energetic costs of reproduction: reliance on postpartum foraging and stored reserves. Behav Ecol. 2016;27:748–56.
21. Lewis RJ, Kappeler PM. Seasonality, body condition, and timing of reproduction in Propithecus verreauxi verreauxi in the Kirindy Forest. Am J Primatol. 2005;67:347–64.
22. Côté SD, Festa-Bianchet M. Birthdate, mass and survival in mountain goat kids: effects of maternal characteristics and forage quality. Oecologia. 2001;127:230–8.
23. Atkinson S, Ramsay M. The effects of prolonged fasting of the body composition and reproductive success of female polar bears (Ursus maritimus). Funct Ecol. 1995;9:559–67.
24. Christiansen F, Víkingsson GA, Rasmussen MH, Lusseau D. Female body condition affects foetal growth in a capital breeding mysticete. Funct Ecol. 2014;28:579–88.
25. Schneider JE, Wade GN. Effects of maternal diet, body weight and body composition on infanticide in Syrian hamsters. Physiol Behav. 1989;46:815–21.
26. Bowen WD, Oftedal OT, Boness DJ. Mass and energy transfer during lactation in a small phocid, the harbor seal (Phoca vitulina). Physiol Zool. 1992;65:844–66.
27. Short HL, Duke WB. Seasonal food consumption and body weights of captive tree squirrels. J Wildl Manag. 1971;35:435–9.
28. Zhang Z-Q, Wang D-H. Seasonal changes in thermogenesis and body mass in wild Mongolian gerbils (Meriones unguiculatus). Comp Biochem Physiol A. 2007;148:346–53.
29. Batzli GO, Pitelka FA. Condition and diet of cycling populations of the California vole, Microtus californicus. J Mammal. 1971;52:141–63.
30. Bronson FH. Mammalian reproductive biology. Chicago: University of Chicago Press; 1989.
31. Tyler NJC, Blix AS. Survival strategies in Arctic ungulates. Rangifer. 1990;10:211–30.
32. Réale D, McAdam AG, Boutin S, Berteaux D. Genetic and plastic responses of a northern mammal to climate change. Proc R Soc B. 2003;270:591–6.
33. van Schaik CP, van Noordwijk MA. Interannual variability in fruit abundance and the reproductive seasonality in Sumatran long-tailed macaques (Macaca fascicularis). J Zool. 1985;206:533–49.
34. Ekelund U, Aman J, Yngve A, Renman C, Westerterp K, Sjostrom M. Physical activity but not energy expenditure is reduced in obese adolescents: a case-control study. Am J Clin Nutr. 2002;76:935–41.
35. Garby L, Garrow JS, Jorgensen B, Lammert O, Madsen K, Sorensen P, et al. Relation between energy expenditure and body composition in man - specific energy expenditure in vivo of fat and fat-free tissue. Eur J Clin Nutr. 1988;42:301–5.
36. Taylor CR, Heglund NC, Maloiy GMO. Energetics and mechanics of terrestrial locomotion. 1. Metabolic energy consumption as a function of speed and body size in birds and mammals. J Exp Biol. 1982;97:1–21.
37. Peyrot N, Thivel D, Isacco L, Morin J-B, Duche P, Belli A. Do mechanical gait parameters explain the higher metabolic cost of walking in obese adolescents? J Appl Physiol. 2009;106:1763–70.
38. Pond CM. Morphological aspects and ecological and mechanical consequences of fat deposition in wild vertebrates. Annu Rev Ecol Evol Syst. 1978;9:519–70.
39. Dietz MW, Piersma T, Hedenstrom A, Brugge M. Intraspecific variation in avian pectoral muscle mass: constraints on maintaining manoeuvrability with increasing body mass. Funct Ecol. 2007;21:317–26.
40. Zamora-Camacho FJ, Reguera S, Rubino-Hispan MV, Moreno-Rueda G. Effects of limb length, body mass, gender, gravidity, and elevation on escape speed in the lizard Psammodromus algirus. Evol Biol. 2014;41:509–17.
41. Gosler AG, Greenwood JJD, Perrins C. Predation risk and the cost of being fat. Nature. 1995;377:621–3.
42. West DB, York B. Dietary fat, genetic predisposition, and obesity: lessons from animal models. Am J Clin Nutr. 1998;67:505S–12S.
43. Dittus WPJ. Arboreal adaptations of body fat in wild toque macaques (Macaca sinica) and the evolution of adiposity in primates. Am J Phys Anthropol. 2013;152:333–44.
44. Heldstab SA, van Schaik CP, Isler K. Being fat and smart: a comparative analysis of the fat-brain trade-off in mammals. J Hum Evol. 2016;100:25–34.
45. Snowdon CT. Infant care in cooperatively breeding species. Adv Stud Behav. 1996;25:643–89.
46. Silk JB. The adaptive value of sociality in mammalian groups. Proc R Soc B. 2007;362:539–59.
47. Gittleman JL, Oftedal OT. Comparative growth and lactation energetics in carnivores. In: Gittleman JL, editor. Carnivore behavior, ecology, and evolution, vol. 1. Dordrecht: Springer; 1989. p. 355–79.
48. Moehlman PD, Hofer H. Cooperative breeding, reproductive suppression, and body mass in canids. In: Solomon NG, French JA, editors. Cooperative breeding in mammals. Cambridge: Cambridge University Press; 1997. p. 76–127.
49. Mitani JC, Watts D. The evolution of non-maternal caretaking among anthropoid primates: do helpers help? Behav Ecol Sociobiol. 1997;40:213–20.
50. Ross C, MacLarnon A. The evolution of non-maternal care in anthropoid primates: a test of the hypotheses. Folia Primatol. 2000;71:93–113.
51. West HE, Capellini I. Male care and life history traits in mammals. Nat Commun. 2016;7:11854.
52. Crick HQP. Load-lightening in cooperatively breeding birds and the cost of reproduction. Ibis. 1992;134:56–61.
53. Scantlebury M, Russell AF, McIlrath GM, Speakman JR, Clutton-Brock TH. The energetics of lactation in cooperatively breeding meerkats Suricata suricatta. Proc R Soc B. 2002;269:2147–53.
54. Garber PA, Leigh SR. Ontogenetic variation in small-bodied new World primates: implications for patterns of reproduction and infant care. Folia Primatol. 1997;68:1–22.
55. Bales K, Dietz J, Baker A, Miller K, Tardif S. Effects of allocare-givers on fitness of infants and parents in callitrichid primates. Folia Primatol. 1999;71:27–38.

56. Lappan S. The effects of lactation and infant care on adult energy budgets in wild siamangs (*Symphalangus syndactylus*). Am J Primatol. 2009;140:290–301.

57. Solomon NG. Current indirect fitness benefits associated with philopatry in juvenile prairie voles. Behav Ecol Soc. 1991;29:277–82.

58. Powell RA, Fried JJ. Helping by juvenile pine voles (*Microtus pinetorum*), growth and survival of younger siblings, and the evolution of pine vole sociality. Behav Ecol. 1992;3:325–33.

59. Schradin C, Pillay N. The influence of the father on offspring development in the striped mouse. Behav Ecol. 2005;16:450–5.

60. Dickinson JL, Koenig WD, Pitelka FA. Fitness consequences of helping behavior in the western bluebird. Behav Ecol. 1996;7:168–77.

61. Packer C, Lewis S, Pusey A. A comparative analysis of non-offspring nursing. Anim Behav. 1992;43:265–81.

62. Roulin A. Why do lactating females nurse alien offspring? A review of hypotheses and empirical evidence. Anim Behav. 2002;63:201–8.

63. Baldovino MC, Di Bitetti MS. Allonursing in tufted capuchin monkeys (*Cebus nigritus*): milk or pacifier? Folia Primatol. 2008;79:79–92.

64. Franco-Trecu V, Tassino B, Soutullo A. Allo-suckling in the south American fur seal (*Arctocephalus australis*) in Isla de lobos, Uruguay: cost or benefit of living in a group? Ethol Ecol Evol. 2010;22:143–50.

65. Bartos L, Vanková D, Hyánek J, Siler J. Impact of allosucking on growth of farmed red deer calves (*Cervus elaphus*). Anim Sci. 2001;72:493–500.

66. Zapata B, González BA, Ebensperger LA. Allonursing in captive guanacos, *Lama guanicoe*: milk theft or misdirected parental care? Ethology. 2009;115:731–7.

67. MacLeod KJ, Nielsen JF, Clutton-Brock TH. Factors predicting the frequency, likelihood and duration of allonursing in the cooperatively breeding meerkat. Anim Behav. 2013;86:1059–67.

68. Clutton-Brock TH. The evolution of parental care. Princeton: Princeton University Press; 1991.

69. König B. Non-offspring nursing in mammals: general implications from a case study on house mice. In: Kappeler PM, van Schaik CP, editors. Cooperation in primates and humans. Berlin: Springer; 2006. p. 191–205.

70. MacLeod KJ, McGhee KE, Clutton-Brock TH. No apparent benefits of allonursing for recipient offspring and mothers in the cooperatively breeding meerkat. J Anim Ecol. 2015;84:1050–8.

71. Clutton-Brock T, Russell A, Sharpe L, Young A, Balmforth Z, McIlrath G. Evolution and development of sex differences in cooperative behavior in meerkats. Science. 2002;297:253–6.

72. Nichols HJ, Amos W, Bell MB, Mwanguhya F, Kyabulima S, Cant MA. Food availability shapes patterns of helping effort in a cooperative mongoose. Anim Behav. 2012;83:1377–85.

73. Brouwer L, van de Pol M, Atema E, Cockburn A. Strategic promiscuity helps avoid inbreeding at multiple levels in a cooperative breeder where both sexes are philopatric. Mol Ecol. 2011;20:4796–807.

74. Zöttl M, Chapuis L, Freiburghaus M, Taborsky M. Strategic reduction of help before dispersal in a cooperative breeder. Biol Lett. 2013;9: 20120878.

75. Wells JC. The evolutionary biology of human body fatness: thrift and control. Cambridge: Cambridge University Press; 2010.

76. Isler K, van Schaik CP. Allomaternal care, life history and brain size evolution in mammals. J Hum Evol. 2012;63:52–63.

77. Pond CM. The fats of life. Cambridge: Cambridge University Press; 1998.

78. Alexander RM. Principles of animal locomotion. Princeton: Princeton University Press; 2003.

79. Marino L. A comparison of encephalization between odontocete cetaceans and anthropoid primates. Brain Behav Evol. 1998;51:230–8.

80. Lukas D, Clutton-Brock T. Cooperative breeding and monogamy in mammalian societies. Phil Trans R Soc B. 2012;279:2151–6.

81. Lukas D, Clutton-Brock TH. The evolution of social monogamy in mammals. Science. 2013;341:526–30.

82. Leigh SR. Relations between captive and noncaptive weights in anthropoid primates. Zoo Biol. 1994;13:21–43.

83. Gittleman JL. Carnivore brain size, behavioral ecology, and phylogeny. J Mammal. 1986;67:23–36.

84. Meier PT. Relative brain size within the north American Sciuridae. J Mammal. 1983;64:642–7.

85. The Animal Diversity Web. Myers P, Espinosa R, Parr C, Jones T, Hammond G, Dewey T. 2006. http://animaldiversity.org/accounts/Mammalia/. Accessed 5 Sept 2016.

86. All the World's Primates. Rowe N, Myers M. 2011. http://www.alltheworldsprimates.org. Accessed 5 Sept 2016.

87. SAS Institute Inc. JMP version 10.0. SAS Institute Inc Cary, North Carolina; 1989.

88. R Core Team. R: a language and environment for statistical computing. Vienna: R Foundation for Statistical Computing; 2015.

89. Dormann CF, Elith J, Bacher S, Buchmann C, Carl G, Carré G, et al. Collinearity: a review of methods to deal with it and a simulation study evaluating their performance. Ecography. 2013;36:27–46.

90. Quinn GP, Keough MJ. Experimental design and data analysis for biologists. Cambridge: Cambridge University Press; 2002.

91. Fox J, Weisberg S. An {R} companion to applied regression. Vol. Second. Thousand Oaks: Sage; 2011.

92. Rogerson P. Statistical methods for geography. London: Sage; 2001.

93. Capellini I, Baker J, Allen WL, Street SE, Venditti C. The role of life history traits in mammalian invasion success. Ecol Lett. 2015;18:1099–107.

94. Freckleton RP, Harvey PH, Pagel M. Phylogenetic analysis and comparative data: a test and review of evidence. Am Nat. 2002;160:712–26.

95. Pagel M. Inferring the historical patterns of biological evolution. Nature. 1999;401:877–84.

96. Orme D. The caper package: comparative analysis of phylogenetics and evolution in R. R package version 2013;5.

97. Fritz SA, Bininda-Emonds ORP, Purvis A. Geographical variation in predictors of mammalian extinction risk: big is bad, but only in the tropics. Ecol Lett. 2009;12:538–49.

98. Hurvich CM, Tsai C-L. Regression and time series model selection in small samples. Biometrika. 1989;76:297–307.

99. Grueber C, Nakagawa S, Laws R, Jamieson I. Multimodel inference in ecology and evolution: challenges and solutions. J Evol Biol. 2011;24:699–711.

100. Burnham KP, Anderson DR, Huyvaert KP. AIC model selection and multimodel inference in behavioral ecology: some background, observations, and comparisons. Behav Ecol Sociobiol. 2011;65:23–35.

101. Symonds MR, Moussalli A. A brief guide to model selection, multimodel inference and model averaging in behavioural ecology using Akaike's information criterion. Behav Ecol Sociobiol. 2011;65:13–21.

102. Garamszegi LZ, Mundry R. Multimodel-inference in comparative analyses. In: Garamszegi LZ, editor. Modern Phylogenetic comparative methods and their application in evolutionary biology. Berlin Heidelberg: Springer; 2014. p. 305–31.

103. Charmantier A, Keyser AJ, Promislow DE. First evidence for heritable variation in cooperative breeding behaviour. Proc R Soc B. 2007;274:1757–61.

104. Russell A, Brotherton P, McIlrath G, Sharpe L, Clutton-Brock T. Breeding success in cooperative meerkats: effects of helper number and maternal state. Behav Ecol. 2003;14:486–92.

105. Heinsohn RG. Parental care, load-lightening and costs. In: Koenig W, Dickinson J, editors. Ecology and evolution of cooperative breeding in birds. Cambridge: Cambridge University Press; 2004. p. 67–80.

106. Klauke N, Segelbacher G, Schaefer H. Reproductive success depends on the quality of helpers in the endangered, cooperative El Oro parakeet (*Pyrrhura orcesi*). Mol Ecol. 2013;22:2011–27.

107. Paquet M, Covas R, Chastel O, Parenteau C, Doutrelant C. Maternal effects in relation to helper presence in the cooperatively breeding sociable weaver. PLoS One. 2013;8:e59336.

108. Walton JM, Wynne-Edwards KE. Paternal care reduces maternal hyperthermia in Djungarian hamsters (*Phodopus campbelli*). Physiol Behav. 1997;63:41–7.

109. McInroy JK. Energetic constraints during reproduction in a harsh environment: leptin and adipose tissues in dwarf hamsters, *Phodopus*. Kingston: Queen's University; 2000.

110. Marshall HH, Sanderson JL, Mwanghuya F, Businge R, Kyabulima S, Hares MC, et al. Variable ecological conditions promote male helping by changing banded mongoose group composition. Behav Ecol. 2016;27:978–87.

111. Ellis KJ. Human body composition: in vivo methods. Physiol Rev. 2000; 80:649–80.

112. Brambell F. The reproduction of the wild rabbit *Oryctolagus cuniculus* (L.). J Zool. 1944;114:1–45.

113. Trillmich F, Wolf JB. Parent–offspring and sibling conflict in Galápagos fur seals and sea lions. Behav Ecol Sociobiol. 2008;62:363–75.

114. French JA, Brewer KJ, Schaffner CM, Schalley J, Hightower-Merritt D, Smith TE, et al. Urinary steroid and gonadotropin excretion across the reproductive cycle in female Wied's black tufted-ear marmosets (*Callithrix kuhli*). Am J Primatol. 1996;40:231–45.

115. Ziegler T, Widowski T, Larson M, Snowdon C. Nursing does affect the duration of the post-partum to ovulation interval in cotton-top tamarins (*Saguinus oedipus*). J Reprod Fertil. 1990;90:563–70.

116. Cantoni D, Brown RE. Paternal investment and reproductive success in the California mouse, *Peromyscus californicus*. Anim Behav. 1997;54:377–86.

117. Lukas D, Clutton-Brock T. Life histories and the evolution of cooperative breeding in mammals. Proc R Soc B. 2012;279:4065–70.

118. Hahn DA. Two closely related species of desert carpenter ant differ in individual-level allocation to fat storage. Physiol Biochem Zool. 2006;79:847–56.

119. Muehlenbein MP, Campbell BC, Richards RJ, Watts DP, Svec F, Falkenstein KP, et al. Leptin, adiposity, and testosterone in captive male macaques. Am J Phys Anthropol. 2005;127:335–41.

120. Harlow HJ. Fasting biochemistry of representative spontaneous and facultative hibernators: the white-tailed prairie dog and the black-tailed prairie dog. Phys Zool. 1995;68:915–34.

121. Clutton-Brock T. Mammal societies. 1st ed. Chichester: John Wiley & Sons; 2016.

122. Lukas D, Clutton-Brock T. Climate and the distribution of cooperative breeding in mammals. R Soc Open Sci. 2017;4:160897.

123. van Woerden JT, van Schaik CP, Isler K. Brief communication: seasonality of diet composition is related to brain size in new World monkeys. Am J Phys Anthropol. 2014;154:628–32.

124. van Woerden JT, Willems EP, van Schaik CP, Isler K. Large brains buffer energetic effects of seasonal habitats in catarrhine primates. Evolution. 2012; 66:191–9.

125. van Woerden JT, van Schaik CP, Isler K. Effects of seasonality on brain size evolution: evidence from strepsirrhine primates. Am Nat. 2010;176:758–67.

126. Maddison WP, Maddison DR. Mesquite: a modular system for evolutionary analysis. Version 3.11 http://mesquiteproject.org. 2017. Accessed 30 Mar 2017.

5

Modelling the range expansion of the Tiger mosquito in a Mediterranean Island accounting for imperfect detection

Giacomo Tavecchia[1]* [ID], Miguel-Angel Miranda[2], David Borrás[2], Mikel Bengoa[3], Carlos Barceló[2], Claudia Paredes-Esquivel[2] and Carl Schwarz[4]

Abstract

Backgrounds: *Aedes albopictus* (Diptera; Culicidae) is a highly invasive mosquito species and a competent vector of several arboviral diseases that have spread rapidly throughout the world. Prevalence and patterns of dispersal of the mosquito are of central importance for an effective control of the species. We used site-occupancy models accounting for false negative detections to estimate the prevalence, the turnover, the movement pattern and the growth rate in the number of sites occupied by the mosquito in 17 localities throughout Mallorca Island.

Results: Site-occupancy probability increased from 0.35 in the 2012, year of first reported observation of the species, to 0.89 in 2015. Despite a steady increase in mosquito presence, the extinction probability was generally high indicating a high turnover in the occupied sites. We considered two site-dependent covariates, namely the distance from the point of first observation and the estimated yearly occupancy rate in the neighborhood, as predicted by diffusion models. Results suggested that mosquito distribution during the first year was consistent with what predicted by simple diffusion models, but was not consistent with the diffusion model in subsequent years when it was similar to those expected from leapfrog dispersal events.

Conclusions: Assuming a single initial colonization event, the spread of *Ae. albopictus* in Mallorca followed two distinct phases, an early one consistent with diffusion movements and a second consistent with long distance, 'leapfrog', movements. The colonization of the island was fast, with ~90% of the sites estimated to be occupied 3 years after the colonization. The fast spread was likely to have occurred through vectors related to human mobility such as cars or other vehicles. Surveillance and management actions near the introduction point would only be effective during the early steps of the colonization.

Keywords: Tiger mosquito, Site-occupancy model, Population dynamics, Invasion, Range expansion

Background

Measuring species range expansion and the pattern of dispersal is a central theme in animal ecology and of particular importance in the management or control of invasive species [39]. Most mathematical models for range expansion assume no false negative for detection of a species, that is to say, if a species is present at a given site, it will always be detected [42]. However, cryptic species or species at the initial phase of the expansion process, might not be detected under a given density threshold [6, 22], which would lead to underestimation of the species prevalence, i.e. number of sites occupied, and the pace of range expansion, i.e. species growth rate. MacKenzie et al. (2006; [26]) proposed an approach based on repeated surveys on sites to estimate the detection probability and the likelihood of species presence accounting for a detection probability <1. In contrast to classical models of range expansion [42], site-occupancy models are discrete in space and time. However, their flexibility permits modelling species occurrence as a function of a continuous spatial or temporal

* Correspondence: g.tavecchia@uib.es
[1]Population Ecology Group, IMEDEA (CSIC-UIB), c. Miquel Marqués 21, 07190 Esporles, Spain
Full list of author information is available at the end of the article

covariates [16] allowing the comparison of predictions of the pattern of colonization similar to those that characterise classical diffusion models. In a simple diffusion model [10] range dynamic is driven only by the intrinsic population growth rate and by the random short-distance movements of individuals [11, 42]. This model predicts that colonization probability covaries negatively with the distance from the central point or the observed initial site of occupation [27]. In many species, however, random short-distance movements are coupled with long-distance dispersal events, leading to a second type of models characterized by multiple centres of diffusion, an expansion process often referred to as 'hierarchical diffusion' or 'stratified dispersal' [14, 42]. At a small spatial scale, a negative association between colonization probability and distance from the site of first colonization does not necessarily occur during stratified dispersal because the species can be absent at intermediate distances. However, a diffusion process from each new colonised site would still exist. A third pattern of range expansion is the one resulting from 'leapfrog' dispersal movements, with no or little subsequent diffusion [7]. This model predicts high colonization and extinction probability but low diffusion. In this model, the overall occupancy probability would increase as a result of range expansion but with no apparent relationship with the distance from the initial occupied site and without a clear diffusion process.

We used dynamic site-occupancy models [27] to measure the rate of expansion of the Asian Tiger mosquito *Aedes (Stegomya) albopictus* (Skuse, 1894) (Diptera; Culicidae) in the Island of Mallorca (Balearic Islands, Spain). We contrasted models consistent with different types of range expansion patterns to investigate the underlying dispersal process. The Asian Tiger mosquito is a daytime-active mosquito native to the tropical and subtropical region of southern Asia [13], and is considered to be one of the most invasive species in the world [23]. Its current distribution includes all continents except Antarctica [20]. In Europe, the species was first detected in Albania in 1979 [1], with no records reported in the rest of the continent until 1990s, when it appeared in Italy [19] from where it rapidly spreads to Southern and Central Europe [29, 41]. The first detection in mainland Spain occurred in 2004 in Sant Cugat del Vallés (Catalonia, Spain; [2]). Currently its distribution in Spain includes most of the Mediterranean coast as well Northern areas of the Iberian Peninsula ([8], 2016). In Mallorca (Balearic Islands) the species was first detected in 2012 in 5 municipalities [30] and it rapidly spread to Ibiza in 2014 (Barceló et al., 2015) and Menorca in 2016 [3]. Despite being able to feed

upon different hosts depending on their availability [45], *Ae. albopictus* adults obtain blood preferably from humans [32]. The expansion of the Tiger mosquito in Europe has recently created public concerns for its possible role in the transmission of the Zika virus, responsible of microcephaly in newborns of infected mothers (ECDPC, 2016) and as a potential vector for Dengue and Chikungunya viruses [34, 46]. Understanding the spread of invasive mosquito species would thus provide important information needed for a successful control and prevention campaigns. At a large spatial scale, the presence of *Ae. Albopictus* is associated with the level of rainfall and day time surface temperature and its dispersal is facilitated by human activities [44]. However, pattern of dispersal and distribution at small spatial scale and from the early steps of the colonization process are largely unknown. The isolated character of recently colonized Balearic Islands offers the unique opportunity to follow the invasion process and to determine the mechanisms underlying species diffusion.

Our first aim was to estimate the prevalence of the occupation, i.e. the proportion of sites occupied, and the annual rate of spread, i.e. the proportional change in the number of sites occupied per year. We subsequently investigated the expansion processes by modelling the colonization probability as a linear function of the distance from the site of first reported occupancy. If the expansion followed a diffusion process, we expected the probability of occupancy to abate with the distance from the first reported occupied site. Indeed, diffusion is a slow process for *Ae albopictus* [28] even when compared with other *Aedes* species [12]. Also, under random short-distance dispersal, models in which the probability of occupancy at a given site is a function of the occupancy of the neighbourhood would provide a good description of the data [4, 47]. Alternatively, if range expansion occurred mainly by leapfrog dispersal, we expected neither the distance from the initial points nor models depending on neighbourhood covariates to be adequate.

Methods

Mosquito prevalence and occupancy rate

Mallorca Island is the largest and most populated island of the Balearic archipelago, Eastern Spain, with a surface of 3640 km^2 and about 860×10^3 inhabitants (in 2015). Since the first record of *Ae. albopictus* in Mallorca in 2012, a network of 784 oviposition traps (described in [30]) was deployed in 40 municipalities to monitor the species distribution and range expansion. The monitoring scheme changed over the years resulting in data sparseness. As a consequence, we first restricted the analysis to data collected during the 3 months of maximum abundance of *Ae. Albopictus* (September–November) during the period 2012–2015. To further reduce data sparseness, we used a cluster-by-distance analysis to group neighboring traps into 70 clusters ('sites', hereafter; Fig. 1). The clustering distance threshold was arbitrarily chosen as a good compromise between data richness and number of sites monitored. Besides reducing data sparseness, the clustering allowed a more straightforward interpretation of the yearly occupancy rate because the number and the identity of clusters remained roughly constant throughout the study (Table 1). Nevertheless the dataset was unbalanced and information gaps persisted for some locations (e.g. 23 locations have been sampled in only 1 year). For each site in the dataset we recorded the distance from the location of first observation in the municipality of Bunyola, about 15 Km north the main city of Palma. Clustering and distance analyses were conducted using program R v3.3.1 [38].

Modelling site-occupancy accounting for detection probability

The occupancy dynamic at each site was investigated using dynamic site-occupancy models [26, 18] in which the occurrence of mosquitoes at site i on a given time t, $z_{i,t}$, is considered as a latent state governed by the occupancy probability, ψ_t. Changes in the occupancy over

time can be described as in metapopulation dynamics by the extinction, ε, and colonization, γ, probabilities. Hence the initial occupancy state, at time 1, is assumed to be

$$z_{i,1} \sim \text{Bernoulli}(\Psi_1)$$

whereas in subsequent period is:

$$z_{i,t} \mid z_{i,t-1} \sim \text{Bernoulli}\left(z_{i,t-1}\left[1-\varepsilon_{t-1}\right] + \left[1-z_{i,t-1}\right]\gamma_{i,t-1}\right)$$

The actual observations, $y_{i,t}$, on a site i at time t are treated as conditional on the occurrence probability and the probability, p_t, to detect the species when present as:

$$y_{i,t} \mid z_{i,t} \sim \text{Bernoulli}(z_{i,t}p_t)$$

By combining the yearly estimates of γ and ε it is possible to calculate several derived quantities such as i) the probability of occupancy in any given year, $\psi_t = \psi_{t-1}(1-\varepsilon_{t-1}) + (1-\psi_{t-1})\ \gamma_{t-1}$ and ii) the proportional increase in the probability of occupancy $\lambda_t = \psi_{t+1}/\psi_t$ [25]. The proportion of sites occupied at equilibrium, ψ_{eq}, that leads to $\psi_t = \psi_{t+1}$, can be calculated using the average colonization and extinction probabilities as $\psi_{eq} = \gamma/(\gamma + \varepsilon)$. In an increasing population typically $\psi_1 < \psi_{eq}$ while $\psi_1 > \psi_{eq}$ when population is decreasing [16, 37].

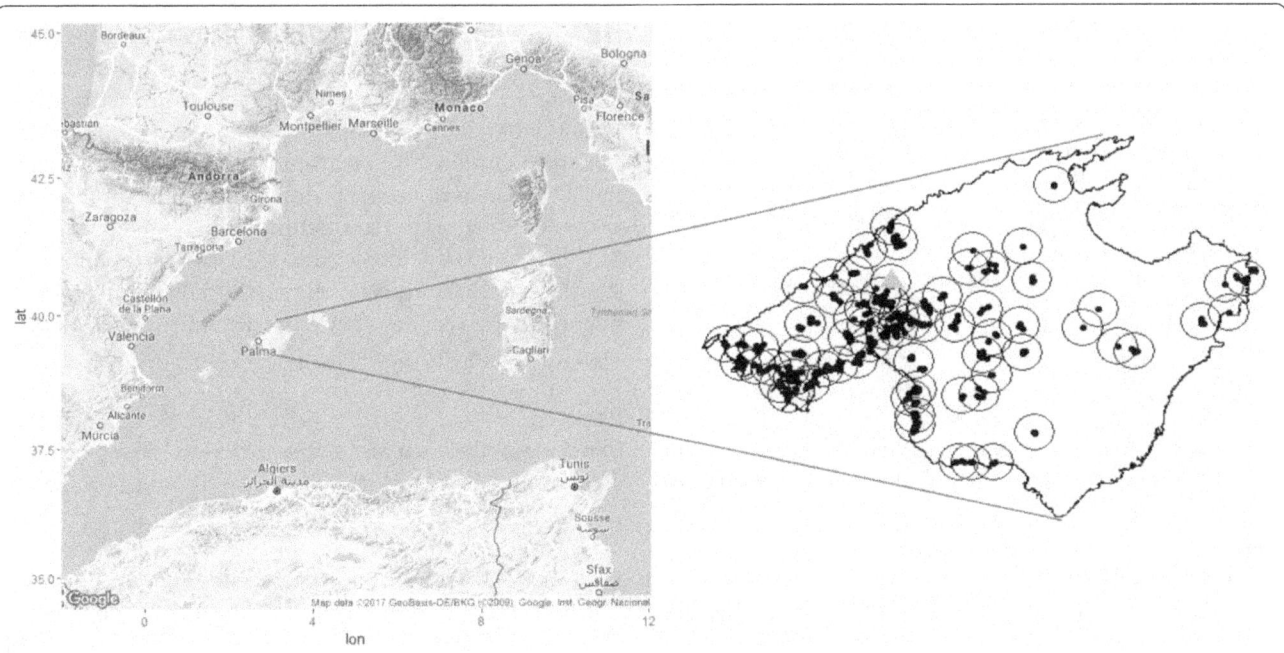

Fig. 1 The Island of Mallorca with the location of the 70 sites (*black dots*) monitored for the presence of *Ae. Albopictus*. The *circles* indicate an area of diameter equal to the average nearest neighbor distance between the sites (3.6 Km). The *grey triangle* is the site of first observation in 2012. Note that the more eastern locations have been monitored in 2012 only when the species was first reported in Mallorca

Table 1 Presence-absence data of tiger mosquitoes from 70 unique sites monitored during autumn 2012 to 2015

Year	Number of unique sites monitored	Observed occupancy rate (%)	Maximum distance from first reported observation (km)
2012	38	26	21.3
2013	39	26	30.1
2014	44	57	31.6
2015	44	93	41.5

Modelling the pattern of expansion through site-occupancy models

The presence of *A. albopictus* at each site was first modelled by assuming the initial occupancy, ψ_1, the extinction, ε, colonization, γ, and detection, p, probabilities varied over time (years), denoted by the model $\psi(t)\gamma(t)\varepsilon(t)p(t)$. This general model was used to estimate the occupancy rate, the turnover, the occupancy growth rate and the extinction probability over time. The model $\psi(t)\gamma(t)\varepsilon(t)p(t)$ does not assume any diffusion process and it is consistent with a leapfrog dispersal pattern in which colonization and extinction change over time but without a particular spatial pattern. We then considered a set of models assuming that the observed mosquito range resulted from a diffusion process. We first modelled the initial occurrence, ψ_1, and the colonization, γ, probabilities as:

$$\text{logit}(\theta_{i,t}) = \alpha_t + \beta_t X_{i,t} \tag{1}$$

where θ refers to the probabilities ψ_1 or γ and X to the distance (standardized) from the site of fist observation (noted '*dist*' in model notation). This approach was used by MacKenzie et al. (2006) to model the expansion of the House finch *Carpodacus mexicanus* (Müller) in North America. Although the model would be consistent with a diffusion process, it does not include the mechanism itself [47]. Following [47], we constrained the colonization probability, γ, at a given site, i, to be dependent on the yearly occupancy rate, $\bar{\psi}_t$, within the site neighbourhood, as $\text{logit}(\gamma_{i,t}) = \alpha_t + \beta_t \bar{\psi}_t$. The autocovariate ψ_t is:

$$\bar{\psi}_t^{\,n} = \frac{1}{l} \sum\nolimits_{j\varepsilon\{n_i\}} \psi_{j,t} \tag{2}$$

and it is defined as the probability of occupancy of the neighbouring patch j, and l is the number of patches located in the neighbourhood. The neighbourhood can be made by the adjacent patches [4] or by the total patches in the study area as an average measure of the overall occupancy rate [47]. Although at different scales, both models are consistent with either a gradual range expansion through diffusion or a stratified diffusion processes. In theory, if all adjacent cells contribute equally, the two models would only differ in the definition of the neighborhood. In our case, however, this similarity does not hold because the sites monitored were unequally spaced and a further clustering was necessary to define the adjacent sites. To do this, we considered a grid made of 4×4 km cells ($n = 263$), a rounded measure of the average nearest neighbour distance between sites. In the majority of the 70 sites monitored ($n = 56$), there was a single site per cell, while 7 cells contained two sites. In this respect the two autoregressive functions cannot be compared because they refer to a different number of sites.

We used a Bayesian framework to estimate model parameters [17]. Bayesian analyses were conducted in WinBUGS [24] using uninformative priors for model parameters (uniform distribution from −20 to +20 for linear predictors and 0 to 1 for probabilities). The posterior distributions of parameters were sampled, using 3 chains and 25,000 simulations (the first 5000 discarded as a burnin period). Model selection in Bayesian analyses is not straightforward [43]. Across nested models, selection can be done using the Deviance Information Criterion (DIC), a generalisation of the Akaike's Information Criterion (AIC; [5]) for hierarchical models. However, comparisons between models with and without an autoregressive structure cannot be performed using the DIC because the DIC is computed at different levels in the hierarchy of data and parameters. Model adequacy was thus assessed by inspecting the estimates and their standard deviations. We report the DIC of all models, but we warn readers that this should not be taken as a strict criterion for model explanatory power.

Results

Mosquito prevalence and occupancy rate

The observed edge of the mosquito distribution, i.e. the site at the greatest distance from the initial reported occupancy, was expanding during the study period at an average rate of 6.1 km per year (Fig. 2). Observed site occupancy rate assuming a detection probability of 1.00 were 0.26 ($n = 38$), 0.26 ($n = 39$), 0.57 ($n = 44$) and 0.93 ($n = 44$) in 2012, 2013, 2014 and 2015, respectively (Table 1). However, the site-occupancy model, $\psi(t)\gamma(t)\varepsilon(t)p(t)$, assuming all parameters time-dependent

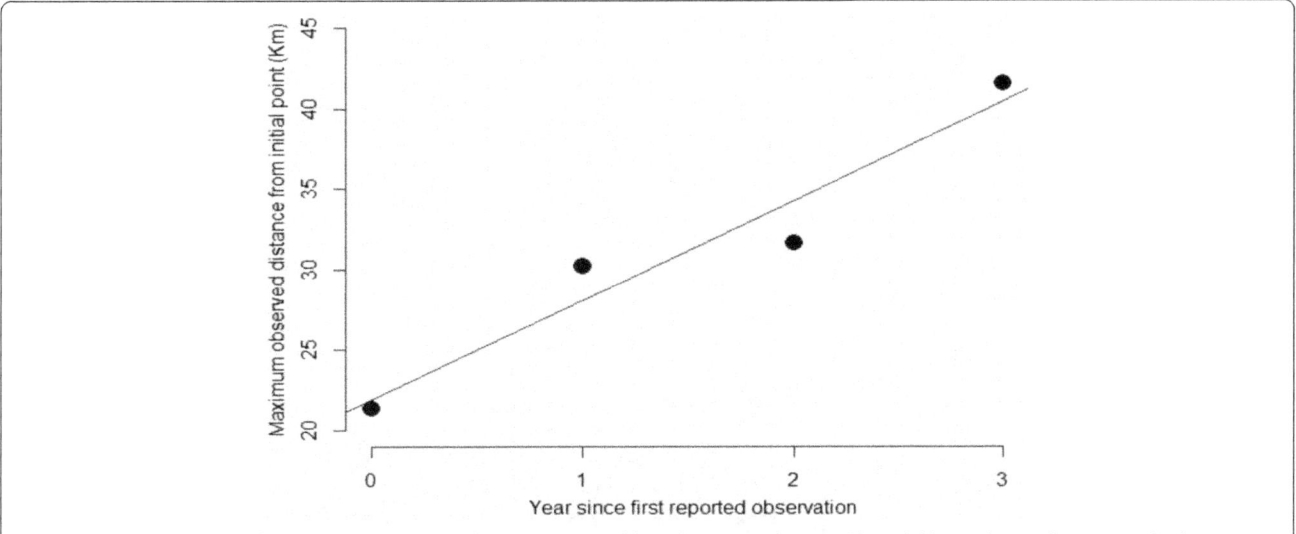

Fig. 2 Maximum distance of a observed occurrence from the point of first observation by year. The *solid line* indicates the expected values assuming a linear diffusion. The speed of the observed expansion is 6.2 km per year (slope of the regression line)

revealed that the detection probability varied from 0.23 in 2013 to 0.85 in 2015, being less than 1 in all years (Table 2). As expected, this model led to estimated occupancy probabilities higher than those observed (0.35, 0.59, 0.57 and 0.90 in 2012, 2013, 2014 and 2015, respectively; Table 2). The probability of local extinction was generally high (average 0.38), but it dropped to 0.07 in 2015 (Table 2). The initial occupancy rate in 2012 ($\psi_{2012} = 0.351$) was lower than the expected occupancy rate at equilibrium ($\psi_{eq} = 0.661$) confirming the observed range expansion over the study area. The average growth rate in the occupancy probability was 1.50, equivalent to a 50% increase per year in the number of sites occupied by the species per year. However, this rate of increase was not constant and the range expansion greatly increased after the initial colonization and during the last year (periods 2012–2013 and 2014–2015, Table 2).

Modelling the pattern of expansion through site-occupancy model

The DIC of the general model (DIC model 2 $\psi(t)\gamma(t)\varepsilon(t)p(t) = 353.89$) improved slightly when the initial probability of occupancy was model as a function of the distance from the point of first reported observation (DIC model 1 $\psi(dist)\gamma(t)\varepsilon(t)p(t) = 352.29$; Tab. 3). The estimates of α and β ($\alpha = -3.581$, $\beta = -5.537$) with the upper 95% credible intervals of β lower than 0.00 (+95%CI = −1.978) indicate that the probability of occupancy in the first year declined as a function of distance. Estimates showed a sharp decline in the occupancy rate, with no occupancy expected at mid-distance between the first reported observation and the farthest site monitored (c. 23 km; Fig. 3). However, the probability of colonization of empty patches in subsequent years did not depend on the distance from the first reported observation (DIC $\psi(dist)\gamma(dist)\varepsilon(t)p(t) = 362.03$; Table 3-

Table 2 Modelling the occupancy dynamics of the tiger mosquito in Mallorca Island. ψ = occupancy probability, γ = colonization probability, ε = extinction probability. Effects: t = time effect, dist = distance from the site of first observation, m.ψ = autocovariate based on the average occupancy rate in the whole area, D = autocovariate based on the adjacent occupancy rate (see details in 'Methods'). Note that ψ(covariate) refers to occupancy rate in 2012 only, the occupancy probabilities for the subsequent years are calculated as derived parameters (see text for details)

Model	Type of dispersal	Autocovariate	Notation	DIC	Reference
1	Initial diffusion + leapfrog	No	$\psi(dist)\gamma(t)\varepsilon(t)p(t)$	352.29	[26]
2	Leapfrog	No	$\psi(t)\gamma(t)\varepsilon(t)p(t)$	353.89	[26]
3	Diffusion	No	$\psi(dist)\gamma(dist)\varepsilon(t)p(t)$	362.03	[26]
4	Diffusion	No	$\psi(t)\gamma(dist)\varepsilon(t)p(t)$	363.84	[26]
5	Leapfrog / Diffusion	Yes	$\psi(t)\gamma(m.\ \psi)\varepsilon(t)p(t)$	374.32	[47]
6	Stratified Diffusion	Yes	$\psi(t)\gamma(D)\varepsilon(t)p(t)$	406.18	[4]

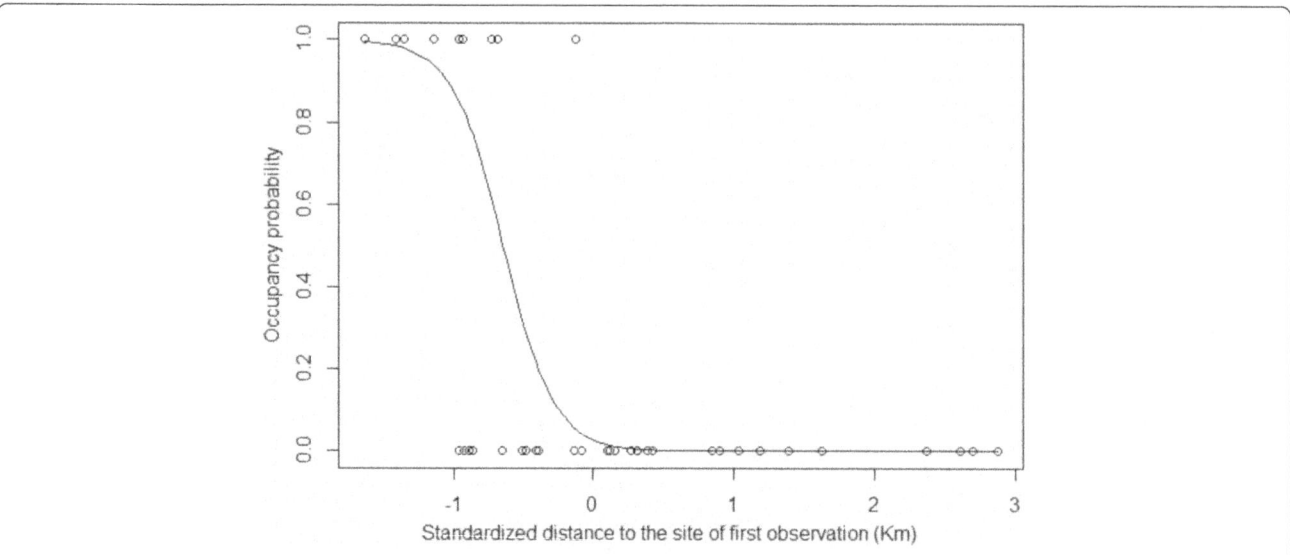

Fig. 3 Predicted (*solid line*) and observed (*circle*) site occupancy probability in relation to the distance from the first reported observation (estimates from model assuming an initial diffusion followed by leapfrog movements, noted ψ(dist)γ(t)ε(t)p(t))

Table 3 Estimates from model ψ(t)γ(t)ε(t)p(t), assuming all parameters variable over time. Credible interval (CI) at 2.5% and 97.5% are reported. Parameters: ψ_i = occupancy probability at time i, γ_i = colonization probability, e.g. the probability that an empty site is occupied between i and $i + 1$, ε_i = extinction probability, e.g. the probability that an occupied size at i is not-occupied at $i + 1$,, λ_i = the growth rate of occupied size between i and $i + 1$, ψ_{eq} = occupation probability at equilibrium (see text for details)

Parameter	Mean	Sd	2.5% quantile	97.5% quantile
ψ_{2012}	0.351	0.102	0.184	0.58
ψ_{2013}	0.587	0.159	0.291	0.887
ψ_{2014}	0.568	0.071	0.428	0.704
ψ_{2015}	0.899	0.044	0.800	0.968
γ_{2012}	0.667	0.201	0.258	0.980
γ_{2013}	0.669	0.182	0.251	0.974
γ_{2014}	0.862	0.075	0.686	0.973
ε_{2012}	0.566	0.24	0.059	0.935
ε_{2013}	0.509	0.133	0.252	0.769
ε_{2014}	0.073	0.051	0.007	0.197
p_{2012}	0.63	0.118	0.384	0.839
p_{2013}	0.26	0.103	0.112	0.510
p_{2014}	0.80	0.048	0.693	0.882
p_{2015}	0.85	0.033	0.782	0.909
λ_{2012}	1.846	0.833	0.685	3.867
λ_{2013}	1.057	0.388	0.580	2.047
λ_{2014}	1.610	0.220	1.261	2.113
ψ_{eq}	0.661	0.066	0.548	0.804

Fig. 4). Models including an autoregression structure in which colonization probabilities were modelled as a linear function of the average neighbouring occupancy rate in the form $\alpha + \beta_{1t}\bar{\psi}_t^{\,n}$ delivered positive but unrealistic standard deviations for the βs parameters (mean ± sd: $\beta_{1,2013}$ = 1.91 ± 11.3, $\beta_{1,2014}$ = 4.637 ± 9.51 and $\beta_{1,2015}$ = −6.667 ± 9.46). Similar imprecise estimates were obtained when only adjacent cells were considered (Fig. 5).

Discussion

The expansion of the tiger mosquito

The range expansion of the Tiger mosquito in the island of Mallorca has been rapid, with an estimated occupancy of monitored sites probability that increased from 0.35 in 2012 to nearly 0.90 in 2015 and an average annual growth rate in the occupancy of 1.50. Interestingly, local extinction probability was relatively high (except from 2014 to 2015) suggesting a high turnover in the occupied sites. During the first year, when the number of mosquitos was presumably small, treatments with insecticides by private citizens and local administration might have caused temporary extinction of the species in some monitored sites. At present we ignore the intensity and influence of these actions. Extinctions can also have occurred naturally because newly colonized locations are expected to be occupied by a small number of mosquitos. However, colonization probability was also high, leading to a fast re-colonization of locations from which the species disappeared. Under a simple diffusion model [42], the colonization probability should negatively covary with the distance from the site of first observation (MacKenzie et al. 2006). Our results indicated that the

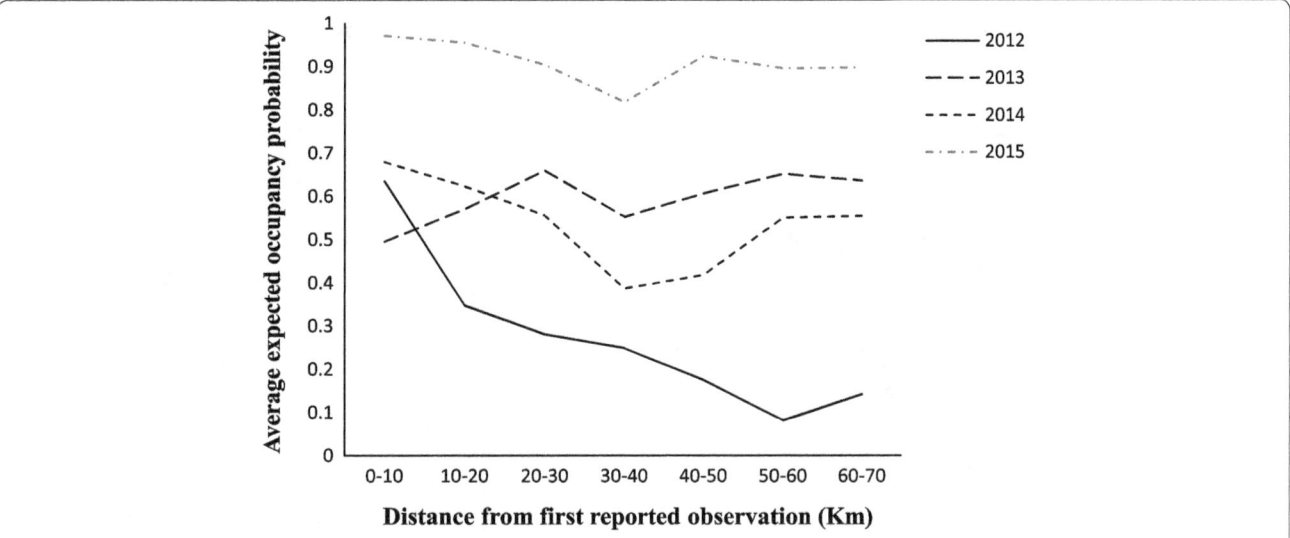

Fig. 4 Average site occupancy probability in relation to the distance from the first reported observation (estimates from the model in which all parameters were time-dependent, noted ψ(t)γ(t)ε(t)p(t))

distribution of occupied sites during the first year responded to what predicted from a simple diffusion model with random short dispersal movements. In contrast, from 2013 to 2015 the colonization of new sites did not occur as a diffusion process, at least not at the spatio-temporal scale considered here. The fast range expansion is better described by leapfrog dispersal movements, in which an increasing number of sites are occupied each year, but without a clear relationship with the distance from the initial colonization. Autoregressive models delivered unrealistic standard deviations of parameter estimates indicating that a gradual diffusion process is unlikely to have shaped the current distribution at least at the spatial scale considered here. Collantes et al. 2015 [8] mentioned a possible diffusion process of dispersal of A. albopictus from its first detection site in mainland Spain in 2004. However, no analyses were conducted for demonstrating such type of dispersal. Other authors also reported leapfrog dispersal movements at larger spatial scale (>500 km) from the site of first detection in Catalonia to the Valencia region (Bueno-Marí et al., 2013). The same pattern of progressive and gradual invasion since the initial point combined with sporadic "jumps" has been also proposed by Roche et al. [35] on a study of the distribution of Ae. albopictus in continental France and Corsica. In comparison to other mosquitos, Ae. albopictus show a low dispersal capability [12]. According to Marini et al. [28], for example, the average flight distance of Ae. albopictus is 119 m per day. However, passive transportation of eggs, through translocations of used tires [33] and adults mosquitoes in vehicles [32] are probably the mechanisms of leapfrog dispersal movements. Despite our conclusions are drawn on a smaller spatial scale than the

one previously considered, they are in agreement with what is known of the colonization pattern of the species. However, they are based on the assumption of a single initial colonization event in 2012. We cannot exclude that subsequent colonization (i.e. independent introductions; see for example [15, 36]) occurred after 2012. At the moment it is unknown whether or where this happened, but multiple introductions from mainland through the local airport or the two main ports of Alcudia and Palma would be consistent with a stratified diffusion as predicted by models with auto-covariates The change in colonization and extinction probability in the last year of the study might partly be due to natural causes. For example the amount of rainfall during the summer 2015 was particularly high and can partially explain the high colonization and recapture probabilities (see below). The short period of the study does not permit to fully investigate a relationship between colonization and rainfall but an important role of weather variables has been found in the oviposition dynamics of Ae. aegypti in Northwestern Argentina [9] and in the abundance of Ae albopictus in the French Riviera (Tar et al. 2013).

Site occupancy models and species range expansion

The pattern of range expansion of any given species depends on several characteristic such as landscape heterogeneity [11], interspecific competition ([47], 2014), species life-history traits [22], climatic and human-related factors [36, 40].

It is not surprising that analytical models of range expansion are a necessary oversimplification of the underlying biological processes. They allow, nevertheless, some generalizations and the estimates of important

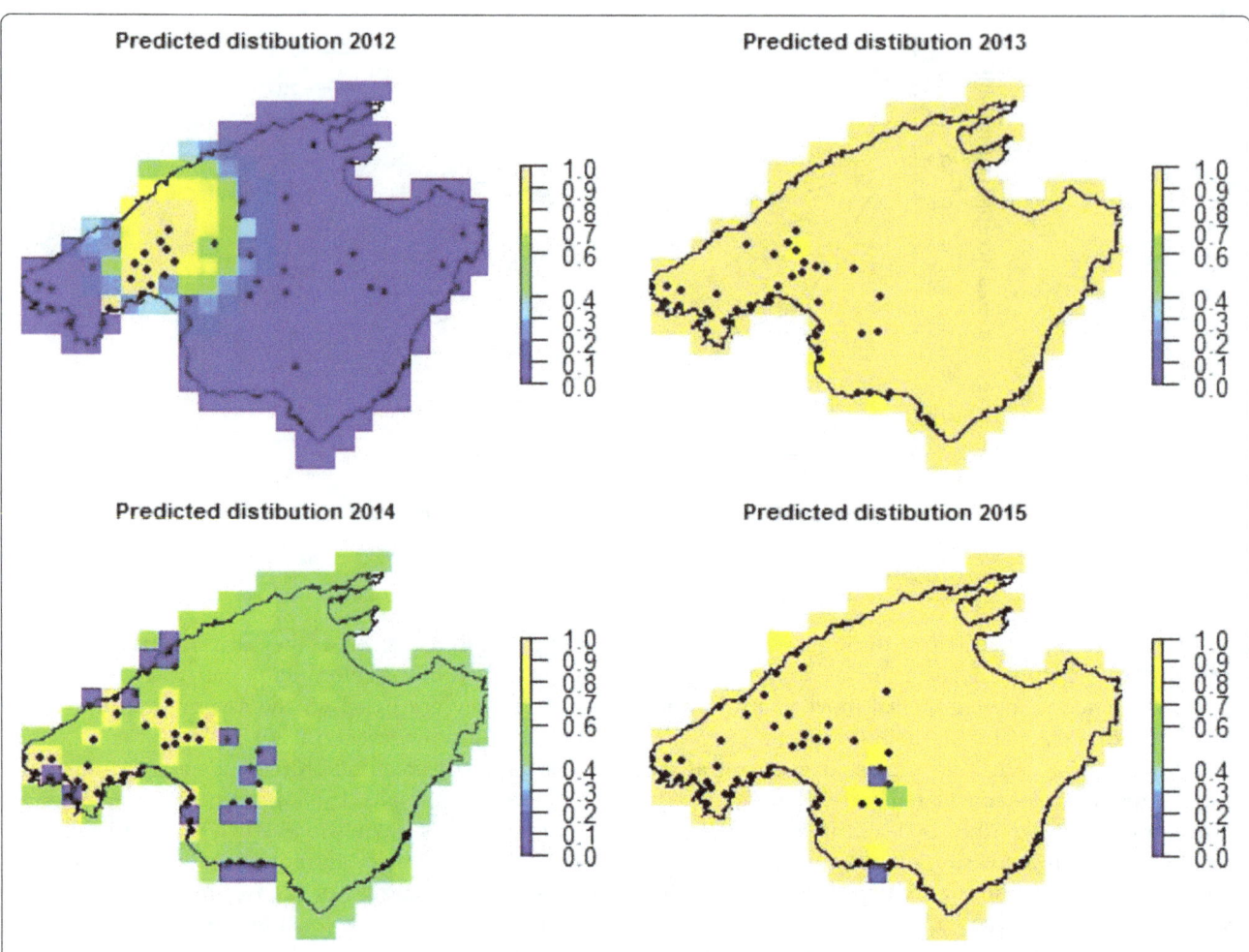

Fig. 5 Predicted site occupancy probabilities according to the autoregressive model as in eq. 3 assuming an effect of the neighboring sites (see text for details). *Black dots* are the monitored sites. Note that areas far away from a monitored site display the average estimate of site occupancy. The estimated average occupancy probability by a non-autoregressive model was 0.35, 0.59, 0.57 and 0.90 in 2012, 2013, 2014 and 2015, respectively (Table 2)

parameters that modulate the range expansion [22]. Here we used dynamic site-occupancy models to estimate occupancy rate and colonization speed of the Tiger mosquito accounting for imperfect detection. In contrast to classical models [14], site-occupancy models are discrete in space and time. However, we constrained parameter variability as a function of a site-dependent covariate to deliver predictions consistent with different patterns of colonization as in classical continuous models, i.e. 'leapfrog' versus 'diffusion'. A clear limitation of our work was the difficulty in finding criteria for model selection. However, the problem of contrasting hierarchical models is not only limited to the present study and it is a topic under study in statistical theory and numerical ecology [43]. Additional problems derived from the fast expansion of the mosquito, the limited number of sites monitored and/or a possible spatially consisted driver of the colonization probability. These factors contribute to reduce the variability in

occupancy rate among sites leading to numerical problems in model fitting when estimating the effect of the covariate. Beside these limitations, we showed the potential of site-occupancy models in estimating range expansion parameters [26] and can be very useful in the study of disease prevalence and vector dynamics [21, 31]. For example, Padilla-Torres et al. [31] used site-occupancy models to study the prevalence of *Ae. aegypti* and *Ae. albopictus*. They concluded that routine surveillance based on rapid larval surveys led to a lower prevalence of both species and suggest a combined used of ovitrap-based surveillance with analytical methods based on imperfect detection. Finally, MacKenzie and Nichols [27] treated occupancy as a surrogate of abundance. In our case mosquito abundance is more likely to be reflected in the probability of detection, which can be seen as the probability of a trap being used by a gravid female. This is because the conditional probability to detect mosquito larvae in the oviposition traps given that a female has used the trap is equal to 1.00. This would explain why in 2015, when the probability of recapture was high (0.85), the extinction probability was low (0.07). However, the link between abundance and detection is not straightforward because it would depend on multiple factors that have not been considered here, i.e. the habitat characteristics or the availability of alternative breeding sites. The present work is more descriptive than predictive and future research should incorporate additional site-dependent covariates in the models such as habitat type and site attractiveness. This can be done with static (opposite to 'dynamic') single-season occupancy-model. Single season models would not allow investigating the expansion process as we did here, but they will avoid trap-clustering and would permit to model mosquito presence using fine scale habitat covariates to predict future distributions.

Conclusions

Assuming a single colonization event in 2012, we concluded that the rapid expansion of *Ae. Albopictus* in Mallorca Island occurred in two phases. In a first phase the distribution appeared consistent with a diffusion process. This was rapidly followed by leap-frog dispersal events that resulted in an estimated occupancy probability of 90% 3 years after the colonization. The two distinct phases imply that surveillance and management actions near the introduction point would only be effective during the early steps of the colonization. The lowest extinction probability was recorded in the year with the highest amount of summer rainfall suggesting a role of weather covariates on the paste of the expansion. Dynamic site-occupancy models offer a robust analytical framework for the study of range expansion. They are particularly suitable for the study of cryptic species with high turnover as they permit to frame imperfect detections.

Abbreviations
AIC: Akaike's Information Criterion; DIC: Deviance Information Criterion

Acknowledgements
We are grateful to all volunteers who helped site monitoring from 2012 to 2015. Thanks to M. Kery and F. Bled for their help in modelling the data and to A. Sanz-Aguilar for her comments on an early draft of the manuscript. Many thanks to G. Guillera-Arroita and J. Lahoz-Monfort for an early discussion about the analysis. We thank two anonymous referees for the constructive comments that have helped to improve the clarity and quality of the work. GT is grateful to I. Hendriks for her support. The occupancy data used in this analysis are available at http://cedai.imedea.uib-csic.es/geonetwork/srv/es/main.home under the terms and conditions specified by the CEDAI database platform.

Funding
GT has been supported by a "Salvador de Madariaga" fellowship (Spanish Minister of Education, Culture and Sport, Ref.: PRX16/00101) for the mobility of researchers.

Authors' contributions
MAM, DB, MB, CPE collected the data and contributed to discussion. GT and CS prepared the data and performed the statistical analyses. All authors have contributed to the editing of the manuscript, read and approved the final version of the text.

Competing interests
Authors do not have any competing interests concerning this study.

Author details
[1]Population Ecology Group, IMEDEA (CSIC-UIB), c. Miquel Marqués 21, 07190 Esporles, Spain. [2]Laboratory of Zoology, Department of Biology, University of the Balearic Island, c. Valldemossa s/n, Palma de Mallorca, Spain. [3]Consultoria Moscard Tigre, c. Gremi Passamaners 24, Local 15, 07009 Palma de Mallorca, Spain. [4]Department of Statistics and Acutarian Science, Simon Fraser University, Burnaby, BC, Canada.

References
1. Adhami J, Reiter P. Introduction and Establishment of Aedes (Stegomyia) Albopictus Skuse (Diptera: Culicidae) in Albania article. J Am Mosq Control Assoc. 1998;14(3):340–3.
2. Aranda C, Eritja R, Roiz D. First record and Establishment of the mosquito *Aedes albopictus* in Spain. Med Vet Entomol. 2006;20:150–2. doi:10.1111/j.1365-2915.2006.00605.x.
3. Bengoa M, Delacour-Estrella S, Barceló C, Paredes-Esquivel C, Leza M, Lucientes J, Molina R, Ángel Miranda M. First Record of *Aedes albopictus* (Skuse, 1894) (Diptera; Culicidae) from Minorca (Balearic Islands, Spain). J Am Mosq Control Assoc. 2016;34:5–9.
4. Bled F, Andrew Royle J, Cam E. Hierarchical Modeling of an invasive spread: the Eurasian collared-dove *Streptopelia decaocto* in the United States. Ecol Appl. 2011;21(1):290–302. doi:10.1890/09-1877.1.
5. Burnham KP, Anderson DR. Model selection and inference. A practical information-theoretic approach. 2nd ed. New York: Springer; 2002.
6. Carey JR. The Incipient Mediterranean Fruit Fly Population in California: Implications for Invasion Biology. Ecology. 1996;77(6):1690–97.
7. Clobert J, Danchin E, Nichols JD, Dhondt AA. Dispersal. New York: Oxford University Press; 2001.
8. Collantes F, Delacour S, Alarcón-Elbal PM, Ruiz-Arrondo I, Delgado JA, Torrell-Sorio A, et al. Review of ten-years presence of *Aedes albopictus* in Spain 2004–2014: known distribution and public Health concerns. Parasit Vectors. 2015;8(1):655. doi:10.1186/s13071-015-1262-y.
9. Estallo EL, Ludueña-Almeida FF, Introini MV, Zaidenberg M, Almirón WR. Weather variability associated with Aedes (Stegomyia) Aegypti (Dengue vector) Oviposition dynamics in northwestern Argentina. PLoS One. 2015; 10(5):e0127820. doi:10.1371/journal.pone.0127820.
10. Fisher RA. The wave of advance of advantageous genes. Ann Hum Genet. 1937;7:355–69.
11. Fraser EJ, Lambin X, Travis JMJ, Harrington LA, Palmer SCF, Bocedi G, et al. Range expansion of an invasive species through a heterogeneous landscape – the case of American mink in Scotland. Divers Distrib. 2015; 21(8):888–900. doi:10.1111/ddi.12303.
12. Goubert C, Minard G, Vieira C, Boulesteix M. Population genetics of the Asian Tiger mosquito *Aedes albopictus*, an invasive vector of human diseases. Heredity. 2016;117(3):125–34.
13. Hawley WA. The Biology of *Aedes albopictus*. J Am Mosq Control Assoc. 1988;1:1–39.
14. Hengeveld R. Dynamics of biological invasions. Springer Science & Business Media. 1989.
15. Kamgang B, Brengues C, Fontenille D, Njiokou F, Simard F, Paupy C. Genetic structure of the Tiger mosquito, *Aedes albopictus*, in Cameroon (Central Africa). PLoS One. 2011;6(5):e20257.
16. Kéry M, Guillera-Arroita G, Lahoz-Monfort JJ. Analysing and mapping species range dynamics using occupancy models. J Biogeogr. 2013;40(8):1463–74. doi:10.1111/jbi.12087.
17. Kery M, Royle JA. Applied Hierarchical Modeling in Ecology: Analysis of Distribution, Abundance and Species Richness in R and BUGS: Volume 1: Prelude and Static Models. London: Academic Press; 2015.

18. Kéry M, Royle JA. Hierarchical Modelling and Estimation of Abundance and Population Trends in Metapopulation Designs'. *Journal of Animal Ecology*. 2010;79(2):453–61.

19. Knudsen AB, Romi R, Majori G. Occurrence and spread in Italy of *Aedes albopictus*, with implications for its introduction into other parts of Europe. J Am Mosq Control Assoc. 1996;12(2):177–83.

20. Kraemer MUG, Sinka ME, Duda KA, Adrian Q N Mylne, Freya M. Shearer, Christopher M. Barker, Chester G. Moore, et al. The global distribution of the Arbovirus vectors *Aedes aegypti* and Ae. Albopictus'. *eLife*. 2015; doi:10.7554/eLife.08347.

21. Lachish S, Gopalaswamy AM, Knowles SCL, Sheldon BC. Site-occupancy modelling as a novel framework for assessing test sensitivity and estimating wildlife disease prevalence from imperfect diagnostic tests. Methods Ecol Evol. 2012;3:339–48.

22. Lockwood JL, Hoopes MF, Marchetti MP. *Invasion Ecology*. John Wiley & Sons, 2009

23. Lowe S, Browne M, Boudjelas S, De Poorter M. 100 of the World's Worst Invasive Alien Species A selection from the Global Invasive Species Database. Published by The Invasive Species Specialist Group (ISSG) a specialist group of the Species Survival Commission (SSC) of the World Conservation Union (IUCN). 2000. 12pp.

24. Lunn DJ, Thomas A, Best N, Spiegelhalter D. WinBUGS – a Bayesian Modelling framework: concepts, structure, and extensibility. Stat Comput. 2000;10:325–37. doi:10.1023/A:1008929526011.

25. MacKenzie DI, Nichols JD. 'Occupancy as a Surrogate for Abundance Estimation'. *Animal Biodiversity and Conservation*. 2004;27:461–67

26. MacKenzie DI, Nichols J, Royle JA, Pollock KH, Bailey LL, Hines JH. Occupancy Estimation and Modeling: Inferring Patterns and Dynamics of Species Occurrence. New York: Academic Press; 2006.

27. MacKenzie DI, Nichols JD, Hines JE, Knutson MG, Franklin AB. Estimating site-occupancy colonization and local extinction when a species is detected imperfectly. Ecology. 2003;84(8):2200–7. doi:10.1890/02-3090.

28. Marini F, Caputo B, Pombi M, Tarsitani G, Della Torre A. Study of *Aedes albopictus* dispersal in Rome, Italy, using sticky traps in mark-release-recapture experiments. Med Vet Entomol. 2010;24(4):361–8. doi:10.1111/j.1365-2915.2010.00898.x.

29. Medlock JM, Hansford KM, Schaffner F, Versteirt V, Hendrickx G, Zeller H, et al. A review of the invasive mosquitoes in Europe: ecology, public Health risks, and control options. Vector Borne Zoonotic Dis. 2012;12(6): 435–47. doi:10.1089/vbz.2011.0814.

30. Miquel M, Río R, Borràs D, Barceló C, Paredes-Esquivel C, Lucientes J. First Detection of *Aedes albopictus* (Diptera: Culicidae) in the Balearic Islands (Spain) and Assessment of Its Establishment according to the ECDC Guidelines. J Am Mosq Control Assoc. 2013;31:8-11

31. Padilla-Torres, Samael D., Gonçalo Ferraz, Sergio L. B. Luz, Elvira Zamora-Perea, and Fernando Abad-Franch. 2013. 'Modeling Dengue Vector Dynamics under Imperfect Detection: Three Years of Site-Occupancy by *Aedes aegypti* and *Aedes albopictus* in Urban Amazonia'. PLOS ONE 8 (3).

32. Paupy C, Delatte H, Bagny L, Corbel V, Fontenille D. *Aedes albopictus*, an Arbovirus vector: from the darkness to the light. Microbes Infect. 2009; 11(14):1177–85. doi:10.1016/j.micinf.2009.05.005.

33. Reiter P, Sprenger D. The used tire trade: a mechanism for the worldwide dispersal of container breeding mosquitoes. J Am Mosq Control Assoc. 1987;3(3):494–501.

34. Rezza G. *Aedes albopictus* And the reemergence of Dengue. Public Health. 2012;12:72.

35. Roche B, Léger L, L'Ambert G, Lacour G, Foussadier R, BEsnard G, et al. The spread of *Aedes albinopictus* in metropolitan France: contribution of environmental drivers and human activities and predictions for a near future. PLoS One. 2015;10(5):e0125600.

36. Rochlin I, Ninivaggi DV, Hutchinson ML, Farajollahi A. Climate change and range expansion of the Asian Tiger mosquito (*Aedes albopictus*) in northeastern USA: implications for public Health practitioners. PLoS One. 2013;8(4):e60874.

37. Royle JA, Nichols JD, Kéry M. Modelling occurrence and abundance of species when detection is imperfect. Oikos. 2005;110(2):353–9. doi:10.1111/j.0030-1299.2005.13534.x.

38. R Core Team. R: A language and environment for statistical computing. R Foundation for Statistical Computing, Vienna, Austria. 2016. http://www.R-project.org/.

39. Sakai AK, Allendorf FW, Holt JS, Lodge DM, Molofsky J, With KA, et al. The population Biology of invasive specie. Annu Rev Ecol Syst. 2001;32:305–32.

40. Semenza JC. Climate Change and Human Health, *International Journal of Environmental Research and Public Health*. 2014;11:7347–53.

41. Scholte EJ, Schaffner F. 'Waiting for the Tiger: Establishment and Spread of the *Aedes albopictus* Mosquito in Europe'. In Emerging Pests and Vector-Borne Diseases in Europe. Ecology and Control of Vector-Borne Diseases (1), edited by Taken W. and Knols B. Wageningen Academic publishers. 2007. 241–60.

42. Shigesada N, Kawasaki K. Biological invasions: theory and practice. UK: Oxford University Press; 1997.

43. Tenan S, O'Hara RB, Hendriks I, Tavecchia G. Bayesian model selection: the Steepest Mountain to climb. Ecol Model. 2014;283(July):62–9. doi:10.1016/j.ecolmodel.2014.03.017.

44. Tran A, L'Ambert G, Lacour G, Benoît R, Demarchi M, Cros M, et al. A rainfall-and temperature-driven abundance model for *Aedes albopictus* populations. Int J Environ Res Public Health. 2013;10(5):1698–719. doi:10.3390/ijerph10051698.

45. Valerio L, Marini F, Bongiorno G, Facchinelli L, Pombi M, Caputo B, Maroli M, della Torre A. Host-feeding patterns of *Aedes albopictus* (Diptera: Culicidae) in urban and rural contexts within Rome Province, Italy. Vector Borne Zoonotic Dis. 2009;10(3):291–4.

46. Vanlandingham D, Higgs S, Huang Yan-Jang S. *Aedes albopictus* (Diptera: Culicidae) and Mosquito-Borne Viruses in the United States. J Med Entomol. 2016;53(5):1024-8. doi:10.1093/jme/tjw025.

47. Yackulic CB, Reid J, Davis R, Hines JE, Nichols JD, Forsman E. Neighborhood and habitat effects on vital rates: expansion of the barred owl in the Oregon coast ranges. Ecology. 2012;93(8):1953–66. doi:10.1890/11-1709.1.

6

Quantifying phenotype-environment matching in the protected Kerry spotted slug (Mollusca: Gastropoda) using digital photography: exposure to UV radiation determines cryptic colour morphs

Aidan O'Hanlon[1]* (iD), Kristina Feeney[1], Peter Dockery[2] and Michael J. Gormally[1]

Abstract

Background: Animal colours and patterns commonly play a role in reducing detection by predators, social signalling or increasing survival in response to some other environmental pressure. Different colour morphs can evolve within populations exposed to different levels of predation or environmental stress and in some cases can arise within the lifetime of an individual as the result of phenotypic plasticity. Skin pigmentation is variable for many terrestrial slugs (Mollusca: Gastropoda), both between and within species. The Kerry spotted slug *Geomalacus maculosus* Allman, an EU protected species, exhibits two distinct phenotypes: brown individuals occur in forested habitats whereas black animals live in open habitats such as blanket bog. Both colour forms are spotted and each type strongly resembles the substrate of their habitat, suggesting that *G. maculosus* possesses camouflage.

Results: Analysis of digital images of wild slugs demonstrated that each colour morph is strongly and positively correlated with the colour properties of the background in each habitat but not with the substrate of the alternative habitats, suggesting habitat-specific crypsis. Experiments were undertaken on laboratory-reared juvenile slugs to investigate whether ultraviolet (UV) radiation or diet could induce colour change. Exposure to UV radiation induced the black (bog) phenotype whereas slugs reared in darkness did not change colour. Diet had no effect on juvenile colouration. Examination of skin tissue from specimens exposed to either UV or dark treatments demonstrated that UV-exposed slugs had significantly higher concentrations of black pigment in their epithelium.

Conclusions: These results suggest that colour dimorphism in *G. maculosus* is an example of phenotypic plasticity which is explained by differential exposure to UV radiation. Each resulting colour morph provides incidental camouflage against the different coloured substrate of each habitat. This, to our knowledge, is the first documented example of colour change in response to UV radiation in a terrestrial mollusc. Pigmentation appears to be correlated with a number of behavioural traits in *G. maculosus*, and we suggest that understanding melanisation in other terrestrial molluscs may be useful in the study of pestiferous and invasive species. The implications of colour change for *G. maculosus* conservation are also discussed.

Keywords: Camouflage, Mollusc, Phenotypic plasticity, Pigmentation, UV radiation, Animal colouration, Digital photography, Disruptive patterning, Gastropoda, Polyphenism, Slug, Terrestrial mollusc, Visual predation

* Correspondence: a.ohanlon4@nuigalway.ie
[1]Applied Ecology Unit, School of Natural Sciences, National University of Ireland Galway, Galway, Ireland
Full list of author information is available at the end of the article

Background

The ability of many animal species to accurately match surrounding habitat features (i.e. camouflage) has been held up as strong evidence of natural selection since Darwin's time [1]. Intraspecific phenotypic variation is common within many animal populations and a consistent pattern of phenotype-environment matching and disruptive markings can be considered an indication that a particular phenotype is adaptive [2]. Such patterns of phenotype-environment matching in animal colouration are most often discussed in the context of predator avoidance and it has been shown that disruptive patterns, as well as background colour matching, can prove highly effective in concealing an animal [3]. The role of predation, therefore, in maintaining variation in cryptic prey colour morphs and patterns is relatively well-understood [4]. Apart from concealment from predators, however, animal colours and patterns may serve other functions such as thermoregulation or social signalling. Thus, not all apparently cryptic morphs are the sole result of directional selection by predators over many generations and, for some species, external cues can stimulate developmental or behavioural mechanisms which lead to rapid adaptation within the lifetime of an individual (i.e. phenotypic plasticity). Plasticity in animal crypsis has been demonstrated in response to a wide range of cues such as diet [5, 6], seasonal temperature changes [7], illumination and visual background properties [8, 9]; detection of a possible predation threat [10], and ultraviolet (UV) radiation [11, 12].

Colour variation appears to be maintained by frequency-dependent selection for many gastropod species (e.g. in land snails [13, 14]; in littorinid snails [15, 16]). Differences in gastropod colour morphs can also arise as adaptations to climatic selection where relatively darker or lighter shell and skin pigmentation evolves within populations exposed to cooler or warmer climates, respectively (e.g. as in littorinids [17, 18]; in Western Irish *Cepea nemoralis* L [19]; and in horn snails [20]). Skin colouration in terrestrial slugs can be highly variable, even within populations [21]. The mechanisms involved in determining slug colouration most likely vary with species and local adaptations may arise within populations of slugs exposed to different environmental conditions. Pigmentation has been explained as a simple Mendelian-inherited trait for some species of slug [22, 23]. Skin colouration and mottling (possibly functioning as crypsis) has been shown to be under polygenic control for *Limax flavus* L and *Limacus maculatus* Kaleniczenco [24]. Colour variation can also arise as a result of hybridization with closely related species (e.g. as in *Arion ater* L and *Arion rufus* L [25]). However, there is also evidence that different colour morphs in slug species can arise as local adaptations to different

environmental conditions. While Evans [25], for example, found similar isozyme profiles between colour morphs of *Arion ater* agg., Taylor [21], in a robust survey of the malacofauna of Ireland and Britain, observed that dark colour morphs were associated with higher altitudes and cooler, wetter climates, suggesting climatic selection for pigmentation. Chevallier [26] provided further evidence of climatic selection for *A. ater* agg. and observed a similar pattern for *Arion lusitanicus* Mabille, with darker morphs prevailing at altitudes above 500 m. Jordaens et al. [27], on the other hand, demonstrated that diet can also influence skin pigmentation in F_1 offspring of three species of arionid slug, resulting in a loss of 'species-specific' colour characteristics. The Kerry spotted slug *Geomalacus maculosus* (Allman) is unusual in that it appears to possess disruptive patterning as well as background-matching colouration which may provide different degrees of camouflage in alternative environments. This EU-protected species is associated with forested and open habitats (blanket bog and mountain heath) in Ireland where it occurs in one of two distinct colour morphs. In forested habitats *G. maculosus* generally possesses a hazel brown to ginger brown body colour with white and yellowish spots which appear to accurately match the moss and lichen-covered bark of trees. In open habitats, on the other hand, the slug generally possesses a dark blue-grey to black body colour with white spots where it seems to accurately match lichen-covered boulder outcrops which it uses for shelter and feeding [21, 28–31]. Little is known about population structure of *G. maculosus* and whether this affects colouration. Reich et al. [32] studied the population genetics of *G. maculosus* on which basis they suggested that the species originated in northern Spain 15Myr ago. No differences in 16S rRNA or COI genes were found between black and brown colour morphs (I. Reich, *pers. comm.*). Furthermore, newly hatched juveniles are brown in both forested and open habitats [31], suggesting that body colouration in this species may be a plastic trait.

This study was undertaken to test the hypothesis that each colour morph of *G. maculosus* is habitat-specific and provides camouflage. We used standardized digital photography to quantify widespread phenotype-environment matching in this internationally important mollusc species. Experiments were also conducted with laboratory-reared juvenile slugs to determine whether different environmental conditions (UV exposure or diet) could induce colour-change and whether this may help to elucidate the function of each colour morph observed in the wild. The study will also help inform the debate regarding potential translocation of the species as a possible mitigation measure within the Environmental Impact Assessment process.

Methods

Study sites and animal sampling

Free-living slugs were photographed from six sites in Ireland across the western counties of Galway, Kerry and Cork (Fig. 1). Sites were selected on the basis that previous surveys had found high numbers of Kerry slugs in these areas [30, 33, 34, 35]. Forested sites were a combination of conifer plantations and oak woodlands (Fig. 1: sites 1, 3 and 5), and open habitats surveyed were blanket bog areas (Fig. 1: sites 2, 4 and 6). Refuge traps (De Sangosse, France) were used to collect slugs. These traps are 50 cm × 50 cm sheets of absorbent material covered with a reflective upper surface and a perforated dark lower surface which maintains a damp, cool environment beneath the trap. This method has been shown to be an effective technique for live-trapping *G. maculosus* [36]. Traps were placed on *Q. petraea* Liebl tree trunks (*n* = 4) at breast height (approx. 1.5 m) at site 5; on sandstone boulder outcrops at sites 4 (*n* = 4) and 6 (*n* = 4); and on granite outcrops at site 2 (site descriptions given in Additional file 1: Table S1). Each trap was checked for slugs one month after they had been set. Traps on sitka spruce *Picea sitchensis* Carr. trees at sites 1 and 3 were set previously by other researchers in the Applied Ecology Unit, NUI Galway, as part of separate projects [34, 35]. Each trap was only checked once for slugs to avoid pseudoreplication, except for site 1. Slugs in this site were removed from tree trunks after they had been photographed for use in a separate behavioural study, so multiple trips were possible since we could be positive that we were not taking pictures of the same individuals. Any additional slugs found near the traps were also photographed and included in colour analyses and subsequently removed from the site for use in a separate study. Slugs were collected with permission from the National Parks and Wildlife Service, Department of Arts, Heritage and the Gaeltacht (Licence No. C097/2015).

Quantification of G. maculosus colour using digital photography

Digital photography was used to estimate the degree of phenotype-environment matching of adult slugs at each site, following a simplified version of the suggested methodology outlined in Stevens et al. [37]. A colour-checker card (X-Rite, Munsell Color Laboratories) was used to standardize the reflectance values obtained from digital photographs. The card consists of 24 coloured squares which are manufactured to represent common natural colours and a six-step greyscale from white to black. The 'adjacent' method validated by Bergman and Beenher [38] was used, whereby slugs were photographed in the same image as the colour-checker card. The camera used was a Nikon D3000 digital single-lens reflex camera with a pixel count of 10.2 megapixels and full control over metering and exposure. All photos of wild slugs in this study were taken at F-stop: f/5.6 with a shutter speed of 1/100 s, ISO 800. Images were saved to the camera memory card as uncompressed Nikon Electronic Format (NEF) raw image files. After transferring all files to a computer, Adobe Photoshop was used to convert the raw NEF files to 8-bit Tagged Image File Format (TIFF) files, for compatibility with GIMP 2.0 image processing software. White balance was corrected in GIMP 2.0 to standardize each photo with reference to the white square of the colour-checker card, such that this white square was equal to a reflectance score of 255 in R, G and B colour channels (i.e. 'true' white) for each image.

To quantify colouration of individual slugs, a 1 cm × 1 cm square was drawn over the mantle of the slug in each image. Mean R, G and B reflectance values (calculated in-program by GIMP 2.0) were then recorded per square. To measure substrate colouration, three additional 3 cm × 3 cm squares were placed over background substrate in the photographs along-side each animal and the mean R, G and B reflectance scores of these three squares was calculated in-program. This was to determine whether *G. maculosus* phenotypes match a random sample of their background substrate (Fig. 2). The same method was used to quantify colour of juvenile slugs in UV-exposure and feeding experiments (outlined below). Due to the small body size of newly-hatched slugs, a 0.5 cm × 0.5 cm square was instead used to calculate mean R, G and B values over the mantle of juvenile slugs. Juvenile slugs were photographed once per month in the laboratory at F-stop f/4 with a shutter speed of 1/100 s, ISO 800.

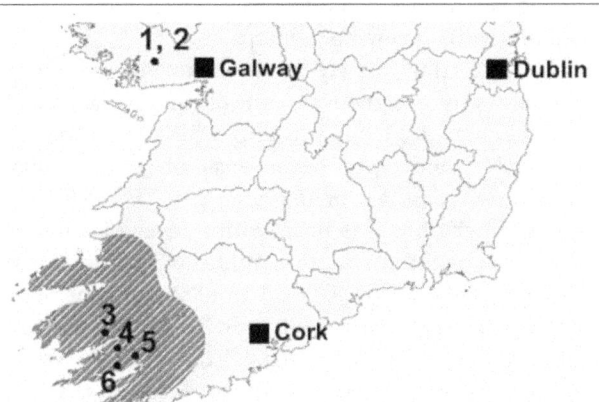

Fig. 1 Partial map of Ireland showing survey sites (black circles) inside the distribution range of *G. maculosus* (shaded area). Conifer (1) and blanket bog (2) habitats at Oughterrard, Co. Galway; conifer habitat at Tooreenafersha, Co. Kerry (3); blanket bog habitat adjacent to Uragh woods, Co. Kerry (4); Oak forest habitat at Glengarriff Nature Reserve, Co. Cork (5) and blanket bog at Leahill Bog, Co. Cork (6). Black squares show the locations of Galway, Cork and Dublin cities for reference. Map modified from G. Kindermann, 2016©

Fig. 2 a Each colour morph of *G. maculosus* was photographed on its natural substrate on tree trunks in forested sites (**b**) or on boulder outcrops in blanket bog (**c**) alongside a colour standard card. Reflectance in R, G and B colour channels was calculated from a 1 cm × 1 cm square drawn over the mantle to estimate slug colouration, and a further three squares measuring 3 cm × 3 cm were drawn over patches of the substrate to calculate background R, G and B reflectance. Scale bars were measured from original photographs

Effect of UV-exposure, darkness and diet on pigmentation
In addition to studying habitat-phenotype matching in wild-caught adult slugs, experiments were performed with newly hatched juveniles to investigate whether UV-exposure or diet can influence body colouration.

To investigate whether UV radiation could affect pigmentation in *G. maculosus*, hatchlings (n = 30) were randomly assigned to either a UV or 'darkness' treatment. For UV treatments, juveniles (n = 15) were kept in the laboratory in a clear plastic container (37 cm × 30 cm × 25 cm) fitted with a cold 13-watt fluorescent UVB bulb (Exo Terra, Canada). The bulb was fixed in the container lid at a distance of 25 cm from the container floor, resulting in a mean UVB irradiance value of approximately 25 µW/cm^2 (estimated from information provided with purchase of the bulb) on the container floor. The floor of each container was lined with a sheet of wet cotton wool covered with tissue paper and each container lid was sealed with ParaFilm® to prevent the slugs from dehydrating. A small shelter constructed from laminated cardboard was placed on one side of the container and food (porridge oats) was placed on the opposite side. This was to ensure that the slugs had shelter from excessive UV radiation and that they would be forced to periodically leave the shelter to feed. The UV light was set on a 14:10 dark: light cycle which approximated with the natural photoperiod when experiments began in February 2016 in a laboratory which also received ambient light. For 'darkness' treatments, juvenile slugs (n = 15) were kept in identical conditions but the containers were not fitted with a UV light source and were kept in constant darkness. Juveniles for each treatment came from two egg-batches laid within the same week which were split in half with each hatchling assigned at random to one of two treatments. Both egg batches were laid in February 2016 by two captive adults of the brown phenotype.

In feeding trials, newly hatched juveniles (n = 45) were randomly assigned to one of three plastic containers (17 cm × 11 cm × 6 cm) where they were fed one of three food types: organic carrot (n = 15 slugs), spinach (n = 15 slugs) or porridge oats (n = 15 slugs). The container floor was lined with a layer of moist tissue paper to maintain damp conditions and protect the slugs from dehydration. Cotton wool was not necessary to maintain adequate moisture in feeding trial boxes (as in the UV and darkness trials) due to their smaller size. The tissue was re-misted three times per week and decaying food was replaced as necessary (typically once per week). The containers were housed in the laboratory on a shelf approximately 5 m from a SW-facing window providing identical natural photoperiod cues to each diet group. Juveniles used in feeding trials originated from egg batches laid in the laboratory in September 2015 by two captive adults of the brown phenotype.

Stereological estimation of epithelial pigment

At the end of the UV / dark trials, all remaining slugs (n = 6 of each colour morph) were sacrificed using chloroform vapour and 1 cm × 1 cm skin samples from the slug mantle (effectively all the mantle) were removed following Rowson et al.[31]. Skin samples were fixed in 4% paraformaldahide before being dehydrated and embedded in paraffin wax. Sections (5 μm thick) were then cut on a Leica RM2125RT microtome and stained with hematoxylin/eosin. Transverse sections (1 per individual: 6 UV-exposed and 6 darkness-reared slugs) were imaged under a Leica DM500 microscope. Simple point-counting methods were used to estimate the volume fraction of pigment to the epithelium (mean of 16 grid samples per individual: mag. ×800). Mean epithelial thickness was also estimated (from 10 measuring points per individual: mag. ×200). Estimates of the volume of pigment per unit projected area of skin were then obtained by multiplying these two parameters [39].

Statistical analyses

To investigate whether animal colouration matches background colouration in free-living slugs in natural habitats, the mean R, G and B reflectance values from each slug were compared with those from the substrate upon which they were photographed. A one-way ANOVA was used to compare colour scores of slugs and substrate between sites of the same habitat type. Colour data of slugs and substrate were pooled for sites of the same habitat type (after it was determined that there were no significant differences between sites; Additional file 1: Table S1) and then tested for bivariate correlation using Pearson's r. Students t-tests were used to test whether RGB reflectance values differed significantly between woodland and blanket bog substrate and between each of the two slug colour morphs. Students t-tests were also used to examine whether RGB reflectance scores, at the end of the experiment, differed between UV-treated slugs and slugs kept in darkness. Paired t-tests were used to test whether juvenile slugs differed in RGB reflectance scores after feeding trials were concluded. Data from stereological estimation of the % volume fraction of black pigment in epithelial sections were not normally distributed. A Mann-Whitney U test was therefore used to examine whether the % volume fraction of black pigment differed significantly between UV-irradiated and dark-reared slugs. A one-way ANOVA was used to test whether juvenile slugs differed in reflectance of R, G and B colour scores at the beginning of UV-exposure trials and feeding trials (to account for any potential variation between newly-hatched individuals). Graphs were prepared and statistical tests were carried out using SPSS (IBM, USA).

Results

Quantification of G. maculosus colour using digital photography

In total, 124 slugs were sampled from forested sites and 71 slugs were sampled from blanket bog sites. Colour scores did not differ significantly for slugs or substrate between sites of the same habitat type (with the exception of the B reflectance from forest slugs; Additional file 1: Table S2). Slugs from forest site 3 showed significantly lower mean B reflectance scores than slugs from both of the other forested sites. Given that subsequent analysis involved pooling of data, this site was removed from the data set – an explanation of why the B reflectance was significantly different in slugs from forest site 3 is given in the discussion. Colour scores of slugs and of substrates were pooled by habitat type (forest or bog). The R, G and B reflectance values of slugs and substrate were strongly and positively correlated for both forest and blanket bog habitats (Fig. 3). Mean R, G and B reflectance values of slugs differed significantly between habitats as did substrate (Table 1). R, G and B reflectance values did not differ significantly between slugs and substrate from the same habitat type but did differ significantly between slugs and substrate from the alternative habitat (i.e. between slugs from forest habitats and substrate from bog habitats, and between slugs from bog habitats and substrate from forest habitats; Table 2).

Effect of UV-exposure, darkness and diet on pigmentation

The R, G and B reflectance scores did not differ significantly between newly-hatched slugs before each experimental treatment (i.e. between UV and darkness treatments, and between different diet treatments; Additional file 1: Table S3).

After a period of 140 days, UV-irradiated slugs displayed significantly lower R, G and B reflectance scores than when they first hatched (Fig. 4), whereas slugs reared in darkness did not differ significantly in colour reflectance scores after the same time period (Table 3). Slugs reared on different diets under laboratory conditions also exhibited significantly lower colour reflectance scores at the end of feeding trials (84 days) than when they first hatched (Table 3). However, colour reflectance scores did not differ significantly between diet treatments after a period of 84 days (comparable only for carrot and oat diets since juveniles reared on spinach died after 56 days; Fig. 5).

Stereological estimation of epithelial pigment

Estimates of the volume fraction of black pigment (Fig. 6) were significantly greater in epithelial sections of UV-irradiated slugs (Mean rank: 9.33) than in darkness-reared slugs (Mean rank: 3.67; U (12) = 35, p = 0.004). There was no significant difference in mean epithelial thickness between slugs from each treatment group (U (12) = 30, p = 0.0649).

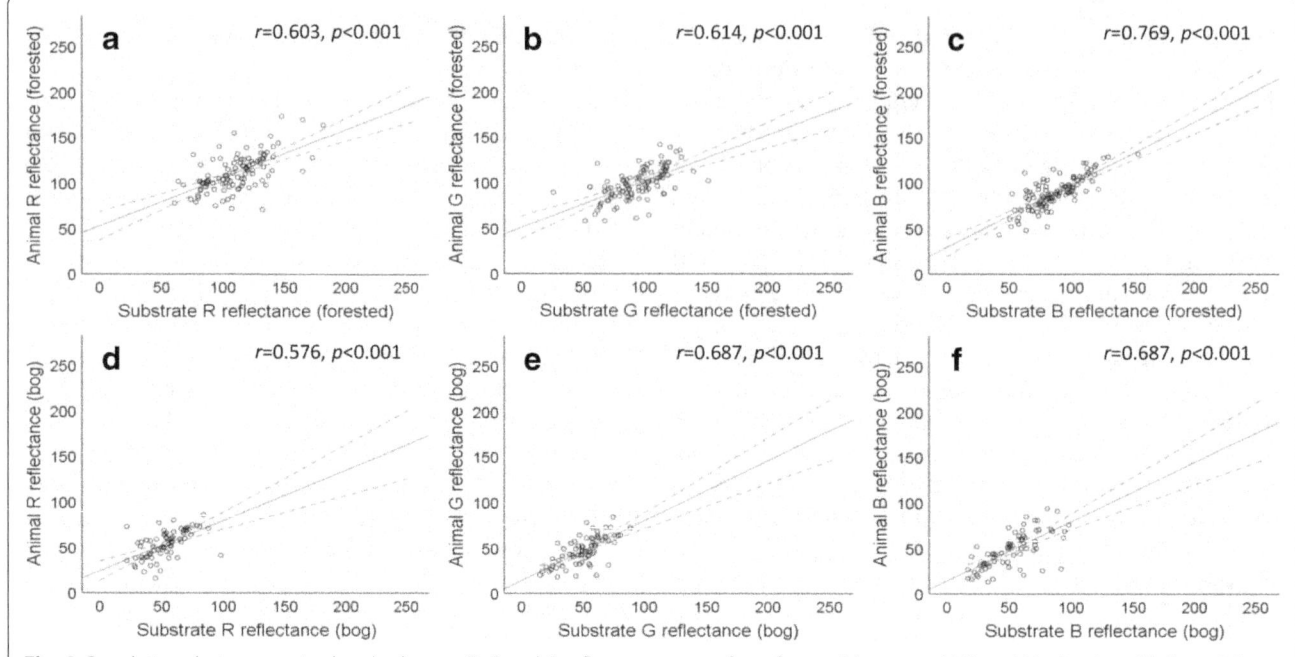

Fig. 3 Correlations between animal and substrate R, G and B reflectance scores from forested (**a–c**; n = 104) and blanket bog (**d–f**; n = 71) habitats. Solid lines show Pearson's r correlation; dotted lines show 95% CI

Discussion

Environment-phenotype matching

The colour values of adult slugs as estimated using digital photography are strongly and positively correlated with the colour properties of the substrate upon which they were found, in both forest and bog habitats. Furthermore, these colour values differed significantly between colour morphs and substrate from each habitat, and when animal colouration was compared to substrate colouration from the alternative habitat. This suggests a 'mismatch' between animal and substrate in alternative habitats and demonstrates that colouration in adult *G. maculosus* is habitat-specific. Brown 'forest' slugs may be mismatched on black and white 'bog' substrates and vice versa, suggesting that mismatched individuals would possess a lower degree of crypsis in the 'wrong' habitat type, possibly increasing their susceptibility to predation.

A consistent pattern of background-phenotype matching is a good indication that a phenotype is adaptive [2]. The correlations found between R, G and B colour channels measured from *G. maculosus* and their substrate in both habitat types is consistent with the hypothesis that this species possesses camouflage which enables both colour morphs of *G. maculosus* to accurately match a random sample of their respective background. Hall et al. [40] demonstrated how disruptive markings can provide effective crypsis in multiple habitats. Spotted patterning, as well as background colour matching therefore likely provides strong, habitat-specific camouflage in *G. maculosus*.

Camouflage implies selective pressure from a visual predator. Although passerine birds are known to prey upon a number of medium-large arionid slug species [41], only one published record of predation exists for *G.*

Table 1 Mean reflectance scores and t test statistics comparing RGB values of slug and substrate images from forested (N = 104 images) and blanket bog (N = 71 images) habitats

	Channel	Forested habitats		Blanket bog habitats		Between-habitats t-tests		
		Mean	SD	Mean	SD	Mean Diff.	t(173)	p
Slug	R	111.68	20.31	53.54	15.34	58.16	21.53	<0.001
	G	98.45	17.69	47.44	15.72	50.98	19.56	<0.001
	B	88.95	17.96	48.66	19.71	40.27	13.99	<0.001
Substrate	R	111.12	23.09	53.21	16.04	57.95	19.62	<0.001
	G	94.14	21.34	50.74	16.41	43.38	15.18	<0.001
	B	86.89	20.04	52.78	20.85	34.09	10.87	<0.001

Table 2 Results from t tests comparing mean RGB values of slugs against RGB values of substrate from their natural habitat, and against substrate from alternative habitats (sample sizes and means ± SD for each colour channel are presented in Table 1)

	Channel	Forested substrate			Bog substrate		
		Mean diff.	t	p	Mean diff.	t	p
Forested slug	R	0.56	0.18	0.852	−57.60	−19.80	<0.001
	G	4.30	1.58	0.115	−46.68	−16.64	<0.001
	B	2.05	0.78	0.437	−38.22	−12.46	<0.001
Bog slug	R	58.51	21.28	<0.001	0.35	0.13	0.894
	G	47.69	18.03	<0.001	−3.29	−1.22	0.224
	B	36.15	11.91	<0.001	−4.13	−1.21	0.227

maculosus by the larval stage of the sciomyzid fly *Tetanocera elata* Fabricius, which appeared to initiate feeding upon reception of tactile or olfactory cues [42]. It is currently unknown whether the spotted patterns on both *G. maculosus* morphs act as disruptive markings against avian predators. Many birds are potentially tetrachromatic [43] and may therefore perceive colour in wavelengths undetected by this study (photography and image analysis methods used in this study provide a basic assessment of colouration in a trichromatic colour space - more accurate colour analysis would involve measuring differences in regions of a light spectrum, and mapping these to models representing the visual capacity of known predators [37]). A possible alternative explanation for spotted markings might be that they function in social signalling, as is known to be the case for other invertebrates (e.g. in wasps: [44, 45]). However, slugs are believed to have poor visual systems relative to many other invertebrate taxa and may be only capable of detecting the overall distribution of light and dark [46]. Furthermore, *G. maculosus* is hermaphroditic which makes a social signalling hypothesis for the spotted patterning unlikely. Allen [41] stated that visual predation by birds is undoubtedly the most important selective

force in the evolution of cryptic colour morphs in prey animals and suggested that this explains why terrestrial gastropods in general living in forests tend to be brown. In addition to the pattern of background matching demonstrated by the results of this study, an unusual startle-response unique among gastropods to *G. maculosus* may also suggest selection by avian predators. Many slugs contract into a humped posture when disturbed [31]. However, *G. maculosus* curls up into a ball shape by contracting its foot completely in half, and secretes a low-viscosity mucus causing the animal to become more slippery (*pers. obs.*). This behaviour is perhaps most easily explained as an adaptation by increasing the difficulty of an avian predator to hold the prey in its bill.

Colour change

Images of juvenile slugs used in feeding trials showed lower reflectance scores in all colour channels after 84 days. However, there was no statistically significant difference in R, G and B reflectance values between food treatments after the feeding trial period was concluded. Thus, even though juvenile slugs were darker after feeding trials, it seems to be independent of diet. Although the lichen composition found in the field differs between

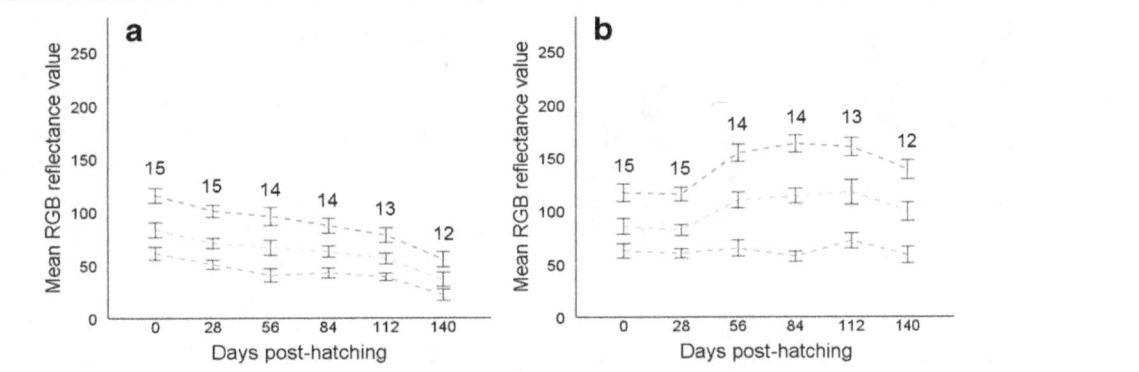

Fig. 4 Juvenile slugs irradiated under UV-lighting (**a**) showed less reflectance in R (*top line*), G (*middle line*) and B (*bottom line*) colour channels with each month, becoming significantly darker than juveniles maintained in darkness (**b**). Colour reflectance scores differed significantly between UV-exposed and darkness reared slugs at the end of the experimental period (R: $t = −14.461$, $p < 0.001$; G: $t = −11.391$, $p = <0.001$; B: $t = −7.700$, $p = <0.001$). Numbers above means show group n at each sampling month; error bars show ±SE for means

Table 3 Effect of different lighting and diet treatments on juvenile slug colouration from the time of hatching (start) to the end of the trials

| | | Channel | Start[a] | | End[b] | | t | p |
			Mean	SD	Mean	SD		
		R	114.80	13.09	55.05	12.48	9.175	<0.001
	UV	G	83.26	11.31	36.26	11.65	7.879	<0.001
Lighting		B	60.64	10.44	21.86	9.40	7.455	<0.001
(fed on oats)								
		R	119.18	15.36	114.94	17.23	0.603	0.559
	Darkness	G	88.02	12.92	92.97	16.89	−1.179	0.263
		B	65.52	11.08	58.28	13.42	2.172	0.053
		R	124.07	16.35	91.68	13.21	6.27	<0.001
	Carrot	G	95.31	11.73	69.51	11.15	5.75	<0.001
		B	71.21	11.09	57.65	6.72	3.06	0.013
Diet		R	119.81	17.11	96.04	11.48	4.378	0.001
(natural photoperiod)	Oats	G	95.88	18.18	69.65	9.23	4.828	<0.001
		B	67.26	17.26	67.26	5.98	2.227	0.043

[a]Start = day of hatching, [b]End = day 140 UV-exposure experiments, and day 84 for diet experiments

blanket bog and forested habitats, adult *G. maculosus* specimens from blanket bog habitats will eat similar foods to specimens from forested habitats (E. Johnston, *pers. comm.*) in captivity. While Jordaens et al. [27] showed how diet can influence body colour in slugs of the subgenus *Carinarion* Hesse, the slight darkening of juveniles observed after 84 days of feeding trials in this study is likely to be a natural result of growth. As in some other slug species (e.g. in *Limax flavus* [24]), newly

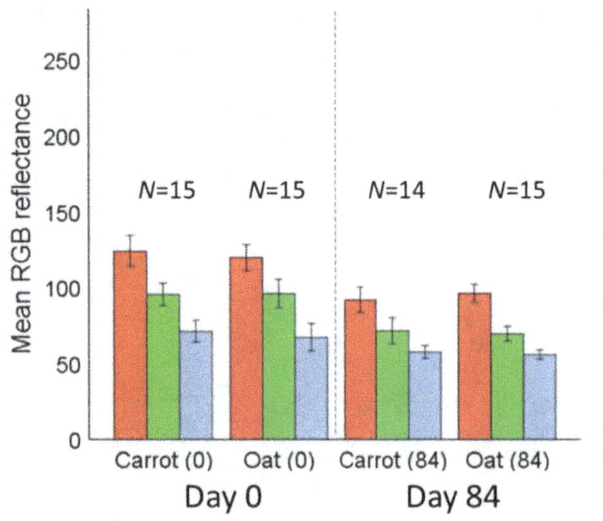

Fig. 5 Juveniles reared on different diets showed significantly less reflectance in R, G and B colour channels (*shown as red, green and blue coloured bars*) after 84 days (*right of the dotted line*) than when they first hatched, but colour reflectance scores not differ significantly between diets after this period; (R: *t* = −0.876, *p* = 0.390; G: *t* = 0.409, *p* = 0.686; B: *t* = 0.612, *p* = 0.546). Error bars show ±2SE for group means; means ± SD given in Table 3

hatched *G. maculosus* juveniles tend to be light brown in colour and somewhat translucent, so the darker colour values recorded after feeding trials were concluded are probably the result of increased overall size and thickening of the integument. Furthermore, the juvenile slugs used in feeding trials were maintained in plastic containers on a shelf in the laboratory approximately 5 m from a SW-facing window and, as such may have been exposed to a natural day/light cycle, which may also have influenced this slight darkening to some degree.

Slugs irradiated under UV-lighting became consistently darker at each sampling month and exhibited the black 'bog' morph after these trials were concluded, showing less reflectance in R, G and B colour channels. Slugs kept in darkness, however, remained lighter than UV-exposed juveniles at the end of the experimental period, showing higher reflectance scores in R, G and B colour channels. This result demonstrates that ultraviolet radiation can induce a change in pigmentation such that slugs exposed to UV radiation become darker in colour. Stereological examination of skin tissue from darkness-reared and UV-exposed juveniles demonstrated that UV-irradiated slugs contained significantly greater amounts of black pigment in their epithelium than darkness-reared slugs. This black pigment is most likely melanin, which has been detected in a number of other slug species (e.g. in *A. ater* [47]; *Arion hortensis* Férussac [48]; *Deroceras reticulatum* Müller [49]) and is known to develop in response to UV-exposure in a wide range of other animal taxa [50] but until this study, has not been demonstrated in slugs. Since juvenile *G. maculosus* from blanket bog habitats tend to be brown [31], it is likely that the black morph of adults develops in response to

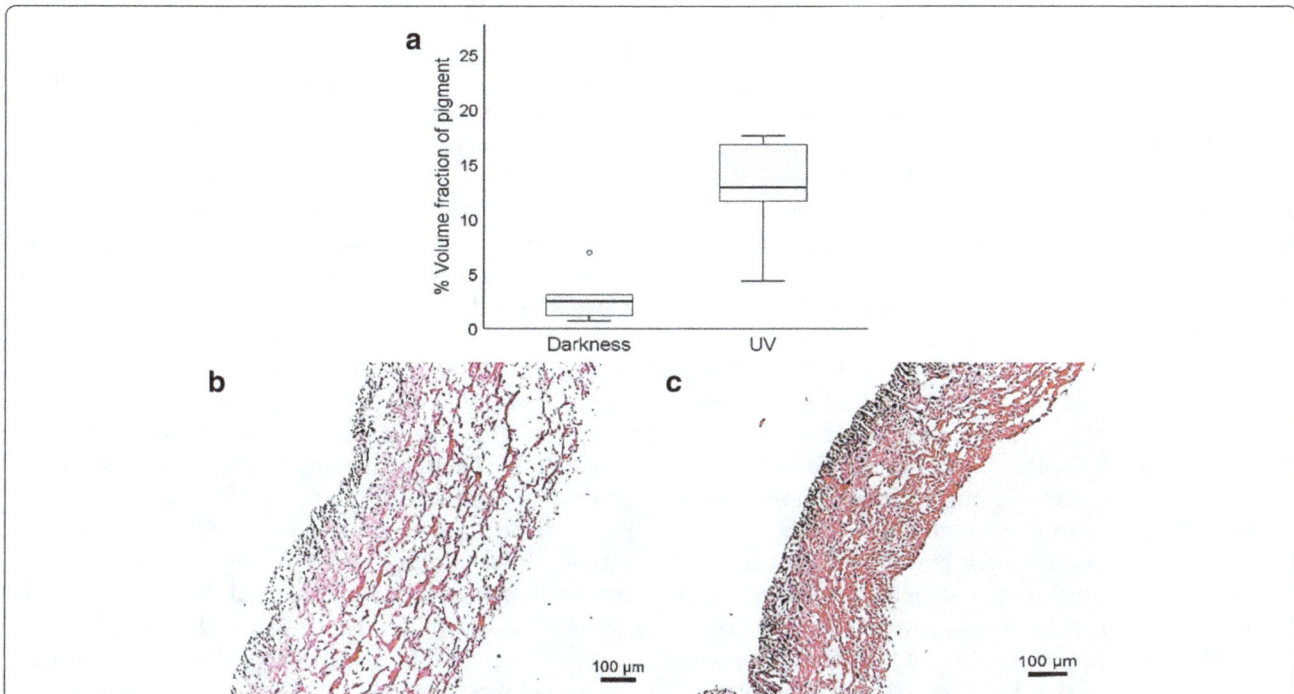

Fig. 6 a UV-exposed slugs contained a significantly greater estimated volume of black pigment than darkness-reared slugs. Volume fraction estimates are expressed as percentages. TS of integument from **b** a darkness-reared (*brown*) juvenile and **c** a UV-exposed (*black*) juvenile (mag. ×200). Black pigment is concentrated in the outer epithelia of both slug colour morphs

higher levels of UV exposure in these habitats, whereas juveniles which develop in relatively darker and more sheltered forested habitats remain brown. McCrone and Sokolove [51] showed that photoperiod was responsible for producing a maturation hormone in *Limax maximus* L., with long photoperiod exposures resulting in the development of male-phase, and shorter photoperiod exposures resulting in the development of female-phase reproductive morphologies upon maturation. A similar hormonal pathway may be present in *G. maculosus*, where black or brown pigmented phenotypes are expressed in response to different levels of UV radiation. Previously, adult *G. maculosus* specimens collected from open habitats appeared to lose some of their black pigment after a period of several weeks in the laboratory, with skin tissue in the grooves between tubercles becoming a paler brown colour (*pers. obs.*), suggesting that colour change may be reversible to some degree. However, it currently remains unclear whether colour change is completely reversible in adults or whether black pigmentation in *G. maculosus* is produced during a key period of early development in response to UV cues. UV radiation reaching the ground in bog habitats, particularly on overcast days and during sunrise/sunset hours when the slugs are most active, is likely to be lower than the UVB radiation emitted during laboratory experiments (25 µW/cm^2). As such it remains unknown how long it takes juveniles to develop into the black morph in the field, and whether this change in

pigmentation is completely or partially reversible. Slugs from forest site 3 were omitted from pooled analyses because they exhibited slightly lower mean reflectance scores in R and G channels, and significantly lower reflectance scores in B colour channels than slugs photographed from the other two forested sites. These slugs, although of the brown phenotype, appear to possess darker skin than slugs from the other forested sites. The area from which slugs were photographed from forest site 3 was noticeably brighter and had a luminosity value three times greater than the other two forest sites surveyed (Additional file 1: Table S1). This site also borders a blanket bog habitat, so it is possible that the slightly darker brown slugs photographed here were exposed to higher levels of UV light than slugs from the other two forested sites. This is consistent with a statement by Rowson et al. [31], that the distinction between brown and black *G. maculosus* colour forms becomes substantially blurred where wooded areas border open habitats.

Origin of colour dimorphism
Reich et al. [32] demonstrated that *G. maculosus* originated during the middle Miocene, approx. 15Myr ago, probably arriving in Ireland from Iberia during the middle ages. The ability of *G. maculosus* to alter the degree of melanin-like pigment in its skin in response to UV exposure could have evolved relatively recently, during the Quarternary period. The Quarternary is characterised by

the periodic growth and retreat of ice sheets across the northern hemisphere [52]. Hewitt [53] has suggested that animal and plant populations survived through several glacial cycles by migrating up and down mountains. Reich et al. [32] suggested that Iberian *G. maculosus* populations may have survived in mountain valleys during periods of glaciation when northern Spain's mountain peaks would have been covered by ice. During warming phases when ice sheets were in retreat, *G. maculosus* populations would again have access to higher mountain altitudes, allowing them to increase their range altitudinally. Exposure to UV radiation can damage DNA and melanin can act as a protective filter in skin against the effects of UV exposure [54–56]. Migration by *G. maculosus* up and down mountain ranges over thousands of years as ice sheets periodically expanded and retreated may therefore have fixed in ancestral populations the ability to cope with different exposures to UV intensity. Dark populations of some slug species have previously been reported to prevail at high altitudes, possibly as a result of climatic selection (e.g. *A. ater* [21]; *A. rufus* and *A. lusitanicus* [26]; and *Lehmannia marginata* Müller [31]). However, pigmentation in these species is not known to be plastic. Since *G. maculosus* can self-fertilize, the capacity of an individual to express either black or brown phenotype in response to differential UV exposure may have been selected over non-plastic melanic morphs, which are known to occur as adaptations to altitudinal gradients in UV intensity in some lizards [12] and insects [57–59]. Skin colour in *G. maculosus* most likely plays a photoprotective role, leading to incidental camouflage against the different substrate types of each habitat. The spotted patterning present in both colour morphs likely affords *G. maculosus* with a degree of bet-hedging by increasing the effectiveness of this incidental camouflage in whichever habitat type it develops. Ahlgren et al. [60] found that the development of black skin pigmentation in the freshwater gastropod *Radix balthica* L. was induced by UV radiation and by the detection of kairomones from predatory fish, demonstrating how pigmentation may serve more than one function in cryptic animals. It is also possible that each *G. maculosus* colour morph has important implications for thermoregulation – with black slugs absorbing heat more efficiently than paler brown slugs, as has been shown for other ectothermic invertebrates on a wide geographic scale [61].

Implications for other species
Dark colour morphs have been reported to prevail at high altitudes in many other terrestrial gastropod species, and this phenomenon has most often been explained as an example of climatic selection. It is possible that colouration is also a plastic trait in at least some of these species. Melanisation is correlated with a suite of behavioural traits in vertebrates: typically, darker vertebrates tend to be more aggressive, sexually active and resistant to stress than lighter vertebrates [62]. Although melanin development pathways differ significantly between vertebrates and invertebrates [63], further research into the links between pigmentation and behaviour may reveal many analogues from invertebrate systems – particularly in species with melanin-based colour polyphenisms. Results from behavioural studies with *G. maculosus* have shown that black slugs collected from bog habitats exhibit a significantly faster escape response (data to be published elsewhere), as well as a greater degree of sinuosity in food-searching behaviour than brown individuals collected from forests (E. Johnston, *pers. comm.*), demonstrating greater levels of boldness and exploratory behaviour, respectively. Such consistent intraspecific behavioural types may also be common to other slugs, and studying their occurrence could have useful practical implications. For example, the grey field slug *D. reticulatum* is a major agricultural pest which can be difficult to identify due to its highly variable skin colour - it occurs on a spectrum of very pale to deep brown-coloured individuals, even within populations [31]. Luther [22] demonstrated that pigmentation in *D. reticulatum* is genetically controlled, with melanised forms dominant to unpigmented individuals. However, the degree of melanisation may well be influenced by UV exposure in darkly pigmented *D. reticulatum*, as has been presently demonstrated for *G. maculosus*. Furthermore, Chevallier [26] reported that populations of the highly invasive slug *A. lusitanicus* are darker at high altitudes, citing it as a case of climatic selection. It is likely that darker forms are at least in part melanised due to higher UV-exposure for *A. lusitanicus*, *D. reticulatum* and a host of other terrestrial slugs in which colour polyphenisms were previously believed to be the sole result of climatic selection or putatively non-adaptive genetic inheritance. It may therefore be useful for researchers interested in developing pest control protocols to investigate whether melanisation could also be used as a predictor of boldness or exploratory behaviour in pestiferous and invasive slugs.

Conclusions
Colour dimorphism in *G. maculosus* is consistent with the idea of habitat-specific crypsis, with a likely additional function for photoprotection. Adult pigmentation is a plastic trait determined by differential exposure to UV radiation and colour dimorphism in this species may have initially originated as an adaptation to clinal differences in UV radiation. Phenotypic plasticity in *G. maculosus* pigmentation could represent a generalist strategy to reduce detectability by visual predators by providing incidental crypsis in different habitat types and experiments are planned to test this hypothesis with passerine

birds. To our knowledge, the results from this study provide the first evidence of plastic colour change in a terrestrial mollusc in response to UV radiation. Recently, *G. maculosus* populations been recorded from a number of plantation forests throughout its Irish distribution [33–35, 64], and forest clear-felling may significantly reduce population sizes in these habitats. The ability of *G. maculosus* to change colour with UV-exposure may also have important implications for the conservation and management of this protected species. Slugs which remain in clear-felled areas should develop the black 'bog' phenotype upon exposure to the relatively higher levels of UV post-felling. We would expect these black phenotypes to provide ineffective camouflage against a tree-stump background, which may reduce their fitness by making the slugs more conspicuous to visually-foraging predators. Careful consideration therefore needs to be given to site and habitat selection where translocation of *G. maculosus* populations may be used as a mitigation measure for forestry activities.

Additional files

Additional file 1: Table S1. Description of study sites. **Table S2**. Results of a one-way ANOVA comparing mean slug and substrate RGB reflectance values between sites of the same habitat type. **Table S3**. Results of a one-way ANOVA comparing mean RGB reflectance values between groups prior to diet experiments; and results of an independent samples *t* test comparing RGB reflectance values between groups prior to UV and darkness experiments. (DOCX 18 kb)

Additional file 2: Epithelial stereology. (XLSX 28 kb)

Additional file 3: Feeding trials. (XLSX 14 kb)

Additional file 4: Lighting trials. (XLSX 18 kb)

Additional file 5: Slug and Substrate RGB reflectance values. (XLSX 21 kb)

Acknowledgements
We are grateful to Dr. Kerry Thompson and Mark Canney for their help in staining and sectioning tissue slides, Eoin MacLoughlin for providing dissection trays and pins, Dr. Claire Heardman and Dr. Rory McDonnell for their advice when selecting survey sites, Dr. Gesche Kindermann for providing a map of *G. maculosus* distribution and Dr. Erin Johnston for helpful discussion about the diets of each *G. maculosus* colour morph.

Funding
This project was funded by an Irish Research Council government of Ireland postgraduate scholarship (GOIPG/2014/657) and is part of a larger project investigating the behavioural ecology of *G. maculosus*.

Authors' contributions
Study design was by AO'H and MJG, field surveys were by AO'H and KF, feeding trials were conducted by AO'H and KF, UV-rearing experiments were conducted by AO'H, analyses were by AO'H and MJG, stereology work was designed by PD. Writing and manuscript production were by AO'H and MJG with input by PD and KF. All authors approved the final manuscript.

Competing interests
The authors declare that they have no competing interests.

Author details
[1]Applied Ecology Unit, School of Natural Sciences, National University of Ireland Galway, Galway, Ireland. [2]Centre for Microscopy and Imaging, National University of Ireland Galway, Galway, Ireland.

References
1. Darwin C. On the origin of species by means of natural selection. London: John Murray; 1859. Chapter IV 'Natural Selection'
2. Endler JA. A predator's view of animal colour patterns. Evol Biol. 1978;11:319–64.
3. Cuthill I, Stevens M, Sheppard J, Maddocks T, Párraga CA, Troscianko TS. Disruptive colouration and background pattern matching. Nature. 2005;434:72–4.
4. Stevens M. Predator perception and the interrelation between different forms of protective colouration. P Roy Soc Lond B Bio. 2007;272:1457–64.
5. Manríquez PH, Lagosb NA, Jaraa ME, Castillac JC. Adaptive shell color plasticity during the early ontogeny of an intertidal keystone snail. P Natl Acad Sci Usa. 2009;106:16298–303.
6. Cranfield MR, Chang S, Pierce NE. The double cloak of invisibility: phenotypic plasticity and larval decoration in a geometrid moth, *Synchlora frondaria*, across three diet treatments. Ecol Entomol. 2009;34:412–4.
7. Brakefield PM, Reitsma N. Phenotypic plasticity, seasonal climate and the population biology of bicyclus butterflies (Satyridae) in Malawi. Ecol Entomol. 1991;16:291–303.
8. Vroonen J, Vervust B, Fulgione D, Maselli V, Van Damme R. Physiological colour change in the Moorish gecko, *Tarentola mauritanica* (Squamata: Gekkonidae): effects of background, light, and temperature. Biol J Linn Soc. 2012;107:182–91.
9. Kang C, Kim YE, Jang Y. Colour and pattern change against visually heterogeneous backgrounds in the tree frog *Hyla japonica*. Sci Rep. 2016;6:22601.
10. Tollrian R, Heible C. Phenotypic plasticity in pigmentation in *Daphnia* induced by UV radiation and fish kairomones. Funct Ecol. 2004;18:497–502.
11. Hannson LA. Plasticity in pigmentation induced by conflicting threats from predation and UV radiation. Ecology. 2004;85:1005–16.
12. Reguera S, Zamora-Camacho F, Moreno-Rueda G. The lizard *Psammodromus algirus* (Squamata: Lacertidae) is darker at high latitudes. Biol J Linn Soc. 2014;112:132–41.
13. Cook LM. Polymorphic snails on varied backgrounds. Biol J Linn Soc. 1986;29:89–99.
14. Cameron RAD. Change and stability in *Cepaea* populations over 25 years – a case of climatic selection. P Roy Soc Lond B Bio. 1992;248:181–7.
15. Wilbur AK, Steneck RS. Polychromatic patterns of *Littorina obtusata* on *Ascophyllum nodosum*: are snails hiding in intertidal seaweed? North Eastern Nat. 1999;6:189–98.
16. Johannesson K, Ekendahl A. Selective predation favouring cryptic individuals of marine snails (*Littorina*). Biol J Linn Sco. 2002;76:137–44.
17. Cook LM, Freeman PM. Heating properties of morphs of the mangrove snail *Littoraria pallescens*. Biol J Linn Soc. 1986;29:295–300.
18. Miller LP, Denny MW. Importance of behavior and morphological traits for controlling body temperature in Littorinid snails. Biol Bull. 2011;220:209–23.
19. Burke DPT. Variation in body colour in western Irish populations of *Cepaea nemoralis* (L.). Biol J Linn Soc. 1989;36:55–63.
20. Miura O, Nishi S, Chiba S. Temperature-related diversity of shell colour in the intertidal gastropod *Batillaria*. J Mollus Stud. 2007;73:235–40.
21. Taylor JW. *Monograph of the Land and Freshwater Mollusca of the British Isles* (Testacellidae, Limacidae, Arionidae). Leeds: Taylor Brothers; 1901;Parts 8–13,
22. Luther A. Zuchtversuche an Ackerschnecken (*Agriolimax reticulatus* Müll. Und *A. agrestis* L.). Acta Soc pro Fauna et Flora Fenn. 1915;40:1–42.
23. Reise H. *Deroceras juranum* – A Mendelian colour morph of *D. rodnae* (Gastropoda: Agriolimacidae). J Zool. 1997;241:103–15.
24. Evans NJ. Observations on variation in body colouration and patterning in *Limax flavus* and *Limax pseudoflavus*. J Nat Hist. 1982;16:847–58.
25. Evans NJ. Studies on the variation and taxonomy in two species aggregates of terrestrial slugs: *Limax flavus*, L. agg. and *Arion ater* L. agg. 1977; PhD thesis, University of Liverpool.

26. Chevallier H. Observations sur le polymorphisme des limaces rouges (*Arion rufus* Linne et *Arion lusitanicus* Mabille) et de l'escargot petit-gris (*Helix aspersa* Müller). Haliotis. 1976;6:41–8.

27. Jordaens K, Van Riel P, Geenen S, Verhagen R, Backeljau T. Food-induced body pigmentation questions the taxonomic value of colour in the self-fertilizing slug *Carinarion* spp. J Molluscan Stud. 2001;67:161–7.

28. Oldham C. Notes on *Geomalacus maculosus*. Proc Malac Soc Lond. 1942;25:10–1.

29. Platts EA, Speight MCD. The taxonomy and distribution of the Kerry slug *Geomalacus maculosus* Allman, 1843 (Mollusca: Arionidae) with a discussion of its status as a threatened species. Ir Nat J. 1988;22:417–30.

30. McDonnell RJ, Gormally MJ. Distribution and population dynamics of the Kerry slug Geomalacus maculosus (Arionidae). Irish Wildlife Manuals, 2011;54. National Parks and Wildlife Service, Department of Arts, Heritage and the Gaeltacht, Dublin, Ireland.

31. Rowson B, Turner J, Anderson R, Symondson B. Slugs of Britain and Ireland. Telford: FSC Publications; 2014. p. 60–1.

32. Reich I, Gormally MJ, Allcock AL, McDonnell RJ, Castillejo J, Iglasias J, Quinteiro J, Smith CJ. Genetic study reveals close links between Irish and northern Spanish specimens of the protected Lusitanian slug *Geomalacus maculosus*. Biol J Linn Soc. 2015;116:156–68.

33. Reich I, McDonnell RJ, McInerney C, Callanan S, Gormally MJ. EU-protected slug *Geomalacus maculosus* and sympatric *Lehmannia marginata* in conifer plantations: what does mark-recapture method reveal about population densities? J Moll Stud. 2016;82:doi:10.1093/mollus/eyw039

34. Johnston E, Kindermann G, O'Callaghan J, Burke D, McLoughlin C, Horgan S, McDonnell RJ, Williams C, Gormally MJ. 2016. Monitoring the EU protected *Geomalacus maculosus* (Kerry Slug): what are the factors affecting catch returns in open and forested habitats? Ecol Res. 2016; doi:10.1007/s11284-016-1412-5

35. Kearney J. Kerry slug (*Geomalacus maculosus* Allman 1843) recorded at Lettercraffroe. Co Galway Ir Nat J. 2010;31:68–9.

36. McDonnell RJ, Gormally MJ. A live trapping method for the protected European slug, *Geomalacus maculosus* Allman 1843 (Arionidae). J Conchol. 2011;40:483–5.

37. Stevens M, Parraga CA, Cuthill IC, Partridge JC, Troscianko TS. Using digital photography to study animal colouration. Biol J Linn Soc. 2007;90:211–37.

38. Bergman TJ, Beehner JC. A simple method for measuring colour in wild animals: validation and use on chest patch colour in geladas (*Theropithecus gelada*). Biol J Linn Soc. 2008;94:231–40.

39. Howard CV, Reed MG. Unbiased stereology. Three dimensional measurement in microscopy (2nd ed.). Garland Science/Bios Scientific: Abingdon; 2005.

40. Hall JR, Cuthill IC, Baddeley R, Shohet AJ, Scott-Samuel NE. Camouflage, detection and identification of moving targets. Proc Roy Soc B Biol Sci. 2013;doi:10.1098/rspb.2013.0064

41. Allen JA. Avian and mammalian predators of terrestrial gastropods. In: Barker GM, editor. Natural enemies of terrestrial Molluscs. Hamilton: CABI; 2004. p. 1–36.

42. Giordani I, Hynes T, Reich I, McDonnell RJ, Gormally MJ. *Tetanocera elata* (Diptera: Sciomyzidae) larvae feed on protected slug species *Geomalacus maculosus* (Gastrpoda: Arionidae): first record of predation. J Insect Behav. 2014;27:652–6.

43. Cuthill I, Partridge JC, Bennett ATD, Church SC, Hart NS, Hunt S. Ultraviolet vision in birds. Adv Study Anim Behav. 2000;29:159–214.

44. De Souza AR, Mourao CA, do Nascimento FS, Lino-Neto J. Sexy faces in a male paper wasp. PLoS One, 2014;doi:10.1371/journal.pone.0098172

45. Izzo AS, Tibbetts EA. Spotting the top male: sexually selected signals in male *Polistes dominulus* wasps. Anim Behav. 2012;83:839–45.

46. Zieger MV, Vakoliuk IA, Tuchina OP, Zhukov VV, Meyer-Rochow VB. Eyes and vision in *Arion rufus* and *Deroceras agreste* (Mollusca; gastropod; Pulmonata): what role does photoreception play in the orientation of these terrestrial slugs? Acta Zool. 2009;90:189–204.

47. Kennedy GY. A porphyrin pigment in the integument of *Arion ater* (L.). J Mar Biol Assoc UK. 1959;38:27–32.

48. Dyson M. An experimental study of wound-healing in *Arion*. 1964; PhD thesis, University of London, UK.

49. Lainé HA. Some observations on the structure of the skin gland of *Agriolimax reticulatus*. 1971; MSc thesis, University of Keele, UK.

50. Roulin A. Melanin-based colour polymorphism responding to climate change. Glob Change Biol. 2014;20:3344–50.

51. McCrone EJ, Sokolove PG. Brain-gonad axis and photoperiodically-stimulated sexual maturation in the slug *Limax maximus*. J Comp Physiol A Sensory, Neural Behav Physiol. 1979;133:117–24.

52. Denton GH, Anderson RF, Toggweiler JR, Edwards RL, Schaefer JM, Putnam AE. The last glacial termination. Nature. 2010;328:1652–6.

53. Hewitt GM. Some genetic consequences of ice ages, and their role in divergence and speciation. Biol J Linn Soc. 1996;58:247–76.

54. Ahmed FE, Setlow RB. Ultraviolet radiation-induced DNA damage and its photorepair in the skin of the platyfish *Xiphophorus*. Cancer Res. 1993;53:2249–55.

55. Jablonski NG, Chaplin G. The evolution of human skin coloration. J Hum Evol. 2000;39:57–106.

56. Svobodova AR, Galandakova A, Sianska J, Dolezal D, Lichnovska R, Ulrichova J, Vostalova J. DNA damage after exposure of mice skin to physiological doses of UVB and UVA light. Arch Dermatol Res. 2012;304:407–12.

57. Loayza-Muro RA, Marticorena-Ruiz JK, Palomino EJ, Merritt C, De Baat ML, Van Gemert M, Verweij RA, Kraak MHS, Admiraal W. Persistence of chironomids in metal polluted Andean high altitude streams: does melanin play a role? Environ Sci Technol. 2013;47:601–7.

58. Karl I, Hoffmann KH, Fischer K. Cuticular melanisation and immune response in a butterfly: local adaptation and lack of correlation. Ecol Entomol. 2010;35:523–8.

59. Parkash R, Munjal AK. Phenotypic variability of thoracic pigmentation in Indian populations of *Drosophila melanogaster*. J Zool Syst Evol Res. 1999;37:133–40.

60. Ahlgren J, Yang X, Hannson LA, Brönmark C. Camouflaged or tanned: plasticity in freshwater snail pigmentation. Biol Lett. 2013;9:20130464.

61. Pinkert S, Brandl R, Zeuss D. Colour lightness of dragonfly assemblages across North America and Europe. Ecography, 2016;doi:10.1111/ecog.02578

62. Ducrest A, Keller L, Roulin A. Pleiotropy in the melanocortin system, colouration and behavioural syndromes. Trends Ecol Evol. 2008;23:502–10.

63. Sugumaran M. Comparative biochemistry of eumelanogenesis and the protective roles of phenoloxidase and melanin in insects. Pigment Cell Res. 2001;15:2–9.

64. Reich I, O'Meara K, McDonnell RJ, Gormally MJ. An assessment of the use of conifer plantations by the Kerry slug (*Geomalacus maculosus*) with reference to the impact of forestry operations. *Irish Wildlife Manuals*. 2012;64. National Parks and Wildlife Service, Department of Arts, Heritage and the Gaeltacht. Theatr Irel.

7

Patterns and dynamics of neutral lipid fatty acids in ants – implications for ecological studies

Félix B. Rosumek[1,2†], Adrian Brückner[1†], Nico Blüthgen[1], Florian Menzel[3] and Michael Heethoff[1*]

Abstract

Background: Trophic interactions are a fundamental aspect of ecosystem functioning, but often difficult to observe directly. Several indirect techniques, such as fatty acid analysis, were developed to assess these interactions. Fatty acid profiles may indicate dietary differences, while individual fatty acids can be used as biomarkers. Ants are among the most important terrestrial animal groups, but little is known about their lipid metabolism, and no study so far used fatty acids to study their trophic ecology. We set up a feeding experiment with high- and low-fat food to elucidate patterns and dynamics of neutral lipid fatty acids (NLFAs) assimilation in ants. We asked whether dietary fatty acids are assimilated through direct trophic transfer, how diet influences NLFA total amounts and patterns over time, and whether these assimilation processes are similar across species and life stages.

Results: Ants fed with high-fat food quickly accumulated specific dietary fatty acids (C18:2n6, C18:3n3 and C18:3n6), compared to ants fed with low-fat food. Dietary fat content did not affect total body fat of workers or amounts of fatty acids extensively biosynthesized by animals (C16:0, C18:0, C18:1n9). Larval development had a strong effect on the composition and amounts of C16:0, C18:0 and C18:1n9. NLFA compositions reflected dietary differences, which became more pronounced over time. Assimilation of specific dietary NLFAs was similar regardless of species or life stage, but these factors affected dynamics of other NLFAs, composition and total fat.

Conclusions: We showed that ants accumulated certain dietary fatty acids via direct trophic transfer. Fat content of the diet had no effect on lipids stored by ants, which were able to synthesize high amounts of NLFAs from a sugar-based diet. Nevertheless, dietary NLFAs had a strong effect on metabolic dynamics and profiles. Fatty acids are a useful tool to study trophic biology of ants, and could be applied in an ecological context, although factors that affect NLFA patterns should be taken into account. Further studies should address which NLFAs can be used as biomarkers in natural ant communities, and how factors other than diet affect fatty acid dynamics and composition of species with distinct life histories.

Keywords: Direct trophic transfer, Lipid metabolism, Dietary routing, Fatty acid biosynthesis, Trophic enrichment, Trophic ecology, Trophic markers, Formicidae, *Formica fusca*, *Myrmica rubra*

* Correspondence: heethoff@bio.tu-darmstadt.de
†Equal contributors
[1]Ecological Networks, Technische Universität Darmstadt, Schnittspahnstr. 3, 64287 Darmstadt, Germany
Full list of author information is available at the end of the article

Background

Trophic interactions play a central role in ecosystem processes, shaping complex food webs with multiple paths and levels [1]. The complexity of interactions within communities, however, makes it difficult to assess their nature and long-term outcome solely by field observations. Several complementary approaches were developed to address this issue, such as fatty acid analysis [2]. Fatty acids have been used to study trophic ecology of organisms in aquatic and terrestrial ecosystems [3, 4]. Variation in fatty acid profiles can answer basic questions about spatial and temporal variation in diets, as well as niche partitioning among species [3, 5, 6]. Also, fatty acids could be used as biomarkers, indicating qualitative and quantitative trophic relationships between organisms [7, 8]. Many recent studies using fatty acid analysis in terrestrial organisms focused on detritivores, such as Collembola and Nematoda [7, 9–14], which established the technique as a useful tool to analyze their feeding interactions in soil food webs [5, 15–17]. However, fatty acid patterns and dynamics depend on an organism's physiology and composition of its natural diet, which are variable among taxonomic groups. Therefore, basic information on lipid metabolism is needed before the application of fatty acid analyses to study trophic relations of a given animal group.

Ants (Hymenoptera: Formicidae) are among the most abundant groups of invertebrates in terrestrial ecosystems, with a wide variety of feeding habits, nesting sites, and interactions with organisms from all trophic levels [18]. Many ant species have a cryptic behavior, which is difficult to study directly (e.g., living underground, inside the leaf-litter or in tree canopies). Moreover, in diverse ecosystems, dozens of species can coexist simultaneously in a given stratum [19]. Thus, complementary techniques are needed to study their trophic ecology. Stable isotopes, for instance, have been extensively used to address many questions in ant ecology [20–22]. The application of DNA barcoding, another modern technique, is still incipient for ants [23–25]. Surprisingly, no study so far tested the applicability of fatty acids to understand trophic ecology of ants.

Ants in general are regarded as omnivorous, feeding on a combination of living prey, dead arthropods, seeds and plant exudates. Less common are specialized feeding habits such as fungus cultivation and predation exclusively upon certain arthropod groups, as well as use of unusual resources such as pollen, animal excrements or mushrooms [18, 26–29]. Fatty acids from the diet could be incorporated without modification (i.e. through direct trophic transfer), or actively modified in response to environmental factors and physiological needs [4, 30, 31]. Many ant species primarily feed on sugars usually obtained from floral and extra-floral nectar or honeydew

[32]. Like all higher organisms, they can synthesize a set of fatty acids from carbohydrates via a decarboxylative Claisen condensation [33]. Fatty acids are mainly stored as neutral lipid fatty acids (NLFAs), which mostly consist of triglycerides, the principal component of the insect fat body [30, 34]. The biosynthesis of saturated palmitic (C16:0) and stearic acids (C18:0) and monounsaturated oleic acid (C18:1n9) seems to be widespread among insects, and correspondingly these fatty acids are the most abundant in their bodies [30]. On the other hand, the ability to synthesize polyunsaturated fatty acids, such as linoleic acid (C18:2n6), is highly variable among species [35, 36]. However, the details of these physiological processes in ants are poorly understood, and there are no studies specifically addressing dynamics of dietary fatty acids assimilation in this important insect group. Knowing which fatty acids can be unambiguously related to food sources, and how well the overall fat composition of ants reflects their diet after any metabolic modification, are crucial steps to apply fatty acid analysis in an ecological context.

Considering the potential use of fatty acids to understand trophic relations, and the lack of information about lipid metabolism in ants, we aim to elucidate patterns and dynamics of neutral lipid fatty acids in ants. We provided ants with high- and low-fat food in a no-choice feeding experiment, and compared the fatty acid profiles of ant workers and larvae over a period of 8 weeks. We specifically ask: (1) whether NLFA amounts and compositions are affected by a high- and a low-fat diet; (2) whether dietary fatty acids are accumulated in the ants' body via direct trophic transfer; (3) how dietary fatty acids shape NLFA patterns over time; (4) whether these patterns and dynamics are the same in different species and life stages.

Methods

Studied species

The experiment was performed with colonies reared in the laboratory, during November and December 2016. We chose two species, common and widespread in the Northern hemisphere, which represent the largest Formicidae subfamilies: *Formica fusca* Linnaeus 1758 (Formicinae) and *Myrmica rubra* Linnaeus 1758 (Myrmicinae). Both have in nature a similar and generalized diet of living and dead arthropods, nectar and honeydew [37, 38], and can thus be reared in the laboratory with a single artificial diet. Six colonies of each species were purchased from Antstore (Berlin, Germany) where ants were fed on an unstandardized diet of honey and dead flies. All colonies had one queen and between 9 and 12 (*F. fusca*) and 15–20 (*M. rubra*) workers. Colonies of *M. rubra* were reproductive during the whole experiment, with lower numbers of eggs and larvae towards the end. For *F. fusca*, larvae were only observed in two colonies in the last week of the experiment. Colonies were kept at a constant temperature of 25 °C and provided three times per week with water and food *ad libitum*.

Low- and high-fat treatments

Three colonies of each species received a low-fat treatment, whereas the remaining three received a high-fat treatment. As low-fat food we used a standardized recipe, suitable for breeding several ant species [39]. It contained 5 g agar, 1 g table salt (NaCl), 1 g vitamin-mineral mix powder (Altapharma, Burgwedel, Germany), 62 ml honey and 1 chicken egg homogenized in 500 ml hot water. The high-fat food followed the same recipe, with addition of 60 ml linseed oil (organic quality, Alnatura, Bickenbach, Germany). The mixture was stirred until it was cool and solid, to avoid separation of the aqueous and fatty phases. Both food mixtures were stored in a freezer at –20 °C until use, and food samples were taken for chemical analysis.

Experimental design

Before beginning the feeding experiment, we collected one worker per colony for fatty acid analysis (= week 0). Workers were chosen randomly from inside and outside the nest (a glass vial kept inside a plastic box). In addition, one larva of *M. rubra* was collected per colony. After starting to apply the treatments, we sampled one worker and one larva in the same way, every week for 8 weeks. Larva sample sizes were smaller from week 5 onwards, because some colonies were not reproductive anymore. In the last week, we also collected the queens for analysis (6 *F. fusca* and 5 *M. rubra*, since one queen died at the beginning of the experiment). All samples were immediately frozen at –20 °C until extraction.

Fatty acid analysis

Total lipids were extracted from the ants using 1 ml of a chloroform:methanol mixture, 2:1 (v/v) over a period of 24 h [40, 41]. Ants were directly refrozen after extraction and subsequently dried for 48 h at 50 °C and weighed with a microbalance (Mettler Toledo, XS3DU, Columbus, USA). The extracts were purified and separated according to the method described by Frostegård et al. [42]. SiOH-columns (Chromabond®) were washed and conditioned with 6 ml hexane. Subsequently, samples were applied on the column and elution of NLFAs (= mono-, di-, and triglycerides) was accomplished with 4 ml chloroform.

The chloroform fractions were evaporated to dryness under gentle nitrogen gas flow and residuals were redissolved in different concentrations of dichloromethane:methanol 2:1 (v/v) to adjust the samples to comparable concentration ranges: 1 ml for *F. fusca* queens and food

samples, 350 μl for workers of both species and *M. rubra* queens, and 50 μl for larvae. 50 μl aliquots (10 μl for high-fat food) were transferred to new glass vials with a conical inlet (150 μl) and 20 μl of internal standard (C19:0 in methanol; ρ_i = 220 ng/μl) were added. Samples were evaporated to dryness again, and finally derivatized to fatty acid methyl esters (FAMEs) with 20 μl TMSH (trimethylsulfonium hydroxide; 0.25 M in MeOH from Fluka, Sigma-Aldrich, St. Louis, USA).

FAME samples of NLFAs were analyzed with a QP2010 Ultra GC/MS (Shimadzu, Duisburg, Germany). The gas chromatograph (GC) was equipped with a ZB-5MS fused silica capillary column (30 m × 0.25 mm ID, df = 0.25 μm) from Phenomenex (Aschaffenburg, Germany). Sample aliquots of 1 μl were injected by using an AOC-20i autosampler-system from Shimadzu into a PTV-split/splitless-injector (Optic 4, ATAS GL, Eindhoven, Netherlands), which operated in splitless-mode. Injection-temperature was programmed from initial 70 °C up to 300 °C and then an isothermal hold for 59 min, sampling-time was set to 3 min and hydrogen was used as carrier-gas with a constant flow rate of 1.3 ml/min. The temperature of the GC oven was raised from initial 60 °C for 1 min, to 150 °C with a heating-rate of 15 °C/min, to 260 °C with a heating-rate of 3 °C/min, to 320 °C with a heating-rate of 10 °C/min and then an isothermal hold at 320 °C for 10 min. Electron ionization mass spectra were recorded at 70 eV from m/z 40 to 650. The transfer line and ion source were kept at 250 °C.

Methyl esters of the NLFAs were identified by comparing gas chromatographic retention times and m/z fragmentation patterns with those of the Supelco® 37 Component FAME Mix standard and the Bacterial Acid Methyl Ester (BAME) Mix standards as commercially available fatty acids (all Sigma-Aldrich) and published literature data [31, 43, 44]. The identity of γ-linolenic acid was additionally confirmed by an iodine catalyzed dimethyl disulfide derivatization [45].

A technical problem during analysis resulted in the loss of a batch of samples. Therefore, we have no data of week 3 for *M. rubra* larvae, week 4 for *M. rubra* workers and week 5 for *F. fusca*.

Data analysis

In general we used two approaches to analyse our data: (1) linear mixed-effect models (LMM) to assess the trophic transfer of certain fatty acids; and (2) multivariate compositional data analysis to describe total NLFA patterns. Only fatty acids with >1% composition were included in our analyses. Queens were not statistically analyzed, since they were sampled just at the end of the experiment.

We used the absolute amount of NLFAs [µg] standardized by dry weight for ants or fresh weight for food [mg], thus reflecting the relative amounts of NLFAs in comparison to non-lipid components [µg/mg]. We additionally ran the analyses with absolute amounts and dry weight as a cofactor, and results were identical for workers, but different for larvae, due to their distinct dynamics (see S1 in Additional file 1, and results for larvae).

At first, we correlated the relative amounts of all NLFAs combined (= total NLFAs) with dry weights of larvae and workers of both species using Spearman's rank correlation. For adults, body weight reflects size polymorphism among workers. For larvae, body weight is a better indicator of larval development than the week of sampling, because queens lay eggs continuously during the reproductive time. Dry weights for workers did not differ between treatments and over time, while larval dry weight increased over time (see S2 in Additional file 1). Since time and size were correlated for larvae (ρ_S = 0.63, p < 0.001), we ran separated LMMs for each factor, with dry weight normalized by square-root transformation.

We statistically tested relative amounts of total NLFAs and of the three most abundant fatty acids (C16:0, C18:0, C18:1n9). We also tested a specific dietary NLFA (C18:2n6), which occurred in higher concentration in the high-fat diet, and was not conspicuously synthesized by the ants. We did not test the amounts of the other two specific dietary NLFAs (C18:3n3 and C18:3n6) and show their results only in plots, because both were always zero in the low-fat treatment and non-zero in the high-fat treatment. Remaining NLFAs that occurred only in very small amounts in ants and food and were not tested either.

Effects on relative amounts were tested with linear mixed-effect models (command lme) as implemented in the R package "nlme" [46] with feeding treatment and time as fixed factors and colony ID as random factor for each species separately. We checked for the normal distribution of the residuals and the homogeneity of variance prior to the analyses and transformed the data if necessary (see S3 in Additional file 1 for data transformation). We further investigated the total NLFA amount in *M. rubra* workers and larvae using a LMM with the same structure as before, but including life stage as a further fixed factor. The difference between workers and larvae was analyzed with a simultaneous test for general linear hypothesis using Tukey pairwise contrasts (package "multcomp"; [47]) of the previous LMM.

We furthermore analyzed whether the overall NLFA composition (i.e. percentages of all fatty acids) of *F. fusca*, *M. rubra* workers and *M. rubra* larvae changed in the different treatments over time. We tested Bray-Curtis similarities (BCS) based on compositional data using

permutational multivariate analysis of variance (PERMANOVA; [48]) for each species separately. Overall 10,000 permutations were performed with feeding treatment and time as fixed factors and colony ID as random factor. We checked the multivariate homogeneity of group dispersions before with a multivariate Levene's test (PERMDISP; all p values >0.1; [49]). These analyses were performed with PRIMER 7.0.12 [50].

Finally, NLFA compositional data were ordinated using principal component analyses (PCA) and according PCA biplots. We compared the differences of the overall NLFA composition in *F. fusca* and *M. rubra* who received the high-fat diet during the experimental time. We used the centered log-ratio transformation after replacing zero values to deal with the constant sum constraint of compositional data and make it suitable for PCA (R packages "zCompositions" and "compositions" [51, 52]). PCA biplots were constructed by plotting factor loadings of compounds that significantly contributed (p < 0.01) to the group separation onto the PCA scatter plots using the R package "vegan" [53]. For a detailed R script of this analysis, see [54]. LMMs and PCAs were performed with R version 3.3.1 [55].

Results

Fatty acid profiles of food and ants

The neutral lipid fatty acid (NLFA) profiles of ants and their food are summarized in Table 1 (for full dataset and value ranges, see Additional file 2). The high-fat food had about 40 times more total concentration of NLFAs than the low-fat food. The main component of the high-fat food was C18:3n3, but it also had notably higher amounts of C16:0, 18:0 and C18:2n6. Besides, it contained C:18:3n6, which was entirely absent from the low-fat food.

C16:0, C18:0 and C18:1n9 were the main fatty acids in ants (Table 1). C18:1n9 was the main component in all experimental workers and queens. On the other hand, larvae had comparatively high levels of C16:0 and C18:0. Ants from the high-fat treatment exhibited higher amounts of C18:2n6, and were the only ones with detectable levels of C18:3n3 and C18:3n6. Queens had less total NLFAs than workers. Samples were variable, thus the profiles in Table 1 do not exactly reflect temporal and treatment differences (particularly for the highly variable larvae); these effects are analyzed below.

Dynamics of total and individual NLFA amounts

For *F. fusca*, there was no difference between treatments in the total amount of NLFAs (Fig. 1a, Table 2). C16:0, C18:0, C18:1n9 and total NLFAs increased over time, but with no treatment effect (Fig. 1b-c, Table 2). On the other hand, we observed an increasingly higher amount of C18:2n6 in the high-fat treatment, while it remained

Table 1 – Fatty acid profiles of food and ants at the beginning and end of the experiment

NLFA	Formica fusca						Myrmica rubra						Myrmica rubra larvae				Food	
	Week 0		Week 8		Queens		Week 0		Week 8		Queens		Week 0		Week 8			
	+	-	+	-	+	-	+	-	+	-	+	-	+	-	+	-	+	-
C12:0 lauric	0.1 (t)	0.1 (t)	0.1 (t)	0.1 (t)	0.1 (t)	0.1 (t)	0.6 (t)	0.4 (t)	0.2 (t)	0.3 (t)	0.1 (1)	0.2 (t)	3.9 (t)	2.4 (t)	0.4 (t)	1.0 (t)	t (t)	t (t)
C14:0 mystric	0.5 (t)	0.3 (t)	0.3 (t)	0.6 (t)	0.1 (t)	0.3 (t)	1.9 (1)	1.6 (t)	0.6 (1)	0.5 (1)	0.1 (1)	0.4 (1)	8.8 (1)	5.0 (1)	1.2 (1)	2.7 (1)	t (t)	t (t)
C16:0 palmitic	59.0 (25)	40.6 (25)	95.2 (12)	129.8 (18)	25.5 (11)	30.5 (18)	55.7 (18)	54.0 (18)	23.1 (32)	31.3 (30)	3.0 (26)	10.3 (17)	532.7 (54)	397.1 (59)	48.9 (33)	50.1 (25)	9.2 (8)	0.8 (32)
C16:1n9 palmitotelic	4.1 (1)	0.9 (t)	2.6 (t)	4.6 (1)	0.9 (t)	2.4 (1)	6.7 (2)	2.6 (1)	1.1 (1)	1.8 (1)	0.1 (1)	1.6 (3)	11.3 (1)	2.7 (t)	1.2 (1)	8.1 (4)	0.1 (t)	t (1)
C18:0 stearic	11.6 (10)	11.4 (12)	26.5 (3)	22.3 (3)	7.2 (3)	5.3 (3)	13.1 (5)	17.4 (6)	7.6 (15)	7.7 (13)	1.7 (15)	1.3 (3)	242.2 (25)	209.2 (31)	29.8 (20)	23.4 (12)	6.6 (6)	0.1 (4)
C18:1n9 oleic	284.7 (64)	224.6 (62)	523.7 (67)	570.3 (78)	151.4 (66)	135.4 (77)	273.5 (74)	251.7 (74)	80.0 (48)	129.2 (55)	4.4 (26)	46.1 (76)	202.6 (19)	51.5 (8)	39.4 (27)	112.2 (56)	1.8 (2)	1.5 (59)
C18:2n6 linoleic	0.4 (t)	0.4 (t)	4.2 (1)	0.2 (t)	3.0 (1)	0.5 (t)	1.3 (1)	1.6 (1)	2.4 (2)	0.3 (t)	1.1 (7)	0.4 (1)	3.2 (t)	2.1 (t)	3.4 (2)	1.3 (1)	5.3 (4)	0.1 (2)
C18:3n3 α-linolenic	0 (0)	0 (0)	138.8 (17)	0 (0)	30.8 (14)	0 (0)	0 (0)	0 (0)	3.9 (2)	0 (0)	3.1 (18)	0 (0)	0 (0)	0 (0)	20 (14)	0 (0)	80.9 (72)	t (1)
C18:3n6 γ-linolenic	0 (0)	0 (0)	22.5 (3)	0 (0)	9.1 (4)	0 (0)	0 (0)	0 (0)	1.3 (1)	0 (0)	0.9 (5)	0 (0)	0 (0)	0 (0)	1.7 (1)	0 (0)	8.9 (8)	0 (0)
C20:0 arachidic	0.1 (t)	0.1 (t)	0.1 (t)	0.1 (t)	0.1 (t)	0 (t)	0.2 (t)	0.2 (t)	0.1 (t)	0.1 (t)	0.1 (t)	0.1 (t)	0.7 (t)	0.6 (t)	0.2 (t)	0.1 (t)	t (t)	t (t)
Total	360.3	278.4	813.9	728.0	228.0	174.5	353.1	329.7	120.3	171.2	14.5	60.3	1005.5	670.6	146.3	198.9	112.8	2.6
Sample size	3	3	3	3	3	3	3	3	3	3	2	3	3	3	1	1	3	2

Average amounts are given in μg of NLFA/mg of dry weight (fresh weight for food). Values in brackets are average percentages of the total composition of NLFAs per sample. +: high-fat treatment; –: low-fat treatment; t: detected in trace amount (less than 0.1 μg/mg or 1% of composition)

small in the low-fat treatment (Fig. 1d, Table 2). Similarly, C18:3n3 and C18:3n6 increased remarkably in the high-fat treatment, but were never recorded in the low-fat treatment (Figs. 1e-f). *Formica fusca* presented considerable polymorphism (coefficient of variation [= CV] of dry weights = 41%), but there was no correlation between body size and total NLFA amount ($\rho_S = -0.06$, $p = 0.66$).

For *M. rubra*, the amounts of C18:2n6, C18:3n3 and C18:3n6 also increased in the high-fat treatment, and the last two NLFAs were completely absent in the low-fat treatment (Fig. 2d-f, Table 2). No time effect was observed for C18:2n6 in this species. There was no treatment effect in C16:0, C18:0, C18:1n9 and total NLFAs, but, opposite to *F. fusca*, we observed an overall decrease over time (Figs. 2a-b, Table 2). *Myrmica rubra* workers varied less in size (CV of dry weights = 17%) and, again, no correlation was found between body size and total NLFA amount ($\rho_S = 0.19$, $p = 0.19$).

Myrmica rubra larvae presented more complex dynamics, because they were influenced both by experimental time effect and their developmental stage. Nevertheless, since these variables were correlated, LMM results were similar, except for C18:1n9 (Table 2). The increasing trends for C18:2n6, C18:3n3 and C18:3n6 were the same

as in workers (Fig. 3d-f, Table 2). Total NLFAs also decreased with time (Fig. 3b, Table 2), but in a higher rate than in workers (Tukey pairwise contrasts, $z = 4.70$, $p < 0.001$, for full model see S4 in Additional file 1). Larvae from the high-fat treatment had more total NLFAs and C18:1n9 overall during the experiment (Figs. 3a-c, Table 2 [A]). However, as larvae increased in dry weight, C18:1n9 actually was higher in the low-fat treatment compared to the high-fat treatment (Table 2 [B]). There was a strong negative correlation between larval dry weight and relative NLFA amount (Fig. 4; Table 2 [B], $\rho_S = -0.72$, $p < 0.001$). The absolute amount of fat slightly increased with body size, but did not follow the growth in other body components, which resulted in lower concentration of NLFAs in larger and older larvae (Fig. 4). This decrease was mostly due to a decline on saturated fatty acids (C16:0 and C18:0, Table 2, see S5 in Additional file 1). Therefore, young larvae had relatively large fat storages and high ratios of saturated:unsaturated fatty acids, which both decreased during development.

Dynamics of overall fatty acid composition

The overall NLFA composition of the ants changed over time (Table 3). Treatment and time affected the composition of *F. fusca* and *M. rubra* larvae. For *M.*

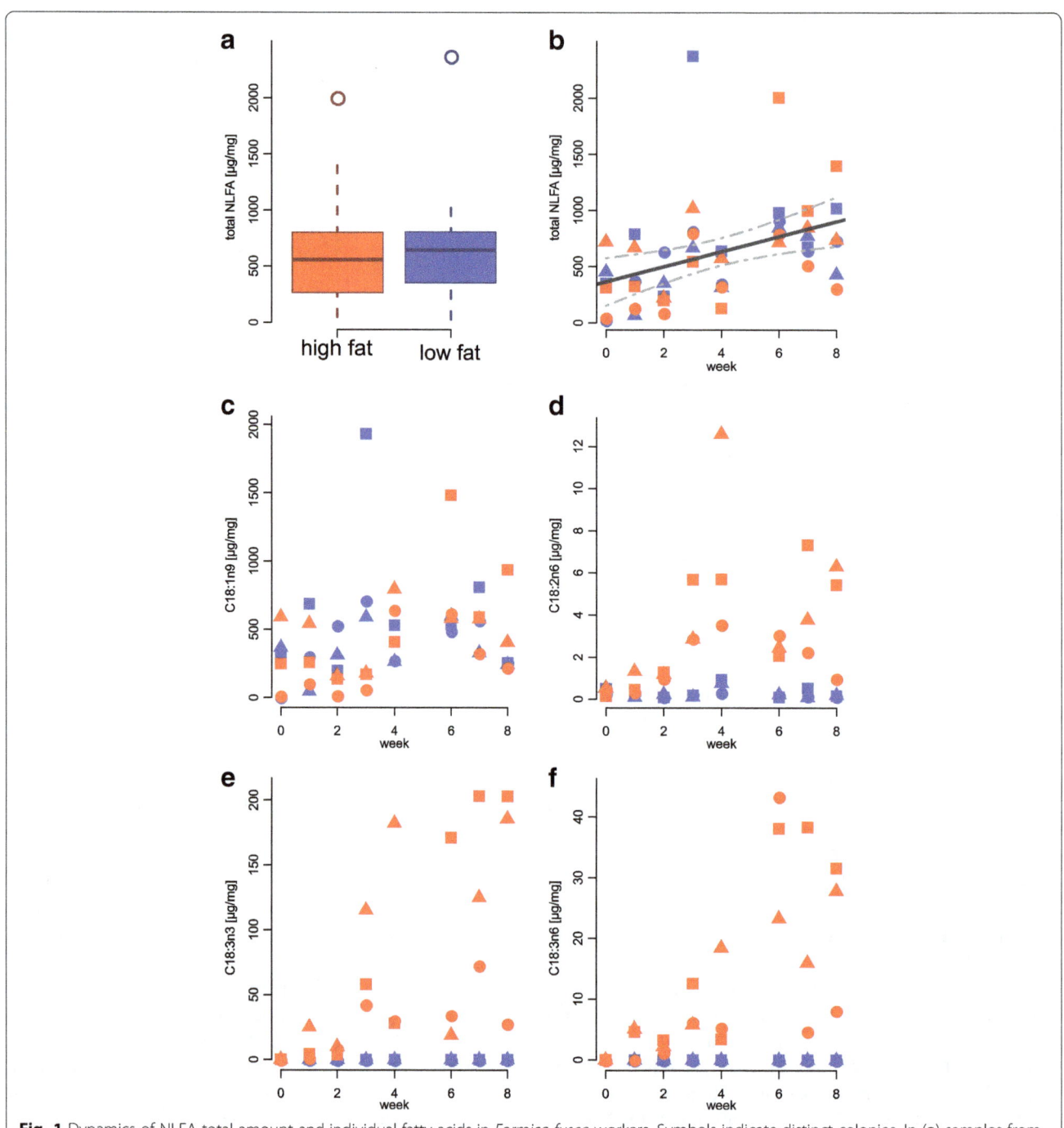

Fig. 1 Dynamics of NLFA total amount and individual fatty acids in *Formica fusca* workers. Symbols indicate distinct colonies. In (**a**) samples from all weeks and colonies are pooled, (**b**) total NLFA, (**c**) C18:1n9, (**d**) C18:2n6, (**e**) C18:3n3, (**f**) C18:3n6

rubra workers, no effect was found. However, this could be understood when the profile change of the high-fat colonies was analyzed with PCA (Fig. 5). For both species, we noticed a shift in composition over time, mainly driven by the dietary fatty acids. For *F. fusca*, C18:2n6, C18:3n3 and C18:3n6 altogether had a statistically significant effect on this shift. For *M. rubra*, only C18:3n3 (the main dietary fatty acid) had a

significant effect. The samples from week 8 were particularly odd, showing small proportions of C18:3n3 and C18:1n9 and relatively high proportions of C16:0 and C18:0. One individual from each treatment had unusually low amounts of total fat and oleic acid (below 20 µg/mg and 10% of composition, respectively; see Additional file 2), which added significant variation to the results. When week 8 was removed from

Table 2 Effects of time, treatment and larval dry weight on relative total amount [µg/mg] and individual amounts of fatty acids

	Total NLFAs				C16:0				C18:0				C18:1n9				C18:2n6			
	df	F	trend	p	df	F	trend	p	df	F	trend	p	df	F	trend	p	df	F	trend	p
F. fusca (n = 48)																				
Treatment	1	0.12		0.74	1	1.39		0.303	1	0.80		0.422	1	0.18		0.69	1	58.44		**0.002**
Time	1	15.42	↑	**< 0.001**	1	22.30	↑	**< 0.001**	1	25.51	↑	**< 0.001**	1	10.52	↑	**0.002**	1	11.66		**0.002**
Treatment x Time	1	0.29		0.60	1	0.16		0.687	1	1.94		0.171	1	0.84		0.37	1	14.97	↑high	**< 0.001**
Residuals	44				44				44				44				44		↓low	
M. rubra (n = 48)																				
Treatment	1	0.02		0.90	1	0.59		0.484	1	0.02		0.893	1	0.26		0.64	1	29.06	↑	**0.006**
Time	1	6.34	→	**0.016**	1	4.60	→	**0.038**	1	9.85	→	**0.003**	1	4.36	→	**0.043**	1	2.98		0.09
Treatment x Time	1	0.07		0.79	1	0.23		0.634	1	0.01		0.956	1	0.02		0.89	1	2.12		0.15
Residuals	44				44				44				44				44			
M. rubra larvae [A] (n = 38)																				
Treatment	1	11.08	↑	**0.029**	1	1.36		0.308	1	3.20		0.148	1	9.93	↑	**0.034**	1	22.55	↑	**0.009**
Time	1	12.27	→	**0.015**	1	9.62	→	**< 0.001**	1	37.08	→	**< 0.001**	1	0.46		0.502	1	0.09		0.765
Treatment x Time	1	0.34		0.56	1	0.20		0.656	1	2.64		0.115	1	1.23		0.276	1	2.17		0.151
Residuals	33				33				33				33				33			
M. rubra larvae [B] (n = 38)																				
Treatment	1	25.78	↑	**< 0.001**	1	2.05		0.161	1	2.75		0.107	1	10.80	↑	**0.003**	1	21.57	↑	**0.001**
Dry weight	1	46.35	→	**< 0.001**	1	53.59	→	**< 0.001**	1	69.50	→	**< 0.001**	1	0.15		0.700	1	0.62		0.434
Treatment x Dry weight	1	3.36		0.076	1	1.42		0.241	1	0.05		0.809	1	4.68	↑low	**0.038**	1	0.06		0.802
Residuals	33				33				33				33		↓high		33			

Results of linear mixed-effect models. Trends indicate the direction of significant effects (α < 0.05, in bold). For larvae, [A] = time as a factor, [B] = dry weight as a factor

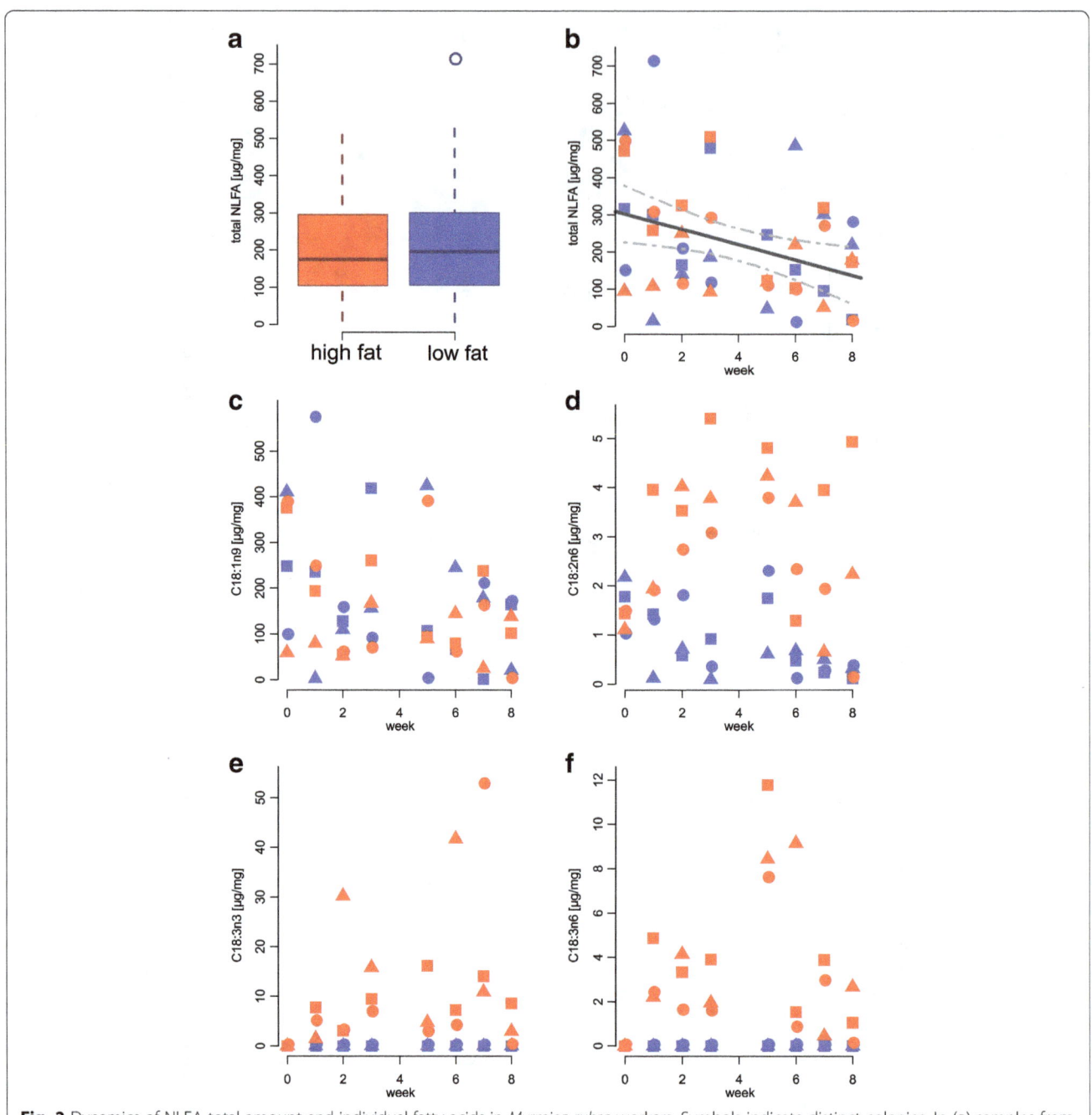

Fig. 2 Dynamics of NLFA total amount and individual fatty acids in *Myrmica rubra* workers. Symbols indicate distinct colonies. In (**a**) samples from all weeks and colonies are pooled, (**b**) total NLFA, (**c**) C18:1n9, (**d**) C18:2n6, (**e**) C18:3n3, (**f**) C18:3n6

the PERMANOVA, the treatment effect was notice-able (Table 3).

Discussion
Fatty acid profiles of ants
Several factors influence the fatty acid composition of insects, such as flying activity, life stage, growth, re-productive status, environmental temperature, and diet [4, 30, 56]. Due to this complexity, Stanley-Samuelson et al. [30] argued against a "typical" insect profile, and

indeed a high variation is found among orders, families, and species [56, 57]. Just a few ant profiles are available in literature: *Myrmica incompleta* Provancher, 1881 (worker and pupae; [58]), *Lasius claviger* (Roger, 1862) (only pupae; [59]), *Myrmica rubra* (only the free fatty acid fraction from head extracts; [60, 61]) and *Polyrha-chis dives* Smith, 1857 (sun-dried workers cultivated as food; [62]). These fatty acid profiles are not entirely comparable due to the multitude of goals and methods, but, together with our results, they indicate C18:1n9 as

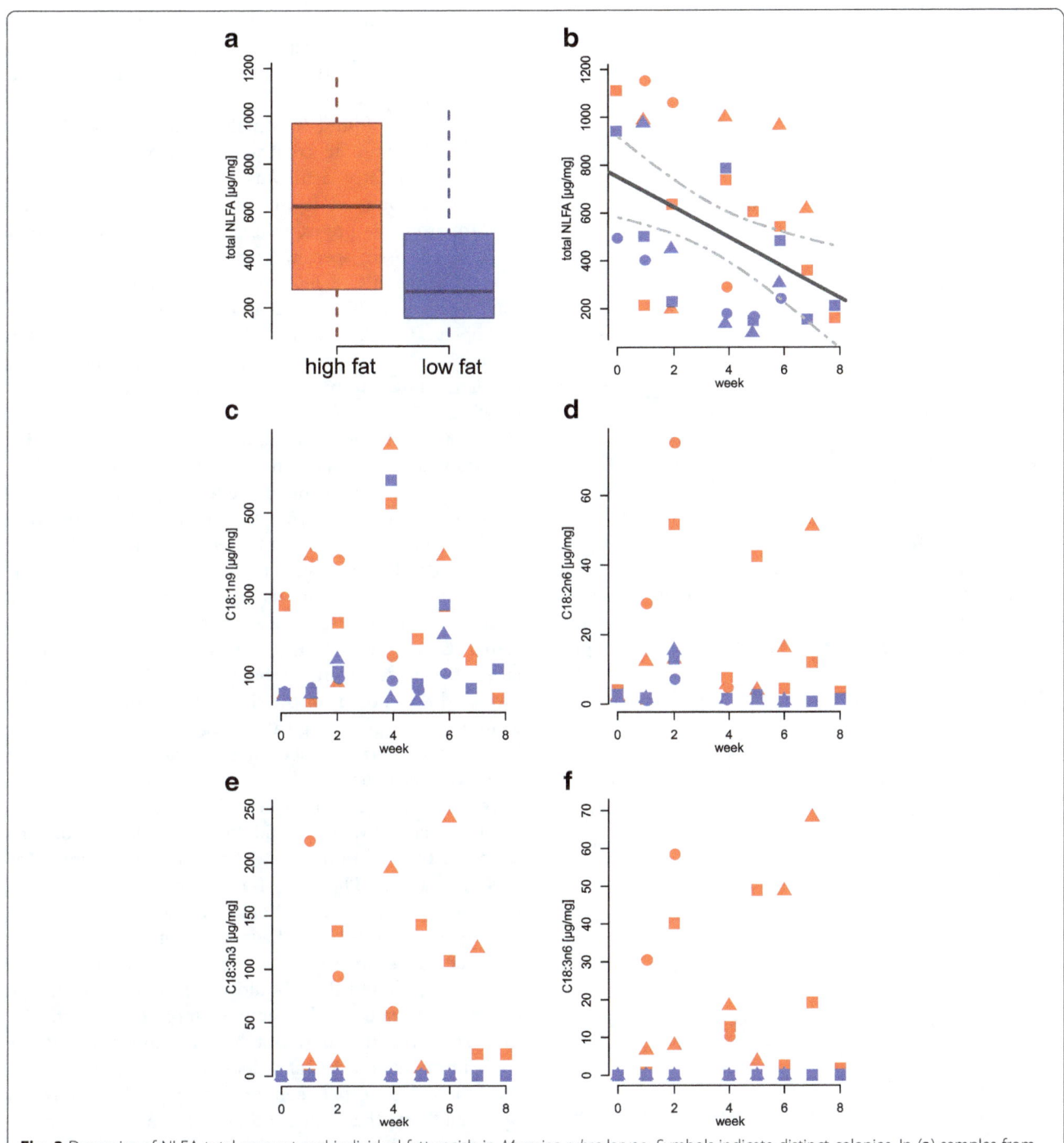

Fig. 3 Dynamics of NLFA total amount and individual fatty acids in *Myrmica rubra* larvae. Symbols indicate distinct colonies. In (**a**) samples from all weeks and colonies are pooled

the predominant NLFA in ant bodies, followed by C16:0 and C18:0. High levels of C18:1n9 are standard for Hymenoptera, but the abundance of other fatty acids varies within the order [56].

Dynamics of individual NLFAs and overall composition

Some fatty acids are extensively synthesized de novo by animals, while others are produced in small amounts,

or only by certain taxa [30, 36]. In our experiment, the food enrichment with linseed oil allowed us to observe the influence of diet on NLFAs found a priori in high, low and null amounts in ants' bodies. C18:3n3 and C18:3n6 were absent in week 0, and solely recorded in the high-fat treatment during the experiment. This suggests that ants are not able to synthesize them, or only in small doses which are directly incorporated in

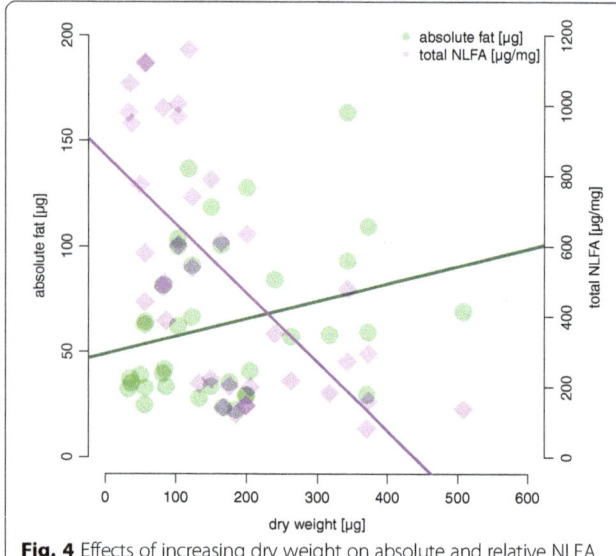

Fig. 4 Effects of increasing dry weight on absolute and relative NLFA amounts of *Myrmica rubra* larvae

the polar lipid fractions (i.e. phospholipids, glycolipids, free fatty acids). The amounts of C18:3n3 and C18:3n6 increased with the time ants fed on the diet, thus their concentration reflects how much/how often the ants consumed a resource. If these NLFAs are neither highly

Table 3 Effects of time and treatment on overall NLFA composition

	df	pseudoF	p
F. fusca			
Treatment	1	8.87	**< 0.001**
Time	7	2.60	**0.015**
Treatment x time	7	0.91	0.543
Residuals	32		
M. rubra (week 8)			
Treatment	1	2.98	0.089
Time	7	1.44	0.197
Treatment x time	7	0.53	0.832
Residuals	32		
M. rubra (week 7)			
Treatment	1	4.70	**0.025**
Time	6	1.46	0.193
Treatment x time	6	0.87	0.871
Residuals	28		
M. rubra larvae			
Treatment	1	22.46	**< 0.001**
Time	7	12.45	**< 0.001**
Treatment x time	7	1.76	0.090
Residuals	23		

PERMANOVA results for overall composition (%) based on Bray-Curtis Similarities. Significant results ($p < 0.05$) are in bold

mobilized nor modified, they should mainly be stored in the fat body when acquired in considerable amounts from the diet, and thus detectable with neutral lipid fatty acid analysis.

C18:2n6 was found in smaller amounts in all samples of the low-fat treatment, but it is not clear whether this fatty acid was synthesized by ants de novo, was obtained from the small amounts in the food, or from the pre-experimental diet. About one third of reported insect species, from five different orders, are able to synthesis C18:2n6, but high interspecific variation was observed within orders [35, 36]. Regarding the Hymenoptera, C18:2n6 biosynthesis was not observed in the mason bee *Osmia lignaria* Say, 1837 (Megachilidae) [35], but it is known from the parasitoid *Nasonia vitripennis* (Walker, 1836) (Pteromalidae) [63]. Regardless of the actual ability of ants to synthesize C18:2n6, its amounts also increased with the diet and, in *F. fusca*, over time as well. In *M. rubra* and its larvae the time effect was not clear.

On the other hand, C16:0, C18:0 and C18:1n9 behaved similarly in both treatments. No treatment effect in C16:0 and C18:0 was noticed, even if they occurred in the high-fat food in levels higher than C18:3n6 and C18:2n6, respectively. Hence, it seems most likely that C16:0, C18:0 and C18:1n9 are synthesized de novo in large amounts from carbohydrates and constantly modified depending on physiological requirements. For example, the physiologically ideal fluidity of the fat body, which changes accordingly with environmental temperature, is achieved through a balanced ratio between saturated and unsaturated fatty acids [4]. Hence, the interplay between β-oxidation and Claisen condensation of these abundant NLFAs should be essential for this mechanism. The lack of a treatment effect on total NLFAs also suggests that, at least under *ad libitum* feeding conditions, ants have no significant energetic loss due to de novo fatty acid biosynthesis. Thus, ants with a sugar-based diet should not have a disadvantage compared to species that acquire most lipids from the diet. However, this may not be true under conditions with limited resources, and detectable differences in ratios could occur between ants that feed directly on lipids and ants that only synthesize them.

Our multivariate analyses showed that a shift in diet results in an equivalent shift in profile, and this difference was more pronounced when the ants fed longer on that resource (Table 3, Fig. 5). The main drivers of this compositional change were specific dietary NLFAs. Therefore, these profiles represent another way to assess dynamics of resource use or detect differences among species [3, 5, 6]. They could be particularly useful when the exact lipid composition of the food is not known, such as in samples collected from the field.

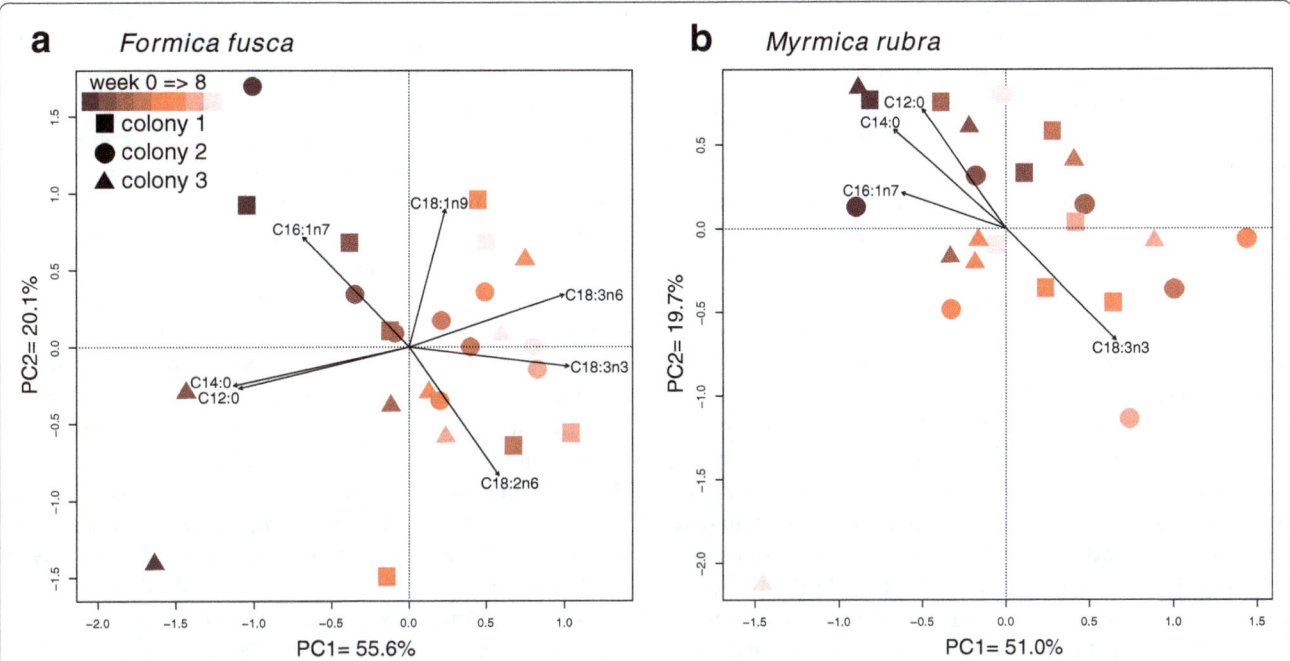

Fig. 5 Principal component analysis for changes in composition over time of colonies under the high-fat treatment. Lighter colors mean later weeks. Arrows show fatty acids with significant effects over compositional changes (for factor loadings, see S6 in Additional file 1). (**a**) *Formica fusca*, (**b**) *Myrmica rubra*

Factors affecting NLFA dynamics

Our results point out to several factors that affect lipid metabolism in ants, and could be important from biological and methodological points of view. First of all, one possible caveat of analytical methods that use ants' full body is that the undigested food stored in their crops could bias the results [20]. If this were the case in our experiment, we would expect higher total amount of lipids in ants of the high-fat treatment, and a conspicuous increase during the first week. Also, higher variance should occur in the high-fat treatment, due to the collection of workers with variable crop filling. However, (1) the amount of NLFAs did not differ between treatments, with the exception of larvae (which do not possess a crop; [64]); (2) we observed linear patterns for total fat and several NLFAs, consistent with lipid storage in the fat body; and (3) variances did not differ between treatments, in all cases (F test; *F. fusca* – F = 1.01, $p = 0.97$, *M. rubra* – F = 1.58, $p = 0.28$, larvae – F = 1.67, $p = 0.28$). Even if ants had undigested food in their crops, its contribution would have been relatively small. Thus, as far as the dietary component of interest does not occur in very high amounts in the food (e.g. ca. 10% NLFAs in our high-fat diet), full body extraction can be used to investigate the effect of diet in ants. In certain research contexts, however, it might be important to fully eliminate this factor, using a methodological alternative such as dissection of the fat body.

The reproductive status of the colonies influenced fatty acid dynamics. Feeding the brood can negatively affect the amount of fat stored by the workers, as observed in *Camponotus festinatus* (Buckley, 1866) [65], potentially explaining the decrease of NLFAs in *M. rubra*. On the contrary, *F. fusca* colonies were getting closer to reproduction mode during the experiment, and effectively we observed larvae in two colonies at the last week (this reproductive timing was also observed in non-experimental colonies kept in the same conditions). These colonies needed to accumulate reserves to fuel upcoming larval feeding and egg laying. Considering this, it is intriguing that queens of both species displayed a very low amount of fat at the end of the experiment.

We also observed an effect of development in compositions and dynamics of *M. rubra* larvae. The young larvae had large fat storages and amounts of saturated fatty acids. Earlier in their growth process, they quickly develop other tissues to build more complex organs [66], resulting in a proportionally smaller amount of NLFAs. The increase in C18:1n9 in the low-fat treatment with development may appear counterintuitive, but this was the only unsaturated NLFA ants were able to synthesize in large amounts. In turn, larvae from the high-fat treatment already received several polyunsaturated NLFAs from the diet. The shift to a more balanced composition between saturated and unsaturated NLFAs might enhance metabolic processes in a more complex body. In contrat to workers, larvae seem to benefit from a high-fat diet from which they accumulate slightly more NLFAs. For *Solenopsis invicta* Buren, 1972 it has been demonstrated that sugars, lipids and proteins are differently allocated among worker subcastes, larvae and queens [67].

The distinct distribution of nutrients among individuals of a colony is not restricted to different life stages, but also among worker subcastes. Several studies observed higher fat storage in workers that stay inside the nest and take care of the brood (= nurses), and less in workers that spend more time in activities outside the nest (= foragers) [67, 68]. However, this pattern may not occur in a few species, and no difference was previously found in field samples of *F. fusca* [69]. In *M. rubra*, nurse and forager subcastes were identified in laboratory colonies smaller than ours, and their role was related to individual age and size [70]. Differences in worker size were unrelated to total amount of fat for both species in our data. Individual variation in fat storages could indicate behavioral subcastes, but it was the same in the reproductive *M. rubra* and the non-reproductive *F. fusca* (CV of NLFA total amounts for all samples = 72% in both species). Thus, we found no evidence for considerable differences in lipid storage across behavioral or morphological subcastes within these species, under our experimental conditions, although these effects may be minute in small colonies and need a specific setup to be detected.

Regardless of the variation across species and life stages in profiles and dynamics, the assimilation of specific dietary NLFAs (C18:2n6, C18:3n3 and C18:3n6) followed the same pattern. Thus, the physiological processes involved in NLFA metabolism should be conserved at least between the subfamilies Formicinae and Myrmicinae, which comprise about three quarters of all valid ant species [71]. It is likely that all ants behave similarly, but this needs to be tested with experiments using species with more diversified feeding behaviors and from more distant branches of the ant tree of life, such as the Ponerinae or Dorylinae [72].

Implications to the study of ant trophic ecology

In trophic ecology, fatty acids can basically be used in two ways: as overall profiles, whose variation indicates differences in diet; and as biomarkers, which indicate specific interactions between organisms [3, 4]. Our results suggest that both applications are suitable for ants. Profiles and individual NLFAs observed in ants changed in response to diet, and these shifts became more pronounced over time. Fatty acid analysis can provide a better resource resolution than stable isotopes, in a more quantitative way and representative timeframe

than barcoding of gut DNA [2]. However, these methods are complementary, rather than opposing, and could be coupled with field observations to provide a comprehensive perspective on ant trophic ecology.

The factors affecting NLFA amounts and composition that we observed should also be considered in an ecological context. A representative sample of castes and life stages is recommended if one is interested in detailed trophic ecology of a particular species. For a study at community level, profiles of forager workers sampled at a similar time may be enough to provide comparative information on resource partitioning, although distinct reproduction times could influence amounts and compositions.

In this study, we did not aim to survey prospective biomarkers for natural resources used by ants. However, the three specific dietary NLFAs (C18:2n6, C18:3n3 and C18:3n6) presented chemical properties of suitable biomarkers, as they were not produced by ants (or only in small amounts) and assimilated through direct trophic transfer, with little or no metabolic modification [4]. They can be found in natural diets of ants, such as in elaiosomes, seeds and other insects, in variable patterns that may allow detection of specific interactions [56, 73, 74]. Thus, they are good candidates for trophic markers. Since their assimilation was not affected by species identity, reproductive status or life stage, the biomarker approach seems to be quite promising for ants. Naturally, the actual relevance of these NLFAs would depend on context and occurrence within a community. On the other hand, since C16:0, C18:0 and C18:1n9 are synthesized from carbohydrates in large amounts, and highly modified to attend physiological needs, it would be difficult to relate their amounts to a particular resource or feeding behavior. Further research can provide more fatty acids useful as biomarkers, related to other resources used by ants, which would likely be distinct from the ones suggested for other groups (e.g. C18:1n9 as an indicator of herbivory in Collembola [5]).

Conclusions

We showed that ants accumulated fatty acids from their diet via direct trophic transfer, and that both, individual NLFAs and overall profiles reflect their diets. The fat content of the diet had no effect in lipids stored by ants, which shows that they are able to synthesize large amounts of NLFAs from sugars. Other factors such as reproductive status and life stage also affected total amounts and profiles of NLFAs. Specific dietary fatty acids were assimilated independent of species or life stage. Fatty acid analysis is a suitable technique to study feeding behavior of ants, and can become a valuable tool to study ant trophic ecology in the field. To this end, central points to be addressed by future research are which biomarkers are most informative of ant diets in natural communities, and how factors other than diet affect fatty acid dynamics and composition of ant species with distinct life histories.

Acknowledgements
We thank Cristina Muñoz Sánchez for assisting at laboratory work. We acknowledge support by the German Research Foundation and the Open Access Publishing Fund of Technische Universität Darmstadt.

Funding
FBR and AB were supported by PhD scholarships from the Brazilian National Council of Technological and Scientific Development (CNPq) and the German National Academic Foundation (Studienstiftung des deutschen Volkes), respectively. This study was partly funded by the German Research Foundation (DFG; HE 4593/5-1).

Authors' contributions
MH provided the initial idea. FBR, AB and MH designed the experiment. FBR and AB collected the data. AB performed data analyses. All authors discussed the data. FBR drafted the manuscript. All authors contributed to the final manuscript. All authors read and approved the final manuscript.

Competing interests
The authors declare that they have no competing interests.

Author details
[1]Ecological Networks, Technische Universität Darmstadt, Schnittspahnstr. 3, 64287 Darmstadt, Germany. [2]Department of Ecology and Zoology, Federal University of Santa Catarina, Campus Trindade, Florianópolis 88040-900, Brazil. [3]Institute of Organismic and Molecular Evolution, Johannes Gutenberg-Universität Mainz, Johannes-von-Müller-Weg 6, 55128 Mainz, Germany.

References
1. Polis GA, Strong DR. Food web complexity and community dynamics. Am Nat. 1996;147:813–46.
2. Birkhofer K, Bylund H, Dalin P, Ferlian O, Gagic V, Hambäck PA, et al. Methods to identify the prey of invertebrate predators in terrestrial field studies. Ecol Evol. 2017;7:1942–53.
3. Budge SM, Iverson SJ, Koopman HN. Studying trophic ecology in marine ecosystems using fatty acids: a primer on analysis and interpretation. Mar Mammal Sci. 2006;22:759–801.
4. Ruess L, Chamberlain PM. The fat that matters: soil food web analysis using fatty acids and their carbon stable isotope signature. Soil Biol Biochem. 2010;42:1898–910.
5. Ruess L, Schütz K, Haubert D, Häggblom MM, Kandeler E, Scheu S. Application of lipid analysis to understand trophic interactions in soil. Ecology. 2005;86:2075–82.
6. Ferlian O, Scheu S, Pollierer MM. Trophic interactions in centipedes (Chilopoda, Myriapoda) as indicated by fatty acid patterns: variations with life stage, forest age and season. Soil Biol Biochem. 2012;52:33–42.
7. Chamberlain PM, Bull ID, Black HIJ, Ineson P, Evershed RP. Collembolan trophic preferences determined using fatty acid distributions and compound-specific stable carbon isotope values. Soil Biol Biochem. 2006;38:1275–81.
8. Pollierer MM, Scheu S, Haubert D. Taking it to the next level: Trophic transfer of marker fatty acids from basal resource to predators. Soil Biol Biochem. 2010;42:919–25.
9. Ruess L, Häggblom MM, Garcia Zapata EJ, Dighton J. Fatty acids of fungi and nematodes – possible biomarkers in the soil food chain? Soil Biol Biochem. 2002;34:745–56.
10. Chamberlain PM, Bull ID, Black HIJ, Ineson P, Evershed RP. Fatty acid composition and change in Collembola fed differing diets: identification of trophic biomarkers. Soil Biol Biochem. 2005;37:1608–24.
11. Haubert D, Häggblom MM, Scheu S, Ruess L. Effects of fungal food quality and starvation on the fatty acid composition of Protaphorura fimata (Collembola). Comp Biochem Physiol B Biochem Mol Biol. 2004;138:41–52.
12. Haubert D, Langel R, Scheu S, Ruess L. Effects of food quality, starvation and life stage on stable isotope fractionation in Collembola. Pedobiologia. 2005;49:229–37.
13. Haubert D, Häggblom MM, Scheu S, Ruess L. Effects of temperature and life stage on the fatty acid composition of Collembola. Eur J Soil Biol. 2008;44:213–9.
14. Haubert D, Pollierer MM, Scheu S. Fatty acid patterns as biomarker for trophic interactions: changes after dietary switch and starvation. Soil Biol Biochem. 2011;43:490–4.
15. Ruess L, Schütz K, Migge-Kleian S, Häggblom MM, Kandeler E, Scheu S. Lipid composition of Collembola and their food resources in deciduous forest stands – implications for feeding strategies. Soil Biol Biochem. 2007;39:1990–2000.
16. Haubert D, Birkhofer K, Fließbach A, Gehre M, Scheu S, Ruess L. Trophic structure and major trophic links in conventional versus organic farming systems as indicated by carbon stable isotope ratios of fatty acids. Oikos. 2009;118:1579–89.
17. Ngosong C, Raupp J, Scheu S, Ruess L. Low importance for a fungal based food web in arable soils under mineral and organic fertilization indicated by Collembola grazers. Soil Biol Biochem. 2009;41:2308–17.
18. Kaspari M. A primer on ant ecology. In: Agosti D, Majer JD, Alonso LE, Schultz TR, editors. Ants: standard methods for measuring and monitoring biodiversity. Washington, DC: Smithsonian Institution Press; 2000. p. 9–24.
19. Agosti D, Majer J, Alonso L, Schultz T, editors. Sampling ground-dwelling ants: case studies from the worlds' rain forests. Perth: Curtin University of Technology; 2000.
20. Blüthgen N, Gebauer G, Fiedler K. Disentangling a rainforest food web using stable isotopes: dietary diversity in a species-rich ant community. Oecologia. 2003;137:426–35.
21. Davidson DW, Cook SC, Snelling RR, Chua TH. Explaining the abundance of ants in lowland tropical rainforest canopies. Science. 2003;300:969–72.
22. Feldhaar H, Gebauer G, Blüthgen N. Stable isotopes: past and future in exposing secrets of ant nutrition (Hymenoptera: Formicidae). Myrmecol News. 2010;13:3–13.
23. Fournier V, Hagler J, Daane K, de León J, Groves R. Identifying the predator complex of Homalodisca vitripennis (Hemiptera: Cicadellidae): a comparative study of the efficacy of an ELISA and PCR gut content assay. Oecologia. 2008;157:629–40.
24. Muilenburg VL, Goggin FL, Hebert SL, Jia L, Stephen FM. Ant predation on red oak borer confirmed by field observation and molecular gut-content analysis. Agric For Entomol. 2008;10:205–13.
25. Penn HJ, Chapman EG, Harwood JD. Overcoming PCR inhibition during DNA-based gut content analysis of ants. Environ Entomol. 2016;45:1255–61.
26. Blüthgen N, Feldhaar H. Food and shelter: how resources influence ant ecology. In: Lach L, Parr CL, Abbott KL, editors. Ant Ecolgy. Oxford: Oxford University Press; 2010. p. 115–36.
27. Brandão CRF, Silva RR, Delabie JHC. Neotropical ants (Hymenoptera) functional groups: nutritional and applied implications. In: Panizzi AR, Parra JRP, editors. Insect bioecology and nutrition for integrated Pest management. Boca Raton: CRC Press; 2012. p. 213–36.
28. von Beeren C, Mair MM, Witte V. Discovery of a second mushroom harvesting ant (hymenoptera: Formicidae) in Malayan tropical rainforests. Myrmecol News. 2014;20:37–42.
29. Sainz-Borgo C. Bird feces consumption by fire ant Solenopsis geminata (Hymenoptera: Formicidae). Entomol News. 2015;124:295–9.
30. Stanley-Samuelson DW, Jurenka RA, Cripps C, Blomquist GJ, de Renobales M. Fatty acids in insects: composition, metabolism, and biological significance. Arch Insect Biochem Physiol. 1988;9:1–33.
31. Brandstetter B, Ruther J. An insect with a delta-12 desaturase, the jewel wasp Nasonia vitripennis, benefits from nutritional supply with linoleic acid. Sci Nat. 2016;103
32. Davidson DW, Cook SC, Snelling RR. Liquid-feeding performances of ants (Formicidae): ecological and evolutionary implications. Oecologia. 2004;139:255–66.

33. Heath RJ, Rock CO. The Claisen condensation in biology. Nat Prod Rep. 2002;19:581–96.

34. Fast PG. Insect lipids: a review. Mem Entomol Soc Can. 1964;37:1–50.

35. Cripps C, Blomquist GJ, de Renobales M. De novo biosynthesis of linoleic acid in insects. Biochim Biophys Acta BBA-Lipids Lipid Metab. 1986;876:572–80.

36. Renobales M, Cripps C, Stanley-Samuelson DW, Jurenka RA, Blomquist GJ. Biosynthesis of linoleic acid in insects. Trends Biochem Sci. 1987;12:364–6.

37. Collingwood CA. The Formicidae (Hymenoptera) of Fennoscandia and Denmark. Fauna Entomol Scand. 1979;8:1–174.

38. Seifert B. Die Ameisen Mittel- und Nordeuropas. Lutra: Tauer; 2007.

39. Bhatkar A, Whitcomb WH. Artificial diet for rearing various species of ants. Fla Entomol. 1970;53:229–32.

40. Folch J, Lees M, Stanley GHS. A simple method for the isolation and purification of total lipides from animal tissues. J Biol Chem. 1957;226:497–509.

41. Gómez-Brandón M, Lores M, Domínguez J. A new combination of extraction and derivatization methods that reduces the complexity and preparation time in determining phospholipid fatty acids in solid environmental samples. Bioresour Technol. 2010;101:1348–54.

42. Frostegård Å, Tunlid A, Bååth E. Microbial biomass measured as total lipid phosphate in soils of different organic content. J Microbiol Methods. 1991;14:151–63.

43. Zelles L. Fatty acid patterns of phospholipids and lipopolysaccharides in the characterisation of microbial communities in soil: a review. Biol Fertil Soils. 1999;29:111–29.

44. Stein SE. Mass Spectra by NIST Mass Spec Data Center. In: Lindstrom PJ, Mallard WG, editors. NIST Chemistry WebBook. National Institute of Standards and Technology. 2015. http://webbook.nist.gov. Accessed 10 Oct 2016.

45. Dunkelblum E, Tan SH, Silk PJ. Double-bond location in monounsaturated fatty acids by dimethyl disulfide derivatization and mass spectrometry: application to analysis of fatty acids in pheromone glands of four Lepidoptera. J Chem Ecol. 1985;11:265–77.

46. Pinheiro J, Bates D, DebRoy S, Sarkar D, R Core Team. nlme: Linear and nonlinear mixed effects models. 2016. http://CRAN.R-project.org/package=nlme. Accessed 15 Dec 2016.

47. Hothorn T, Bretz F, Westfall P. Simultaneous inference in general parametric models. Biom J. 2008;50:346–63.

48. Anderson MJ. A new method for non-parametric multivariate analysis of variance. Austral Ecol. 2001;26:32–46.

49. Anderson MJ. Distance-based tests for homogeneity of multivariate dispersions. Biometrics. 2006;62:245–53.

50. Clarke KR, Gorley RN. PRIMER v7: User Manual/Tutorial. Plymouth: PRIMER-E; 2015.

51. Palarea-Albaladejo J, Martín-Fernández JA. zCompositions — R package for multivariate imputation of left-censored data under a compositional approach. Chemom Intell Lab Syst. 2015;143:85–96.

52. van den Boogaart KG, Tolosana R, Bren M. compositions: Compositional data analysis. 2014. https://CRAN.R-project.org/package=compositions. Accessed 15 Dec 2016.

53. Oksanen F, Blanchet FG, Kindt R, Legendre P, Minchin PR, O'Hara RB, et al. vegan: Community ecology package. 2016. https://CRAN.R-project.org/package=vegan. Accessed 15 Dec 2016.

54. Brückner A, Heethoff M. A chemo-ecologists' practical guide to compositional data analysis. Chemoecology. 2017;27:33–46.

55. R Core Team. R: A language and environment for statistical computing. R Foundation for Statistical Computing. 2016. https://www.R-project.org/. Accessed 15 Dec 2016.

56. Thompson SN. A review and comparative characterization of the fatty acid compositions of seven insect orders. Comp Biochem Physiol Part B Comp Biochem. 1973;45:467–82.

57. Hanson BJ, Cummins KW, Cargill AS, Lowry RR. Lipid content, fatty acid composition, and the effect of diet on fats of aquatic insects. Comp. Biochem. Physiol. Part B Comp. Biochem. 1985;80:257–76.

58. Stanley-Samuelson DW, Howard RW, Akre RD. Nutritional interactions revealed by tissue fatty acid profiles of an obligate myrmecophilous predator, *Microdon albicomatus*, and its prey, *Myrmica incompleta*, (Diptera: Syrphidae) (Hymenoptera: Formicidae). Ann Entomol Soc Am. 1990;83:1108–15.

59. Barlow JS. Fatty acids in some insect and spider fats. Can J Biochem. 1964;42:1365–74.

60. Brian MV, Blum MS. The influence of *Myrmica* queen head extracts on larval growth. J Insect Physiol. 1969;15:2213–23.

61. Brian MV. Caste control through worker attack in the ant *Myrmica*. Insect Soc. 1973;20:87–102.

62. Bhulaidok S, Sihamala O, Shen L, Li D. Nutritional and fatty acid profiles of sun-dried edible black ants (*Polyrhachis vicina* Roger). Maejo Int J Sci Technol. 2010;4:101–12.

63. Blaul B, Steinbauer R, Merkl P, Merkl R, Tschochner H, Ruther J. Oleic acid is a precursor of linoleic acid and the male sex pheromone in *Nasonia vitripennis*. Insect Biochem Mol Biol. 2014;51:33–40.

64. Wheeler WM. Ants - their structure, development and behavior. New York: Columbia University Press; 1910.

65. Rosell RC, Wheeler DE. Storage function and ultrastructure of the adult fat body in workers of the ant *Camponotus festinatus* (Buckley) (Hymenoptera : Formicidae). Int J Insect Morphol Embryol. 1995;24:413–26.

66. Wang YJ, Happ GM. Larval development during the nomadic phase of a nearctic army ant, *Neivamyrmex nigrescens* (Cresson) (Hymenoptera: Formicidae). Int J Insect Morphol Embryol. 1974;3:73–86.

67. Sorensen AA, Busch TM, Vinson SB. Control of food influx by temporal subcastes in the fire ant, *Solenopsis invicta*. Behav Ecol Sociobiol. 1985;17:191–8.

68. Tschinkel WR. Sociometry and sociogenesis of colonies of the harvester ant, *Pogonomyrmex badius*: worker characteristics in relation to colony size and season. Insect Soc. 1998;45:385–410.

69. Silberman RE, Gordon D, Ingram KK. Nutrient stores predict task behaviors in diverse ant species. Insect Soc. 2016;63:299–307.

70. Brian MV. Brood-rearing behaviour in small cultures of the ant *Myrmica rubra* L. Anim Behav. 1974;22:879–89.

71. Bolton B. Bolton B. An online catalog of the ants of the world. 2016. http://antcat.org/. Accessed 27 Feb 2017.

72. Kück P, Hita Garcia F, Misof B, Meusemann K. Improved phylogenetic analyses corroborate a plausible position of *Martialis heureka* in the ant tree of life. Gilbert MTP, editor. PLoS ONE. 2011;6:e21031.

73. Hughes L, Westoby MT, Jurado E. Convergence of elaiosomes and insect prey: evidence from ant foraging behaviour and fatty acid composition. Funct Ecol. 1994;8:358–65.

74. Reifenrath K, Becker C, Poethke HJ. Diaspore trait preferences of dispersing ants. J Chem Ecol. 2012;38:1093–104.

8

Adaptive responses to salinity stress across multiple life stages in anuran amphibians

Molly A. Albecker* [ID] and Michael W. McCoy

Abstract

Background: In many regions, freshwater wetlands are increasing in salinity at rates exceeding historic levels. Some freshwater organisms, like amphibians, may be able to adapt and persist in salt-contaminated wetlands by developing salt tolerance. Yet adaptive responses may be more challenging for organisms with complex life histories, because the same environmental stressor can require responses across different ontogenetic stages. Here we investigated responses to salinity in anuran amphibians: a common, freshwater taxon with a complex life cycle. We conducted a meta-analysis to define how the lethality of saltwater exposure changes across multiple life stages, surveyed wetlands in a coastal region experiencing progressive salinization for the presence of anurans, and used common garden experiments to investigate whether chronic salt exposure alters responses in three sequential life stages (reproductive, egg, and tadpole life stages) in *Hyla cinerea*, a species repeatedly observed in saline wetlands.

Results: Meta-analysis revealed differential vulnerability to salt stress across life stages with the egg stage as the most salt-sensitive. Field surveys revealed that 25% of the species known to occur in the focal region were detected in salt-intruded habitats. Remarkably, *Hyla cinerea* was found in large abundances in multiple wetlands with salinity concentrations 450% higher than the tadpole-stage LC_{50}. Common garden experiments showed that coastal (chronically salt exposed) populations of *H. cinerea* lay more eggs, have higher hatching success, and greater tadpole survival in higher salinities compared to inland (salt naïve) populations.

Conclusions: Collectively, our data suggest that some species of anuran amphibians have divergent and adaptive responses to salt exposure across populations and across different life stages. We propose that anuran amphibians may be a novel and amenable natural model system for empirical explorations of adaptive responses to environmental change.

Keywords: Secondary salinization, Anuran amphibian, Sea level rise, Saltwater tolerance, Climate change, Complex life history

Background

Accumulating greenhouse gas concentrations are increasing the energy retained in the atmosphere, which is in turn causing global mean sea levels to rise through intensified ice sheet and glacier melting and thermal expansion of ocean water [1–4]. Sea levels have already risen 17-21 cm over the past 110 years, and current models forecast that sea levels could rise an additional 40–63 cm over the next century with additions expected if ice sheets on Greenland and West Antarctica collapse [2, 4–8]. Ancillary impacts of climate change on coastal wetlands include alterations in the frequency and intensity of storm surges and coastal flooding, which may compound the effects of coastal erosion and saltwater inundation. The magnitude of sea level rise and impact on coastal ecosystems will vary depending on glacial isostatic adjustment, tectonic processes, oceanic circulation patterns, sediment compaction and accretion, wind patterns, and gravitational changes [4, 9–15], yet many areas are already being affected by sea level rise [16–20].

Rising salinities are broadly anticipated to negatively impact freshwater organisms inhabiting coastal regions by reducing both the quality and quantity of suitable habitat, lowering individual fitness (e.g., increased physiological stress, increased morphological deformities, reduced fecundity, and modifications to growth,

* Correspondence: albeckerm09@students.ecu.edu
Department of Biology, Howell Science Complex, East Carolina University, Greenville, NC, USA

development, and mortality), reducing population carrying capacity, and by altering biological interactions, disease risk, species movement, and community structure [21–24].

Osmoregulators require a wide variety of physiological, morphological, life historical, and behavioral traits to conserve water and expel enough excess ions to survive higher salinities. Although examples of adaptive responses across strong abiotic clines are multiplying quickly [25–30], adaptive responses might be slowed by an organism's life history strategy, amount of standing genetic variation, demographic constraints (e.g., competition), or decoupling of environmental cue from response [31–35]. For example, organisms with complex life cycles, such as amphibians, have different ontogenetic life stages that are typically marked by abrupt shifts in morphology, physiology, behavior, and often distinct changes in habitat use. Therefore, the same stressor may differently impact each life stage, and require multiple adaptive responses across life stages to successfully adapt to an emerging environmental stressor.

Amphibians are a classic model for exploring responses to environmental stressors such as salinity. Amphibians are widely regarded as important indicator species of wetland quality due to a life history tied to freshwater coupled with unique characteristics such as permeable skin, an inability to concentrate and excrete excess salts, and poor dispersal capabilities [36–39]. Additionally, amphibians comprise a significant proportion of the vertebrate biomass in wetland ecosystems [40, 41] and have been classified by the IUCN as "climate change susceptible" [23]. Most amphibians are obligatorily aquatic throughout the egg and larval period and become semi-terrestrial upon metamorphosis. Depending on the species, amphibians typically return to water as adults to breed or rehydrate.

A recent review identified ca. 140 anuran amphibian species that have been observed in saline habitats (ranging from tidal mangrove swamps to inland freshwater habitats contaminated with road deicing salts). Yet these species represent only 2% of all known species [38, 39], supporting the widely held belief that anurans are a generally salt-sensitive, freshwater order. A few notable species of amphibians such as *Fejervarya cancrivora* and *Bufo viridis* are known to tolerate brackish conditions [38, 39, 42–46], but these species still require freshwater habitats to complete their life cycles suggesting differential vulnerability to salt exposure across life stages even in specialist salt-tolerant species [47–50].

In addition to field observations, there are many published studies that experimentally explore embryonic, tadpole, or adult responses to salt stress. These studies typically evaluate how saltwater impacts anuran survivorship and behavior in a single life stage, and in doing so, provide indispensable and informative data on expected responses across a range of salinities. Hopkins and Brodie published an extensive review of saltwater tolerance in amphibians [39], which provides a useful framework to better understand and predict how salinization affects anuran populations. Yet the data contained in these studies has not yet been coalesced to precisely quantify how salt tolerance changes across different life stages. Moreover, to best predict how anurans will respond to progressively increasing salinities, we not only need to define how salinity affects each life stage, but also how labile salt-tolerant responses are across populations.

In this study, we use multiple, complementary strategies to evaluate salt sensitivity in anurans generally, and substitute space for time to explore whether populations that inhabit coastal wetlands with a history of increasing salt exposure demonstrate adaptive responses across multiple life stages. First, we conducted a meta-analysis to establish an empirically derived quantitative framework of expected survivorship following exposure to saltwater in anuran amphibians for different life stages. Second, we performed a field survey of brackish and freshwater wetlands to describe and characterize amphibian distributions along a salt gradient in a coastal location predicted to be among the most impacted by sea level rise. Third, we substitute space for time in common garden experiments to investigate how exposure to saltwater across life stages differs among chronically salt-exposed (coastal) and salt-naïve (inland) anuran populations.

We focus on reproductive behaviors, egg hatching patterns, and post-hatching tadpole survival for our common garden experiments. During breeding events, male frogs amplex females and then she will transport the male to assess potential egg laying sites. Females of some species are highly discriminatory and choose among oviposition sites to avoid a variety of biotic and abiotic stressors [51, 52]. Oviposition site choice behaviors are under strong selection because her choice can considerably impact offspring survival and performance by affecting fertilization success, mortality risk to offspring, as well as resource availability to offspring [46, 51–55]. After eggs have been deposited, developing clutches are vulnerable to aquatic contaminants because frog eggs are enclosed by a permeable, jelly coat and lack a hard, protective shell [56, 57]. Upon hatching, the larvae of many frog species are obligatorily aquatic and cannot survive on land until the completion of metamorphosis. During this period, tadpoles respire and osmoregulate via gills that function similar to freshwater teleosts such that ions and salts are conserved and excess water is expelled [58–60]. We chose reproductive choices, embryo hatching success, and tadpole survival because these stages are key periods in the anuran life cycle that are highly vulnerable to external stressors, including saltwater, and strongly influence individual fitness and population persistence [46, 51, 61–65].

Methods
Study location
We conducted these studies in eastern North Carolina, USA. North Carolina's coastline, barrier islands, and coastal habitats are predicted to be among the most significantly impacted by sea level rise due to the geomorphology of the Northern coastal zone (Albemarle embayment), coastal subsidence rates (–1 mm ± 0.15 mm/yr.), and gently sloped coastal plains [15, 19, 66–68]. Indeed, the North Carolina coast has already seen intensified coastal flooding, and increased saltwater intrusion into coastal lowlands and freshwater aquifers making it an important location for investigating the impacts of sea level rise and increasing salinities on coastal organisms [11, 19, 69].

Meta-analysis
Literature search
We searched Google Scholar and Scopus databases for experimental studies evaluating the survivorship of anuran amphibians after experimental exposure to saltwater. We conducted the primary, exhaustive searches on December 16–20, 2014. Literature was checked again on July 14, 2015, September 23, 2015, February 25, 2016, and February 2, 2017 to ensure recently published work was included. We used the search terms (and all combinations of): "frog" OR "anuran" OR "amphibian" AND "saltwater" OR "salt" OR "salinity" OR "ocean" OR "NaCl" AND "mortality" OR "survivorship". Initial searches returned ~24,500 hits in total. These studies were further refined by scanning titles and abstracts. We excluded studies that did not mention survivorship or mortality of anurans and exposure to saltwater in the abstract. We also cross checked against the list of studies in Hopkins and Brodie's review of amphibian salt tolerance to ensure all appropriate studies were included [39].

Data extraction
After refining our database to 129 studies, each study was read in detail and data were extracted from the text or figures. We extracted data only on studies that experimentally and directly manipulated salt concentrations against known sample sizes (e.g., field observations and studies with incidental, non-targeted salt exposure were excluded). We used studies that exposed frogs to saltwater solutions comprised of sodium chloride (NaCl), (e.g. InstantOcean° or natural seawater) and excluded studies that exposed frogs to mixed salt solutions (e.g., mixed road salt solutions) [70]. In studies where multiple saltwater compositions (e.g., $MgCl_2$, KCl, $CaCl_2$)

were tested, we only used data from the trials that utilized NaCl. See Additional file 1 for detailed list of studies.

We used GraphClick° software version 3.0.3 (Arizona Software) to extract estimates from published figures and graphs. We report the mean survivorship (with error) for studies containing multiple replicates across salinities. For studies that compare survivorship across replicate populations, we present global averages across all populations tested. Although two studies report intra-specific differences in saltwater tolerance across different populations (e.g., [45, 71]), there were too few studies available to permit a meaningful formal analysis on population level differences in saltwater tolerance across studies or species. We recorded species identity, family, life stage (tadpole, egg, or adult), experimental salinity concentrations, sample size (N), survivorship (as proportion), the standard deviation of survivorship (converted from standard error when necessary), location of the study, and length of exposure (in hours) for each study. Because different studies reported salinity using different units, we used standard conversions to transform all salinity measurements to parts per thousand (ppt).

Field survey
Study sites
We monitored wetlands regularly to make sure species that breed at different times could be detected. We surveyed 55 salt and freshwater wetlands in eastern North Carolina between February and September of 2014 for the presence of anuran amphibians. We included bogs, retention areas, marshes, ponds, ditches, and swamps, but excluded estuaries, sea grass beds, and other large, open water habitats. The most southern and eastern location was Cape Hatteras National Seashore and the survey extended northward to the town of Nags Head. Along this transect, we surveyed wetlands along Rodanthe, New Inlet, Bodie Island, Oregon Inlet, and Pea Island National Wildlife Refuge. We also sampled wetlands along an east to west transect spanning from the outer banks of NC, across Roanoke Island, which lies between the inner and outer banks and throughout Alligator River National Wildlife Refuge located on the Albemarle peninsula. The geographic bounds of the study area are 35°55′7″N to 35°14′7″N, and between 75°48′43″W to 75°27′27″W, excluding the Atlantic Ocean and the Pamlico, Croatan, and Roanoke sounds.

Survey techniques
We used standard sampling methods to characterize anuran presence and relative abundance including auditory call surveys, standardized dip netting for larvae, and active searching for adults [72, 73]. Our primary

approach used auditory surveys to identify and locate frog populations, as well as to determine species identities and relative abundances of the anurans present. When frogs were detected via call, the site was geo-referenced using a Garmin® GPSMAP 60CSx GPS navigator (Garmin, Ltd., Olathe, KS) and salinity (in ppt) and the temperatures of the air and water were measured using YSI Professional Plus multiparameter meter (Xylem, Inc., Yellow Springs, OH). We returned the following day (auditory surveys occurred at night) to the geo-referenced sites to determine egg mass/larvae presence using fixed-effort dip netting, and visual transect surveys [72, 73]. To ensure that we thoroughly surveyed all wetlands for the presence of amphibians (and not just wetlands with detectable choruses), we used Google Maps® and visual surveys to identify additional wetlands that were not identified using call surveys, and sampled these wetlands using visual transect surveys and dip-netting for the presence of adult and/or larval anuran species. Tuberville et al. [74] conducted a thorough amphibian field survey along the North Carolina coast that included Cape Hatteras and Cape Lookout National Seashore and documented the current or historic presence of 17 anuran species, and we use the results of this study as a comparison for our own observations. Notably, the Tuberville study did not record salinity of locations in which anurans were observed.

Common garden experiments

We used *Hyla cinerea*, the American green tree frog (average size: 3.2–5.7 cm), for each of our common garden experiments, as this species is common across the Southeastern United States and has been repeatedly documented in saltwater intruded environments [38, 51, 75, 76]. These experiments were conducted between May and August 2015. To characterize and identify how responses to saltwater differ among populations, we compared individuals from chronically salt-exposed *Hyla cinerea* populations (hereafter referred to as "coastal" populations) against individuals from freshwater, salt-naive *Hyla cinerea* populations (hereafter referred to as "inland" populations). We located coastal and inland populations via the field survey. All coastal individuals were collected from sites in which salinities remained at or above 3 ppt over the course of the breeding season, and all inland individuals were collected from populations with salinities below 1 ppt. Coastal populations and inland populations were geographically separated from one another by at least 190 km, so we assume that pairs collected from populations within these locations are sufficiently distant both geographically and environmentally to provide an accurate assessment of population-level differences produced by the different salinity of their habitats.

Oviposition site choice and egg hatching

We tested oviposition site choice by collecting four amplexed pairs of *Hyla cinerea* from either coastal or inland populations. Each pair was placed into an 18-Liter clear bin, the bottom of which was lined with six pint cups. Three of the six cups contained 400 ml tap water (0 ppt) treated with API® Tap Water Conditioner (Chalfont, PA), and the remaining cups contained 400 ml saltwater prepared by mixing treated tap water with InstantOcean Sea Salt® (Blacksburg, VA). Each bin contained a single saltwater concentration that was either 4 ppt, 6 ppt, 8 ppt, or 12 ppt. In doing so, we presented each pair with a binary choice between laying eggs in freshwater or saltwater. The four salt concentration treatments collectively comprised a single replicate (i.e., four bins = one replicate). On nights when multiple replicates were conducted, each replicate was arranged in a spatial block at the site of collection.

Bins were left in situ overnight to allow pairs to complete breeding. The following morning, adult frogs were released, lids fastened to each cup, and bins were transported to the laboratory. Each cup was individually photographed, the salinity measured, and then monitored for hatching. Eggs hatched after 72–96 h, defined as the point in which individuals were no longer retained in egg matrix and have functional gills (Gosner stage 20 [77]). Hatchlings were counted and recorded.

Tadpole survivorship

To determine the effects of salinity on tadpole survival, we utilized the individuals hatched from eggs laid in freshwater during the previous oviposition experiments. Hatchlings were held in the laboratory that was maintained at 26.67 °C (~80 °F) and allowed to develop until reaching Gosner stage 25 (approximately 5 days) [77]. Several studies have indicated that acclimatizing anurans to elevated salinities reduces mortality [50, 52, 78], and natural salinity fluctuations typically do not exceed +/− 2 ppt per day, excluding an extreme event such as storm surge or flooding event. Therefore, to best mimic natural conditions and quantify survival, tadpoles were gradually acclimatized to a specified target salinity over 6 days. We chose five target salinities, 0.5 ppt, 4 ppt, 6 ppt, 8 ppt, and 12 ppt, which are representative of natural salinities observed in coastal wetlands. Freshwater treatments (0.5 ppt) were maintained at 0.5 ppt throughout the six-day acclimatization period. The 4 ppt treatments were raised by 0.67 ppt per day, 6 ppt treatments were raised by 1 ppt per day, 8 ppt treatments raised by 1.33 ppt per day, and 12 ppt treatments were raised by 2 ppt per day, with final target salinities reached on day 6.

We divided each clutch into five groups of fifty tadpoles, which were then randomly assigned to one of the five salinity treatments, replicated 8 times for each location. Each clutch divided into five groups comprised a single replicate block to account for potential parental effects. Groups of tadpoles were placed into 350 mL glass containers containing 300 mL of treated tap water (treated with API® Tap Water Conditioner (Chalfont, PA)) within a laboratory with 12-h light/dark cycle. After acclimatizing overnight, salinity was increased incrementally each day according to treatment. Prior to water changes each day, tadpole mortality in each cup was assessed and recorded, and deceased individuals were removed. Tadpoles were fed 0.01 g of Spirulina fish food flakes (Ocean Star International, Coral Springs, FL) each day following the water change. To perform water changes, tadpoles were carefully poured into a small holding container and returned after 300 mL of new, treated water with experimentally raised saltwater concentrations (InstantOcean Sea Salt® (Blacksburg, VA)) was poured into glass containers.

Statistical analyses

We use a Bayesian approach to analyze our data. For all statistical analyses we used JAGS interfaced with the R statistical programming environment, version 3.2.3 [79] via "R2jags" [80], "rjags" [81], and "coda" [82] packages. For each analysis, we ran 5000 iterations of three separate Markov Chain Monte Carlo (MCMC) chains with starting values that varied by an order of magnitude, each with a burn in of 2500 unless otherwise specified [83]. We used Gelman-Rubin diagnostics to assess model convergence in each analysis [83].

Meta-analysis

To estimate the probability of survival in saltwater for each life stage across anuran taxa and across salinities, we tested how increasing salinity affects anuran survivorship across clades for each life stage (e.g., egg, larvae, adult). We did not use phylogenetically corrected data because a recent review of all instances of amphibians in saline environments revealed no phylogenetic signal [39] and we detected no signal of phylogeny in the unexplained deviance from our analysis. We performed a Bayesian beta regression with an uninformative (relatively flat; mean = 0, std. dev. = 0.001) Gaussian prior. We chose the beta distribution because the data extracted for the meta-analysis were often only reported as "proportion survived" or "proportion killed" and lacked the necessary information (e.g., sample sizes and replicate numbers) required to back-calculate starting densities. In this analysis, survivorship and salinity were considered fixed effects, with individual studies treated as random effects.

Field survey

We utilized the posterior distribution from the meta-analysis of all anuran species to predict the probability of anuran survivorship across several salinities including the salinities where we observed coastal *Hyla cinerea* during field surveys. Specifically, we generated a survival curve (with uncertainty) across salinities ranging from 1 ppt (freshwater) up to 40 ppt, and estimated the expected probability and credible intervals for finding frogs in sites with salinity concentrations we found in our field observations. Although 40 ppt exceeds the salinity of natural seawater (35 ppt), Gordon and colleagues observed *Fejervarya cancrivora* tadpoles in 39 ppt water in 1961 [48]. While this particular observation was not included in our meta-analysis due to its non-experimental nature, we wanted to ensure that all possible salinities were considered in our meta-analysis.

Common garden experiments

We used ImageJ® software to quantify the number of eggs that were laid in each cup. Briefly, photograph files for each container were imported and changed to 8-bit images. The image background was subtracted, images were made binary, and files were converted to a mask. To separate groups of eggs that were clumped together, we used the watershed feature to demarcate individual egg boundaries. Outputs were visually inspected to ensure that all eggs were included and correctly counted.

We ran two-stage tests for both oviposition site-choice and hatching data. In the first step, we analyzed the data in binary form to ask if the probability of egg deposition or hatching changed as a function of the interaction between source population (e.g., coastal vs. inland) and salinity. In the second step, given that egg deposition or hatching occurred (i.e., excluding all cups in which zero eggs were laid or hatched), we analyzed the proportion of eggs deposited into freshwater and the proportion of offspring hatched as a function of the interaction between source population and salinity. These dual approaches answer distinct but complementary questions. Regarding oviposition, the first test asks if the probability of depositing eggs into saltwater or freshwater reflects a choice between salinities, while the second test reveals how parental investment differs according to salinity. Regarding hatching, the first test uncovers differences in the probability of complete loss due to salinity, while the second test reveals thresholds of sensitivity to salt.

To test the probability of oviposition, we ran Bernoulli regression to test for a relationship between egg presence or absence according to salinity and location (step one above). To test whether there were differences in investment (step two above), we ran a binomial regression to examine whether salinity and location affected the proportion of eggs deposited by a female into saltier water. For both of these analyses, we used uninformative Gaussian priors (mean = zero and precision as a decaying power function with exponent = −2). To test the probability of hatching and proportion that hatched, we use informed priors based on the posterior distribution produced by the egg stage meta-analysis. Similar to the oviposition analyses, we ran Bernoulli regression to determine the relationship between egg hatching and salinity and location. We then used a binomial regression to analyze differences in the proportion of eggs that hatched in each salinity and location. Each of these four models considers salinity and location (e.g., coastal or inland) as fixed effects with "bin" nested in location as a random effect to account for parental effects [84].

Tadpole survivorship

To quantify how salinity, location, and time (e.g., day) affect tadpole survivorship, we used a binomial regression with informed priors based on the posterior distribution produced by the tadpole stage meta-analysis. This model considers salinity and location (e.g., coastal or inland) as fixed effects with "clutch" included as a random effect to account for sibship [84]. For this analysis we ran four separate MCMC chains with 50,000 iterations, each with a burn in of 25,000 [83].

Results

Meta-analysis

Effects of salt on amphibian survivorship

We utilized data from 39 papers published between 1961 to early 2017 (see Additional file 1 for detailed information). Overall, the literature uniformly demonstrates that increasing saltwater concentrations lowers anuran survivorship across all three life-stages (Fig. 1). We found that across all studies included in this analysis, the lethal concentration of saltwater required to impose 50% mortality (LC_{50}) to anuran amphibian eggs is 4.15 ppt (95% Bayesian credible interval [BCI] = 2.25 to 6.25 ppt). The LC_{50} for larval anurans is 5.5 ppt (4.24–6.65 ppt BCI), while the LC_{50} for adults is 9.0 ppt (0–19.9 ppt BCI).

Field surveys

Species presence

In coastal freshwater habitats (<3 ppt) with no connection to saltwater influence (e.g., municipal retention ponds), we documented the regular presence of 16 of the 17 anuran species found in the Tuberville study including *Hyla cinerea, Hyla chrysoscelis, Hyla squirella, Hyla femoralis, Anaxyrus fowleri, Anaxyrus quercicus, Anaxyrus terrestris, Lithobates sphenocephalus, Lithobates clamitans, Lithobates virgatipes, Lithobates catesbeianus, Gastrophryne carolinensis, Pseudacris ocularis, Pseudacris crucifer,* and *Acris gryllus.* We did not detect

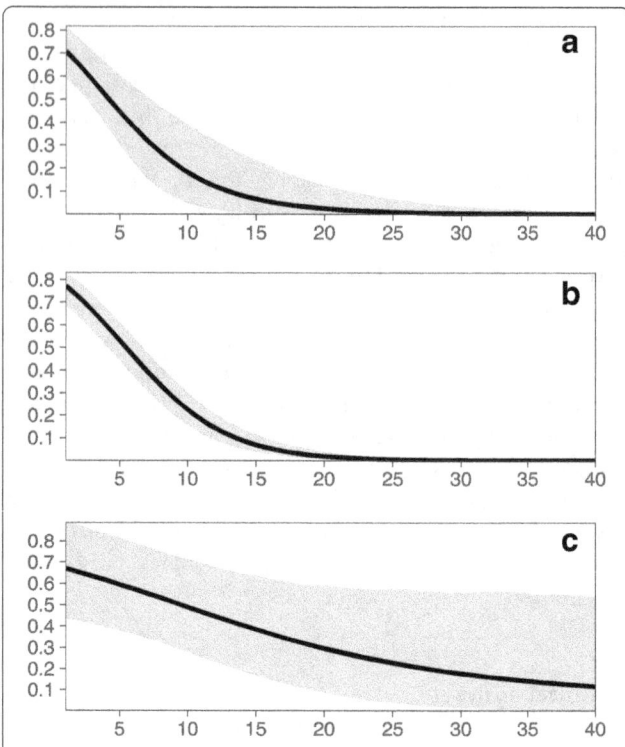

Fig. 1 The predicted survival for each life stage across the anuran amphibian clade as a function of salinity (in parts per thousand). Panel **a** is the predicted survival for the egg stage, panel **b** is predicted tadpole survival, and panel **c** is predicted adult survival

Scaphiopus holbrookii [74]. In salt-invaded wetlands (>3 ppt), we documented the presence of 4 of those 16 species (*Hyla cinerea, Gastrophryne carolinensis, Lithobates catesbeianus,* and *Lithobates sphenocephalus*) (Table 1).

Relative abundance

In general, we noted that relative abundances of all species (except *Hyla cinerea*) declined as wetlands grew more saline. *Hyla cinerea* demonstrated unique distribution patterns along North Carolina's coast as the most abundant species found within salt-invaded habitats along both the inner and outer banks. Notably, in some locations we observed that the relative abundance of *Hyla cinerea* actually increased with increasing salinity, a pattern not shared with any of the other species found in salt-invaded wetlands. We collected early and late stage *Hyla cinerea* tadpoles, metamorphs (between Gosner stages 31–39 [77]), and adults from multiple locations including from ponds and marshes with 3.9 ppt, 8.3 ppt, 11 ppt, 16.8 ppt, and 23.4 ppt water.

Probability of field findings

Using the posterior probability distributions from our meta-analysis we examined the relative probability of finding frogs in the observed salinities: 3.9 ppt, 8.3 ppt,

Table 1 The location and identity of the four anuran species observed in coastal, salt-invaded wetlands along with the highest salinity in which each species was observed

Species	Highest Salinity Observed	Occurrence	Location
Lithobates sphenocephalus	11 ppt	Abundant	Alligator River NWR
Hyla cinerea	23.4 ppt	Abundant	Cape Hatteras National Seashore
Gastrophryne carolinensis	3.9 ppt	Abundant	Alligator River NWR
Lithobates catesbeianus	6.2 ppt	Rare	Pea Island NWR

11 ppt, 16.8 ppt, and 23.4 ppt saltwater. The expected probability of survival for an individual anuran following exposure to a 3.9 ppt saltwater solution during the egg stage is 0.52 (0.39–0.66 95% BCI), 0.60 (0.51–0.68 BCI) for larval anurans, and 0.62 (0.39–0.84 BCI) for adults. The probability of survival in 8.3 ppt water for eggs is 0.25 (0.09–0.45 BCI), 0.32 (0.23–0.39 BCI) for larvae, and 0.53 (0.32–0.74 BCI) for adult frogs. At 11 ppt, the survivorship for eggs is 0.15 (0.03–0.35 BCI), larval survivorship is 0.18 (0.12–0.25 BCI), with adult survivorship predicted at 0.46 (0.25–0.70 BCI). Wetlands at 16.8 ppt have 0.04 (0.002–0.19 BCI) expected egg survivorship, 0.04 (0.02–0.07 BCI) expected larval survivorship, and 0.35 (0.13–0.61 BCI) expected adult survivorship. In 23.4 ppt wetlands, 0.01 (0.00–0.008 BCI) eggs are expected to survive, larval survivorship is 0.01 (0.002–0.02 BCI), and expected adult survivorship is 0.25 (0.05–0.57 BCI) (Table 2).

Common garden experiments

The oviposition site choice experiment utilized *Hyla cinerea* pairs collected from three geographically discrete populations from inland and coastal locations in eastern North Carolina. The subsequent egg hatching and tadpole survivorship experiments utilized the offspring of the collected pairs. For the coastal locations, we sampled three discrete populations along the inner and outer banks of North Carolina. We collected 1 replicate from a population near New Inlet bridge (35°41′11.5″ N, 75°29′03.92″W), 1 replicate from Coastal Studies

Institute on Roanoke Island (35°52′26.14″ N, 75°39′38.54″ W), and 2 replicates from Point Peter Road, Alligator River National Wildlife Refuge (35°46′13.1″ N, 75°44′30.1″ W). These populations are separated by the Croatan and/or Roanoke Sounds. For the inland locations, we sampled three discrete populations around Greenville, North Carolina. Specifically, we collected 1 replicate from a population near MacGregor Downs Road (35°37′15.8″ N, 77°26′45.29″ W), 1 replicate along Pactolus Highway (35°37′18.9″ N, 77°20′43.8″ W), and 2 replicates from a retention pond on 10th street (35°35′26.49″ N, 77°19′09.89″ W). Each inland population is at least 5 km apart from other populations with the Tar river and multiple highways between populations.

Oviposition site choice

We conducted four replicates in coastal and inland locations. Pairs successfully bred in every bin except one that contained a coastal pair. On average, females laid 1363 eggs (minimum = 713 eggs, maximum = 3039 eggs) per bin. We found that location (e.g., coastal vs. inland) and salinity both affected the probability that a female will lay her eggs in a particular pool (Fig. 2). As salinity increased, pairs from inland populations were less likely to deposit eggs in salinized water, while coastal females maintained a high probability of laying eggs in the higher salinity treatments (Fig. 2). For example, in the lower salinity treatments (4 ppt), females showed no

Table 2 Predicted survivorship (and Bayesian Credible Intervals) of anurans in various salinities based on the findings of the meta-analysis (Fig. 1). Each salinity concentration represents the salinity of a wetland in which frogs were observed along North Carolina's coast

Salinity (ppt) in which anurans were observed:	Predicted Egg Survivorship (+95% BCIs)	Predicted Larval Survivorship (+95% BCIs)	Predicted Adult Survivorship (+95% BCIs)
3.9	0.52 (0.39–0.66)	0.60 (0.51–0.68)	0.62 (0.39–0.84)
8.3	0.25 (0.09–0.45)	0.32 (0.23–0.39)	0.53 (0.32–0.74)
11	0.15 (0.03–0.35)	0.18 (0.12–0.25)	0.46 (0.25–0.70)
16.9	0.04 (0.002–0.19)	0.04 (0.02–0.07)	0.35 (0.13–0.61)
23.4	0.01 (0.00–0.008)	0.01 (0.002–0.02)	0.25 (0.05–0.57)

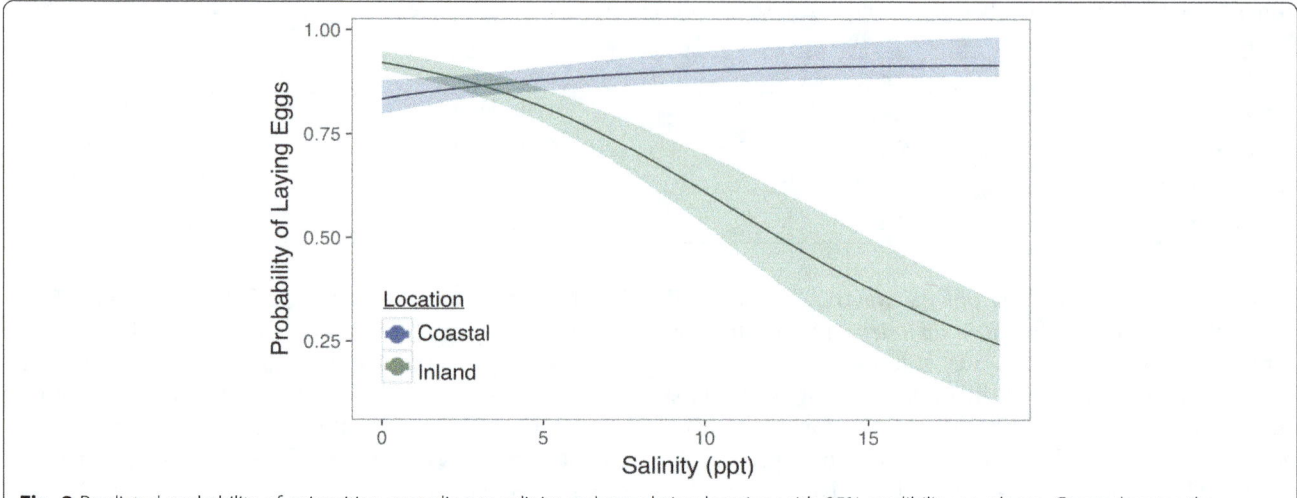

Fig. 2 Predicted probability of oviposition according to salinity and population location with 95% credibility envelopes. Green denotes the oviposition patterns from inland populations; *blue* indicates the oviposition patterns from coastal populations

divergence with inland females having 0.87 (0.85–0.91 BCI) probability of laying any eggs in the 4 ppt water, and coastal females having 0.84 (0.81–0.88 BCI) probability of laying eggs. Yet in the higher salinity treatments in which females chose between fresh or 12 ppt water, inland females had a 0.51 (0.41–0.61 BCI) probability of laying any eggs into 12 ppt water, while coastal females exhibited 0.91 (0.88–0.96 BCI) probability of laying eggs. Source population and salinity both affected the proportion of eggs laid in freshwater (Fig. 3). Pairs from both locations tended to lay the majority of their eggs into freshwater as salinity increased, but at 12 ppt, pairs from inland populations laid only 6% (0.04–0.07 BCI) into the saline water, while coastal pairs laid 16% (0.14–0.18 BCI) of their eggs in the saline water (Fig. 3).

Egg hatching

Salinity and source population affect the probability that any eggs would hatch out of a particular treatment (Fig. 4). At 4 ppt, the probability that an egg sourced from inland parents would hatch is 0.31 (0.24–0.38 BCI), while the probability that an egg laid by coastal parents would hatch is 0.54 (0.47–0.61 BCI). At higher salinities (10 ppt), eggs from both populations had an exceedingly low probability of hatching (inland probability: 0.02 (0.007–0.03 BCI); coastal probability: 0.04 (0.02–0.06 BCI)) (Fig. 4). We also observed that although the proportion of eggs that hatched in 3 ppt was similar across locations (inland proportion hatched: 0.33 (0.27–0.38 BCI); coastal proportion hatched: 0.36 (0.31–0.42 BCI)), 10% (0.07–0.11 BCI) of the coastal-sourced eggs hatched

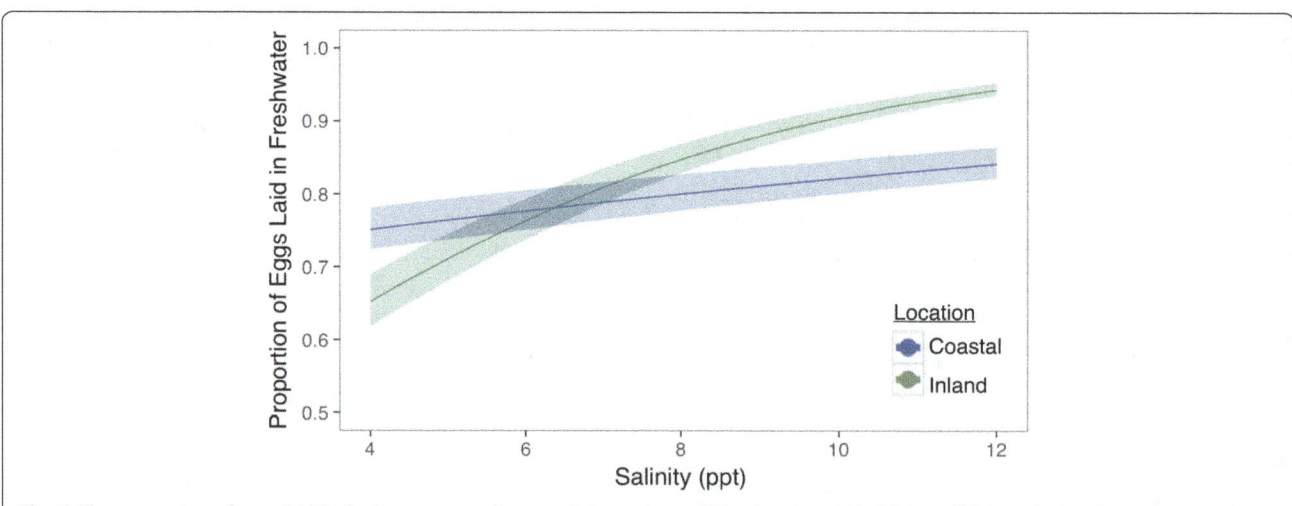

Fig. 3 The proportion of eggs laid in freshwater according to salinity and population location with 95% credible envelopes. Green denotes the proportion of eggs laid in freshwater by inland populations; *blue* indicates the proportion of eggs laid in freshwater from coastal populations

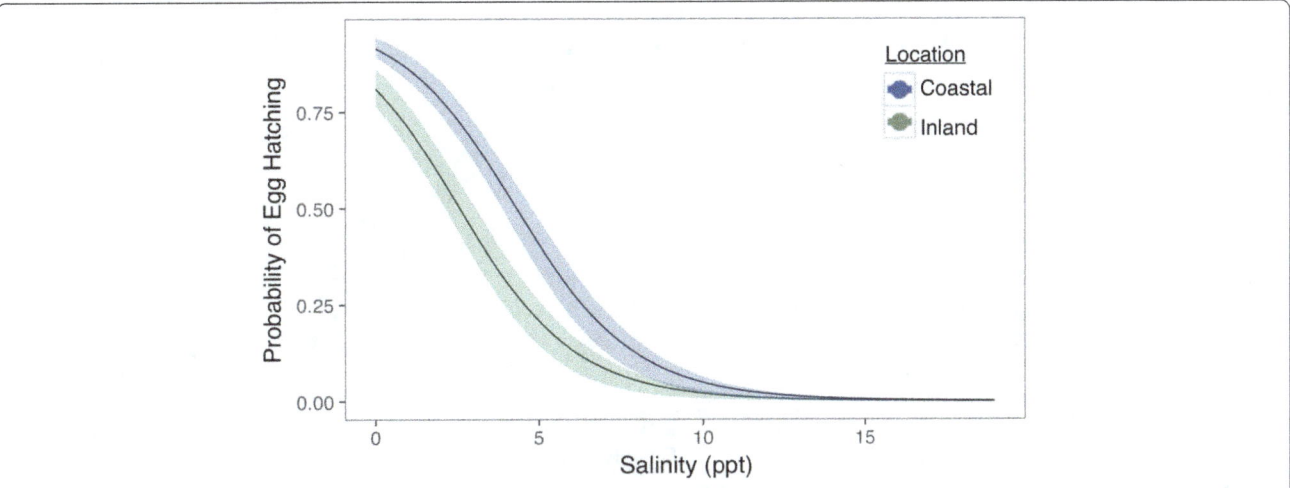

Fig. 4 Predicted probability of egg hatching according to salinity and population location with 95% credible envelopes. Green denotes the hatching patterns from inland populations; *blue* indicates hatching patterns from coastal populations

at 6 ppt compared to 3% (0.02–0.04 BCI) of the eggs sourced from inland populations (Fig. 5).

Tadpole survivorship

The predicted survival probability for coastal and inland *Hyla cinerea* tadpoles following a 6-day acclimation to freshwater (0.5 ppt) for coastal-sourced tadpoles is 0.98 (0.96–0.99 BCI) and 0.98 (0.96–0.99 BCI) for inland-sourced tadpoles (Fig. 6). At 4 ppt, predicted survivorship for coastal offspring is 0.96 (0.92–0.98 BCI) while inland offspring survivorship is 0.97 (0.95–0.99 BCI). Survivorship in 6 ppt treatments is 0.94 (0.90–0.97 BCI) for coastal tadpoles and 0.95 (0.89–0.98 BCI) from inland tadpoles. In the 8 ppt treatments, coastal tadpoles had higher survivorship at 0.97 (0.94–0.99 BCI) than inland tadpoles at 0.84 (0.73–0.92 BCI). At 12 ppt, we

again observed higher survivorship among coastal tadpoles with 0.24 (0.14–0.39 BCI) survivorship compared to inland tadpoles with 0.09 (0.04–0.16 BCI) survivorship. The random effect standard deviation representing parental influence is 0.17. Fixed effect slope and intercept estimates are listed in Additional file 2.

Discussion

We are at the precipice of dramatic environmental transformation as a result of global climate change, which provides the ideal canvas for exploring organismal responses to environmental change. Wetlands in coastal zones around the globe are among those anticipated to be most severely impacted from climate change due to increased frequency and intensity of coastal storms as

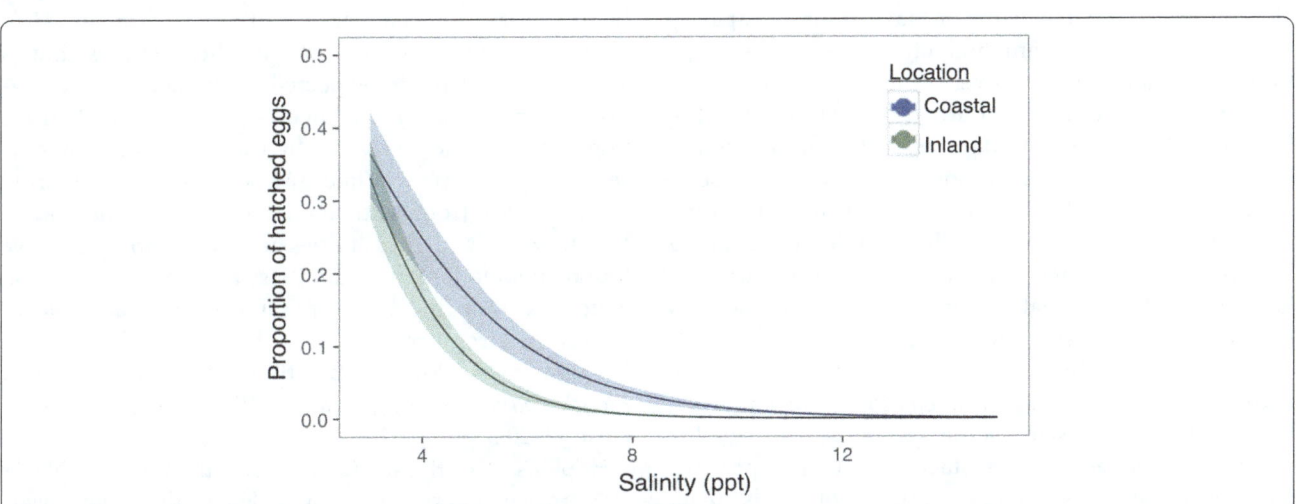

Fig. 5 The proportion of eggs that hatched according to salinity and population location with 95% confidence envelopes. Green denotes the proportion of eggs hatched from inland populations; *blue* indicates the proportion of eggs hatched from coastal populations

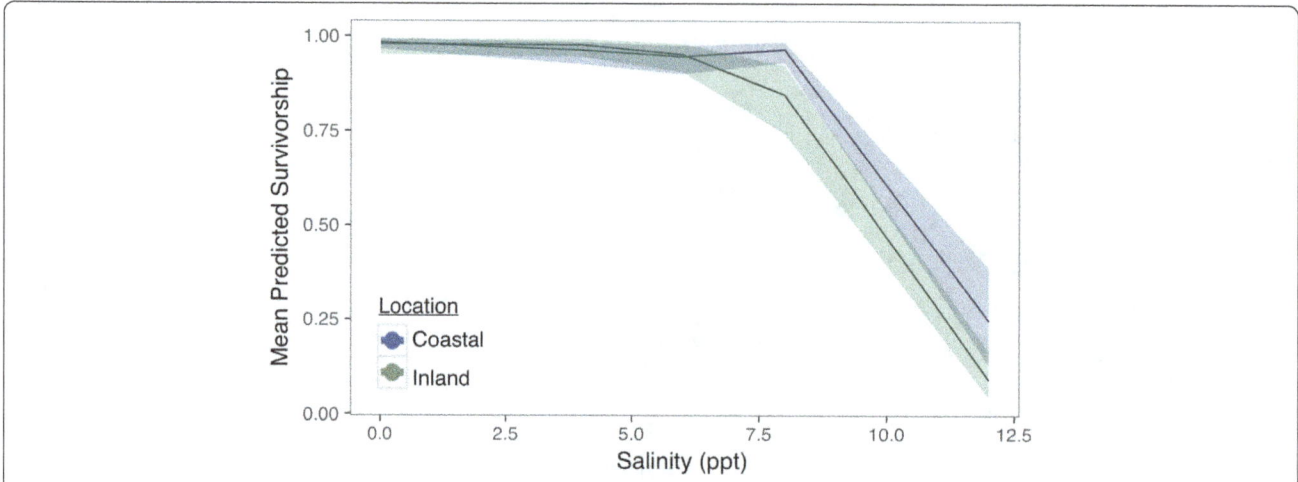

Fig. 6 Mean probability of tadpole survivorship according to salinity and population location with 95% credible envelopes. Green denotes the proportion of tadpoles sourced from inland populations; *blue* indicates tadpoles sourced from coastal populations

well as increased flooding and secondary salinization from sea level rise [1, 2, 8, 19, 66, 68, 85]. Yet despite the amount of cultural and research attention that climate change garners, a distressing deficiency exists in our empirical understanding of how rising salinities will impact coastal freshwater habitats and the animal communities sustained therein.

Ecological niche models aimed at understanding how environmental changes will impact affected populations typically predict that species that cannot emigrate to more suitable habitats are at risk of being locally extirpated as environmental quality degrades [3, 86–94]. This forecast is rational for freshwater organisms (like amphibians) that inhabit coastal wetlands given the lethal nature of osmotic stress [91, 94–100]. However, an important assumption inherent in most model predictions is that species either completely lack or have limited capacity to respond to environmental change – an assumption that can lead to overestimates of extinction rates or expected range contraction [24, 91, 94, 96, 98, 101–103]. Although adaptive evolution is increasingly well appreciated as a potential source of rescue for some, it is unclear whether organisms with complex life history strategies will be able to adapt to environmental change. In amphibians, we currently lack the ability to make more informed predictions that include adaptation for two main reasons. First, we do not know how sensitivity to salt stress varies across different life stages, and second, we know little about whether salt-tolerant responses are evolutionarily labile across life stages. In this paper, we address these gaps using a variety of tools (e.g., meta-analysis, field surveys, and common-garden experiments).

Meta-analysis and field surveys

Studies on amphibian responses to saltwater often begin with some variant of the statement, *it is well accepted that frogs do not belong in saline habitats*. These statements stem from long standing dogma that amphibians are not physiologically equipped to osmoregulate in non-freshwater environments. Nonetheless, we observed *Lithobates catesbeianus*, *Lithobates sphenocephalus*, *Gastrophryne carolinensis*, and *Hyla cinerea* in brackish marshes in coastal North Carolina. These four species have been reported in brackish habitats previously [42, 75, 104–106] and the recurrence of these observations draws attention to the paucity of information explaining why some species are repeatedly observed inhabiting brackish wetlands while other closely related species are absent [39]. A particularly interesting contribution on this subject stems from our repeated field observations of abundant and thriving *Hyla cinerea* populations in salt marshes with salinities 450% higher than the expected larval LC_{50} concentration (as revealed by the meta-analysis). Indeed, these findings were inconceivable by the authors at the outset of the survey. While previous studies reported *Hyla cinerea* from saltmarshes along the Chesapeake Bay in Maryland in salinities up to 15 ppt [104], we found populations in salinities as high as 23 ppt, which is also the highest salinity that any North American frog species has been found to date (though Puerto Rican populations of *Rhinella marina*, *Eleutherodactylus coqui*, and *Lithobates grylio* come close at 20.5 ppt [107]).

Hopkins and Brodie (2015) recently updated Neill's 1958 review and provide a valuable and thorough review of all published observations of amphibians in saltwater [38, 39]. In their review, Hopkins and Brodie present a range of salinity tolerances revealed by experimental and field studies and suggest that the median maximum experimental salinity that can be tolerated by anuran amphibians falls between 9 ppt–12 ppt [39, 48, 108, 109]. Our meta-analysis refines and builds upon these estimates by providing an empirically derived range of survival probability estimates for each salinity and life stage. For example, at 9 ppt we may expect around 21% of eggs to survive, 27% of larvae to survive, and 50% of adults to survive – fundamental information for managing anuran populations across landscapes affected by salinization.

The meta-analysis underlines the fact that amphibians have different abilities to persist in saline environments according to life stage. Though most studies test the effects of salt on a single life stage, our meta-analysis integrates the findings of all of these studies to better understand how salt sensitivity changes through each life stage and provides a quantitative baseline and important context for our common garden experiments and field observations of anurans in salinities as high as 66% seawater. Broadly, all studies examined in our meta-analysis demonstrate declines in survivorship as salinity increased across each life stage, our analyses, which includes studies on 35 species representing 26 different genera across 10 families. We found that eggs are the most sensitive to osmotic stress across the anuran clade, followed by the larval stage, and adults are the least susceptible. The results of the meta-analysis indicate that the lethal experimental salt concentration in which 50% mortality (LC_{50}) is expected for eggs occurs at approximately 4.15 ppt for anuran eggs, 5.5 ppt for larvae, and 9.0 ppt for adults. Although the uncertainty in LC_{50} concentrations identified in the meta-analysis stems largely from differences in sample sizes (only three studies on adult frogs met our criteria for the meta-analysis), they might also reflect greater sensitivity during particular stages among species.

Embryos, for example, are expected to be more sensitive to external stressors than other stages because important developmental pathways are initiated during the early embryonic period and so perturbations at this stage may be teratogenic or fatal [110, 111]. It has been shown that pathogens (e.g., bacteria, endoparasites, or waterborne fungi), predators, ultraviolet radiation, and toxins can all have strong effects on embryonic survival, and induce effects that carry over to affect developmental outcomes in later life stages [57, 112–115]. Tadpoles are also expected to be more sensitive to water quality than are adults because they are obliged to the aquatic habitat. Larvae may be more tolerant of osmotic stress than embryos if they can increase the activity or concentration of ion pumps in the gills. However, tadpoles raised in saltwater tend to have stunted developmental rates and metamorphose at smaller sizes compared to freshwater-raised tadpoles [46, 52, 59, 64, 65, 116, 117], which can affect adult survival and reproductive success [118]. Adults, on the other hand, are less confined to aquatic environments and thus can reduce contact with stressful habitats via behavioral avoidance or dispersal. Additionally, adults can likely physiologically tolerate a greater degree of osmotic stress and/or desiccation by increasing urea in the blood [44, 119], altering cellular ion or water transport [58, 59, 120, 121], or adjusting the permeability of the skin [122, 123].

Common garden experiments

In the oviposition, hatching, and tadpole survivorship experiments, we find evidence for altered and adaptive responses to salinization across multiple life stages in *Hyla cinerea*. Specifically, we report differences in egg deposition patterns, hatching success, and tadpole survivorship between salt-exposed coastal and salt-naïve inland populations of North American green treefrogs (*Hyla cinerea*). We focus on reproductive behaviors, egg hatching, and tadpole viability because they are stages and traits that are highly vulnerable to environmental quality, and directly affect fitness and population viability [39, 48, 65, 124–128]. Female oviposition site selection directly affects the fitness of both the parents and the offspring, so decisions about oviposition sites should reflect an adaptive response. Therefore, we expected strong patterns of saltwater avoidance among both coastal and inland populations if salt were equally lethal to eggs and offspring from both inland and coastal populations [51, 61, 129, 130]. However, we found that coastal and inland frogs exhibited different patterns of oviposition site selection across the experimental salt gradient. Both inland and coastal pairs increasingly avoided saline water as salinity increased but inland frogs had greater response and did not deposit any eggs in salinities above ~12 ppt, whereas coastal pairs laid approximately 24% of their eggs in the highest salinities (Fig. 3). Additionally, eggs laid by coastal parents have higher probabilities of hatching in higher salinities and more coastal tadpoles survive in higher salinities when compared to inland-sourced conspecifics. Our inferences are based on experiments on three coastal and three inland populations and so should be extrapolated more broadly with caution. However, collectively our results provide evidence that some coastal populations of *Hyla cinerea* are responding adaptively to saltwater exposure across multiple life stages, which is contrary to expected outcomes given the general reputation of anuran amphibians as a highly salt sensitive order. Gomez-Mestre and Tejado report similar findings in *Bufo calamita*, the Natterjack Toad, in which embryos and tadpoles from brackish populations demonstrate higher survival compared to tadpoles from inland, freshwater populations

[45, 131]. Together these studies suggest that the ability to respond adaptively to saltwater exposure may be more possible than previously appreciated, and future studies may consider using comparative, common-garden approaches to not only determine how salt-exposure affects various endpoints, but also whether other species also exhibit population-level differences in salt tolerance across species and life stages.

The physiological mechanisms that explain why coastal pairs have relaxed salt avoidance behaviors, higher hatching success, and higher tadpole survivorship are likely to be numerous and spread across the different life stages. In adults, coastal male *Hyla cinerea* may have more viable and motile sperm in saline water. A recent study examined sperm survivorship and motility in *Hyla cinerea* located in Charleston, South Carolina (a coastal location) and found that 4 ppt saltwater reduced ability of sperm to survive and swim, but that study did not compare coastal and inland populations [51]. Alternatively, adult coastal females may increase the partitioning of yolk resources into eggs, or alter the egg coat matrix to provide additional protection against osmotic stressors compared to inland eggs. In tadpoles, coastal individuals may have an increased abundance of water channels (AQPs) and ion pumps (e.g., Na^+/K^+-ATPase) in the gills that enhance the ability to maintain internal water and ion balance, thus improving survival. Several studies have demonstrated that exposure to saltwater can increase the quantity and activity of sodium-potassium pumps in tadpole gills [58, 65, 132]. These hypotheses remain to be tested in coastal *Hyla cinerea*, leaving the exact mechanisms explaining the observed patterns undefined. Moreover, the adaptive processes that produce advantageous physiological responses also remain largely unknown.

There are three possible overlapping adaptive processes that may explain the divergence in responses that we observed between coastal and inland anuran populations; local adaptation, phenotypically plastic responses, and/or maternal effects. Local adaptation occurs when populations have higher fitness in their local environmental conditions compared to populations from other environments, and our results are consistent with expected outcomes if coastal populations are becoming locally adapted to tolerate elevated salt concentrations across different life stages [133]. Adaptive evolution is a well-appreciated process that can sustain or rescue populations facing strong selection gradients [134–138]. Yet several criteria must be met before local adaptation can be confirmed. "Adaptive" phenotypes must be shown to correlate positively with fitness, and the production of putatively adaptive phenotypes should be directly linked to specific environmental drivers, and studies on adaptive responses must demonstrate a genetic basis for differences observed among populations [139]. Our results are consistent with the expectations of the first two criteria, but we are not yet able to deduce whether there is a genetic basis for such changes.

Phenotypic plasticity (defined here as the ability to modulate phenotype in response to environmental cues) can also produce phenotypes that appear different and adaptive, yet may be genetically indistinguishable from other populations [139–141]. Because plasticity can promote adaptation, inhibit adaptation, or be the adaptive response itself, uncovering the role of phenotypic plasticity remains one of the most important challenges for understanding and predicting adaptive responses to climate change [31–34, 140, 142–145]. Indeed, some degree of phenotypic plasticity has been observed in nearly every trait that has been measured to date, which underlines the importance of examining the contribution of plasticity in studies of adaptive responses [32–34, 140, 144–146].

Maternal effects induced by environmental conditions experienced by the parents are also emerging as important factors that influence offspring fitness in different environments [147, 148]. Increased prevalence of maternally affected traits are expected when the environment experienced by the mother matches the environment experienced by the offspring [149], and in such situations, can explain up to 96% of the variation in improved offspring fitness in stressful environments [150].

The divergent responses that we present in this paper may be the production of either maternal effects, phenotypic plasticity, or local adaptation alone. However, some blend of these mechanisms is more likely. For example, exposure to saltwater during the ontogeny of coastal individuals may have initiated cascades of plastic responses that predisposed females from coastal populations toward salt tolerant responses. These responses may have transferred to offspring, which mixes plasticity with maternal effects. Alternatively, coastal individuals with increased ability to tolerate salt through enhanced plasticity may have been favored by selection. Presumably, selecting for more plastic individuals would gradually increase the overall amount of plasticity observed in coastal populations, which blends plasticity with genetic adaptation (*sensu* Baldwin effect) [151]. In reality, there are a multitude of possible mechanistic combinations as plasticity, local adaptation, and maternal effects can be reciprocal processes that serve as both the product and raw material for selection and adaptation. Future research should prioritize discerning how adaptive evolution, phenotypic plasticity, and maternal effects are interwoven to produce different responses to environmental stressors especially in organisms with complex life cycles. A more complete understanding of all contributing processes will help managers identify thresholds of tolerance, detect vulnerable populations, and determine which organisms are likely to successfully tolerate novel stressors and persist in their environments.

Despite the consistent differences in behavior, embryo, and larval survivorship we observed between inland and coastal populations, our results indicate that all populations and life stages of *Hyla cinerea* (coastal and inland populations) are salt-sensitive. Frog pairs laid the majority of eggs into freshwater in all populations; saltwater negatively affected hatching rates across all populations, and saltwater reduced survivorship for both coastal and inland tadpoles. While we have focused on the degree to which these responses differed among populations as indications of adaptive responses, we believe that it should be noted that anurans on the whole, remain an osmotically sensitive group of organisms even in chronically salt-exposed populations. The continued preference for, and higher performance in, freshwater, even among coastal populations, may indicate that thresholds of saltwater tolerance exist.

Conclusions

This study provides the following insights: First, our meta-analysis offers a quantitative baseline for salt tolerance in anurans and provides important context for future field observations and experimental studies exploring saltwater tolerance in anurans. The meta-analysis also shows that generally, anurans are salt-sensitive across species and across life stages and are therefore likely to be adversely affected by progressive salinization of freshwater systems. Second, we show different sensitivities and responses to salt stress across life stages and across populations, significant information for future studies and management. Third, we provide initial evidence that despite their sensitivity, some anuran species (*Hyla cinerea*) have populations that are able to respond adaptively to salt stress across different life stages. Though these findings are an encouraging indication that some frog populations may persist through salinization, our results also illuminate that much more remains to be known. Key unknowns include the physiological mechanisms and adaptive processes that underlie salt tolerance in anurans, determining whether we can expect adaptive responses to match the pace and intensity of environmental change (i.e., define the limits of tolerance and rates of adaptation), and exploring the factors that govern amphibian distributions across salt-invaded landscapes (i.e., why only 4 out of the 17 possible species occur in brackish wetlands).

Testing multiple mechanistic hypotheses about adaptive processes (e.g., maternal effects, genetic evolution, and phenotypic plasticity) in ecological time in wild macro-organisms has remained an empirical challenge. Yet identifying populations with complex life cycles that

demonstrate divergent responses to an environmental stressor across life stages (such as coastal frog populations adapting to saline environments) may provide unique and valuable opportunities to empirically address questions about the etiology of adaptive and non-adaptive responses, how novel adaptive phenotypes emerge, and how population and demographic dynamics interact with adaptive processes.

Acknowledgements
We thank members of McCoy lab group, K. McCoy, A. Stuckert, and J. Touchon for their thoughtful insights during the development of this work. We also extend thanks to A. Stuckert, T. McFarland and C. Thaxton for field and laboratory assistance. This manuscript was improved by helpful reviews by Gareth Hopkins and Ilaria Bernabò.

Funding
This work was funded by North Carolina Sea Grant (Project No. 2014-R/14-HCE-3), National Science Foundation Doctoral Dissertation Improvement Grant (#1701690) awarded to McCoy and Albecker, the Graduate Women in Science Nell Mondy Fellowship, as well as research grants from the North Carolina Herpetological Society, East Carolina University's Coastal Maritime Council, and Explorer's Club.

Authors' contributions
MAA and MWM conceived the study. MAA carried out experimentation and data extraction. MAA and MWM analyzed and interpreted results, and contributed to writing the manuscript. Both authors read and approved the final manuscript.

Competing interests
The authors declare that they have no competing interest.

References
1. Meehl GA, Washington WM, Collins WD, Arblaster JM, Hu A, Buja LE, Strand WG, Teng H. How much more climate change and sea level rise? Science. 2005;307:1769–72.
2. Domingues CM, Church JA, White NJ, Gleckler PJ, Wijffels SE, Barker PM, Dunn JR. Improved estimates of upper-ocean warming and multi-decadal sea-level rise. Nature. 2008;453:1090–3.
3. Nicholls RJ, Tol RS. Impacts and responses to sea-level rise: a global analysis of the SRES scenarios over the twenty-first century. Philos Trans A Math Phys Eng Sci. 2006;364:1073–95.
4. Church JA, Clark PU, Cazenave A, Gregory JM, Jevrejeva S, Levermann A, Merrifield MA, Milne GA, Nerem RS, Nunn PD, et al. IPCC fifth assessment report (AR5), climate change. Phys Sci Basis. 2013;2013:1–124.
5. Scavia D, Field JC, Boesch DF, Buddemeier RW, Burkett V, Cayan DR, Fogarty M, Harwell MA, Howarth RW, Mason C, et al. Climate change impacts on U. S. coastal and marine ecosystems. Estuaries. 2002;25:149–64.
6. Senior CA, Jones RG, Lowe JA, Durman CF, Hudson D. Predictions of extreme precipitation and sea-level rise under climate change. Philos Trans A Math Phys Eng Sci. 2002;360:1301–11.
7. Rahmstorf S, Perrette M, Vermeer M. Testing the robustness of semi-empirical sea level projections. Climate Dynam. 2012;39:861–75.
8. Kemp AC, Horton BP, Donnelly JP, Mann ME, Vermeer M, Rahmstorf S. Climate related sea level variations over the past two millennia. PNAS. 2011; 108:11017–22.
9. DaLaune RD, Pezeshki SR. The influence of subsidence and saltwater intrusion on coastal marsh stability: Louisiana gulf coast. J Coast Res. 1994;12:77–89.

10. Abrams PA. Implications of dynamically variable traits for identifying, classifying, and measuring direct and indirect effects in ecological communities. Am Nat. 1995:112–34.

11. Michener WK, Blood ER, Bildstein KL, Brinson MM, Gardner LR. Climate change, hurricanes and tropical storms, and rising sea level in coastal wetlands. Ecol Appl. 1997;7:770–801.

12. Loaiciga HA. Climate change and ground water. Ann Assoc Am Geogr. 2003;93:30–41.

13. Day JW, Christian RR, Boesch DM, Yáñez-Arancibia A, Morris J, Twilley RR, Naylor L, Schaffner L, Stevenson C. Consequences of climate change on the Ecogeomorphology of coastal wetlands. Estuar Coasts. 2008;31:477–91.

14. Meyssignac B, Cazenave A. Sea level- a review of present day and recent past changes and variability. J Geodyn. 2012;58:96–109.

15. Williams SJ. Sea level rise implications for coastal regions. J Coast Res. 2013; 63:184–96.

16. Baldwin AH, Mendelssohn IA. Effects of salinity and water level on coastal marshes: an experimental test of disturbance as a catalyst for vegetation change. Aquat Bot. 1998;61:255–68.

17. Williams K, Ewel KC, Stumpf RP, Putz FE, Workman TW. Sea-level rise and coastal forest retreat on the west coast of florida, USA. Ecology. 1999;80: 2045–63.

18. Geddes NA, Mopper S. Effects of environmental salinity on vertebrate florivory and wetland communities. Nat Areas J. 2006;26:31–7.

19. Kopp RE, Horton BP, Kemp AC, Tebaldi C. Past and future sea-level rise along the coast of North Carolina, USA. Clim Change. 2015;132:693–707.

20. Knighton AD, Mills K, Woodroffe CD. Tidal-creek extension and saltwater intrusion in northern Australia. Geology. 1991;19:831–4.

21. Morris JT, Sundarshwar PV, Nietch CT, Kjerfve B, Cahoon DR. Responses of coastal wetlands to rising sea level. Ecology. 2002;83:2869–77.

22. Hamer AJ, McDonnell MJ. Amphibian ecology and conservation in the urbanising world- a review. Biol Conserv. 2008;141:2432–49.

23. Foden WB, Mace GM, Vié J-C, Angulo A, Butchart SHM, DeVantier L, Dublin HT, Gutsche A, Stuart SN, Turak E. Species susceptibility to climate change impacts. In: Vie J-C, Hilton-Taylor C, Stuart SN, editors. Wildlife in a changing world: an analysis of the 2008 IUCN red list of threatened species. Volume 1. Barcelona, Spain; 2009. p. 77–88.

24. Reed TE, Schindler DE, Waples RS. Interacting effects of phenotypic plasticity and evolution on population persistence in a changing climate. Conserv Biol. 2011;25:56–63.

25. Anderson JT, Perera N, Chowdhury B, Mitchell-Olds T. Microgeographic patterns of genetic divergence and adaptation across environmental gradients in Boechera stricta (Brassicaceae). Am Nat. 2015;186:S60–73.

26. Brady SP. Road to evolution? Local adaptation to road adjacency in an amphibian (Ambystoma maculatum). Sci Rep. 2012;2

27. Fraser DJ, Weir LK, Bernatchez L, Hansen MM, Taylor EB. Extent and scale of local adaptation in salmonid fishes: review and meta-analysis. Heredity. 2011;106:404–20.

28. Lamichhaney S, Barrio AM, Rafati N, Sundström G, Rubin C-J, Gilbert ER, Berglund J, Wetterbom A, Laikre L, Webster MT. Population-scale sequencing reveals genetic differentiation due to local adaptation in Atlantic herring. Proc Natl Acad Sci. 2012;109:19345–50.

29. Mopper S, Strauss SY. Genetic structure and local adaptation in natural insect populations: effects of ecology, life history, and behavior: Springer eScience & Business Media; 2013.

30. Reznick DN, Ghalambor CK. The population ecology of contemporary adaptations: what empirical studies reveal about the conditions that promote adaptive evolution. Genetica. 2001;112:183–98.

31. Pfennig DW, Wund MA, Snell-Rood EC, Cruickshank T, Schlichting CD, Moczek AP. Phenotypic plasticity's impacts on diversification and speciation. Trends Ecol Evol. 2010;25:459–67.

32. Wund MA. Assessing the impacts of phenotypic plasticity on evolution. Integr Comp Biol. 2012;52:5–15.

33. Nonaka E, Svanbäck R, Thibert-Plante X, Englund G, Brännström Å. Mechanisms by which phenotypic plasticity affects adaptive divergence and ecological speciation. Am Nat. 2015;186:E126–43.

34. Hendry AP. Key questions on the role of phenotypic plasticity in eco-evolutionary dynamics. J Hered. 2015; esv060

35. Reed TE, Waples RS, Schindler DE, Hard JJ, Kinnison MT. Phenotypic plasticity and population viability: the importance of environmental predictability. Proc R Soc Lond B Biol Sci. 2010;277

36. Vitt LJ, Caldwell JP, Wilbur HM, Smith DC. Amphibians as harbingers of decay. Bioscience. 1990;40:418.

37. Carignan V, Villard M-A. Selecting indicator species to monitor ecological integrity: a review. Environ Monit Assess. 2002;78:45–61.

38. Neill WT. The occurrence of amphibians and reptiles in saltwater areas, and a bibliography. Bull Mar Sci. 1958;8:1–97.

39. Hopkins GR, Brodie JED. Occurrence of amphibians in saline habitats: a review and evolutionary perspective. Herpetol Monogr. 2015;29:1–27.

40. Gibbons JW, Winne CT, Scott DE, Willson JD, Glaudas X, Andrews KM, Todd BD, Fedewa LA, Wilkinson L, Tsaliagos RN, et al. Remarkable amphibian biomass and abundance in an isolated wetland: implications for wetland conservation. Conserv Biol. 2006;20:1457–65.

41. McCoy MW, Barfield M, Holt RD. Predator shadows: complex life histories as generators of spatially patterned indirect interactions across ecosystems. Oikos. 2009;118:87–100.

42. Christman SP. Geographic variation for salt water tolerance in the frog Rana sphenocephala. Copeia. 1974;1974:773–8.

43. Gibbons JW, Coker JW. Herpetofaunal colonization patterns of atlantic coast barrier islands. Am Midl Nat. 1978;99:219–33.

44. Balinsky JB. Adaptation of nitrogen metabolism to hyperosmotic environment in amphibia. J Exp Zool. 1981;215:335–50.

45. Gomez-Mestre I, Tejado M. Local adaptation of an anuran amphibian to osmotically stressful environments. Evolution. 2003;57:1889–99.

46. Wu C-S, Kam Y-C. Effects of salinity on the survival, growth, development, and metamorphosis of Fejervarya limnocharis tadpoles living in brackish water. Zoolog Sci. 2009;26:476–82.

47. Gordon MS, Tucker VA. Osmotic regulation in the tadpoles of the crab-eating frog (Rana cancrivora). J Exp Biol. 1965;42:437–45.

48. Gordon MS, Schmidt-Nielsen K, Kelly HM. Osmotic regulation in the crab-eating frog (Rana Cancrivora). J Exp Biol. 1961;38:659–78.

49. Gordon MS. Intracellular osmoregulation in skeletal muscle during salinity adaptation in two species of toads. Biol Bull. 1965;128:218–29.

50. Gordon MS. Osmotic regulation in the green toad (Bufo viridis). J Exp Biol. 1962;39:261–70.

51. Wilder AE, Welch AM. Effects of salinity and pesticide on sperm activity and Oviposition site selection in green Treefrogs, Hyla cinerea. Copeia. 2014; 2014:659–67.

52. Hsu W-T, Wu C-S, Lai J-C, Chiao Y-K, Hsu C-H, Kam Y-C. Salinity acclimation affects survival and metamorphosis of crab-eating frog tadpoles. Herpetologica. 2012;68:14–21.

53. Wu CS, Gomez-Mestre I, Kam YC. Irreversibility of a bad start: early exposure to osmotic stress limits growth and adaptive developmental plasticity. Oecologia. 2012;169:15–22.

54. Sanzo D, Hecnar SJ. Effects of road de-icing salt (NaCl) on larval wood frogs (Rana sylvatica). Environ Pollut. 2006;140:247–56.

55. Li N, Phummisutthigoon S, Charoenphandhu N. Low salinity increases survival, body weight and development in tadpoles of the Chinese edible frog Hoplobatrachus rugulosus. Aquacult Res. 2015;

56. Haramura T. Salinity tolerance of eggs of Buergeria japonica (Amphibia, Anura) inhabiting coastal areas. Zoolog Sci. 2007;24:820–3.

57. Touchon JC. Hatching plasticity in two temperate anurans: responses to pathogen and predation cues. Can J Zool. 2006;84:556–63.

58. Wu CS, Yang WK, Lee TH, Gomez-Mestre I, Kam YC. Salinity acclimation enhances salinity tolerance in tadpoles living in brackish water through increased Na(+), K(+) -ATPase expression. J Exp Zool A Ecol Genet Physiol. 2014;321:57–64.

59. Uchiyama M, Yoshizawa H. Salinity tolerance and structure of external and internal gills in tadpoles of the crab-eating frog, Rana Cancrivora. Cell Tissue Res. 1992;267:35–44.

60. Dietz TH, Alvarado RH. Na and Cl transport across gill chamber epithelium of Rana catesbeiana tadpoles. Am J Physiol. 1974;226:764–70.

61. Haramura T. Use of oviposition sites by a Rhacophorid frog inhabiting a coastal area in Japan. J Herpetol. 2011;45:432–7.

62. Smith MJ, Shreiber ESG, Scroggie MP, Kohout M, Ough K, Potts J, Lennie R, Turnbull D, Jin C, Clancy T. Associations between anuran tadpoles and salinity in a landscape mosaic of wetlands impacted by secondary salinisation. Freshw Biol. 2006;52:75–84.

63. Dougherty CK, Smith GR. Acute effects of road de-icers on the tadpoles of three anurans. Appl Herpetol. 2006;3:87–93.

64. Christy MT, Dickman CR. Effects of salinity on tadpoles of the green and golden bell frog (Litoria aurea). Amphibia-Reptilia. 2002;23:1–11.

65. Bernabò I, Bonacci A, Coscarelli F, Tripepi M, Brunelli E. Effects of salinity stress on Bufo Balearicus and *Bufo bufo* tadpoles: tolerance, morphological gill alterations and Na(+)/K(+)-ATPase localization. Aquat Toxicol. 2013;132-133:119–33.

66. Craft C, Clough J, Ehman J, Joye S, Park R, Pennings S, Guo H, Machmuller M. Forecasting the effects of accellerated sea-level rise on tidal marsh ecosystem services. Front Ecol Environ. 2009;7:73–8.

67. Kemp AC, Horton BP, Culver SJ, Corbett DR, Ovd P, Gehrels WR, Douglas BC, Parnell AC. Timing and magnitude of recent accelerated sea-level rise. Geology. 2009;37:1035–8.

68. Titus JG, Richman C. Maps of lands vulnerable to sea level rise: modeled elevations along the US Atlantic and gulf coasts. Climate Res. 2001;18:205–28.

69. Parkinson RW. Sea-level rise and the fate of tidal wetlands. J Coast Res. 1994;10:987–9.

70. Hintz WD, Relyea RA. Impacts of road deicing salts on the early-life growth and development of a stream salmonid: salt type matters. Environ Pollut. 2017;

71. Crother BI, Fontenot CL. Amphibian and reptile monitoring in the Ponchartrain-Maurepas region. In: Lake Pontchartrain Basin research program (PBRP); 2006. p. 35.

72. Heyer R, Donnelly MA, Foster M, Mcdiarmid R. Measuring and monitoring biological diversity: standard methods for amphibians: Smithsonian Institution; 2014.

73. Rader RB, Batzer DP, Wissinger SA. Bioassessment and management of north American freshwater wetlands: Wiley; 2001.

74. Tuberville TD, Willson JD, Dorcas ME, Gibbons JW. Herpetofaunal species richness of southeastern national parks. Southeast Nat. 2005;4:537–69.

75. Brown ME, Walls SC. Variation in salinity tolerance among larval anurans: implications for community composition and the spread of an invasive, non-native species. Copeia. 2013;2013:543–51.

76. Wells KD. The ecology and behavior of amphibians. Chicago: The University of Chicago Press; 2007.

77. Gosner KL. A simplified table for staging anuran embryos and larvae with notes on identification. Herpetologica. 1960:183–90.

78. Gordon MS, Tucker VA. Further observations on the physiology of salinity adaptation in the crab-eating frog (Rana Cancrivora). J Exp Biol. 1968;49:185–93.

79. R: a language and environment for statistical computing. In: R Core development team. 3.2.3 ed. Vienna: R Foundation for Statistical Computing; 2014.

80. Su Y-S, Yajima M. R2jags: using R to run 'JAGS': R package version 0.5–7; 2015.

81. Plummer M. Rjags: Bayesian graphical models using MCMC: R package version 3–15; 2015.

82. Plummer M, Best N, Cowles K, Vines K. CODA: convergence diagnosis and output analysis for MCMC. R News. 2006;6(1):7–11.

83. Gelman A, Carlin JB, Stern HS, Rubin DB. Bayesian data analysis. Texts in statistical science series. Boca Raton: Chapman & Hall/CRC; 2004.

84. Bennett JE, Racine-Poon A, Wakefield JC. MCMC for nonlinear hierarchical models. London, UK: Chapman and Hall; 1996.

85. Nicholls RJ, Cazenave A. Sea-level rise and its impact on coastal zones. Science. 2010;328:1517–20.

86. Bradshaw WE, Holzapfel CM. Climate change. Evolutionary response to rapid climate change. Science. 2006;312:1477–8.

87. Chen IC, Hill JK, Ohlemuller R, Roy DB, Thomas CD. Rapid range shifts of species associated with high levels of climate warming. Science. 2011;333:1024–6.

88. Davis MB, Shaw RG, Etterson JR. Evolutionary responses to changing climate. Ecology. 2005;86:1704–14.

89. Dawson TP, Jackson ST, House JI, Prentice IC, Mace GM. Beyond predictions: biodiversity conservation in a changing climate. Science. 2011;332:53–8.

90. Harley CD. Climate change, keystone predation, and biodiversity loss. Science. 2011;334:1124–7.

91. Moritz C, Agudo R. The future of species under climate change: resilience or decline? Science. 2013;341:504–8.

92. Parmesan C. Ecological and evolutionary responses to recent climate change. Annu Rev Ecol Evol Syst. 2006:637–69.

93. Walther G-R, Post E, Convey P, Menzel A, Parmesan C, Beebee TJC, Fromentin J-M, Hoegh-Guldberg O, Bairlein F. Ecological responses to recent climate change. Nature. 2002;416:389–95.

94. Thomas CD, Cameron A, Green RE, Bakkenes M, Beaumont LJ, Collingham YC, Erasmus BF, De Siqueira MF, Grainger A, Hannah L. Extinction risk from climate change. Nature. 2004;427:145–8.

95. Chown SL. Trait-based approaches to conservation physiology: forecasting environmental change risks from the bottom up. Philos Trans R Soc Lond B Biol Sci. 2012;367:1615–27.

96. Lewis OT. Climate change, species-area curves and the extinction crisis. Philos Trans R Soc Lond B Biol Sci. 2006;361:163–71.

97. Maclean IM, Wilson RJ. Recent ecological responses to climate change support predictions of high extinction risk. Proc Natl Acad Sci U S A. 2011;108:12337–42.

98. Schwartz MW, Iverson LR, Prasad AM, Matthews SN, O'Connor RJ. Predicting extinctions as a result of climate change. Ecology. 2006;87:1611–5.

99. Stuart SN, Chanson JS, Cox NA, Young BE, Rodrigues AS, Fischman DL, Waller RW. Status and trends of amphibian declines and extinctions worldwide. Science. 2004;306:1783–6.

100. Traill LW, Lim ML, Sodhi NS, Bradshaw CJ. Mechanisms driving change: altered species interactions and ecosystem function through global warming. J Anim Ecol. 2010;79:937–47.

101. Davis MB, Shaw RG. Range shifts and adaptive responses to quaternary climate change. Science. 2001;292:673–9.

102. Holt R, Gomulkiewicz R. Conservation implications of niche conservatism and evolution in heterogeneous environments. In: Evolutionary conservation biology. Volume 2004: Cambridge University Press; 2004. p. 244–64.

103. Lawler JJ, Shafer SL, Bancroft BA, Blaustein AR. Projected climate impacts for the amphibians of the western hemisphere. Conserv Biol. 2010;24:38–50.

104. Hardy JD. Notes on the distribution of *Mycrohyla carolinensis* in southern Maryland. Herpetologica. 1953;8:162–6.

105. Hardy JDJ. Amphibians of the Chesapeake Bay region. Chesapeake Sci. 1972;13:S123–8.

106. Gunzburger MS. Reproductive ecology of the green treefrog (*Hyla cinerea*) in northwestern Florida. Am Midl Nat. 2006;155:321–8.

107. Rios-López N. Effects of increased salinity on tadpoles of two anurans from a Caribbean coastal wetland in relation to their natural abundance. Amphibia-Reptilia. 2008;29:7–18.

108. Munsey LD. Salinity tolerance of the african Pipid frog, *Xenopus laevis*. Copeia. 1972;1972:584–6.

109. Ruibal R. The ecology of a brackish water population of *Rana pipiens*. Copeia. 1959;1959:315–22.

110. Wilbur HM. Complex life cycles. Annu Rev Ecol Syst. 1980;11:67–93.

111. Meteyer CU, Cole RA, Converse KA, Docherty DE, Wolcott M, Helgen JC, Levey R, Eaton-Poole L, Burkhart JG. Defining anuran malformations in the context of a developmental problem. J Iowa Acad Sci. 2000;107:72–8.

112. Burkhart JG, Helgen JC, Fort DJ, Gallagher K, Bowers D, Propst TL, Gernes M, Magner J, Shelby MD, Lucier G. Induction of mortality and malformation in *Xenopus laevis* embryos by water sources associated with field frog deformities. Environ Health Perspect. 1998;106:841.

113. Grant KP, Licht RL. Effects of ultraviolet radiation on life-history stages of anurans from Ontario, Canada. Can J Zool. 1995;73:2292–301.

114. Kiesecker JM, Blaustein AR. Influences of egg laying behavior on pathogenic infection of amphibian eggs. Conserv Biol. 1997;11:214–20.

115. Rohr JR, McCoy KA. A qualitative meta-analysis reveals consistent effects of atrazine on freshwater fish and amphibians. Environ Health Perspect. 2010;2010:20–32.

116. Wood L, Welch AM. Assessment of interactive effects of elevated salinity and three pesticides on life history and behavior of southern toad (*Anaxyrus terrestris*) tadpoles. Environ Toxicol Chem. 2015;34:667–76.

117. Langhans M, Peterson B, Walker A, Smith GR, Rettig JE: Effects of salinity on survivorship of wood frog (*Rana sylvatica*) tadpoles. 2009.

118. Berven KA. Factors affecting population fluctuations in larval and adult stages of the wood frog (*Rana sylvatica*). Ecology. 1990;71:1599–608.

119. Shoemaker V, Hillman S, Hillyard S, Jackson D, McClanahan L, Withers P, Wygoda M. Exchange of water, ions, and respiratory gases in terrestrial amphibians: Environmental physiology of the amphibians; 1992. p. 125–50.

120. Uchiyama M, Konno N. Hormonal regulation of ion and water transport in anuran amphibians. Gen Comp Endocrinol. 2006;147:54–61.

121. Konno N, Hyodo S, Matsuda K, Uchiyama M. Effect of osmotic stress on expression of a putative facilitative urea transporter in the kidney and urinary bladder of the marine toad, *Bufo marinus*. J Exp Biol. 2006;209:1207–16.

122. McClanahan L Jr, Stinner JN, Shoemaker VH. Skin lipids, water loss, and energy metabolism in a south American tree frog (*Phyllomedusa sauvagei*). Physiol Zool. 1978;51:179–87.

123. Lillywhite HB. Water relations of tetrapod integument. J Exp Biol. 2006;209:202–26.

124. Roberts J: Variations in salinity tolerance in the Pacific Treefrog, *Hyla regilla*. Oregon [dissertation] Corvallis, OR: Oregon State University 1970.

125. Chinathamby K, Reina RD, Bailey PC, Lees BK. Effects of salinity on the survival, growth and development of tadpoles of the brown tree frog, *Litoria ewingii*. Aust J Zool. 2006;54:97–105.

126. Brand AB, Snodgrass JW, Gallagher MT, Casey RE, Van Meter R. Lethal and sublethal effects of embryonic and larval exposure of *Hyla versicolor* to stormwater pond sediments. Arch Environ Contam Toxicol. 2010;58:325–31.

127. Petranka JW, Doyle EJ. Effects of road salts on the composition of seasonal pond communities: can the use of road salts enhance mosquito recruitment? Aquat Ecol. 2010;44:155–66.

128. Thirion J-M. Salinity of the reproduction habitats of the western spadefoot toad *Pelobates cultripes* (cuvier, 1829), along the atlantic coast of France. Herpetozoa. 2014;27:13–20.

129. Rieger JF, Binckley CA, Resetarits WJ Jr. Larval performance and oviposition site preference along a predation gradient. Ecology. 2004;85:2094–9.

130. Refsnider JM, Janzen FJ. Putting eggs in one basket: ecological and evolutionary hypotheses for variation in oviposition-site choice. Annu Rev Ecol Evol Syst. 2010;41:39–57.

131. Gomez-Mestre I, Tejado M. Adaptation or Exaptation? An experimental test of hypotheses on the origin of salinity tolerance in *Bufo calamita*. J Evol Biol. 2005;18:847–55.

132. Havird JC, Henry RP, Wilson AE. Altered expression of Na(+)/K(+)-ATPase and other osmoregulatory genes in the gills of euryhaline animals in response to salinity transfer: a meta-analysis of 59 quantitative PCR studies over 10 years. Comp Biochem Physiol Part D Genomics Proteomics. 2013;8:131–40.

133. Savolainen O, Lascoux M, Merila J. Ecological genomics of local adaptation. Nat Rev Genet. 2013;14:807–20.

134. Martin G, Aguile R, Ramsayer J, Kaltz O, Ronce O. The probability of evolutionary rescue: towards a quantitative comparison between theory and evolution experiments. Philos Trans R Soc B. 2013;368:20120088.

135. Bourne EC, Bocedi G, Travis JM, Pakeman RJ, Brooker RW, Schiffers K. Between migration load and evolutionary rescue: dispersal, adaptation and the response of spatially structured populations to environmental change. Proc R Soc Lond B Biol Sci. 2014;281:20132795.

136. Gonzalez A, Ronce O, Ferriere R, Hochberg ME. Evolutionary rescue: an emerging focus at the intersection between ecology and evolution. Philos Trans R Soc Lond B Biol Sci. 2013;368:20120404.

137. Carlson SM, Cunningham CJ, Westley PAH. Evolutionary rescue in a changing world. Trends Ecol Evol. 2014;29:521–30.

138. Bell G. Evolutionary rescue and the limits of adaptation. Philosophical Transactions of the Royal Society of London B: Biological Sciences. 2013;368: 20120080.

139. Merilä J, Hendry AP. Climate change, adaptation, and phenotypic plasticity: the problem and the evidence. Evol Appl. 2014;7:1–14.

140. Urban MC, Richardson JL, Reidenfelds NA. Plasticity and genetic adaptation mediate amphibian and reptile responses to climate change. Evol Appl. 2014;7:88–103.

141. Urban MC, Bocedi G, Hendry AP, Mihoub J-B, Pe'er G, Singer A, Bridle JR, Crozier LG, De Meester L, Godsoe W, et al. Improving the forecast for biodiversity under climate change. Science. 2016;353

142. Chevin LM, Lande R, Mace GM. Adaptation, plasticity, and extinction in a changing environment: towards a predictive theory. PLoS Biol. 2010;8: e1000357.

143. Lande R. Adaptation to an extraordinary environment by evolution of phenotypic plasticity and genetic assimilation. J Evol Biol. 2009;22:1435–46.

144. Whitman DW, Agrawal AA. What is phenotypic plasticity and why is it important? 2009. p. 1–63.

145. Murren CJ, Auld JR, Callahan H, Ghalambor CK, Handelsman CA, Heskel MA, Kingsolver J, Maclean HJ, Masel J, Maughan H. Constraints on the evolution of phenotypic plasticity: limits and costs of phenotype and plasticity. Heredity. 2015;115:293–301.

146. Forsman A. Rethinking phenotypic plasticity and its consequences for individuals, populations and species. Heredity. 2015;115:276–84.

147. Marshall DJ, Uller T. When is a maternal effect adaptive? Oikos. 2007;116: 1957–63.

148. Räsänen K, Kruuk L. Maternal effects and evolution at ecological timescales. Funct Ecol. 2007;21:408–21.

149. Kirkpatrick M, Lande R. The evolution of maternal characters. Evolution. 1989;1989:485–503.

150. Chirgwin E, Marshall DJ, Sgrò CM, Monro K. The other 96%: can neglected sources of fitness variation offer new insights into adaptation to global change? Evol Appl. 2016;10:267–75.

151. Crispo E. The Baldwin effect and genetic assimilation: revisiting two mechanisms of evolutionary change mediated by phenotypic plasticity. Evolution. 2007;61:2469–79.

9

No speed dating please! Patterns of social preference in male and female house mice

Miriam Linnenbrink*⬡ and Sophie von Merten

Abstract

Background: In many animal species, interactions between individuals of different sex often occur in the context of courtship and mating. During these interactions, a specific mating partner can be chosen. By discriminating potential mates according to specific characteristics, individuals can increase their evolutionary fitness in terms of reproduction and offspring survival. In this study, we monitored the partner preference behaviour of female and male wild house mice (*Mus musculus domesticus*) from populations in Germany (G) and France (F) in a controlled cage setup for 5 days and six nights. We analysed the effects of individual factors (e.g. population origin and sex) on the strength of preference (selectivity), as well as dyadic factors (e.g. neutral genetic distance and major histocompatibility complex (MHC) dissimilarity) that direct partner preferences.

Results: Selectivity was stronger in mice with a pure population background than mixed individuals. Furthermore, female mice with a father from the German population had stronger selectivity than other mice. In this group, we found a preference for partners with a larger dissimilarity of their father's and their partner's MHC, as assessed by sequencing the H2-Eß locus. In all mice, selectivity followed a clear temporal pattern: it was low in the beginning and reached its maximum only after a whole day in the experiment. After two days, mice seemed to have chosen their preferred partner, as this choice was stable for the remaining four days in the experiment.

Conclusions: Our study supports earlier findings that mate choice behaviour in wild mice can be paternally influenced. In our study, preference seems to be potentially associated with paternal MHC distance. To explain this, we propose familial imprinting as the most probable process for information transfer from father to offspring during the offspring's early phase of life, which possibly influences its future partner preferences. Furthermore, our experiments show that preferences can change after the first day of encounter, which implies that extended observation times might be required to obtain results that allow a valid ecological interpretation.

Keywords: Social preference, Decision making process, Mate choice, *Mus musculus domesticus*, MHC, Familial imprinting

Background

In social animals, individuals interact with each other in a broad range of different situations. Interactions between individuals of different sex often occur in the context of courtship, pair bonding, and mating. A preference for some possible social partners over others can ultimately lead to mate choice. The evolution of mate choice is assumed to be driven by several mechanisms [1], such as preferences for direct or indirect phenotypic benefits and genetic correlations between mating preferences and preferred traits [2].

Selective mating is meant to increase the evolutionary fitness of individuals in terms of reproduction and offspring survival [1, 3–6]. Consequently, different mating strategies have evolved. Assortative mating results from the reproduction of phenotypically or genotypically matching mates and promotes population differentiation [7, 8]. In contrast, disassortative mating occurs when individuals prefer dissimilar mates compared to neutral expectations [9–12]. The latter strategy maintains or even increases genetic variability and counteracts possible disadvantages due to inbreeding depression (e.g. [13]).

An important basis for mate choice behaviour lies in the evolution of mechanisms to recognize conspecifics' characteristics and thus identify potential mates. In mice (*Mus musculus sp.*), the recognition of individuals,

* Correspondence: linnenbrink@evolbio.mpg.de
Max-Planck Institute for Evolutionary Biology, Plön, Germany

families, and populations is mainly regulated by two sensory systems: olfaction and vocalisation. Chemical signals used are volatiles (i.e. pheromones), peptides (e.g. of the major histocompatibility complex, MHC), and proteins (e.g. major urinary proteins) [14–16]. Acoustic communication occurs via ultrasonic vocalisation [16–18].

The MHC (officially termed "H2" in mice [19]) is a highly polymorphic gene complex that encodes many proteins with key roles in the adaptive immune system. Since Yamazaki and colleagues [20] detected MHC-related mating preferences in laboratory mice, many studies have reported an influence of MHC loci on mate choice in nearly all classes of vertebrates [21]. Potential reasons for MHC being involved in mate discrimination are kin recognition and the enhancement of the offspring's immune competence, which occurs by increasing either MHC diversity or dissimilarity by choosing a compatible mate [22]. An important pre-requisite for MHC-based mate choice in mice is the ability to identify and discriminate potential partners based on MHC alleles. Several studies have analysed the influence of MHC on partner choice for house mice. They mostly support the hypothesis of disassortative mating [23, 24] and raised evidence for familial imprinting [23, 25]. Familial imprinting is the non-genetical transmission (i.e. learning is involved) of preferences (e.g. preference for food, home area or mates (see Immelmann [26] for review) during the early phase of life from mostly one parent as reference to its offspring.

Wild mice offer a perfect model system to study the behavioural and genetic basis of mate choice, since they are still far more natural than common laboratory mouse strains in both aspects. Due to the vast amount of studies on lab mice, numerous genetic tools are available and can be also applied for studies in wild mice. Examples of such study populations are two originally wild caught populations of *M. m. domesticus* (one population from France and one from Germany), which separated about 3000 years ago. This separation is reflected in the divergence of the nuclear genome and gene expression differences [27–29], as well as in ultrasonic vocalisation [30]. Montero and colleagues [31] studied the degree of mutual mate recognition according to population origin under semi-natural conditions and identified complex mating patterns in these two populations. Assortative mating according to population background was only observed when mice of the single populations could get familiar with each other before individuals of both populations had the chance to meet. Further, the mating patterns observed were based on paternally influenced mate preferences, such that mice with a father from the French population preferred mating with a partner from that population, individuals fathered by a German male preferred mating with an individual from the German

population. The authors suggested genomic or familial imprinting as being involved.

The aim of the present study is to reveal possible factors that determine the mating patterns found by Montero et al. [31] in more detail. We studied the exact two mouse populations mentioned to shed light on the evolution of mate choice in the early phase of population differentiation. By using a controlled cage setup, we were able to follow single (focal) individuals during their decision making processes and analyse the persistency of their choices. To offer the same possible mates as in the study by Montero and colleagues [31], focal mice were allowed to choose between four partners of either the same or different population origins (France and Germany) or reciprocal crosses of both. Focal mice were females and males of the same four genotypes. We aimed to confirm the findings of preferences for paternally matching population backgrounds. In order to do so we investigated three different aspects of mouse behaviour in our setup, (i) general activity of the focal mice, (ii) their degree of selectivity (i.e. strength of preference independent of its direction) and underlying temporal changes when several potential partners where presented, and (iii) factors correlating with the direction of preference, including the effect of MHC on preference behaviour, which might possibly serve as evidence for MHC-driven partner preference.

Methods
Animals

The mice used for this study originate from two *M. m. domesticus* populations. The ancestors of the experimental individuals were originally caught in France in the Massif Central (2005) and in Germany around Cologne/Bonn (2006), and were kept in the mouse facility at the MPI Plön under outbred conditions to maintain genetic and behavioural variability. At the time of the experiment, mice from the French population were in the 8th, mice from the German population in the 6th and 7th generation. A specific breeding has been set up for this experiment and we obtained experimental mice from 15 breeding pairs. Pure offspring of both populations ("German" GG and "French" FF) and reciprocal crosses between mice of the two populations ("mixed individuals"), either with a mother from the German and a father from the French population (GF) or with a mother from the French and a father from the German population (FG) have been bred. All mice were kept and raised under standard conditions together with both parents until weaning at the age of four weeks. Female and male offspring were kept separated after weaning to ensure no sexual experience before the experiment. We had a balanced system of experimental mice: six females and males per genotype (GG, FF, GF, FG), which results in a

No speed dating please! Patterns of social preference in male and female house mice

103

total number of 48 mice in the behavioural experiment. For genetic analyses, we additionally included the parents of the focal mice and some siblings that had not been used in the behavioural experiment, resulting in genetic samples from 76 individuals. All information on mice including population background, sex, microsatellite data and experimental information can be found in Additional file 1: Table S1.

Experimental setup and procedure

The experimental setup consisted of five standard macrolon cages (Techniplast). One central cage (40.5 × 28.0 × 20.0 cm) and four Type III satellite cages, which were connected via Plexiglas tubes to each one of the four sides of the central cage (Fig. 1). Each Plexiglas connection was equipped with two RFID ring antennae (TSE Industries Inc.), one close to the central cage and one close to the respective satellite cage. Each mouse was equipped with an RFID tag (Iso FDX-B, Planet ID), which is read every time the mouse passes one of the antennae.

The satellite cages were divided into two parts by a metal grid. In each of the four satellite cages, a so-called satellite mouse was placed in the larger, outer part. The smaller inner part could be accessed by the so-called focal individual from the central cage, in which it was

placed at the beginning of the experiment. This setup allowed the focal mouse to move between the central and all outer cages. Through the separation mice could interact (i.e. via smell and vocalisation), but copulation and thus, from an animal ethical viewpoint unwanted, offspring was prevented. Water and food as well as bedding were provided for all mice ad libitum, for the satellite mice in their part of the respective cage, for the focal mouse in the central cage. The order a mouse participated in the experiment (first as focal vs. first as satellite mouse, which we randomised as much as possible within genotype and equal between genotypes) had no effect on their behaviour in the experiment (Additional file 1: Table S1).

Each group of satellite mice, consisting of four mice of similar sex and differing genotype (FF, GG, FG, and GF), formed a satellite-quad. The mice in each quad stayed the same over experiments to ensure a comparable measure of preference across cage systems. Each quad participated four times as satellite mice, one time for each of the four possible genotypes of focal mice.

Each run of the experiment was started by placing the focal and satellite mice in their respective cages and starting the computer program monitoring the RFID antennae. Each experiment ran over five days and six nights, always starting around 16:00 h on a Thursday

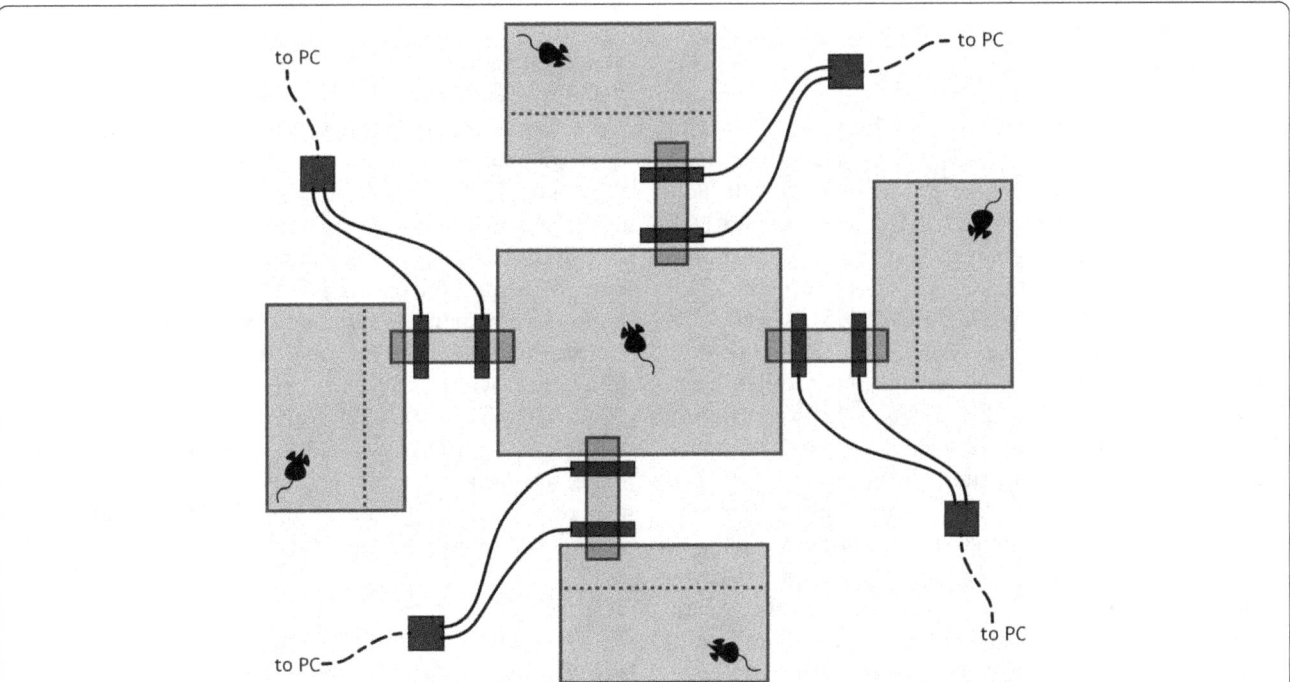

Fig. 1 Schematic top view of the experimental setup. One central cage is connected via Plexiglas tubes with four satellite cages. Satellite cages are divided by a metal grid (*dotted lines*) to prevent mating of the focal mouse (with access to the central cage and the smaller inner parts of the satellite cages) and the four satellite mice (with access only to the larger outer part of their respective satellite cage). Each tube is fitted with a double RFID ring antennae system, all connected to a PC to record the movements of the the focal mouse, which is equipped with an RFID tag. Food, water and shelter were provided for all mice, but are not shown in the figure for clarity

and finishing the following Wednesday around 8:00 h. The rooms were set with a dark-light cycle of 12:12 h with lights on at 7:00 and off at 19:00 h, both for the breeding before the experiments, as well as during the experiments.

We chose such a comparatively long duration of the experiment as we were interested in the decision making process. Further, in this time frame, each female mouse can be assumed to have concluded at least one full oestrus cycle [32]. We decided against a daily control of the females' oestrus state as a two week pilot experiment showed a very high sensitivity to any disturbances in the experimental room. Females also usually enter oestrus synchronised as soon as they perceive certain pheromones of a nearby male [33], which was the case in our study.

Behavioural data

For each focal individual a text-file was generated with time stamps for every antenna read, and the identification number of the respective antenna. By this, we could gain information about the number of antenna reads as a proxy for activity. Using a self-written script in R [34] we calculated the duration of time spent in the four satellite cages, which served a proxy for preference behaviour of the focal mouse.

Genetic data

Microsatellite genotyping and analysis

To determine if genetic distance is an important factor of partner preference we chose 13 unlinked microsatellite markers published by Teschke et al. [27]. These markers are Chr3_24R, Chr16_21R, CHr12_05R, Chr10_45R, Chr 01_25R, Chr17_09R, Chr05_45R, Chr13_22R, Chr19_08R, Chr14_16R, Chr09_20R, Chr01_23, and Chr02_02R. Forward primers were labelled with FAM or HEX, and PCR was performed using 5 ng/µl DNA template together with the Multiplex PCR kit (QIAGEN). After processing PCR products with HiDi formamide and 500 ROX size ROX standard, samples were run on an ABI 3730 sequencer (Applied Biosystems). Raw alleles have been called using GeneMapper 4.0 (Applied Biosystems). The proportion of shared alleles [35] and the pairwise genetic distance (Cavalli-Sforza distance; in the present paper referred to as CAS) between all individuals were calculated with the program MSA [36] and visualized using MEGA 6 [37]. As a paternal influence on partner choice was proposed by Montero and colleagues [31], we also included the genetic distance from the mother and father of the focal mouse to each of the satellite mice (referred to as CASmat or CASpat, respectively).

Sequencing and analysis of the H2-Eß locus

We chose one locus of the MHC Class II complex (H2-Eß) for Sanger Sequencing and, more specifically,

decided to focus on Exon 2 as this exon is known for determining the antigen binding groove and thus might be most directly involved in pathogen resistance and thus interesting for partner choice. Primers used for PCR and sequencing are: Forward 5′CGG GCA TCT TGT CGG CAG AGA AGA AG 3′ and Reverse 5′CAC CGT GGT TCC GCC CCA GCC ACC 3′. Sequences were edited manually with Seqman (included in DNASTAR, Inc., Madison, USA) and aligned with the algorithm Clustal-W [38], included in the program MEGA 6 [37]. The phase of diploid sequences was estimated following Stephens and colleagues [39, 40], implemented in DNASp [41]. To assess whether MHC-dissimilarity between two potential mating partners were important, the number of amino acid differences per site (p-distance) between sequences of the focal individual to the four satellite individuals was calculated, as well as the p-distance between the mother and father of the focal individual to each of the satellite mice (in this paper referred to as MHC, MHCmat and MHCpat, respectively). The latter has been done, to identify a possible influence of the parents on the offspring's choice. Furthermore, we addressed the question if MHC-diversity of the potential mate is influencing mate choice. Therefore, we again calculated the number of amino acid differences per site (p-distance) between the two haplotypes within each satellite individual. Individual H2-Eß sequences have been submitted to Dryad.

Statistical analysis

Patterns of activity

To estimate the activity patterns of focal mice, we analysed the number of antenna reads per hour for each focal mouse. We tested for changes in activity over time using a generalised linear model (with Poisson error distribution, fitting to our count data) with experimental day, genotype and sex as fixed factors, and activity per hour as response variable. We used square-root transformed data to improve distribution of residuals (checked with QQ-plot and Shapiro-Wilk test: $W = 0.9948$, $p = 0.3522$). For the analysis of temporal behavioural patterns, we had to exclude one of the focal mice, as during the run of one individual (a GF male) the recording was stopped unintentionally after three days due to power failure.

Selectivity of focal mice

To analyse the selectivity (i.e. the strength of preference that focal mice show for some satellite mice over others, independent on the direction of preference), we calculated a selectivity index SI with the formula SI = SD/SD(max), where SD is the standard deviation of the four proportional durations (in %) that the respective focal mouse spent in each of the four satellite cages, and

SD(max) is the maximal standard deviation theoretically possible, which is in our case (4 possibilities) 50. The resulting SI values range from 0 (25% of time in each of the four satellite cages = no selectivity) to 1 (100% of time in one of the four satellite cages, no time in the three others = highest selectivity). SI was calculated in ten minute intervals for each focal mouse to get measures for the change of selectivity over time, and over the whole period to get a measure of overall selectivity for each individual.

To calculate which factors might influence the selectivity, we applied a generalised linear model with the overall SI per focal mouse as response variable. We used the best fitting error distribution, which was a beta-distribution. The tested factors were the sex of the focal individual, its maternal and paternal population background, and whether it is of pure or mixed population background. The t- haplotype is a selfish genetic element that has previously been shown to affect mate choice in house mice. Even though some experimental animals carried the t-haplotype, we could not detect any influence of t-haplotype on selectivity and thus retained all individuals in the analysis, irrespective of their t-haplotype status, and included the t-haplotype status of the focal mouse as a random effect in the model testing for selectivity.

Factors correlating with preference

To test if the direction of preference of focal mice depends on characteristics of the satellite mice and/or dyadic factors between focal and satellite mice (e.g. genetic distance), we applied a generalised linear mixed effects model (function glmmadmb from the R package glmmADMB which allows for a beta error distribution, which was the best fit for our data). The proportion of duration spent in each of the satellite cages served as response variable, and identity of the focal mouse, satellite-quad and t-haploytpe of the satellite mouse as random factors. The following characteristics of satellite mice served as fixed factors: maternal and paternal population background (i.e. F or G), information whether the satellite individual is of pure or mixed population origin (i.e. FF and GG vs. FG and GF), information whether the population background is matching with that of the focal individual (i.e. paternal matching (e.g. **FG** chooses **GG**), maternal matching (e.g. **FG** chooses **FF**), exact match (e.g. **GF** chooses **GF**), no match (e.g. GF chooses FG), MHC diversity (p-distance of the two amino acid alleles of the satellite mouse), MHC (p-distances of each satellite mouse to the focal individual), MHCmat and MHCpat (p-distances of each satellite mouse to the focal individual's mother or father, respectively), CAS (genetic distances based on microsatellites between each satellite mouse and the focal

individual), CASmat and CASpat (genetic distance of each satellite mouse to the focal individual's mother or father, respectively). We further included cage position as an additional factor to test if the preference of focal mice was influenced by the position of the cage inside the experimental room. During experiments, we had already aimed to minimise a possible position effect by randomising the position of satellite mice (by genotype) over trials.

In a second step, following the results from the analysis of selectivity (see Table 1) and the results from the models analysing the direction of preference (see Table 2), we performed correlations to analyse the influence of MHCpat on the direction of preference, separated by the groups that differ in selectivity. We calculated Spearman rank correlation coefficients between the duration spent in each of the satellite cages and the factor MHCpat, grouped by the type of population background (pure vs. mixed), and by sex and paternal population background. We used square-root transformed data to improve distribution of residuals (checked with a QQ-plot and Shapiro-Wilk test: $W = 0.9818$, $p = 0.0915$).

All statistical tests were carried out using R 2.14.1 [34].

Stability of preference

We used two different approaches to estimate the degree to which a preference is stable over time: First, we used the overall duration of time each focal mouse spent in the most preferred cage. Second, we determined in 10 min intervals the preferred cage for each focal mouse. We calculated how often this preference changes between cages, giving us the number of "preference blocks" as an estimator on how stable or unstable the choice of a focal mouse is: Few long preference blocks are a sign for a comparatively stable choice, many short preference blocks hint at a rather unstable choice. We further calculated how long the preference for the last preferred cage lasted, to estimate after how much time in the experiment on average the choice is established and how stable it is (see Additional file 2: Figure S1 for

Table 1 Influence of focal mouse characteristics on selectivity

Factor (characteristics of focal mice)	df	X^2	p
Maternal population background	1	0.6581	0.41723
Paternal population background	1	2.9311	0.08689
Sex	1	0.6974	0.40365
Pure or mixed population background	**1**	**5.3001**	**0.02132**
First as focal or first as satellite mouse	1	0.0154	0.90127
Maternal population background: Sex	1	0.0173	0.89533
Paternal population background: Sex	**1**	**5.4789**	**0.01925**

Presented are the results of all fixed effects of the generalised linear model, degrees of freedom (df), Chi-Square values (X^2) and p-values. T-haplotype status was included as random effect. Significant results are printed in bold

Table 2 Effect of different parameters on the direction of preference

Factor (characteristics of satellite mice)	df	X²	P
Maternal population background	1	0.0346	0.8525
Paternal population background	1	0.9560	0.3282
Pure or mixed population background	1	1.2095	0.2714
Relative matching population background	3	1.9556	0.5817
MHC diversity	1	0.0001	0.9906
MHC distance to the focal mouse	1	0.2943	0.5875
MHC distance to the mother of the focal mouse	1	0.7897	0.3742
MHC distance to the father of the focal mouse	**1**	**4.3466**	**0.0371**
Genetic distance to the focal mouse	1	2.5150	0.1128
Genetic distance to the mother of the focal mouse	1	0.0001	0.9904
Genetic distance to the father of the focal mouse	1	0.0726	0.7876
Position of the respective satellite cage in the room	3	2.3073	0.5111

Presented are the results of all fixed effects of the generalised linear mixed model: degrees of freedom (df), Chi-Square values (X²) and p-values. Individual identity of focal mice, t-haplotype status and quad-number of satellite mice were included as random effects. Significant results are printed in bold

two examples). We calculated these preference block parameters over all mice and separated by sex.

Additionally we tested if the initial preferences (preferred satellite mouse after the first 10 min, 90 min and 24 h of the experiment) matched the final stable choice, using Chi-Square tests.

Results

We monitored the partner preference behaviour of wild house mice (*M. m. domesticus*). Over five days and six nights, we allowed each of 24 males and 24 females to associate with their preferred partner among four individuals of opposite sex inside a controlled cage setup, which allowed sensory interaction via smell and vocalisation, but no full physical contact or mating (Fig. 1). Both the focal individuals and the potential partners differed in their genetic background with respect to population background (pure FF: population originating from France; pure GG: population originating from Germany; FG and GF: mixed population background, maternal origin given first). An allele-sharing tree shows genetic differentiation between FF and GG, even though the branch lengths are small and the level of differentiation is thus low (Additional file 3: Figure S2a). FG and GF individuals share alleles with both parent populations. When focusing on the MHC locus *H2-Eß*, this separation is not evident, and both populations share alleles (Additional file 3: Figure S2b). We used the duration that focal mice spent in each of the satellite cages as a proxy for their preference.

Patterns of activity

Over the whole time in the setup, each mouse was registered 86 times per hour on average. The activity changed throughout the days of the experiment (LRT = 110.216, $p < 0.001$; Additional file 4: Figure S3), decreasing by an average of 19 antenna reads per hour on each day. Mice were most active during the first 24 h (including the complete first night), followed by a steep drop in activity and a steady but low, continuous decrease until the end of the experiment. This pattern is most likely due to habituation to the surroundings. Furthermore, we observed a daily rhythm in activity, with peaks occurring shortly after both lights-on and lights-off and highly reduced activity around mid-day (Additional file 4: Figure S3). A difference in activity could not be observed between sexes or genotypes (genotype: LRT = 4.767, $p = 0.1897$; sex: LRT = 0.609, $p = 0.4351$).

When we compared the time spent in any of the satellite cages ("social time") to the time spent in the central cage, we found that focal mice spent an average of 68.5% of their time close to another mouse rather than being alone in the central cage. The time that focal mice spent in the cage of the preferred mouse was approximately equal to the time spent in the central cage (central cage mean (sd) = 49.04 h (32.24 h), preferred cage: 49.31 h (27.55 h); Fig. 2). In contrast, most mice spent less time in each of the three non-preferred cages than the central cage (Fig. 2).

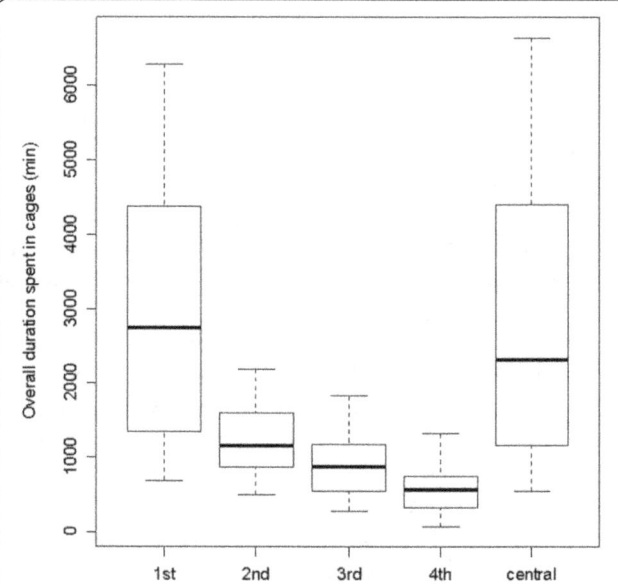

Fig. 2 Duration of time spent in each of the four satellite cages and the central cage. The overall duration of time spent in the cage of the most preferred satellite mouse (1st) was higher than the duration of time spent in the cages of the three less preferred mice (2nd – 4th) and slightly higher than the duration of time spent in the central cage

No speed dating please! Patterns of social preference in male and female house mice

107

Selectivity of focal mice

To analyse the selectivity of focal mice (i.e. the strength of preferences for some satellite mice over others, independent on the direction of preference), we calculated a selectivity index (SI) based on the time spent in satellite cages. SI ranges from 0 (no selectivity) to 1 (highest selectivity).

The selectivity changes over time (Fig. 3). In the first minutes of the experiment, the average values of SI are very high ($SI_{average(10min)} = 0.50$), followed by a steep drop. The lowest point occurs at about 90 min (duration: $SI_{average(90min)} = 0.19$). The high initial values of SI do not reflect selectivity itself, but rather the lack of time to explore other cages, which also explains the drop of selectivity when mice start to explore. After 90 min, SI increases again and reaches a maximum after 22.8 h among females ($SI_{average(max-females)} = 0.45$), while a local maximum occurs after 23.7 h among males ($SI_{average(local-max-males)} = 0.33$). The total maximum of males occurs only after 120.3 h ($SI_{average(max-males)} = 0.34$). Overall, females and males show a slight difference in their selectivity ($SI_{average(females)} = 0.38$ vs. $SI_{average(males)} = 0.31$). The temporal daily variation in SI reflects active vs. inactive phases of individuals and is congruent with the observed activity pattern (Additional file 4: Figure S3).

Further, we aimed to identify factors involved in determining the selectivity of focal mice. Selectivity was significantly higher in mice with pure genotypes (FF, GG) as opposed to mixed individuals (FG, GF) (Fig. 4a, Table 1). Furthermore, the interaction between sex and paternal population background had a significant effect on selectivity (Fig. 4b, Table 1): males with a father from the French population showed higher selectivity than females with a father from the French population, while individuals with a father from the German population showed the opposite effect. Paternal population

background and sex did not show significant effects on their own. Age and maternal population background also had no significant effects on selectivity.

Factors correlating with preference

Using a generalised linear mixed effects model, we identified the paternal MHC as potential factor correlated with preference behaviour. The more distant the paternal MHC of the focal mouse is from that of the satellite mouse (MHCpat), the more time the focal mouse spent with the satellite mouse (Fig. 5a, Table 2). None of the other tested factors had any significant effects on preference.

As described before, we found significant differences in selectivity between pure and mixed mice, as well as a significant interactive difference in selectivity depending on the sex and the paternal population background of focal mice (see Results section, Selectivity of focal mice. Therefore, we tested for a correlation between MHCpat and the direction of preference using data separated according to these groups. We found no difference in the influence of MHCpat on preference between pure and mixed individuals (Fig. 5b, Table 3). When separated by sex and the paternal population background, MHCpat correlated with preference in only female mice with a father from the German population (xG females Fig. 5c, Table 3). Interestingly, the mean MHCpat was slightly higher (but not significantly) in females with a father from the German population (xG females, mean: 0.167, standard deviation: 0.039) compared to the other three groups (xF females: 0.154 ± 0.037; xG males: 0.157 ± 0.030; xF: 0.156 ± 0.034).

Stability of preference

The overall proportion of time that each focal individual spent in its preferred satellite cage was 78.8% on average (minimum: 43.9%, maximum: 99.8%) of the total social time (the time spent in any of the satellite cages).

The mean number of preference blocks over all focal individuals is 7.5, and the duration of the last preference block is 66.7% on average. Results separated by females and males reflect the patterns of selectivity described above: females had a lower number of preference blocks on average (7.0) and a higher proportional duration of the last preference block (72.6) than males (number of blocks: 8.0; duration of last block: 60.8). This reflects a more stable preference in females than males. Considering the whole duration of the experiment (136 h), females did not change their preferences after an average of 37.3 h and males after 53.3 h.

The initial "preference" after 90 min (when selectivity was very low) and the final preference of focal mice matched in 13 of the 47 cases, which is not different from the matches expected by chance (11.75 of 47;

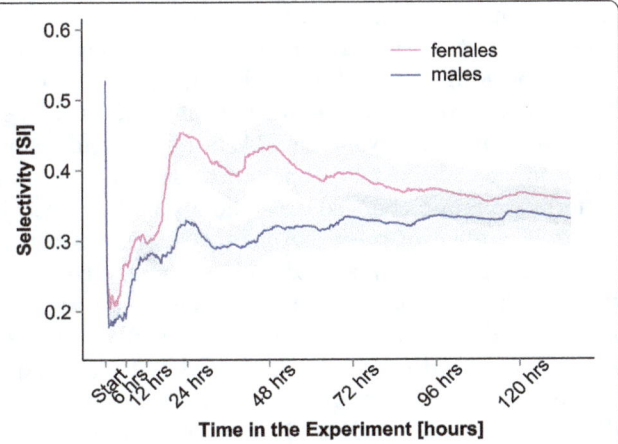

Fig. 3 Temporal patterns of selectivity. Selectivity (given as SI, see Methods) measures the strength of preference. The average SI for males (*blue*) and females (*pink*) over the whole duration of the experiment is presented, including 95% confidence intervals

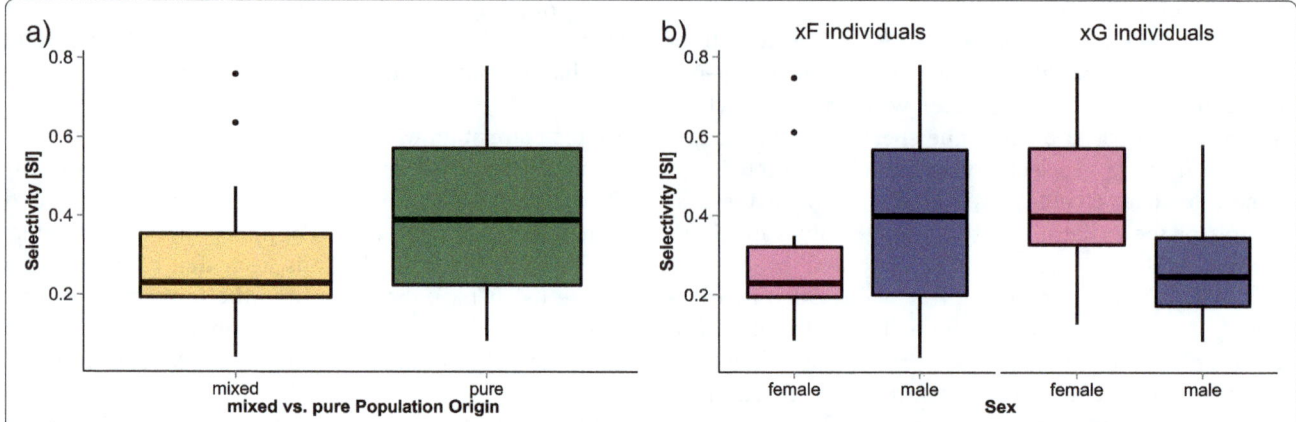

Fig. 4 Three factors influencing selectivity. **a** Individuals with a pure population background (FF and GG, shown in *green*) choose stronger than those with a mixed population background (FG and GF, shown in *yellow*). **b** Paternal population background (F or G) and sex had a significant interactive effect on selectivity. xF individuals (mice with a father from the French population, irrespective of the mother's population): xF males had a higher selectivity than xF females. xG individuals (mice with a father from the German population, irrespective of the mother's population): xG females had a higher selectivity than xG males

$X^2(1) = 0.0034$, $p = 0.953$). The preference after 24 h in the experiment matched the final preference in 23 of 47 cases, which is significantly higher than the expectation by chance ($X^2(1) = 5.713$, $p = 0.025$). The "preference" after 10 min when the selectivity index SI was high and the final preference of focal mice only matched in 12 of the 47 cases, which is at chance level ($X^2(1) = 0.0000$, $p = 1$). This confirms that our conjecture that high initial SI values do not reflect selectivity but the lack of time to explore all cages (see Patterns of activity).

Discussion

We monitored the partner preference behaviour of wild house mice to characterize individual factors (e.g. population background) as well as dyadic factors (e.g. MHC dissimilarity between pairs of mice) correlating with preference behaviour and possibly mate choice. Moreover, to investigate the process that leads to the choice of a specific partner, we continuously tracked the behaviour of focal mice for nearly one week, which also assured that each female entered oestrus at least once.

Patterns of activity

All mice exhibited similar activity patterns during the experiments, with reduced activity around midday and highest activity levels around "sun-down" and "sun-up". Consistent with this activity pattern, we observed small daily changes in the strength of preference. As in most animals, daily activity patterns in mice are strongly influenced by light [42, 43], but cage enrichment, feeding schedule, and social factors also play a role [44, 45]. Our mice were most likely mainly entrained to the artificial day-night schedule maintained in our keeping facilities, but they also could have been influenced by social factors such as the activity of the four satellite mice. In line

with this activity pattern, the duration that the focal mice spent in satellite cages peaked at regular intervals during midday (the time of lowest activity), when the focal mice were resting in one of the satellite cages with a chosen satellite mouse instead of being alone in the central cage. This resulted in the seemingly increased selectivity during midday.

Decision making process in partner preference

To our knowledge, our study is the first to test the degree to which individuals prefer social partners over others, how this selectivity varies over an extended period of time, and at which time a stable preference is established. Most experiments on partner preference and mate choice have been based on short tests and usually last less than 10 min (e.g. [5, 46–49]). However, we chose to run our experiment for five days and six nights. Thonhauser et al. [50] used a long-term setup of 18 days to investigate the mate choice behaviour of female mice, but they only recorded the position of the focal mouse once per day. They were thus likely not able to follow the complete behavioural patterns, as some of the patterns we found changed on an hourly basis, especially in the beginning of the experiment. Our results clearly show a temporal change in selectivity over the course of the experiment. This change over time is consistent in all mice and might be part of a decision making process. Also Manser and colleagues [51] used RFID technique to monitor preference behaviour over several days. They did, however, not analyse their data with respect to temporal changes of selectivity. Instead, their analysis was based on the total time focal mice spent with the potential partners, comparable to our analysis on the direction of preference.

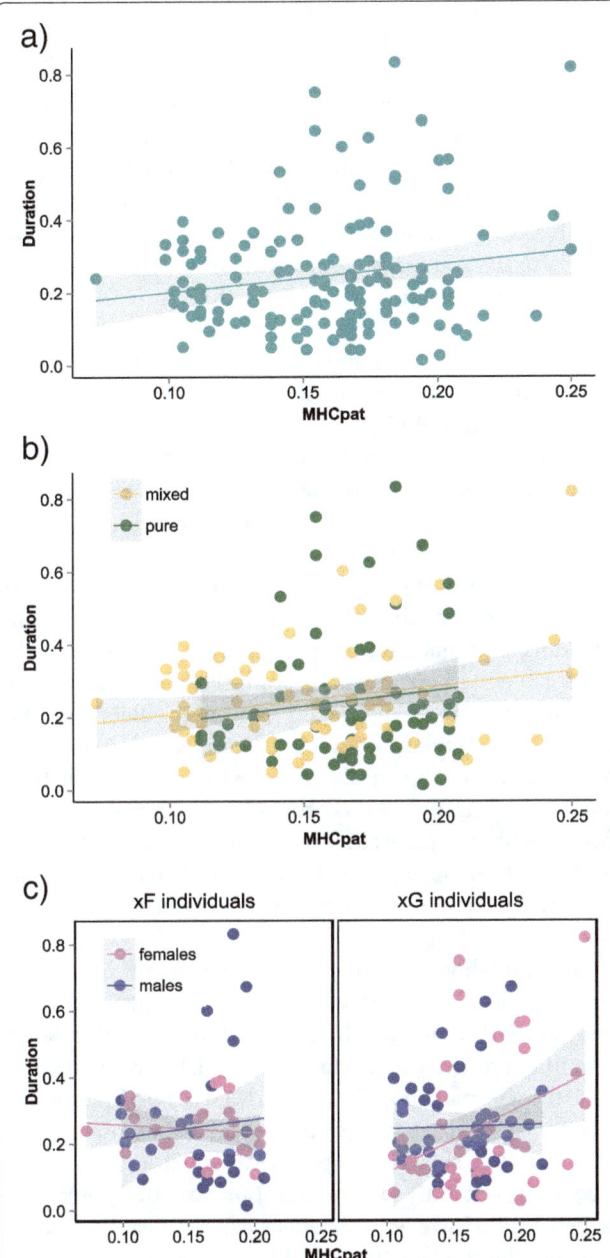

Table 3 Spearman correlation between the duration spent in satellite cages and the distance between the paternal MHC and the MHC of the respective satellite mouse

Group of data	R	T	df	P
All mice	0.1187	1.4042	138	0.1625
Separated by population background				
Pure population background	0.0907	0.7284	64	0.4690
Mixed population background	0.1896	1.6385	72	0.1057
Separated by sex and paternal population background				
xF females[a]	−0.1368	−0.6906	25	0.4962
xG females	**0.3656**	**2.4532**	**39**	**0.0187**
xF males	−0.0017	−0.0086	26	0.9932
xG males	−0.0017	−0.0110	42	0.9913

Values for spearman correlation are presented for all mice and separated by the groups differing in their strength of choice (i.e. mice with a pure population background vs. mice with a mixed population background; and mice differing in sex and their paternal population background). Significant results are printed in bold
[a]xF: individuals with a father from population F; xG: individuals with a father from population G

Fig. 5 Correlation between the duration spent in satellite cages and the distance between the paternal MHC and the MHC of the respective satellite mouse (MHCpat). **a** Correlation between duration and MHCpat shown for all mice. **b** Separate correlations for focal individuals of pure or mixed population background. **c** Separate correlations for focal individuals depending on their sex and their paternal population background (i.e. if the father of the respective focal mouse belongs to the French or German population). For better visualisation of the data we used the linear model based smoothing function from the R package "ggplot2"

cages and thus met all possible companions. At this time, selectivity started to increase and peaked after about 24 h, and this maximum was more pronounced in females (see section Selectivity depends on sex and population background). At this time, the mice seemed to have explored their environment sufficiently to preferentially remain near one of the satellite mice. Indeed, final preferences were established and relatively constant after about 37.3 (females) and 53.3 h (males).

The oestrus cycle of female mice might potentially have an influence on selectivity, with a possibly increased selectivity in both sexes when entering oestrus. We did not control for oestrus and can thus not exclude that the change in oestrus status might have influenced selectivity in both sexes. The females of this study were housed without males before the onset of the experiment. Females housed in this way often enter an anoestrus state and only come into oestrus about three to four days after introduction of a male [52]. As selectivity in our experiment reached its peak after 24 h, the females have most likely not yet been in oestrus, which would make an influence of oestrous on selectivity in our study rather unlikely.

In only about one-quarter of the focal mice, the final preference of social partner was the same as the initial choice, which is what is expected by chance. We conclude that the aparent "choice" at the beginning of the experiment is differentiated from the final stable preference after about two days and might just be an artefact resulting from a lack of time to assess all given possibilities. The stability in preference for a given partner is in line with findings by Montero and colleagues [31], who detected a high degree of mate fidelity in a mate choice experiment under semi-natural conditions.

Selectivity values were low after the first minutes, as focal mice had just started to explore the new environment and to meet all four satellite mice. After 10 min, only half of the individuals had visited all satellite cages, and only after 90 min had all of them visited all satellite

These results clearly indicate that the duration of the experiment can influence the measured selectivity and preference behaviour. Apart from possible influences of the oestrus cycle, a decrease of selectivity may occur due to habituation to the setup and learning that actual mating is not possible. Further, the time to establish a stable preference might change depending on the number of choices given. In a comparable mate choice experiment that was also conducted in a four-choice setup lasting only 10 min, not all of the focal mice visited all possible partners before the end of the experiment [53]. In such short experiments, it is thus possible that mice had no chance to aquire all information necessary to make an informed decision. Thus we suggest to run pilot experiments with an extended period of time to deterime the duration most suitable to answer the question asked.

Association between sexes – a measure for mate choice?

House mice live in socially sub-structured populations and form small reproductive units. In such units, one dominant male usually sires most of the offspring with one or several females [54, 55]. Dominant males defend their territories by frequent urinary marking and fighting with intruders [55–58]. It was shown, however, that not just male territories are important in the social structure of wild mice, but also the membership of males in a family group [31]. Montero and colleagues further observed that males sometimes shared nests with females and found repeated matings with the same partner within the same nest box (multiple nest boxes were offered in a large semi-natural setup). They suggest that familiarity is an important factor in repeated mating with the same partner [31].

We used a controlled cage setup to balance between the advantage of allowing sensory cues (which mice could use for individual assessment and potential mate choice [16, 46]) and the aim of not producing unwanted offspring for ethical reasons. It has been shown that social preferences measured in lab experiments do not necessarily lead to a higher number of matings between social partners [50, 51, 59, 60]. Thus, social preference cannot always be interpreted as mate choice. However, in the wild, there is usually not a direct choice between two or more possible partners like in a laboratory experiment. Under natural conditions, we expect selective attention, resulting in a higher likelihood to locate a preferred type of partner. This can eventually lead to a higher chance of mating.

Following these considerations, the behaviour observed (i.e. individuals of different sex spending time close to each other) can be an important step in mate choice and does not simply reflect a general social preference. Spending time with a potential partner not only increases the likelihood of mating with this partner (as a result of the increased chances to do so), but might also be interpreted as a form of securing a social resource.

Selectivity depends on sex and population background

A clear difference in the strength of preference could be observed between pure and mixed individuals, with pure individuals being more selective. This supports the idea that homozygous individuals show stronger selectivity than heterozygous individuals, which was described previously by Yamazaki et al. [20]. Furthermore, we found a difference in the preference behaviour between females and males, which was mostly apparent during the first half of the experiment. Females generally show higher selectivity and stability than males. Stronger selectivity in female individuals compared to males is in line with hypotheses considering higher female investment in producing and raising offspring in species in which paternal care is not common [61].

The mate choice of males is supposedly driven by two strategies: either choosing one individual female and providing paternal care, or mating with as many females as possible without further commitments [62]. While the males in our study indeed had lower selectivity than females, this tendency was not significant when taking the average over the whole experimental period. Furthermore, males also have costs in reproduction, even in species with no paternal care, so the chance of having more viable offspring can be increased if males do not just arbitrarily mate with several females but make choices.

Over the whole experiment, the two sexes differed reciprocally in the strength of preference, depending on their fathers' population background: males with a father from the French population and females with a father from the German population showed a stronger preference than the other two groups. We have no obvious explanation for such an interactive effect on selectivity depending not only on the sex but also the population background. However, this effect can at least partially be explained by the fact that the only factor correlating with the preference behaviour, i.e. MHCpat, was strongest in female mice with a father from the German population.

Paternal influences on mate preference

In the mate choice experiment by Montero et al. [31], hybrid mice preferred paternally matching mates. Furthermore, assortative mating could be detected when mice had the chance to become familiar with individuals of their own population first. No assortative mating was observed when mice were unfamiliar with their own population before encountering individuals of the other population. In our experiments, all mice were unfamiliar, and like Montero et al. [31], we did not find assortative mating. Concerning paternal versus maternal matching, we found a stronger effect for paternal matching, but both values were not significant (Table 2). This suggests that there was not enough power in the experiment to resolve this issue.

Nevertheless, our results reveal another non-random partner preference in the form of a stronger preference for partners in which the MHC (H2-Eß locus) of the focal individual's fathers and the MHC of the chosen satellite mouse show increased dissimilarity. This effect was only apparent in females with a father from the German population. In other words, female mice with a father from the German population spent more time with partners whose MHC locus was more different from that of the focal individual's father than with those that had a smaller difference. This might be due to the higher mean dissimilarity between the paternal MHC of these mice and their satellite individuals' MHC. We cannot exclude a similar effect for the other groups (females with a father from the French population and males independent of their father's population) if more dissimilar individuals had been available.

There are two hypotheses regarding why MHC-driven mate choice might occur (selected reviews: [26, 63, 64]): the good genes hypothesis [65, 66], which corresponds to a partner preference based on MHC-diversity, and the complementary allele hypothesis [7, 67], which corresponds to a partner preference based on MHC-dissimilarity. Our findings support the complementary allele hypothesis for two reasons. First, we detected an effect of MHC-dissimilarity, expected under the complementary allele hypothesis, and not MHC-diversity, which would be more likely under the good gene hypothesis. Second, a detailed look at the haplotype distribution of the most and second most preferred individuals vs. never preferred individuals showed that neither individuals within the preferred individuals, nor within the non preferred individuals share certain haplotypes. Preferred and unpreferred individuals even share haplotypes. Thus, we found no evidence for certain MHC-alleles being favoured.

Several studies support mate choice based on MHC-complementary alleles, even though complementary alleles may include different degrees of dissimilarity (e.g. [68–70]). The effect of MHC-dissimilarity detected in our study is based on the MHC of the focal individual's father. This effect does not involve self-referencing the individual's own MHC during partner choice. It rather includes the information of the father's MHC, which can be transmitted by either genomic or familial (sexual) imprinting from the father to its offspring in the first days in the nest while the father is still present (as shown by Montero et al. [31]).

Isles and colleagues support the hypothesis that preference behaviour is driven by genomic imprinting [49, 71]. They suggest that genes inherited with a parental bias are in linkage disequilibrium with odour determination and odorous information processing (e.g. MHC). There is no evidence that MHC genes themselves are inherited in a parent-of-origin manner (i.e. genomic imprinting [72]). The effects of familial imprinting and early learning are well known [73, 74]. Familial imprinting has been described to play a role in various situations, like kin recognition [75], mate choice [76, 77], and recognition of environmental cues [76, 78].

Penn and Potts [23] and Yamazaki et al. [25] have already suggested evidence of familial (olfactory) imprinting in mice. We thus propose that olfactory imprinting (i.e. early learning of the fathers smell) during the early days of life might be the mechanism underlying the fathers' influence on their offspring's preference behaviour in our study. It has already been described that familial imprinting takes place in the early nest phase [79–81]. Tramm and Servedio [82] suggest that paternally driven familial imprinting is more likely to evolve than maternal influences on mate choice. Furthermore, paternally driven familial imprinting might be a compromise between the two strategies of male mating behaviour described above (one partner, few offspring, and parental care vs. several partners, many offspring, and no parental care). Olfactory imprinting of MHC-information would enable the father to provide some care for all of its offspring (from several females) without being bound to a single nest.

Conclusions

This study revealed two patterns of paternal influence on partner choice in wild mice: First, paternal population background in interaction with sex of the focal mouse had an influence on selectivity, i.e. strength of preference. Second, the preference for a certain partner seems to be driven by the distance between the paternal MHC to the satellite mouse' MHC, an effect we only observed in the group with the strongest preference, females with a German father. As MHC genes are not expressed in a parent-of-origin manner (genetic imprinting) we support the hypothesis of familial (olfactory) imprinting as most probable process for information transfer from father to offspring during the early days in the nest. Further studies under more natural conditions are needed to elucidate this process, with social interactions such as territoriality or the hierarchy structure of competing individuals taken into account. Finally, we wish to highlight the importance of an appropriate duration of behavioural experiments as the formation of preference is based on a real decision making process.

Additional files

Additional file 1: Table S1. All individual information (microsatellite data, experimental information). (CSV 27 kb)

Additional file 2: Figure S1. Two examples visualising how to assess the consistency of choice (upper panel the "German" male "Hermann" and lower panel the "French" male "Jacques"). Two measures were used to define the consistency of choice. 1. Block-Count and 2. Block Size. Preference is based on the selectivity index SI calculated in intervals increasing by 10 minutes (the first interval being 10 minutes, the second 20 minutes, and so forth). For each time point the preferred cage (1-4) is plotted. "Hermann" changed his preference over time 4 times (= 5 blocks), compared to "Jacques" who changed his mind only once (= 2 blocks). After 38% of "Hermann" being in the experiment (as the last block is 62%), he chose the mouse from cage 4 to be his preferred partner, "Jacques" already decided after just 5% of the total time in the experiment that mouse 1 is the best. "Hermann" never chose mouse 3 and "Jacques" never chose mice 1 and 2. (PDF 35 kb)

Additional file 3: Figure S2. a) Allele sharing tree (Bowcock *et al.* 1994) for all animals based on microsatellites. Even though a separation between populations is evident, small branch lengths reflect a still close relationship between individuals of all breeding types. b) Neighbour-joining tree of H2-Eß locus Exon 2 haplotype sequences. Bootstrap values > 50 are shown. No pattern of population divergence can be detected, both populations share several alleles. (PNG 301 kb)

Additional file 4: Figure S3. Mouse activity patterns over the whole experimental period and diurnal rhythm of activity averaged over all days. a) Average activity measured as number of antenna hits per hour for each day of the experiment. After a steep drop of activity in the beginning of the experiment, activity steadily drops unitl the end. Boxplots include outliers (black dots) and additionally all individual data points (grey dots, slightly jittered along the x-axis for clarity). b) Diurnal rhythm of mouse activity. Each light grey line shows the number of antenna hits per hour of individual as a proxy for its activity. The open circle represent the average over all mice of the given time of the day. The light yellow box indicates the phase of lights-on in the experimental room. Mice clearly are active all over the day, with a siesta in the early afternoon (12:00 – 15:00h). (PDF 618 kb)

Abbreviations

CAS: Cavalli-Sforza-distance between a satellite mouse and the focal individual; CASmat: Cavalli-Sforza-distance between a satellite mouse and the focal individual's mother; CASpat: Cavalli-Sforza-distance between a satellite mouse and the focal individual's father; F: French population; FF: Mouse with pure French population background; FG: Mouse with a mother from the French and a father from the German population; G: German population; GF: Mouse with a mother from the German and a father from the French population; GG: Mouse with pure German population background; LRT: Likelihood Ratio Test; MHC: p-distance between a satellite mouse and the focal individual; MHCmat: p-distance between a satellite mouse and the focal individual's mother; MHCpat: p-distance between a satellite mouse and the focal individual's father; SI: Selectivity index; xF: Mouse with a father from the French population and a mother from any of the two populations; xG: Mouse with a father from the German population and a mother from any of the two populations

Acknowledgements

We would like to thank Diethard Tautz for the opportunity to conduct our MiSoMate experiment in his facilities, as well as discussions and comments on the manuscript. Many thanks also to Alexandra Matt, Rafik Neme, Anja Schunke, Leslie Turner and Bernhard Haubold for discussion of statistical questions. We thank Heike Harre, Heinke Buhtz, Hans-Joachim Hamann, Harald Deiwick and Christine Pfeifle for a lot of technical support and Till Sckerl for help in running the experiments. We also wish to thank two anonymous reviewers for a fair and helpful evaluation of the manuscript.

Funding

This study was financed by institutional funds of the Max-Planck Society.

Authors' contributions

ML and SvM conceived and performed research, analysed the data and wrote the manuscript. Both authors read and approved the final manuscript.

Competing interests

The authors declare that they have no competing interests.

References

1. Kokko H, Brooks R, Jennions MD, Morley J. The evolution of mate choice and mating biases. Proc R Soc Lond B Biol Sci. 2003;270:653–64.
2. Andersson M, Simmons LW. Sexual selection and mate choice. Trends Ecol Evol. 2006;21:296–302.
3. Sardell RJ, Kempenaers B, DuVal EH. Female mating preferences and offspring survival: testing hypotheses on the genetic basis of mate choice in a wild lekking bird. Mol Ecol. 2014;23:933–46.
4. Ihle M, Kempenaers B, Forstmeier W. Fitness benefits of mate choice for compatibility in a socially monogamous species. PLoS Biol. 2015;13:e1002248.
5. Raveh S, Sutalo S, Thonhauser KE, Thoß M, Hettyey A, Winkelser F, Penn DJ. Female partner preferences enhance offspring ability to survive an infection. BMC Evol Biol. 2014;14:14.
6. Drickamer LC, Gowaty PA, Holmes CM. Free female mate choice in house mice affects reproductive success and offspring viability and performance. Anim Behav. 2000;59:371–8.
7. Tregenza T, Wedell N. Genetic compatibility, mate choice and patterns of parentage: invited review. Mol Ecol. 2000;9:1013–27.
8. Dieckmann U, Metz JAJ, Doebeli M, Tautz D. Adaptive Speciation. Cambridge: Cambridge University Press; 2004.
9. Crow JF, Kimura M. An introduction to population genetics theory. 1970.
10. Lewontin RC. The genetic basis of evolutionary change. New York: Columbia University Press; 1974.
11. Jiang Y, Bolnick DI, Kirkpatrick M. Assortative mating in animals. Am Nat. 2013;181:E125–38.
12. Burley N. The meaning of assortative mating. Ethol Sociobiol. 1983;4:191–203.
13. Sherborne AL, Thom MD, Paterson S, Jury F, Ollier WE, Stockley P, Beynon RJ, Hurst JL. The genetic basis of inbreeding avoidance in house mice. Curr Biol. 2007;17:2061–6.
14. Mucignat-Caretta C, Caretta A. Message in a bottle: major urinary proteins and their multiple roles in mouse intraspecific chemical communication. Anim Behav. 2014;97:255–63.
15. Leinders-Zufall T, Brennan P, Widmayer P, Maul-Pavicic A, Jäger M, Li X-H, Breer H, Zufall F, Boehm T. MHC class I peptides as chemosensory signals in the vomeronasal organ. Science. 2004;306:1033–7.
16. Asaba A, Hattori T, Mogi K, Kikusui T. Sexual attractiveness of male chemicals and vocalizations in mice. Front Neurosci. 2014;8:231.
17. Holy TE, Guo Z. Ultrasonic songs of male mice. PLoS Biol. 2005;3:e386.
18. Hoffmann F, Musolf K, Penn DJ. Ultrasonic courtship vocalizations in wild house mice: spectrographic analyses. J Ethol. 2012;30:173–80.
19. Klein J. The major histocompatibility complex of the mouse. Science. 1979;203:516–21.
20. Yamazaki K, Boyse E, Mike V, Thaler H, Mathieson B, Abbott J, Boyse J, Zayas Z, Thomas L. Control of mating preferences in mice by genes in the major histocompatibility complex. J Exp Med. 1976;144:1324–35.
21. Ziegler A, Kentenich H, Uchanska-Ziegler B. Female choice and the MHC. Trends Immunol. 2005;26:496–502.
22. Milinski M. The major histocompatibility complex, sexual selection, and mate choice. Annu Rev Ecol Evol Syst. 2006;37:159–86.
23. Penn DJ, Potts WK. The evolution of mating preferences and major histocompatibility complex genes. Am Nat. 1999;153:145–64.
24. Penn D, Musolf K. The evolution of MHC diversity in house mice. In: Machol An M, Baird SJE, Munclinger P, Pi Alek J, editors. Evolution of the house mouse. Cambridge: Cambridge Univ Press; 2012. p. 221–52.
25. Yamazaki K, Beauchamp GK, Kupniewski D, Bard J, Thomas L, Boyse E. Familial imprinting determines H-2 selective mating preferences. Science. 1988;240:1331–3.
26. Immelmann K. Ecological significance of imprinting and early learning. Annu Rev Ecol Syst. 1975;6:15–37.
27. Teschke M, Mukabayire O, Wiehe T, Tautz D. Identification of selective sweeps in closely related populations of the house mouse based on microsatellite scans. Genetics. 2008;180:1537–45.
28. Ihle S, Ravaoarimanana I, Thomas M, Tautz D. An analysis of signatures of selective sweeps in natural populations of the house mouse. Mol Biol Evol. 2006;23:790–7.
29. Staubach F, Lorenc A, Messer PW, Tang K, Petrov DA, Tautz D. Genome patterns of selection and introgression of haplotypes in natural populations of the house mouse (*Mus musculus*). PLoS Genet. 2012;8:e1002891.

30. von Merten S, Hoier S, Pfeifle C, Tautz D. A role for ultrasonic vocalisation in social communication and divergence of natural populations of the house mouse (Mus musculus domesticus). PLoS One. 2014;9:e97244.

31. Montero I, Teschke M, Tautz D. Paternal imprinting of mating preferences between natural populations of house mice (Mus musculus domesticus). Mol Ecol. 2013;22:2549–62.

32. Bingel AS, Schwartz NB. Pituitary LH content and reproductive tract changes during the mouse oestrous cycle. J Reprod Fertil. 1969;19:215–22.

33. Whitten WK. Modification of the oestrous cycle of the mouse by external stimuli associated with the male. J Endocrinol. 1956;13:399–404.

34. R Core Team. R: A Language and Environment for Statistical Computing. Vienna: R Foundation for Statistical Computing; 2016.

35. Bowcock A, Ruiz-Linares A, Tomfohrde J, Minch E, Kidd J, Cavalli-Sforza LL. High resolution of human evolutionary trees with polymorphic microsatellites. Nature. 1994;368:455–7.

36. Dieringer D, Schlötterer C. Microsatellite analyser (MSA): a platform independent analysis tool for large microsatellite data sets. Mol Ecol Notes. 2003;3:167–9.

37. Tamura K, Stecher G, Peterson D, Filipski A, Kumar S. MEGA6: molecular evolutionary genetics analysis version 6.0. Mol Biol Evol. 2013;30:2725–9.

38. Thompson JD, Higgins DG, Gibson TJ. CLUSTAL W: improving the sensitivity of progressive multiple sequence alignment through sequence weighting, position-specific gap penalties and weight matrix choice. Nucleic Acids Res. 1994;22:4673–80.

39. Stephens M, Donnelly P. Ancestral inference in population genetics models with selection (with discussion). Aust N Z J Stat. 2003;45:395–430.

40. Stephens M, Smith NJ, Donnelly P. A new statistical method for haplotype reconstruction from population data. Am J Hum Genet. 2001;68:978–89.

41. Rozas J. DNA Sequence Polymorphism Analysis using DnaSP. In Posada, D. (ed.) Bioinformatics for DNA Sequence Analysis; Methods in Molecular Biology Series Vol. 537. NJ, USA: Humana Press; 2009. p. 337–350.

42. Dallmann R, Mrosovsky N. Scheduled wheel access during daytime: A method for studying conflicting zeitgebers. Physiol Behav. 2006;88:459–65.

43. Aschoff J. Circadian system properties. In Environmental Physiology 1981: Proceedings of the 28th International Congress of Physiological Sciences, Budapest, 1980. pp 1–17.

44. de Groot MH, Rusak B. Housing conditions influence the expression of food-anticipatory activity in mice. Physiol Behav. 2004;83:447–57.

45. de Visser L, van den Bos R, Spruijt BM. Automated home cage observations as a tool to measure the effects of wheel running on cage floor locomotion. Behav Brain Res. 2005;160:382–8.

46. Musolf K, Hoffmann F, Penn DJ. Ultrasonic courtship vocalizations in wild house mice, Mus musculus musculus. Anim Behav. 2010;79:757–64.

47. Latour Y, Perriat-Sanguinet M, Caminade P, Boursot P, Smadja CM, Ganem G. Sexual selection against natural hybrids may contribute to reinforcement in a house mouse hybrid zone. Proc R Soc Lond B Biol Sci. 2014;281:20132733.

48. Asaba A, Okabe S, Nagasawa M, Kato M, Koshida N, Osakada T, Mogi K, Kikusui T. Developmental social environment imprints female preference for male song in mice. PLoS One. 2014;9:e87186.

49. Isles AR, Baum MJ, Ma D, Keverne EB, Allen ND. Genetic imprinting: urinary odour preferences in mice. Nature. 2001;409:783–4.

50. Thonhauser KE, Raveh S, Hettyey A, Beissmann H, Penn DJ. Scent marking increases male reproductive success in wild house mice. Anim Behav. 2013;86:1013–21.

51. Manser A, König B, Lindholm AK. Female house mice avoid fertilization by t haplotype incompatible males in a mate choice experiment. J Evol Biol. 2015;28:54–64.

52. Whitten WK. Occurrence of anoestrus in mice caged in groups. J Endocrinol. 1959;18:102–7.

53. Roberts SC, Gosling LM. Genetic similarity and quality interact in mate choice decisions by female mice. Nat Genet. 2003;35:103–6.

54. Crowcroft P, Rowe FP. Social organization and territorial behaviour in the wild house mouse (Mus musculus L.). In Proceedings of the Zoological Society of London. London; 1963;(140):517–31.

55. Reimer J, Petras M. Breeding structure of the house mouse, Mus musculus, in a population cage. J Mammal. 1967;48:88–99.

56. Poole T, Morgan H. Aggressive behaviour of male mice (Mus musculus) towards familiar and unfamiliar opponents. Anim Behav. 1975;23:470–9.

57. Poole TB, Morgan H. Social and territorial behaviour of laboratory mice (Mus musculus L.) in small complex areas. Anim Behav. 1976;24:476–80.

58. Wolff RJ. Mating behaviour and female choice: their relation to social structure in wild caught House mice (Mus musculus) housed in a semi-natural environment. J Zool. 1985;207:43–51.

59. Zala SM, Bilak A, Perkins M, Potts WK, Penn DJ. Female house mice initially shun infected males, but do not avoid mating with them. Behav Ecol Sociobiol. 2015;69:715–22.

60. Rolland C, MacDonald DW, De Fraipont M, Berdoy M. Free female choice in house mice: leaving best for last. Behaviour. 2003;140:1371–88.

61. Trivers R. Parental investment and sexual selection In Sexual selection and the descent of man edited by B. Campbell, 1871–1971. Chicago, Illinois: Aldine Press; 1972.

62. Clutton-Brock TH. Review lecture: mammalian mating systems. Proc R Soc Lond B Biol Sci. 1989;236:339–72.

63. Mays HL, Hill GE. Choosing mates: good genes versus genes that are a good fit. Trends Ecol Evol. 2004;19(10):554–9.

64. Piertney S, Oliver M. The evolutionary ecology of the major histocompatibility complex. Heredity. 2006;96:7–21.

65. Kamiya T, O'Dwyer K, Westerdahl H, Senior A, Nakagawa S. A quantitative review of MHC-based mating preference: the role of diversity and dissimilarity. Mol Ecol. 2014;23:5151–63.

66. Von Schantz T, Wittzell H, Goransson G, Grahn M, Persson K. MHC genotype and male ornamentation: genetic evidence for the Hamilton-Zuk model. Proc R Soc Lond B Biol Sci. 1996;263:265–71.

67. Hamilton WD, Zuk M. Heritable true fitness and bright birds: a role for parasites? Science. 1982;218:384–7.

68. Zeh JA, Zeh DW. The evolution of polyandry I: intragenomic conflict and genetic incompatibility. Proc R Soc Lond B Biol Sci. 1996;263:1711–7.

69. Huchard E, Baniel A, Schliehe-Diecks S, Kappeler PM. MHC-disassortative mate choice and inbreeding avoidance in a solitary primate. Mol Ecol. 2013;22:4071–86.

70. Jacob S, McClintock MK, Zelano B, Ober C. Paternally inherited HLA alleles are associated with women's choice of male odor. Nat Genet. 2002;30:175–9.

71. Reusch TB, Häberli MA, Aeschlimann PB, Milinski M. Female sticklebacks count alleles in a strategy of sexual selection explaining MHC polymorphism. Nature. 2001;414:300–2.

72. Isles AR, Baum MJ, Ma D, Szeto A, Keverne EB, Allen ND. A possible role for imprinted genes in inbreeding avoidance and dispersal from the natal area in mice. Proc R Soc Lond B Biol Sci. 2002;269:665–70.

73. Lorenc A, Linnenbrink M, Montero I, Schilhabel MB, Tautz D. Genetic differentiation of hypothalamus parentally biased transcripts in populations of the house mouse implicate the Prader-Willi syndrome imprinted region as a possible source of behavioral divergence. Mol Biol Evol. 2014;31:3240–9.

74. Svensson EI, Eroukhmanoff F, Karlsson K, Runemark A, Brodin A. A role for learning in population divergence of mate preferences. Evolution. 2010;64:3101–13.

75. Gerlach G, Hodgins-Davis A, Avolio C, Schunter C. Kin recognition in zebrafish: a 24-hour window for olfactory imprinting. Proc R Soc Lond B Biol Sci. 2008;275:2165–70.

76. Caspers BA, Hoffman JI, Kohlmeier P, Krüger O, Krause ET. Olfactory imprinting as a mechanism for nest odour recognition in zebra finches. Anim Behav. 2013;86:85–90.

77. Milinski M, Griffiths S, Wegner KM, Reusch TB, Haas-Assenbaum A, Boehm T. Mate choice decisions of stickleback females predictably modified by MHC peptide ligands. Proc Natl Acad Sci U S A. 2005;102:4414–8.

78. Hinz C, Namekawa I, Behrmann-Godel J, Oppelt C, Jaeschke A, Müller A, Friedrich RW, Gerlach G. Olfactory imprinting is triggered by MHC peptide ligands. Sci Rep. 2013;3:1–8.

79. Fillion TJ, Blass EM. Infantile experience with suckling odors determines adult sexual behavior in male rats. Science. 1986;231:729–31.

80. D'Udine B, Alleva E. Early experience and sexual preferences in rodents. In P. Bateson, eds. Mate choice. Cambridge: Cambridge University Press; 1983. p. 311–27.

81. Leon M. Chemical communication in mother-young interactions. Pheromones Reprod Mammals. 1983;39:77.

82. Tramm NA, Servedio MR. Evolution of mate-choice imprinting: competing strategies. Evolution. 2008;62:1991–2003.

10

Sleeping site ecology, but not sex, affect ecto- and hemoparasite risk, in sympatric, arboreal primates (*Avahi occidentalis* and *Lepilemur edwardsi*)

May Hokan[1,2], Christina Strube[2], Ute Radespiel[1] and Elke Zimmermann[1*]

Abstract

Background: A central question in evolutionary parasitology is to what extent ecology impacts patterns of parasitism in wild host populations. In this study, we aim to disentangle factors influencing the risk of parasite exposure by exploring the impact of sleeping site ecology on infection with ectoparasites and vector-borne hemoparasites in two sympatric primates endemic to Madagascar. Both species live in the same dry deciduous forest of northwestern Madagascar and cope with the same climatic constraints, they are arboreal, nocturnal, cat-sized and pair-living but differ prominently in sleeping site ecology. The Western woolly lemur (*Avahi occidentalis*) sleeps on open branches and frequently changes sleeping sites, whereas the Milne-Edward's sportive lemur (*Lepilemur edwardsi*) uses tree holes, displaying strong sleeping site fidelity. Sleeping in tree holes should confer protection from mosquito-borne hemoparasites, but should enhance the risk for ectoparasite infestation with mites and nest-adapted ticks. Sex may affect parasite risk in both species comparably, with males bearing a higher risk than females due to an immunosuppressive effect of higher testosterone levels in males or to sex-specific behavior. To explore these hypotheses, ectoparasites and blood samples were collected from 22 individuals of *A. occidentalis* and 26 individuals of *L. edwardsi* during the dry and rainy season.

Results: *L. edwardsi*, but not *A. occidentalis*, harbored ectoparasites, namely ticks (*Haemaphysalis lemuris* [Ixodidae], *Ornithodoros* sp. [Argasidae]) and mites (*Aetholaelaps trilyssa*, [Laelapidae]), suggesting that sleeping in tree holes promotes infestation with ectoparasites. Interestingly, ectoparasites were found solely in the hot, rainy season with a prevalence of 75% ($N = 16$ animals). Blood smears were screened for the presence and infection intensity of hemoparasites. Microfilariae were detected in both species. Morphological characteristics suggested that each lemur species harbored two different filarial species. Prevalence of microfilarial infection was significantly lower in *L. edwardsi* than in *A. occidentalis*. No significant difference in infection intensity between the two host species, and no effect of season, daytime of sampling or sex on prevalence or infection intensity was found. In neither host species, parasite infection showed an influence on body weight as an indicator for body condition.

(Continued on next page)

* Correspondence: elke.zimmermann@tiho-hannover.de
[1]Institute of Zoology, University of Veterinary Medicine Hannover,
Buenteweg 17, 30559 Hannover, Germany
Full list of author information is available at the end of the article

(Continued from previous page)

Conclusions: Our findings support that sleeping site ecology affects ectoparasite infestation in nocturnal, arboreal mammalian hosts in the tropics, whereas there is no significant effect of host sex. The influence of sleeping site ecology to vector-borne hemoparasite risk is less pronounced. The observed parasite infections did not affect body condition and thus may be of minor importance for shaping reproductive fitness. Findings provide first evidence for the specific relevance of sleeping site ecology on parasitism in arboreal and social mammals. Further, our results increase the sparse knowledge on ecological drivers of primate host-parasite interactions and transmission pathways in natural tropical environments.

Keywords: Parasite, Mites, Ticks, Microfilaria, Seasonality, Sociality, Sex, Ecology, Behavior, Western woolly lemur, Milne-Edward's sportive lemur, Primate, Tropics, Madagascar

Background

The distribution and abundance of parasites are influenced, amongst others, by environmental factors as well as interactions with the host's behavioral ecology. Environmental factors like temperature and rainfall impact patterns of parasitism in the way that warmth and humidity favor hatching of arthropod eggs, usually resulting in higher abundance of temporary ectoparasites and insect vectors, such as mosquitos, in the hot and rainy season [1, 2]. The cattle tick (*Amblyomma variegatum*) becomes more active in the early wet season, when temperature increases [3]. However, in a rainforest-dwelling lemur species, the diademed sifaka (*Propithecus diadema*), ticks (*Haemaphysalis lemuris*) were found to be more prevalent in the dry than in the rainy season [4]. Other etcoparasite infestations do not differ with season. Prevalence of mites (*Spelaeorhynchus praecursor*) and argasid ticks (*Ornithodores* sp.) in bats, for example, did not vary seasonally [5]. Hence, the influence of environmental factors, such as season can vary between the different ectoparasite genera and must therefore be taken into account in studies on parasite infections.

Host behavioral ecology is also described to affect the distribution and abundance of parasites in mammals. Parasite avoidance behaviors, such as auto- and allo-grooming as well as mud baths are suggested to reduce ectoparasite infestation, while defecating outside nests or dens may reduce exposure to endoparasites [6]. In a wide range of animals, social grouping is documented to not only provide protection from predators but also from flying insects, such as flies and mosquitoes by reducing exposure of the animal's body surface [7]. Furthermore, behaviors related to sleeping site ecology are proposed to reduce exposure to insects such as mosquitoes (e.g. *Anopheles* spp.) and the parasites they may transmit. For example, chimpanzees (*Pan troglodytes schweinfurthii*) prefer to build their nests in a tree species (*Cynometra alexandri*) with insect repellant properties, potentially reducing the risk of malaria infection via mosquito bites [8]. Moreover, sleeping in burrows or holes may provide protection from flying insects [9, 10], in addition to conferring essential benefits such as

insulation from unfavorable climate conditions or protection from predators [11–13]. On the other hand, burrows of rodents, for instance, provide an excellent habitat for ectoparasites such as mites, fleas or ticks due to their stable, dark, moist, and warm microclimate. For example, fleas and mites co-occur more often in voles (*Microtus* spp.) using deep and complex burrow systems than in a congeneric species, which sleeps above ground or uses shallow burrows [14]. In addition, the year-round presence of the host in such burrows provides ectoparasites with a regular food supply [15, 16]. In bats, roosting habits have been related to prevalence and species richness of a specific bat ectoparasite, the bat fly, with heavier parasitism found in bats using more permanent, enclosed roosts [17]. Ectoparasite infestation may therefore constitute an important cost of sleeping in regularly revisited, confined spaces. These ectoparasites, especially the flying insects, may function as vectors transmitting hemoparasites such as *Plasmodium* spp., *Babesia* spp. or filarial nematodes. The first two are haemosporidian parasites, that can be detected by microscopy and differentiated by comparing shape and size of the parasitic stages located inside the erythrocytes as well as size and position of their nucleus [18, 19]. Also microfilaria can be well detected by microscopy, situated between erythrocytes. Species can be differentiated using morphological parameter such as body length, size and proportion of the cephalic space, position of the nerve ring and form of the tail [20, 21].

Furthermore, host traits such as sex and body mass can affect patterns of parasitism. Male flying squirrels have been reported to be more susceptible to ecto- and hemoparasites than females [22], which may be due to an immunosuppressive effect of testosterone or to sex-specific differences in behavior [23]. Nevertheless, a number of studies found no sex differences in prevalence or infection intensity of parasites [24, 25] and some even found higher parasite infection rates in females [26]. Body condition, measured by body mass, is discussed to be linked to parasite infection and ultimately may affect fitness [27–29]. In monkeys and apes, animals in a poorer condition are often more heavily parasitized than indivudals in better condition [30]. In contrast, in rufous mouse lemurs (*Microcebus rufus*) it was found that individuals bearing a higher ecto- and endoparasite load had a better body condition [31]. Thus, the evidence for an influence of sex and body mass on parasite infection in wild hosts is ambiguous and requires further attention.

The goal of this study was to evaluate the impact of host sleeping site ecology, season and sex on patterns of ecto- and hemoparasite infections in a tropical seasonal environment, using two Malagasy sympatric nocturnal primate species as models. The Western woolly lemur (*Avahi occidentalis*) and the Milne Edward's sportive lemur (*Lepilemur edwardsi*) are both arboreal, dry deciduous forest-dwelling, folivorous primates, endemic to northwestern Madagascar [32]. Both species are nocturnal, exhibit a comparable body mass of approximately 1 kg with no sex dimorphism, share the same habitat and thereby same climate conditions and are pair-living and thus match in social pattern [33, 34]. However, they differ prominently in their choice of sleeping sites as well as in their sleeping site related behavior: *A. occidentalis* sleeps on open branches or tree forks and changes its sleeping sites frequently, whereas *L. edwardsi* sleeps in tree holes with high sleeping site fidelity [35, 36]. Thus, these two host species present unique models to assess the effect of host sleeping site ecology on parasite risk, while controlling for the factors climate condition, activity, host body size and sociality. The following hypotheses

were explored: Ectoparasite and hemoparasite prevalence, infection intensity and species richness are assumed to be higher in the host species sleeping in tree holes. In addition, both ecto- and hemoparasites are assumed to be more prevalent during the warm, rainy season and more prevalent in males than in females. As an indicator for the host's condition and thereby fitness, we examined the effect of parasitism on host's body mass.

Methods
Study site
The study was conducted in a 30.6 ha forest parcel named Jardin Botanique A (JBA), located at 16° 19′ S, 46° 48′ E in the Ankarafantsika National Park in northwestern Madagascar. The park consists of dry deciduous forest and is subject to pronounced seasonality, with a dry season from May to October and a hot, rainy season from November to March (Fig. 1).

Sample collection
Sample collection took place during two periods, from July to October 2013 and from March to May 2014, representing the dry and rainy season, respectively. Twenty-two individuals of *Avahi occidentalis* (10 male, 12 females) were captured during the study period by remote immobilization with a combination of ketamine (Ketanest®, 25 mg/ml) and xylazine (Rompun®, 20 mg/ml) using a blowpipe and 1 ml cold air pressure darts (Telinject®, Germany). Seven of these individuals were sampled more than once (Additional file 1). Dosages based on estimated body weights were as follows:

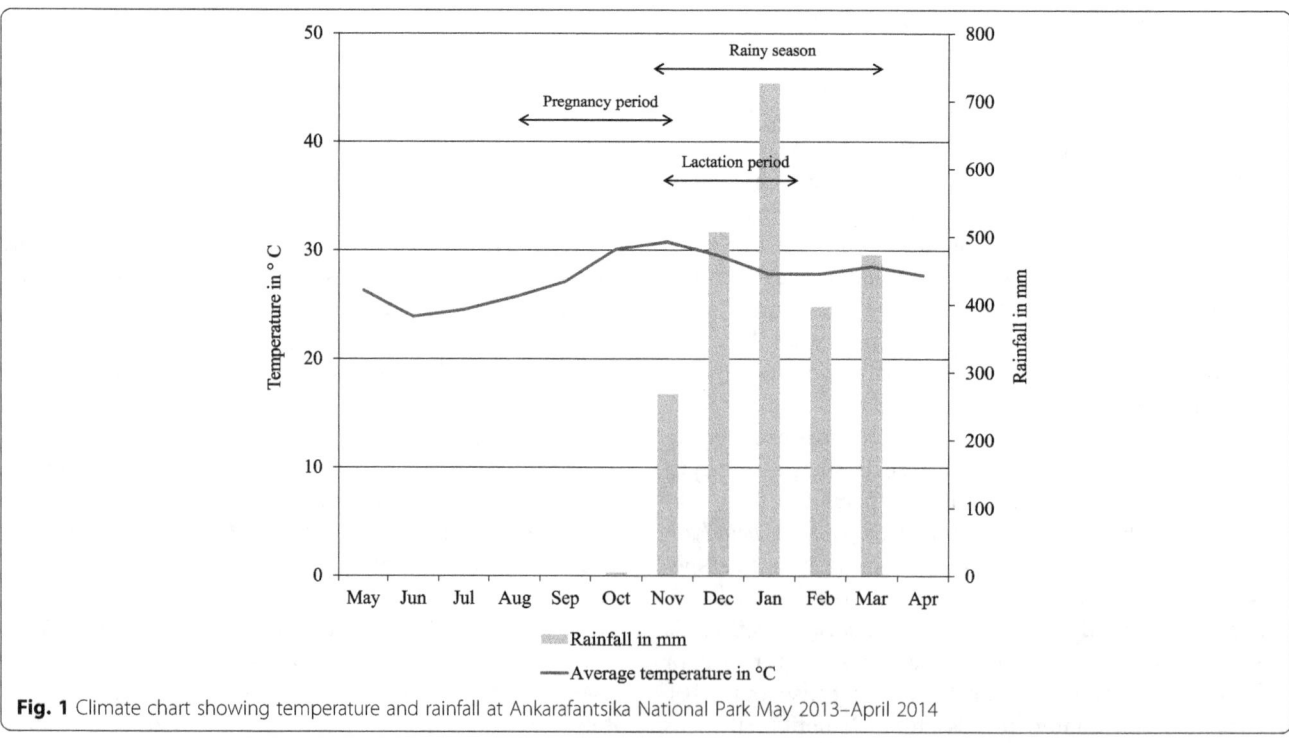

Fig. 1 Climate chart showing temperature and rainfall at Ankarafantsika National Park May 2013–April 2014

10 mg/kg ketamine and 0.5 mg/kg xylazine. Twenty-six individuals of *Lepilemur edwardsi* (12 males, 14 females) were captured directly in their tree holes and sedated with the same drug combination. Twelve of these individuals were sampled more than once (Additional file 1).

All captured individuals were weighed (5 Kg-balance, AEG, precision: 1 g) and macroscopically examined for ectoparasites. All ticks and representative samples of mites were removed and preserved in 90% ethanol. The body parts of the hosts harboring ectoparasites were noted in a protocol. Blood was taken either from the femoral vein or collected opportunistically during the process of ear marking and tissue collection. In total 29 blood samples (17 in the dry and 12 in the rainy season) from *A. occidentalis* and 44 blood samples (27 in the dry and 17 in the rainy season) from *L. edwardsi* were collected. Three to five blood smears per sample were prepared, air dried and fixed with methanol. Additionally, in the rainy season one to three drops of blood from 16 *L. edwardsi* and from 9 *A. occidentalis* were collected into cryotubes with 0.5 ml RNAlater (Qiagen, Hilden, Germany) and frozen at –12 °C.

All procedures were approved by the Ministère de l'Environnement, de l'Ecologie et des Forêts and Madagascar National Parks (MNP) and necessary research permits were obtained from the Malagasy authorities (License N° 167 /13/MEF/SG/DGF/DCB.SAP/SCB obtained on the 13th of July 2013 and N°072/14 obtained on the 12th of March 2014).

Microscopic examination

Ectoparasites were mounted in polyvinyl lactophenol for morphological identification. Blood smears were Giemsa stained and scanned for the presence of hemoparasites. Ecto- and hemoparasites were morphologically identified. All parasites were photographed with an Olympus CAMEDIA C-5050 Zoom digital camera, then visualized and measured with the cell^B Image Acquisition Software (version 3.1; Olympus Soft Imaging Solutions).

Quantifying hemoparasitemia

Initially, the exact quantity of blood on each slide was unknown. However, in order to quantify the level of parasitemia, the quantity of blood on each slide was determined. Therefore, all blood smears were photographed with a Canon EOS 60D under the same conditions with a constant distance between the camera lens and the slide (ISO - 100, aperture F/16, shutter speed 1/30 s). The photos were cropped to the same size and showed only the blood slide without margins. They were then transformed into black and white. The color intensity of the whole image was measured with the image-processing program ImageJ (version 1.48; U.S. National Institutes of Health). The same technique was applied to blood smears from baboons prepared as reference with known blood quantities. These quantities varied between 1 and 30 µl augmenting in steps of 1 µl for the smears with volumes from 1 to 10 µl and in steps of 2 µl for the smears with volumes from 10 to 30 µl. Of each quantity, five slides were prepared, resulting in a total of 100 reference slides. The blood was obtained from baboons from the German Primate Center (Göttingen, Germany), representing residual amounts of blood taken for medical examination. The measured intensities of the baboon blood photos allowed assigning each blood amount to an intensity interval. Thus, the amount of blood on each of the lemur blood smears could be determined with an accuracy of ± 1 µl. Microfilaria present on each blood smear were counted and absolute counts were transformed into microfilaria per µl of blood.

DNA extraction, PCRs and sequence analyses

The RNAlater-preserved blood samples were dissolved with 0.5 ml distilled water and centrifuged at 4000 x *g* for 3 min. After centrifugation, the supernatant containing the RNAlater was removed. For DNA extraction with the NucleoSpin® Tissue kit (Macherey-Nagel, Düren, Germany), 180 µl lysis buffer and 25 µl proteinase K was added to the blood pellet and incubated overnight at 56 °C. The following day, DNA was purified according to the manufacturer's instructions.

To identify the microfilariae observed in a blood sample of *L. edwardsi*, a PCR was performed amplifying the ITS1–5.8S–ITS2 rDNA region using the primer set NC5 and NC2 [37]. The 50 µl reaction mixture contained 5 µl 10× Taq buffer (5 Prime, Hilden, Germany), 1 µl of 10 mM deoxynucleotide triphosphates, 2 µl of each primer (10 µmol each), 1 µl Taq Polymerase (5 Prime, Hilden, Germany) and 2 µl DNA template. PCR was performed using the peqSTAR thermocycler (Peqlab VWR, Erlangen, Germany) under the following conditions: an initial denaturation at 95 °C for 3 min, 30 cycles of 94 °C for 30 s, 55 °C for 30 s (annealing), 71 °C for 30 s (extension), followed by a final elongation step at 72 °C for 10 min. Positive and negative controls were included. The PCR products were visualized by gel electrophoresis on 1% agarose gels.

The amplified fragment was inserted into the pCR4™4-TOPO® vector and cloned into One Shot® TOP10 chemically competent *E. coli* using TOPO® TA cloning kit for sequencing (Invitrogen, Karlsruhe, Germany). Plasmid DNA was obtained using the NucleoSpin® Plasmid Kit (Macherey-Nagel, Dueren, Germany) following the manufacturer's recommendations. Afterwards, the insert was sequenced by Sanger sequencing (Seqlab Sequence Laboratories Göttingen). The obtained sequence was analysed using Clone Manager Professional Edition 9

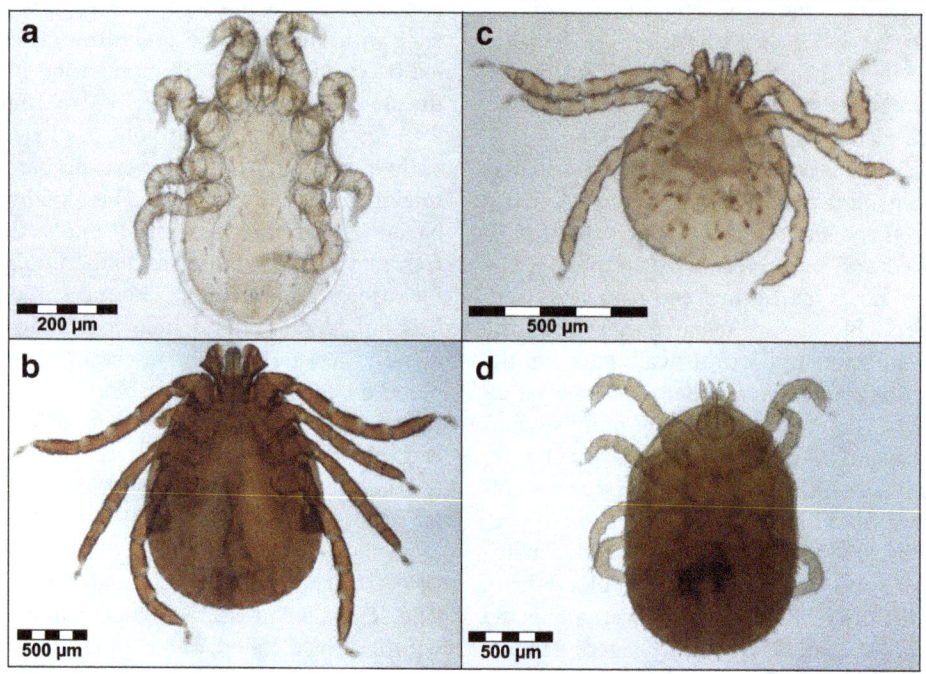

Fig. 2 Ectoparasites collected from *Lepilemur edwardsi* at Ankarafantsika National Park in Madagascar. **a** *Aetholaps trilyssa*, **b** *Haemaphysalis lemuris* (adult male), **c** *H. lemuris* (larva) **d** *Ornithodoros* sp. (nymph)

(Sci-Ed Software, Denver, USA) and compared to publicly available sequences using BLAST [38].

Statistical analyses

To assess the difference in ectoparasite prevalence (= percentage of infected individuals) between the two host species and the two sexes, we used a Fisher's Exact Test (Statistica 6.1, StatSoft. Inc. Tulsa, USA). Regarding hemoparasites, we first analyzed the overall prevalence of microfilaria infection for *A. occidentalis* and for *L. edwardsi*. For this purpose, each host individual was included once per season and a host was considered positive for a season, if at least one of its blood smears from that period was positive. For all subsequent analyses, the information content of multiple sampling was included. In order to assess the influence of host species, sex, season and time of day, when blood samples were collected, on the probability of microfilariae presence, generalized linear mixed models (GLMMs) were constructed with binomial error structure and logit link function. The models contained the variables "species" (*A. occidentalis*, *L. edwardsi*), "sex" (male, female), "season" (dry, rainy) and "time of day" (morning, noon, afternoon, night) as fixed factors. Animal ID was included as random effect, because 21 individuals (28%) of both species contributed more than one sample. We successively tested the different factors one by one by comparing them to a null model containing only the random factor with the Anova

function using the Chi-square distribution for determining the *p*-values. We then tested the full model containing all factors to evaluate their significance. Except when examining the factor host species, all variables were tested separately for the host species *A. occidentalis* and *L. edwardsi* in order to assess whether the observed effect is only present in one or in both species. These analyses were performed in R v.3.2.2 [39] using the package lme4 [40].

Additionally, the influence of the same factors ("species", "sex", "season" and "time of day") on the infection intensity, i.e. the level of microfilaremia (number of microfilariae/μl) in microfilaria-positive samples was determined. This was done with the same fixed and random factors as described above, employing a linear mixed effect model (LMM) using the package nlme [41]. For this purpose, microfilaremia values were log-transformed in order to achieve normality in the distribution.

Furthermore, a LMM was constructed to assess the impact of "species", "sex" and "season" on the length of

Table 1 Number of individuals infected with microfilariae (positives/total number of sampled animals)

Host species	Dry season	Rainy season
A. occidentalis	10/15 (66.67%)	6/9 (66.67%)
L. edwardsi	8/20 (40%)	7/16 (43.75%)

Table 2 GLMMs testing the influence of the different factors on the probability of infection with microfilaria

Measure	Term	Estimate	Standard error	z	p value
Overall prevalence	Intercept	- 7.83	2.14	- 3.66	< 0.001*
	Species	16.15	3.61	4.48	< 0.0001*
Prevalence *A. occidentalis*	Intercept	- 0.16	1.19	- 0.13	0.890
	Sex	- 1.74	1.12	- 1.55	0.122
	Season	0.73	0.00	0.73	0.467
	Time of day	- 0.03	0.63	- 0.08	0.940
Prevalence *L. edwardsi*	Intercept	8.30	3.67	2.26	0.024*
	Sex	- 3.90	5.86	- 0.67	0.506
	Season	3.40	4.36	0.78	0.435
	Time of day	0.32	1.35	0.24	0.81

*Significant p-values (< 0.05)

microfilariae. Systematic length differences between these subsamples might indicate a difference in filarial species.

Finally, we examined the host's body mass as an indicator for the host's condition. Using a Mann-Whitney U Test (IBM SPSS Statistics, version 24), it was tested whether there was a difference in body mass between individuals carrying ecto- and hemoparasites and those which did not.

Results
Ectoparasites

No ectoparasites could be found on *Avahi occidentalis* at any time of the year. *Lepilemur edwardsi* carried no lice, but mites and ticks, and both ectoparasite taxa were restricted to the rainy season. In the latter season, mites had a prevalence of 75% in *L. edwardsi* ($N = 16$ individuals). The difference in mite prevalence between the two lemur species in the rainy season was statistically significant (Fisher's Exact Test: Chi sq. = 13.93, df = 1, $p = 0.0002$).

Mites on *L. edwardsi* were macroscopically visible, crawling through the animal's fur all over the body. They were morphologically identified as *Aetholaelaps trilyssa* (Laelaptidae) (Fig. 2 a) [42]. Mite infestation was not associated with any evidence of skin alterations, such as alopecia, erythema or elevated desquamation.

Ticks were found in the inguinal region with a prevalence of 18.8% in *L. edwardsi* ($N = 16$ individuals) and were identified as *Haemaphysalis lemuris* (Ixodidae) (two larvae, one adult male, Fig. 2 b, c). They were only observed in hosts ($N = 3$ individuals) that were also infected with mites. Additionally, one of the host individuals was co-infected with *Ornithodoros* sp. (Argasidae) (one nymph, Fig. 2 d).

There was no significant difference in ectoparasite prevalence between males and females (Fisher's Exact Test: Chi sq. = 0.76, df = 1, $p = 0.38$). Body mass of individuals infested with ectoparasites was not significantly different from those who were not (Mann-Whitney $U = 51.5$, $n_1 = 10$, $n_2 = 14$ $p = 0.29$).

Hemoparasites

Microscopic examination of blood smears revealed filarial nematodes in both lemur species. The prevalence of microfilariae was 66.7% in *A. occidentalis* ($N = 24$ individuals) and 41.7% in *L. edwardsi* ($N = 36$ individuals). In both species, the prevalence was very similar across seasons (Table 1).

The GLMM revealed a significant difference in prevalence between the two host species (Table 2) with *A. occidentalis* showing a higher prevalence. Season, sex or time of blood collection had no significant effect on infection status, neither in *A. occidentalis* nor in *L. edwardsi* (Table 2).

In the positive samples, the number of microfilaremia (= intensity of infection) ranged from 0.1 to 27.9 microfilariae/µl with a mean of 3.1 ± 6.4 microfilariae/µl in *A. occidentalis* ($N = 19$ positive individuals) and from 0.2 to 10.1 microfilariae/µl with a mean of 2.4 ± 3.0 microfilariae/µl in *L. edwardsi* ($N = 19$ positive individuals) (Table 3).

The linear mixed effect model revealed no difference in microfilaremia between the two host species. In *L. edwardsi*, the factor season had the highest impact on microfilaremia. In contrast, season had no significant effect on microfilaremia in *A. occidentalis* (Fig. 3, Additional file 2).

On average, 8–9 microfilariae per sample were measured. Microfilariae of *A. occidentalis* were 200 ± 29.3 µm long ($N = 100$ microfilariae). Microfilariae of *L. edwardsi* were 190 ± 22.4 µm long ($N = 100$ microfilariae). This

Table 3 Microfilaria intensity in blood samples of *A. occidentalis* and *L. edwardsi*

Host species	Dry season	Rainy season
A. occidentalis	1.42 (± 1.88) mf/µl (N = 10, n = 10)	5.02 (± 9.03) mf/µl (N = 6, n = 9)
L. edwardsi	1.25 (± 1.67) mf/µl (N = 8, n = 10)	3.57 (± 3.73) mf/µl (N = 7, n = 9)

mf microfilariae
N number of sampled individuals
n number of positive samples

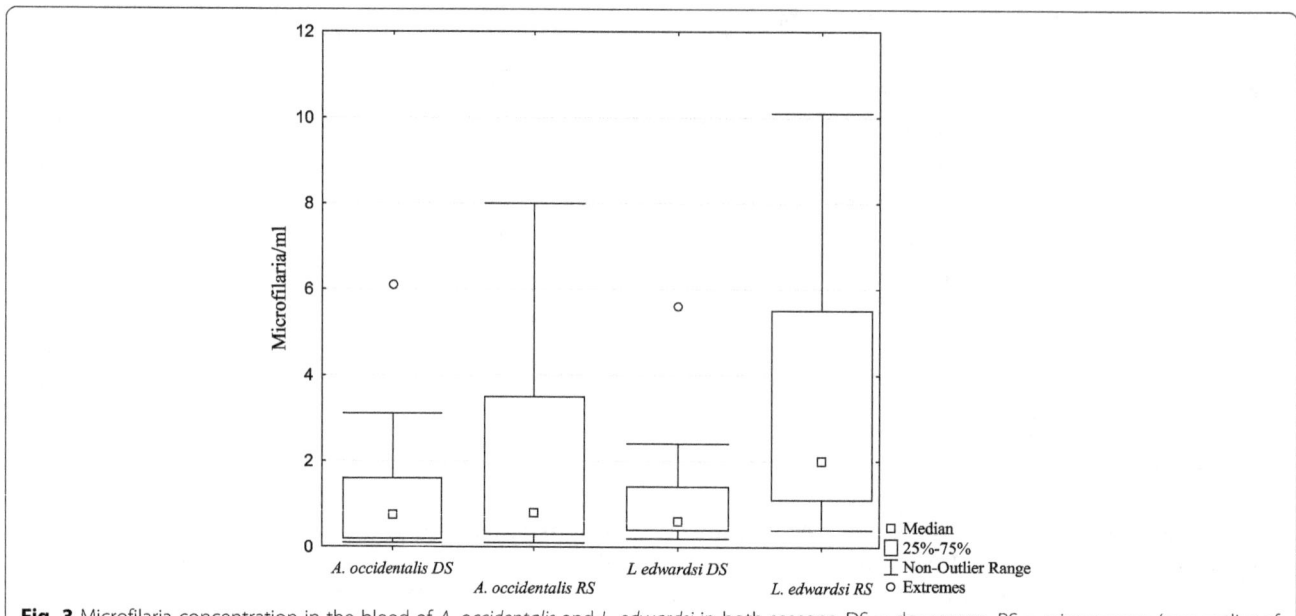

Fig. 3 Microfilaria concentration in the blood of *A. occidentalis* and *L. edwardsi* in both seasons. DS = dry season, RS = rainy season (one outlier of 27.9 from *A. occidentalis* in the rainy season was removed)

difference in length between microfilariae from the different host species was not statistically significant. However, microfilariae of both species were longer in the dry season than in the rainy season (Fig. 4) (Table 4).

The linear mixed effect model found microfilaria in *L. edwardsi* to be significantly longer in the dry than in the rainy season ($p = 0.024$, Additional file 3). In *A. occidentalis*, this factor was not significant in the full model, but a model with the single factor season explained the results better than the null model (L. Ratio: 8.47, df = 5, $p = 0.004$). The seasonal difference in length that is clearly visible at least in *L. edwardsi* might indicate the presence of different filarial species at different times of the year. In both host species body mass of individuals infected with microfilaria was not significantly different from those who were not (*L. edwardsi*: Mann-Whitney $U = 53.0$, $n_1 = 15$, $n_2 = 9$ $p = 0.41$; *A. occidentalis* Mann-Whitney $U = 42.0$, $n_1 = 8$, $n_2 = 12$, $p = 0.68$).

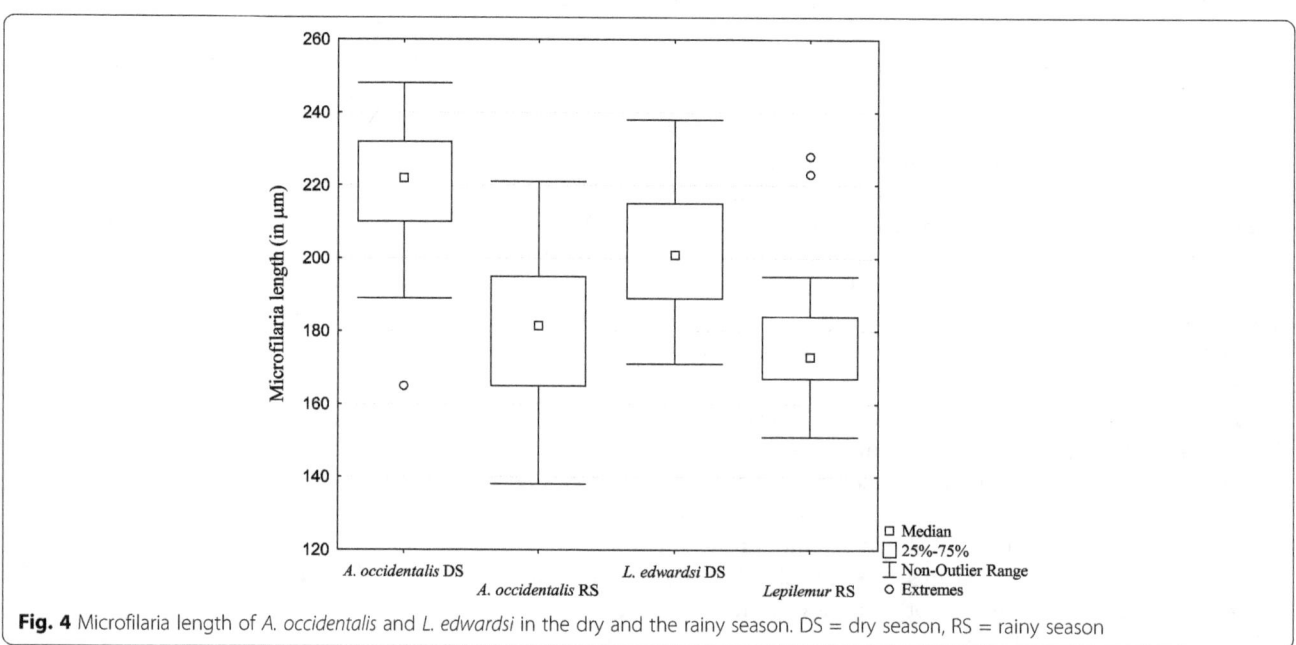

Fig. 4 Microfilaria length of *A. occidentalis* and *L. edwardsi* in the dry and the rainy season. DS = dry season, RS = rainy season

Table 4 Mean length (± SD) of microfilariae in the two sampling seasons

Host species	Dry season	Rainy season
A. occidentalis	219 (± 17.9) μm (N = 7, n = 50)	181 (± 25.8) μm (N = 6, n = 50)
L. edwardsi	202 (± 19.5) μm (N = 6, n = 50)	177 (± 17.6) μm (N = 5, n = 50)

N number of sampled individuals
n number of measured microfilariae

Co-infections with ectoparasites and microfilaria could be observed in *L. edwardsi* (n = 6). However, there were also individuals, which were infested with ectoparasites but did not show microfilariae (n = 6) or were infected with microfilariae but did not show ectoparasites (n = 4).

The ITS1–5.8S–ITS2 rDNA region of microfilariae from a blood sample of *L. edwardsi* could be successfully amplified. The sequence with a length of 1363 bp showed 83–96% identity to different nematodes belonging to the family Onchocercidae such as *Onchocerca* sp., *Brugia* sp. and *Mansonella* sp. However, the percentage of similarity did not allow reliable genus assignment. The top hit was an unnamed filarial nematode, which had been found in another lemur species, the Verreaux's sifaka (*Propithecus verreauxi*) (query coverage: 47%, identity: 96%, E-value: 0; GenBank accession no.: LN869520) [43].

Discussion

We examined whether host sleeping site ecology shapes the pattern of parasitism with regard to infection with ectoparasites and vector-borne hemoparasites by studying two ecologically similar sympatric lemur species, *Avahi occidentalis* and *Lepilemur edwardsi*. In general ectoparasite infestation was low compared to other mammals. Three species of ectoparasites, one mite and two tick species, were collected from *L. edwardsi*, whereas no ectoparasites were found on *A. occidentalis*. *L. edwardsi* showed infestation with ectoparasites only during the wet season. On the other hand, prevalence of microfilaria was significantly higher in *A. occidentalis* than in *L. edwardsi*. Neither host species showed a difference in body mass between individuals carrying mites, ticks or microfilaria and those who did not. This might be an indication that the animals' health and most likely fitness are not affected by the low parasite intensity with ecto- and haemoparasites. However, some individuals with a low body condition may have deceased quickly or fell victim to predators and may therefore not have been detected. Moreover, ectoparasites may also transmit vector-born diseases such as borreliosis [44] which may lead to rapid illness and mortality before being detected in differences in body weight. The absence of ectoparasites in the openly sleeping host *A. occidentalis* and their seasonal presence in the tree hole-sleeping *L. edwardsi* support the hypothesis that tree-holes may constitute a suitable habitat for ectoparasites. *Aetholaelaps trilyssa*, a mite belonging to the family Laelaptidae, showed a prevalence of 75% in *L. edwardsi*. Mites of this family are commonly found on various lemur species but were never associated with clinical disease [45, 46]. Some laelaptid mites are known to be nidicolous temporary parasites, living in the nest of the host but infesting the host for feeding [47]. Apart from the morphological description, there is not much known about the Malagasy endemic genus *Aetholaelaps*, but it is possible that *Aetholaelaps trilyssa* behaves as a nidicolous temporary mite. In that case, this mite's ecology would explain why it was found at high prevalence in *L. edwardsi*, a host that sleeps in regularly revisited tree holes, but not in *A. occidentalis*, a species that sleeps on open branches and rotates its sleeping site more frequently [35, 36]. The fact that some host species adapt their sleeping behavior by changing their sleeping sites in order to avoid ectoparasite infestation supports this explanation [16, 48].

A dark, moist and regularly re-visited tree hole may favor parasitism by ticks, especially with nest-adapted tick species. *Ixodes hexagonus*, for instance, is often found in or around nests of hosts and is often associated with nesting mammals like hedgehogs [49]. Ixodid ticks were observed on three *L. edwardsi* individuals (18.8%) and identified as *Haemaphysalis lemuris* (larva and adults), which has previously been described as parasitizing *Lepilemur* spp. [4, 50, 51]. Furthermore, one of these three individuals was co-infected with *Ornithodoros* sp., an argasid tick. To our knowledge, argasid ticks have not been documented to parasitize lemurs before.

Ticks are known to parasitize several host species and *H. lemuris*, in particular, has been found to parasitize a large variety of lemur species (*Microcebus rufus*, *M. griseorufus*, *Lemur catta*, *Varecia variegata*, *Lepilemur ruficaudatus*, *L. leucopus*, *Propithecus verrauxi*), some of which sleep in tree holes and some openly [52–54]. Given that many lemur species with a similar ecology were reported to be parasitized by this ectoparasite, its lack on *A. occidentalis* was surprising. However, it has to be kept in mind that only a relatively small number of individuals was sampled, and infestation may therefore have remained unnoticed if the overall prevalence was low. Nevertheless, the sampled individuals in this study nearly represent the whole *Lepilemur* and *Avahi* population of the 30.6 ha study site [55, 56]. In general, the number of studies investigating ectoparasitism in *Avahi* spp. is very limited, and so far only the sucking louse *Phtirpediculus avahidis* has been described to parasitize this genus of lemurs [57]. In summary, the pattern of ectoparasitism found in these two lemur species suggests that the choice of sleeping sites may drive the presence of ectoparasites in these primates. However, the benefits provided by tree holes, e.g. protection from predators and insulation [48], probably outweigh the costs that may be imposed by the elevated ectoparasite infestations.

All ectoparasites were only observed in the rainy season, indicating annual dynamics of parasite activity. Survival and development of mites and ticks is directly influenced by temperature, i.e. occurrence and abundance of ticks often increases with high temperatures and after rainfall [3, 58]. The tick *Amblyomma variegatum*, for instance, was even observed to disappear in cattle in the dry season [59]. As for mites, they have repeatedly been documented with invariable prevalence and intensity at different times of the year [60, 61]. Nevertheless, mites were also noted to reproduce more intensively during the reproductive periods of their hosts, i.e. while they were pregnant or lactating. As a consequence, the annual cycle may be influenced by seasonal abiotic conditions as well as seasonal reproductive activities of the host [62]. The pregnancy period of *L. edwardsi* in Ankarafantsika National Park lasts from July to November and lactation starts in October, coinciding with the beginning of the rainy season [63]. In this study, the samples representing the rainy season were collected in March and April, corresponding to the end of the rainy season. No data was collected at the beginning of the rainy season due to heavy rainfalls. However, one could speculate that the mites collected from *L. edwardsi* in March and April are the remnants of a preceding reproductive peak of the mite *A. trilyssa*, which might have occurred during the lactation period at the beginning of the rainy season.

Filarial nematodes were detected microscopically in blood smears of both lemur species. Length measurements revealed that microfilariae of both lemur species were significantly longer in the dry than in the rainy season. This might indicate the presence of two different nematode species, one occurring predominantly in the dry, and the other in the rainy season. So far, four species of microfilariae were described in lemurs, all belonging to the family Onchocercidae [64]. Of these species, the microfilariae found in the dry season correspond in size to *Paulianfilaria pauliani*. However, different methods of blood smear fixation can result in various microfilaria lengths, so that species identification by comparison of microfilariae measurements from different studies alone becomes unreliable [21]. No adult worm could be recovered as dissections of animals would have been necessary. Thus, further morphological identification was not feasible. Unfortunately, only microfilaria DNA from one sample of *L. edwardsi*, taken

in the rainy season, could be successfully sequenced. The generated sequence was most similar to sequences derived from microfilariae found in Verreaux's sifakas (*Propithecus verreauxi*), belonging to the family Onchocercidae with no further assigned name [43]. To date, these are the only two sequences from Malagasy filarial nematodes which are available in the NCBI GenBank. Consequently, BLAST search did not enable genus allocation. No RNAlater-samples were taken in the dry season, so that microfilariae detected in the dry season samples could only be analyzed morphologically.

It is likely that the filarial nematode species have been transmitted by different arthropod vectors which may vary in their abundance between the dry and the rainy season. The vectors of these microfilariae are most likely mosquitoes or other flying insects rather than ticks or lice, since microfilariae were also found in *A. occidentalis*, who did not show any ectoparasites [65, 66]. It was hypothesized that the tree hole-sleeping *L. edwardsi* should contain microfilariae less often than *A. occidentalis* that sleeps openly. In support of the hypothesis that tree holes may confer some degree of protection from flying vectors, such as mosquitoes, the prevalence of microfilariae was significantly higher in *A. occidentalis* than in *L. edwardsi*. A total of 66.7% of the population of *A. occidentalis* carried microfilariae, whereas only 41.7% of *L. edwardsi* were infected. As the lifestyles of these two lemur hosts are very similar in terms of habitat use and group size and they are also comparable in body size, other factors are less likely to cause this difference in filarial prevalence. However, it is possible that the microfilaria species may be better adapted to *A. occidentalis*, enhancing microfilaria survival in this primate host. Furthermore, *A. occidentalis* may be more susceptible to microfilarial infection. Another aspect to be considered is that the two lemur species may carry different microfilaria species, one of which occurring more frequently, explaining the difference in prevalence. In case we are dealing with only one filarial species, which infects both lemur species, *A. occidentalis* might be more susceptible. The findings also suggest that the insect vectors are diurnal, since both lemur species stay in their sleeping sites during daytime and differ then in exposure, whereas they are both nocturnal arboreal foragers. However, no difference in the level of microfilaremia was detected between the two hosts. Nevertheless, these findings may indicate that sleeping in tree holes may affect prevalence of hemoparasites in primate hosts.

In contrast, no effect of season on microfilaria prevalence was found despite presumably higher vector abundance during the wet season, neither in *A. occidentalis* nor in *L. edwardsi*. Filarial nematodes have a long prepatent period and once developed, they persist in their host for several months, possibly concealing a seasonal effect. In Kirindy Forest, a habitat with comparable seasonal conditions to those in Ankarafantsika National Park, Verreaux's sifakas (*Propithecus verreauxi*) hosting microfilaria also showed no effect of seasonality on infection [43]. However, when looking at the level of microfilaremia as a proxy for infection intensity, the model containing the factor season in *L. edwardsi* differed significantly from the null model, indicating higher infection intensity in the rainy season (cf. Fig. 3).

No relation was detected between sex and either presence or infection intensity of filarial nematodes. Some studies have reported higher prevalence in males, e.g. in raccoons (*Procyon lotor*) [67], probably caused by sex-differences in body size or in hormone levels, such as immunosuppressive effects of testosterone [68]. The studied lemur species do not display sexual dimorphism in size, resulting in equal exposure of both sexes to vectors. Additionally both species are pair-living with dominant or at least co-dominant females [34], which may result in a smaller difference in androgen levels between sexes compared to other mammals [69]. Males and females may therefore be equally immunocompetent, although future studies are needed to investigate this hypothesis in more detail.

Since the presence of microfilariae circulating in the blood stream is subjected to a circadian rhythm [70], time of sampling was included as a factor in the statistical models. However, neither the likelihood nor intensity of infection was affected by the time of blood collection. Microfilaraemia levels are dependent on the intravascular distribution of the parasite. Dreyer et al. [71] documented an average microfilaria concentration that was 1.25 times higher in capillary blood than in venous blood obtained at the same time. Moreover, the number of microfilariae was proportionally higher in the capillary system of the skin at the time when biting activity of the local mosquito vector was highest. In our study, blood smears were sometimes derived from venous blood drawn from the femoral vein and at other times from capillary blood that was collected in the process of ear-marking and tissue collection. This sampling disparity might have led to the absence of an effect of time of blood collection on microfilaria concentration, as well as to the lack of difference in infection intensity between *A. occidentalis* and *L. edwardsi* mentioned above.

Conclusions

The findings of the present study support the hypothesis that sleeping site ecology affect patterns of parasitism in nocturnal, arboreal primate hosts in a seasonal environment. *L. edwardsi*, which uses tree holes as sleeping sites, showed infestation with three different ectoparasite taxa during the wet season, whereas no ectoparasites were found on *A. occidentalis*, which sleeps on open branches but is otherwise ecologically similar. Thus, repeatedly revisited tree holes seem to present a driver of ectoparasite infestations. In contrast, prevalence of microfilaria was significantly higher in *A. occidentalis* than in *L. edwardsi*. Hence, sleeping in tree holes might protect from bites of flying insects that transmit hemoparasites, such as filarial nematodes. In conclusion, sleeping site ecology is an important ecological driver of parasite distribution and transmission in wild populations.

Additional files

Additional file 1: Table with the individual blood sampling frequency in the dry and in the rainy season. (DOCX 15 kb)

Additional file 2: Table with the results of the LMMs testing the influence of host species, sex, season and time of day on the number of microfilaremia (= intensity of infection). (DOCX 17 kb)

Additional file 3: Table with the results of the LMMs testing the influence of host species, sex and season on microfilaria length. (DOCX 18 kb)

Acknowledgements

We would like to thank the Ministère de l'Environnement, de l'Ecologie et des Forêts and Madagascar National Parks for granting research permits. We are grateful to Solofonirina Rasoloharijaona for continuous support in Madagascar, Bertrand Andriatsitohaina for help in the sample collection and Jhonny Kennedy for guide services, as well as the Durrell Wildlife Preservation Trust for providing climate data of Ampijoroa. We thank Kerstin Mätz-Rensing (DPZ, Göttingen) for providing the baboon blood. We also thank Sabine Schicht for advice during the genetic work and Andrea Springer for commenting on the manuscript, as well as Sönke von den Berg for technical support.

Funding

The German Academic Exchange Service (DAAD) funded travel expenses and field stay for data collection. The German Society for Primatology (GfP) supported costs for material and Malagasy field assistants.

Author's Contributions

EZ initiated this study and guided the zoological part in Madagascar. EZ, CS and UR conceived and designed the study. MH collected the data, performed laboratory work and drafted the manuscript. CS supervised the parasitological examinations and guided the genetic work regarding the hemoparasites. UR instructed statistical analyses of the data. All authors participated in data analysis and interpretation and revised and approved the final manuscript.

Competing interests

The authors declare that they have no competing interests.

Author details

[1]Institute of Zoology, University of Veterinary Medicine Hannover, Buenteweg 17, 30559 Hannover, Germany. [2]Institute for Parasitology, Centre for Infection Medicine, University of Veterinary Medicine Hannover, Buenteweg 17, 30559 Hannover, Germany.

References

1. Heath A. Seasonality in ectoparasites. New Zeal Entomol. 1978;6:364–5.
2. Kovats R, Campbell-Lendrum D, McMichel A, Woodward A, Cox JSH. Early effects of climate change: do they include changes in vector-borne disease? Phil Trans R Soc B. 2001;356:1057–68.
3. Kaiser M, Sutherst R, Bourne A, Gorissen L, Floyd R. Population dynamics of ticks on Ankole cattle in five ecological zones in Burundi and strategies for their control. Prev Vet Med. 1988;6:199–222.
4. Klompen H, Junge RE, Williams CV. Ectoparasites of *Propithecus diadema* (primates: Indriidae) with notes on unusual attachment site selection by *Haemaphysalis lemuris* (Parasitiformes: Ixodidae). J Med Entomol. 2015;52: 315–9.
5. Gannon MR and Willig MR. Ecology of ectoparasites from tropical bats. Environm Entomol. 1995;24:1495–1503.
6. Kowalewski M, Zunino GE. The parasite behavior hypothesis and the use of sleeping sites by black howler monkeys (*Alouatta caraya*) in a discontinuous forest. Neotrop Primates. 2005;13:22–6.
7. Mooring MS, Hart BL. Animal grouping for protection from parasites: selfish herd and encounter-dilution effects. Behaviour. 1992;123:173–93.
8. Samson DR, Muehlenbein MP, Hunt KD. Do chimpanzees (*Pan troglodytes schweinfurthii*) exhibit sleep related behaviors that minimize exposure to parasitic arthropods? A preliminary report on the possible anti-vector function of chimpanzee sleeping platforms. Primates. 2013;54:73–80.
9. Anderson JR. Sleep, sleeping sites, and sleep-related activities: awakening to their significance. Am J Primatol. 1998;46:63–75.
10. Heymann EW. Sleeping habits of tamarins, *Saguinus mystax* and *Saguinus fuscicollis* (Mammalia; primates; Callitrichidae), in north-eastern Peru. J Zool. 1995;237:211–26.
11. Reichman O, Smith SC. Burrows and burrowing behavior by mammals. J Mammal. 1990;2:197–244.
12. Schmid J. Tree holes used for resting by gray mouse lemurs (*Microcebus murinus*) in Madagascar: insulation capacities and energetic consequences. Int J Primatol. 1998;19:797–809.
13. Radespiel U, Cepok S, Zietemann V, Zimmermann E. Sex-specific usage patterns of sleeping sites in grey mouse lemurs (*Microcebus murinus*) in northwestern Madagascar. Am J Primatol. 1998;46:77–84.
14. Krasnov BR, Matthee S, Lareschi M, Korallo-Vinarskaya NP, Vinarski MV. Co-occurrence of ectoparasites on rodent hosts: null model analyses of data from three continents. Oikos. 2010;119:120–8.
15. Hart BL. Behavioral adaptations to pathogens and parasites: five strategies. Neurosci Biobehav Rev. 1990;14:273–94.
16. Butler J, Roper T. Ectoparasites and sett use in European badgers. Anim Behav. 1996;52:621–9.
17. Patterson BD, Dick CW, Dittmar K. Parasitism by bat flies (Diptera: Streblidae) on neotropical bats: effects of host body size, distribution, and abundance. Parasitol Res. 2008;103:1091–100.
18. Martinsen E, Paperna I, Schall J. Morphological versus molecular identification of avian Haemosporidia: an exploration of three species concepts. Parasitology. 2006;133:279–88.
19. Valkiūnas G, Iezhova TA, Križanauskienė A, Palinauskas V, Sehgal RN, Bensch S. A comparative analysis of microscopy and PCR-based detection methods for blood parasites. J Parasitol. 2008;94:1395–401.
20. Nelson G. The identification of infective filarial larvae in mosquitoes: with a note on the species found in "wild" mosquitoes on the Kenya coast. J Helminthol. 1959;33:233–56.
21. Schacher JF. Morphology of the microfilaria of *Brugia pahangi* and of the larval stages in the mosquito. J Parasitol. 1962;679–92.
22. Perez-Orella C, Schulte-Hostedde AI. Effects of sex and body size on ectoparasite loads in the northern flying squirrel (*Glaucomys sabrinus*). Can J Zool. 2005;83:1381–5.
23. Klein S. Hormonal and immunological mechanisms mediating sex differences in parasite infection. Parasite Immunol. 2004;26:247–64.
24. Sol D, Jovani R, Torres J. Geographical variation in blood parasites in feral pigeons: the role of vectors. Ecography. 2000;23:307–14.
25. Viljoen H, Bennett NC, Ueckermann EA, Lutermann H. The role of host traits, season and group size on parasite burdens in a cooperative mammal. PLoS One. 2011;6:e27003.
26. Christe P, Glaizot O, Evanno G, Bruyndonckx N, Devevey G, Yannic G, Patthey P, Maeder A, Vogel P, Arlettaz R. Host sex and ectoparasites choice: preference for, and higher survival on female hosts. J Anim Ecol. 2007;76: 703–10.
27. Chapman CA, Wasserman MD, Gillespie TR, Speirs ML, Lawes MJ, Saj TL, Ziegler TE. Do food availability, parasitism, and stress have synergistic effects on red colobus populations living in forest fragments? Am J Phys Anthropol. 2006;131:525–34.
28. Coop RL, Holmes PH. Nutrition and parasite interaction. Int J Parasitol. 1996; 26:951–62.
29. Newey S, Thirgood SJ, Hudson PJ. Do parasite burdens in spring influence condition and fecundity of female mountain hares Lepus Timidus? Wildl Biol. 2004;10:171–6.
30. Milton K. Effects of bot fly (*Alouattamyia baeri*) parasitism on a free-ranging howler (*Alouatta palliata*) population in Panama. J Zool Soc London. 239: 39–63.
31. Rafalinirina HA, Aivelo T, Wright PC, Randrianasy J. Comparison of parasitic infections and body condition in rufous mouse lemurs (*Microcebus rufus*) at Ranomafana National Park, southeast Madagascar. Madagascar Conserv Dev. 2015;10:60–6.
32. Mittermeier R, Louis EEJ, Richardson M, Schwitzer C, Langeand O, Rylandy AB, Hawkins F, Rajaobelina S, Ratsimbazafy J, Rasoloarison R, et al. Lemurs of Madagascar. 3rd ed. Washington D.C.: Conservation International; 2010.
33. Méndez-Cárdenas MG, Zimmermann E. Duetting — a mechanism to strengthen pair bonds in a dispersed pair-living primate (*Lepilemur edwardsi*)? Am J Phys Anthropol. 2009;139:523–32.
34. Ramanankirahina R, Joly M, Zimmermann E. Peaceful primates: affiliation, aggression, and the question of female dominance in a nocturnal pair-living lemur (*Avahi occidentalis*). Am J Primatol. 2011;73:1261–8.
35. Rasoloharijaona S, Rakotosamimanana B, Randrianambinina B, Zimmermann E. Pair-specific usage of sleeping sites and their implications for social organization in a nocturnal Malagasy primate, the Milne Edwards' sportive lemur (*Lepilemur edwardsi*). Am J Phys Anthropol. 2003;122:251–8.
36. Ramanankirahina R, Joly M, Zimmermann E. Seasonal effects on sleeping site ecology in a nocturnal pair-living lemur (*Avahi occidentalis*). Int J Primatol. 2012;33:428–39.
37. Gasser R, LeGoff L, Petit G, Bain O. Rapid delineation of closely-related filarial parasites using genetic markers in spacer rDNA. Acta Trop. 1996;62:143–50.
38. Altschul SF, Gish W, Miller W, Myers EW, Lipman DJ. Basic local alignment search tool. J Mol Biol. 1990;215:403–10.
39. R Core Team 2015. R: A language and environment for statistical computing. R Foundation for Statistical Computing, Vienna, Austria. http://www.R-project.org/ last Accessed 16 Aug 2016.
40. Bates D, Mächler M, Bolker B, Walker S. Fitting linear mixed-effects models using lme4. J Stat Softw. 2014;67:1–48.
41. Pinheiro J, Bates D, DebRoy S, Sarkar D. Linear and nonlinear mixed effects models. R package version. 2007;3:57.
42. Domrow R, Taufflieb R. A second species of *Aetholaelaps* from a Malagasy lemur (Acarina, Laelapidae). Acarologia. 1963;
43. Springer A, Fichtel C, Calvignac-Spencer S, Leendertz FH, Kappeler PM. Hemoparasites in a wild primate: infection patterns suggest interaction of plasmodium and Babesia in a lemur species. Int J Parasitol Parasites Wildl. 2015;4:385–95.
44. Larsen PA, Hayes CE, Williams CV, Junge RE, Razafindramanana J, Mass V, Rakotondrainibe H, Yoder AD. Blood transcriptomes reveal novel parasitic zoonoses circulating in Madagascar's lemurs. Biol Lett. 2016;12: 20150829.
45. Junge RE, Dutton CJ, Knightly F, Williams CV, Rasambainarivo FT, Louis EE. Comparison of biomedical evaluation for white-fronted brown lemurs (*Eulemur fulvus albifrons*) from four sites in Madagascar. J Zoo Wildl Med. 2008;39:567–75.
46. Singleton CL, Norris AM, Sauther ML, Cuozzo FP, Youssouf Jacky IA. Ring-tailed lemur (*Lemur catta*) health parameters across two habitats with varied levels of human disturbance at the Bezà Mahafaly special reserve, Madagascar. Folia Primatol. 2015;86:56–65.
47. O'Connor BM. Acariformes: parasitic and commensal mites of vertebrates. In: Goodman SM, Benstead JP, editors. The natural history of Madagascar. Chicago and London: University of Chicago Press; 2003. p. 593–602.
48. Reckardt K, Kerth G. Roost selection and roost switching of female Bechstein's bats (*Myotis bechsteinii*) as a strategy of parasite avoidance. Oecologia. 2007;154:581–8.
49. Arthur DR. The host relationships of Ixodes hexagonus leach in Britain. Parasitology. 1953;43:227–38.
50. Hoogstraal H, Theiler G. Ticks (Ixodoidea, Ixodidae) parasitizing lower primates in Africa, Zanzibar, and Madagascar. J Parasitol. 1959;45:217–22.

51. Uilenberg G, Hoogstraal H and Klein J-M. Les tiques (Ixodoidea) de Madagascar et leur role vecteur (Ticks of Madagascar in their roles as vectors). Arch Inst Pasteur Madagascar. 1979;Numéro Spécial:46.

52. Durden LA, Zohdy S, Laakkonen J. Lice and ticks of the eastern rufous mouse lemur, *Microcebus rufus*, with descriptions of the male and third instar nymph of *Lemurpediculus verruculosus* (Phthiraptera: Anoplura). J Parasitol. 2010;96:874–8.

53. Hoogstraal H. Ticks (Ixodoidea) of the Malagasy faunal region (excepting the Seychelles). Bull Mus Comp Zool Harv Coll. 1953;111:36–113.

54. Rodriguez IA, Rasoazanabary E, Godfrey LR. Multiple ectoparasites infest *Microcebus griseorufus* at Beza Mahafaly special reserve. Madagascar Conserv Dev. 2012;7:45–8.

55. Ganzhorn JU. Food partitioning among Malagasy primates. Oecologia. 1988; 75:436–50.

56. Warren RD, Crompton RH. A comparative study of the ranging behaviour, activity rhythms and sociality of *Lepilemur edwardsi* (primates, Lepilemuridae) and *Avahi occidentalis* (primates, Indriidae) at Ampijoroa. Madagascar J Zool. 1997;243:397–415.

57. Paulian R. Un nouvel anoploure de lémurien malgache. Bull Soc Entomol France. 1960;65:306–8.

58. Shoorijeh SJ, Ghasrodashti AR, Tamadon A, Moghaddar N, Behzadi MA. Seasonal frequency of ectoparasite infestation in dogs from shiraz, southern Iran. Turk J Vet Anim Sci. 2008;32:309–13.

59. Mattioli R, Janneh L, Corr N, Faye J, Pandey V, Verhulst A. Seasonal prevalence of ticks and tick-transmitted haemoparasites in traditionally managed N'Dama cattle with reference to strategic tick control in the Gambia. Med Vet Entomol. 1997;11:342–8.

60. Wright P, Arrigo-Nelson S, Hogg K, Bannon B, Morelli TL, Wyatt J, Harivelo A, Ratelolahy F. Habitat disturbance and seasonal fluctuations of lemur parasites in the rain forest of Ranomafana National Park, Madagascar. In: Huffman MA, Chapman CA, editors. Primate parasite ecology: the dynamics and study of host-parasite relationships. Cambridge: Cambridge University Press; 2009. p. 311–30.

61. Schwitzer N, Clough D, Zahner H, Kaumanns W, Kappeler P, Schwitzer C. Parasite prevalence in blue-eyed black lemurs *Eulemur flavifrons* in differently degraded forest fragments. Endang Species Res. 2010;12:215–25.

62. Lourenço S, Palmeirim JM. Which factors regulate the reproduction of ectoparasites of temperate-zone cave-dwelling bats? Parasitol Res. 2008;104:127–34.

63. Randrianambinina B, Mbotizafy S, Rasoloharijaona S, Ravoahangimalala R, Zimmermann E. Seasonality in reproduction of *Lepilemur edwardsi*. Int J Primatol. 2007;28:783–90.

64. Irwin MT, Raharison J-L. A review of the endoparasites of the lemurs of Madagascar. Malagasy Nat. 2009;2:66–93.

65. Klei T, Rajan T. World class parasites: volume 5. The Filaria. New York, Boston, Dordrecht, London, Moscow: Kluwer Academic Publisher; 2002.

66. Anderson RC. Nematode parasites of vertebrates: their development and transmission. 2nd ed. New York: CABI Publishing; 2000.

67. Telford SR Jr, Forrester DJ. Hemoparasites of raccoons (*Procyon lotor*) in Florida. J Wildlife Dis. 1991;27:486–90.

68. Zuk M, McKean KA. Sex differences in parasite infections: patterns and processes. Int J Parasitol. 1996;26:1009–24.

69. Von Engelhard N, Kappeler PM, Heistermann M. Androgen levels and female social dominance in *Lemur catta*. Proc Biol Sci. 2000;267:1533–9.

70. Hawking F. Microfilaria infestation as an instance of periodic phenomena seen in host-patasite relationships. Ann N Y Acad Sci. 1962;98:940–53.

71. Dreyer G, Pimentael A, Medeiros Z, Beliz F, Moura I, Coutinho A, de Andrade LD, Rocha A, da Silva LM, Piessens WF. Studies on the periodicity and intravascular distribution of *Wuchereria bancrofii* microfilariae in paired samples of capillary and venous blood from Recife, Brazil. Tropical Med Int Health. 1996;1:264–72.

11

Higher resting metabolic rate in long-lived breeding Ansell's mole-rats (*Fukomys anselli*)

Charlotte Katharina Maria Schielke[1], Hynek Burda[1,2], Yoshiyuki Henning[1], Jan Okrouhlík[3] and Sabine Begall[1*] (iD)

Abstract

Background: Reproduction is an energetically expensive process that supposedly impairs somatic integrity in the long term, because resources are limited and have to be allocated between reproduction and somatic maintenance, as predicted by the life history trade-off model. The consequence of reduced investment in somatic maintenance is a gradual deterioration of function, i.e. senescence. However, this classical trade-off model gets challenged by an increasing number of contradicting studies. Here we report about an animal model, which adds more complexity to the ongoing debate. Ansell's mole-rats are long-lived social subterranean rodents with only the founder pair reproducing, while most of their offspring remain in the parental burrow system and do not breed. Despite of a clear reproductive trade-off, breeders live up to twice as long as non-breeders, a unique feature amongst mammals.

Methods: We investigated mass-specific resting metabolic rates (msRMR) of breeders and non-breeders to gain information about the physiological basis underlying the reproduction-associated longevity in Ansell's mole-rats. We assessed the thermoneutral zone (TNZ) for breeders and non-breeders separately by means of indirect calorimetry. We applied generalized linear mixed-effects models for repeated measurements using the msRMR in the respective TNZs.

Results: TNZ differed between reproductive and non-reproductive Ansell's mole-rats. Contrary to classical aging models, the shorter-lived non-breeders had significantly lower msRMR within the thermoneutral zone compared to breeders.

Conclusion: This is the first study reporting a positive correlation between msRMR and lifespan based on reproductive status. Our finding contradicts common aging theories, but supports recently introduced models which do not necessarily link reproductive trade-offs to lifespan reduction.

Keywords: Aging, Reproduction, Mole-rat, Resting metabolic rate, Oxidative stress

Background

Aging is defined as a gradual decline in intrinsic physiological function leading to an increase in morbidity and mortality rate (reviewed in: [1]). However, the mechanisms behind aging, i.e. senescence processes are still poorly understood. The disposable soma theory is a prevailing model of aging, which is based on a trade-off between energy demanding processes, including growth, somatic maintenance, and reproduction due to limited resource availability [2]. Reproduction is energetically expensive [3–6], and is often considered a central force shaping different life histories [7]. As soon as an animal starts reproducing, energy resources are allocated to reproduction. Consequently, less energy is available for somatic maintenance and protection, leading to a gradual accumulation of somatic damage. This process is thought to be even amplified, because reproduction increases the metabolic rate to cover the increased energy demand [8]. This increase in metabolic rate is predicted to result in oxygen radicals, i.e. reactive oxygen species (ROS), highly reactive byproducts which cause oxidative damage to DNA, lipids and proteins [1].

However, high energy turnover does not necessarily increase oxidative damage and mortality. Contrary to earlier expectations, correlational and experimental

---

* Correspondence: sabine.begall@uni-due.de
[1]Faculty of Biology, University of Duisburg-Essen, Essen, Germany
Full list of author information is available at the end of the article

studies published recently show no negative effect of high metabolic rate on lifespan [9–12], or even a positive association [13]. Moreover, in the brown trout, higher metabolic rates were negatively correlated with levels of H_2O_2, a highly potent ROS. These controversial associations between metabolic rate and oxidative damage and / or lifespan can in parts be ascribed to different experimental setups and tissues studied (reviewed in: [8, 14, 15]). Moreover, some studies indicate that at least reproductive females have an increased ability to protect from oxidative damage, termed oxidative shielding, in order to protect the offspring from prenatal somatic damage (reviewed in: [8]). Consequently, more research is needed to gather representative data from animals with different life histories, to gain a comprehensive understanding of how life history trade-offs influence lifespan. Here we present the data of a subterranean mammal with a unique life history to contribute to the discussion stated above.

Ansell's mole-rats (*Fukomys anselli*) are subterranean rodents of the family Bathyergidae with an extraordinary long lifespan (22 years being the maximum recorded age thus far; own observations). They live in multigenerational families where typically only the founder pair (breeders) reproduces. Most of the offspring (non-breeders) forego reproduction and remain in the natal family. Incestuous mating (i.e. between brothers and sisters) usually does not occur, however, adult non-breeders readily mate with unrelated conspecifics if given a possibility [16, 17]. A clear contradiction to the classic trade-off model has been shown in this species: breeding individuals live up to twice as long as their non-breeding counterparts, a feature which is unique amongst mammals [18].

Thus far, proximate factors contributing to this bimodal aging pattern are not known. In contrast to naked mole-rats where reproductive behavior of non-breeders is aggressively suppressed by the mother, Ansell's mole-rats facilitate incest avoidance by individual recognition. Moreover, they exhibit pronounced sociopositive behaviors like grooming and huddling between all family members [19]. Moreover, previous studies showed that daily activity between breeders and non-breeders does not show differences, and social rank does not influence life expectancy [18, 20, 21]. Hence, extrinsic factors like aggression, fighting and higher workload in non-breeders are not likely to influence the lifespan difference in first-place. Thyroid hormone levels, as a possible intrinsic factor, showed no status-dependent difference, as well [22].

Here, we test the hypothesis that breeders and non-breeders of Ansell's mole-rats differ in their mass specific resting metabolic rate (msRMR), as a possible approach to understand the bimodal aging pattern.

Methods

Study animals

We measured oxygen consumption (VO_2) using open flow respirometry in 26 Ansell's mole-rats (six reproductive females and six reproductive males, eight non-reproductive females and six non-reproductive males) in order to determine msRMR for both reproductive states (Table 1). None of the reproductive females were pregnant or lactating during the time of measurements. Of the 26 tested Ansell mole-rats, four non-reproductive (two males, two females) and four reproductive animals (three males, one female) were wild-captured and lived for more than one year in captivity, thus being fully acclimatized to laboratory conditions. Trapping and export of wild Ansell's mole-rats were approved by the Zambian Wildlife Authorities (permit numbers 4790 and 4060, issued June 7th, 2010). Maintenance was approved by the Veterinary Office of the City of Essen (AZ: 32-2-1180-71/328). Housing conditions have been described elsewhere [23]. Briefly, mole-rats were kept on animal litter in glass terraria in the animal facilities of the Department of General Zoology at the University of Duisburg-Essen (Germany). Mole-rats were fed ad libitum with carrots, potatoes, and once per week with apples, salad and cereals. Light conditions were 12D:12L. All experiments were performed in accordance with the guidelines and regulations of the local authorities. All experiments were approved by the North Rhine-Westphalia State Environment Agency (permit no. AZ: 87-51.04.2010.A359/01) and have been performed in accordance with their guidelines and regulations.

Experimental procedure

We first determined the thermoneutral zone (TNZ) for both reproductive states by measuring oxygen consumption (VO_2) of Ansell's mole-rats at 13 different ambient temperatures (T_a) ranging between 10 and 40 °C. For non-reproductive individuals the sample size was 14 at temperatures 10, 15, 20, 25, 26, 28, 30, 32, 33, 34, 35, and 37 °C and 11 at 40 °C. For reproductive animals, the sample size was 12 for temperatures 28, 30, 32, and 33 °C and 4 for temperatures 20, 25, 26, 34, 35, and 37 °C. The tested temperatures did not include extremes for reproductive animals in order to avoid impairments of reproductive animals from established colonies due to hypothermia and/or hyperthermia. Especially high temperatures can be critical, as prolonged exposure of Ansell's mole-rats to ambient temperatures about 42 °C were shown to be potentially lethal (previous own accidental observations).

Before each experimental trial, the animals were food deprived for at least 12 h. To establish an oxygen analyzer baseline, the oxygen analyzer (Servomex Type 5200 Multi Purpose, Crowborough, UK) was calibrated using 99.999% N_2 (Air Liquide, Düsseldorf, Germany) for 0% oxygen and compressed outdoor air for 20.95% oxygen at the beginning

Table 1 Basic parameters and mean mass-specific resting metabolic rate (msRMR) of *Fukomys anselli*

Sex	Status	N	Age (years)	Body mass (g)	Range of msRMR ($ml\ O_2 \times g^{-1} \times h^{-1}$)	Mean msRMR ($ml\ O_2 \times g^{-1} \times h^{-1}$)
M	R	6	6.9 ± 2.4	83.9 ± 10.5	0.77–1.71	1.17 ± 0.13
	NR	6	4.8 ± 2.8	108.8 ± 21.4	0.43–1.46	0.91 ± 0.17
F	R	6	8.9 ± 3.1	87.2 ± 21.8	0.63–1.80	1.17 ± 0.20
	NR	8	2.4 ± 1.2	68.2 ± 8.9	0.42–1.79	0.86 ± 0.21
Grand mean		26	5.8 ± 4.3	85.6 ± 21.4	0.42–1.80	1.02 ± 0.23

Age, body mass and mass-specific resting metabolic rates within the boundaries of their respective thermoneutral zones (non-reproductive animals: 26–30 °C; reproductive animals: 28–33 °C) of studied *Fukomys anselli* (*N* sample size, *M* males, *F* females, *R* reproductive; *NR* non-reproductive). Values are given as mean ± SD

and at the end of each experimental trial. After initial calibration, the animals were placed in a custom-made tight stainless steel metabolic chamber with a volume of 896 ml (16 cm × 7 cm × 8 cm) supplemented with tissue paper as nest litter. The chamber was closed with a clear Perspex lid to enable direct observation of the animal's behavior inside the chamber. The watertight chamber was submerged in a water bath (temperature controlled by HAAKE D1, Type 001–3603, Germany). The animals spent at least 30 min within the chamber to acclimatize before measurements started. Measurements were finished when the animal calmed down within the chamber and stable data were gathered over a period of 10 min, which can be considered a truly resting state. In this state, the animals lay down and breathe regularly. If an animal did not calm down within 60 min, the experiment was stopped and repeated later with a time lag of three days at minimum. Under extreme temperatures (<15 °C and >35 °C) the maximum duration of an experimental trial was reduced to 30 min to avoid hypothermia or overheating, respectively. Temperature within the chamber was measured with a digital thermometer (GMH 3230, Greisinger Electronic, Germany) with its probe placed behind a grid to avoid damage by the animals. Ambient air was pushed through the chamber at a constant rate of 350 ml × min^{-1}. The incurrent airflow was regulated by a flow controller (Model 35830, Analyt-MTC, Müllheim, Germany), following a carbon dioxide (Sodalime) and a water trap (indicating Drierite) before the oxygen content was measured by a paramagnetic oxygen sensor (Servomex Type 5200 Multi Purpose, Crowborough, UK). Software DIAdem 8.0 (National Instruments, Germany) was used to visualize and to record the oxygen content of the excurrent air every second. Immediately after each trial the animals were weighted. Oxygen consumption was calculated following the equation of Lighton [24]:

$$VO_2 = \frac{F_{ri}(F_iO_2 - F_eO_2)}{1 - F_eO_2}$$

with: VO_2 = oxygen consumption (ml O_2 × min^{-1}), F_{ri} = incurrent flow rate (ml × h^{-1}) F_iO_2 = oxygen incurrent fractional concentration (%), F_eO_2 = oxygen excurrent fractional concentration. Data gathered by the

oxygen sensor were corrected to standard temperature and pressure conditions (273.15 K, 101.325 kPa). The RMR was calculated as a 10-min mean of lowest oxygen consumption and expressed in ml O_2 × h^{-1}. msRMR was then calculated as RMR divided by the individual's weight and expressed in ml O_2 × g^{-1} × h^{-1}.

Statistical analysis

The extent of TNZ was assessed separately for reproductive and non-reproductive Ansell's mole-rats by a step-down procedure [25] of permutation version of the Jonckheere-Terpstra test [26]. Briefly, starting from the temperature, where the mean msRMR was the lowest, we tried to detect an increasing trend in msRMR data assorted by increasing ambient temperature. If a trend was detected, the msRMR data for the highest temperature were excluded and tested again until no trend in msRMR data was detected and the upper critical temperature was identified. The determination of the lower critical temperature was similar, but here the lowest temperature was excluded stepwise until no trend could be detected. The metabolic rate data were included in TNZ estimation only when the sample size for the given temperature comprised five or more values. The analyses revealed different ranges of TNZ for non-reproductive and reproductive individuals. The observed range of TNZ was 26–30 °C in non-reproductive and 28–33 °C in reproductive Ansell's mole-rats, respectively (Table 2).

To estimate the effects of various predictors on msRMR (dependent variable) within the TNZ boundaries, data were analyzed with generalized linear mixed-effects models for repeated measurements using the lme4 package [27] and Gamma distribution was assumed (link identity). First, we extended the null model to include one of the independent variables (square-root transformed body mass, ambient temperature as factor, reproductive status) and tested which of them optimally improved the model using the Akaike information criterion and model comparison by χ^2-statistic. Subsequently, we extended this model by inclusion of different variables in the same way as described above and we continued until there was no variable left which would significantly improve the model (forward selection). Finally, interaction model of all significant factors was constructed.

Table 2 Critical temperatures and Jonckheere-Terpstra statistical analysis for reproductive and non-reproductive *Fukomys anselli*

Parameter	Reproductive	Non-reproductive
Upper critical temperature	≥33 °C[a]	30 °C
Temperature range /JT/p	N/A	28...32 °C/368/0.0486
Temperature range /JT/p	32...33 °C/90/0.16	28...30 °C/102/0.43
Lower critical temperature	28 °C	26 °C
Temperature range /JT/p	26...33 °C/421/0.0002	25...30 °C/471/0.043
Temperature range /JT/p	28...33 °C/410/0.346	26...30 °C/246/0.1364

Upper critical temperature and lower critical temperature together with parameters of statistical analysis in reproductive and non-reproductive *F. anselli*. [a]minimal value only, because of lack of RMR data for reproductive animals at temperatures above 33 °C, see Methods for more detail. JT represents Jonckheere-Terpstra-statistics, p is the probability of trend in data

Individual identity was treated as a random factor. Since the age of some individuals was not known, the same procedure as described above with an additional factor age was applied to all animals with known age. This analysis showed that age is not a significant factor (results not shown). All calculations and statistical analyses were conducted using R 3.0.2 [28].

Results

msRMR at different temperatures are depicted in Fig. 1. Both, TNZ and mean msRMR differed between reproductive and non-reproductive Ansell's mole-rats (Tables 1 and 2). The mean msRMR of reproductive Ansell's mole-rats (1.17 ± 0.17 ml $O_2 \times g^{-1} \times h^{-1}$) within their TNZ of 28–33 °C was significantly higher than the mean msRMR of non-reproductive animals (0.89 ± 0.19 ml $O_2 \times g^{-1} \times h^{-1}$) within their TNZ of 26–30 °C (Fig. 2, Table 1). The lowest mean msRMR (0.81 ± 0.23 ml and 1.1 ± 0.23 $O_2 \times g^{-1} \times h^{-1}$) was observed at ambient temperatures of 28 °C and 32 °C in non-reproductive and reproductive mole-rats, respectively. The optimal model of msRMR determined by generalized linear mixed models for repeated

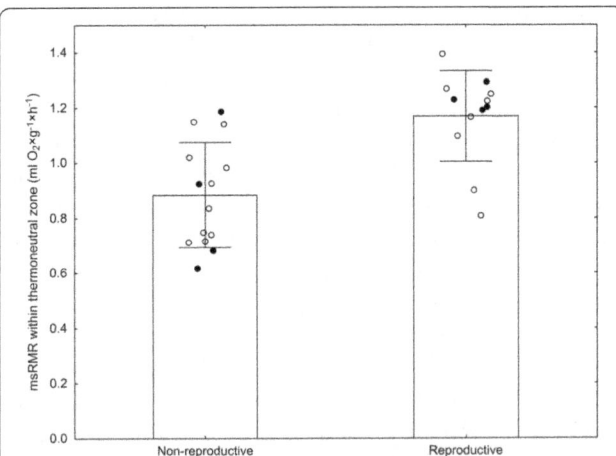

Fig. 2 Mass specific resting metabolic rate (msRMR) of non-reproductive ($N = 14$) and reproductive ($N = 12$) *Fukomys anselli* within the boundaries of their respective thermoneutral zones (non-reproductive: 26–30 °C; reproductive: 28–33 °C). The mean msRMRs of each individual (open circles) throughout its TNZ are presented together with the mean of the respective reproductive status (bars) and its SD (whiskers). Full symbols indicate wild-derived animals

measurements consisted only of the reproductive status (AIC = 8.83; comparison with null model $\chi^2 = 6.72$, df = 1, $p = 0.01$) and showed fixed effects of 0.89 ± 0.07 and 1.17 ± 0.09 for non-reproductive and reproductive factor level, respectively. The difference itself (0.28 ± 0.09) is statistically significant ($t = 2.995$, $p < 0.003$) suggesting that the msRMR of reproductive individuals is about 30% higher than that of non-reproductive ones.

Discussion

Metabolic rate with respect to reproduction

Low msRMR is a common trait in bathyergid rodents interpreted as an ecophysiological adaptation to the subterranean habitat [29], and our measurements generally confirm previous studies. However, our finding that long-lived breeders of *F. anselli* have higher metabolic rates compared to shorter-lived non-breeders is novel. The higher metabolic rate observed in breeders is not surprising per se, since reproduction is energy demanding, especially for lactating females [8, 30, 31], albeit none of the reproductive females were pregnant or lactating during the time of measurements. In Damaraland mole-rats, breeding females also live significantly longer than non-breeding females [32], but no significant differences in msRMR of female breeders and non-breeding individuals were found. However, changes in daily energy expenditure were observed depending on seasonal changes in rainfall [33]. Higher msRMR in breeders compared to non-breeders of *F. anselli* are in line with findings in pregnant naked mole-rats, another long-lived bathyergid species (>30 years), which have a higher body temperature compared to their non-breeding counterparts [34]. To the best

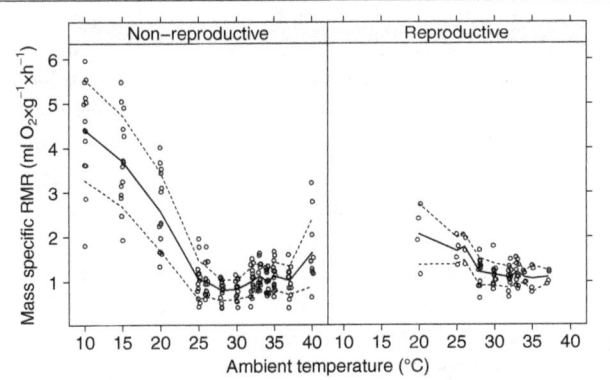

Fig. 1 Mass specific resting metabolic rate of non-reproductive (left) and reproductive (right) *Fukomys anselli* at different ambient temperatures. Means are connected by a solid line, interrupted lines connect mean ± SD

of our knowledge, metabolic data of reproductive naked mole-rats is not available, but higher body temperature suggests a higher metabolic rate. This assumption is also in line with most mammals, where energy intake and/or daily energy expenditure was shown to be increased in lactating females [35]. This aspect is most interesting since investment in reproduction was long thought to impair somatic maintenance according to the classical trade-off model, but recent findings refer to the trade-off model as being too simplistic [14]. Especially in terms of female reproduction, a meta-analysis from different homeothermic vertebrates by Blount et al. [8] has shown that in intraspecific comparisons between breeders and non-breeders, breeders had lower levels of oxidative damage in certain tissues. This effect could be attributed to upregulation of antioxidant defense mechanisms, such as glutathione or superoxide dismutase activity, which shows a tissue-dependent upregulation in several species during reproduction [36–39]. This oxidative shielding hypothesis, even if not consistent across different studies (reviewed in: [8]), suggests a reproduction-induced protection of mothers and offspring. Ansell's mole-rats are continuously reproducing once they achieve the reproductive status. Oxidative shielding might protect the animals from detrimental pregnancy effects due to a higher energy turnover in female breeders compared to non-breeders. However, the bimodal lifespan in Ansell's mole-rats is not sex-dependent, indicating a general effect in terms of reproductive status, msRMR, and lifespan rather than just a pregnancy effect restricted to females.

The mechanisms underlying the higher msRMR in male Ansell's mole-rat breeders which do not lactate cannot be derived from the present results. Even so, it could be argued that reproduction in monogamous species includes parental care in male and female, thus it could be that a closer look at monogamous male breeders reveals a higher energy turnover during the reproductive phase as well. Furthermore, breeders are sexually active throughout the year, which could increase the overall msRMR in these individuals.

Higher msRMR in long-lived breeders is not in line with classical aging theories

Oxidative stress as a main factor contributing to life history trade-offs is getting challenged by increasing contradictory studies [14]. Hence, higher msRMR in breeders is in line with those studies, which did not find any correlation [9] or even support a positive correlation between RMR and lifespan [11, 13, 40–43]. In a large comparative study by de Magalhães et al. [9], data from 300 mammalian species were analyzed, but after correcting for body mass and phylogeny, no influence of metabolic rate on lifespan, at least in eutherians, could be found. On the other hand, the developmental schedule of a species, i.e. age of sexual maturity and postnatal growth rate, were both correlated with longevity. Mole-rats have a relatively long gestational period of about 3 months (100 days), slow postnatal growth rates, and reach sexual maturity at about 1 year of age [44, 45]. These features are in line with developmental schedule-associated implications reported in de Magalhães et al. [9]. Although these developmental features might account for overall longevity in mole-rats, the distinct intraspecific lifespan difference between breeding and non-breeding Ansell's mole-rats is still a special case which must be discussed in light of studies that support a positive correlation between msRMR and lifespan [18].

The uncoupling-to-survive hypothesis [46] complements simplistic theories of senescence by explaining apparent exceptions. It suggests that elevated oxygen consumption, a measure for msRMR in the present study, could be also observed due to uncoupling of proton flux in the mitochondria. This process, also referred to as inducible proton-leak, is facilitated by uncoupling proteins and increases RMR. On the other hand, inducible proton-leak is known to reduce ROS production by reducing mitochondrial membrane potentials [47]. Hence the higher msRMR measured in breeders of Ansell's mole-rats could be due to higher rates of mitochondrial uncoupling compared to non-breeders. Several studies found higher rates of uncoupling in those laboratory mice that lived longer compared to other individuals with shorter lifespans [13, 41, 42]. A recent study even showed that mitochondrial H_2O_2 levels (which is a ROS) were negatively correlated to the metabolic rate in the brown trout [43]. However, in case of mole-rats this model should be considered carefully, since in naked mole-rats, surprisingly high levels of oxidative damage to DNA, lipids and proteins were found, which contrasts with the proposed benefit of mitochondrial uncoupling [48]. Nevertheless, it would be interesting to establish a protocol to investigate mitochondrial ROS in *F. anselli* in future studies. Mitochondrial uncoupling should be taken into account as a possible proximate mechanism contributing to the bimodal lifespan of *F. anselli*, but the adaptive background of such a selective protective mechanism on species level is still puzzling. It may be that the upregulation of protective, or in general, repair mechanisms are just a side effect of sexual activity in this species. For instance, long-lasting pair bonding could lead to higher oxytocin levels, known to decrease stress hormones and promote immunity (reviewed in [49]). Hence, further research should illuminate the linkage between different physiological processes, and how the different processes impacts an animal's lifespan.

Conclusions

Bathyergid species with their exceptionally long lifespan are already interesting animal models in current aging research, but this is the first study to report a positive correlation between msRMR and lifespan based on reproductive status. The bimodal lifespan of our animal model, the Ansell's mole-rat, provides an exceptional opportunity to investigate protective mechanisms such as oxidative shielding on species level without the usual shortcomings related to experimental manipulation of the study animals. In general, our finding stresses the complexity of currently discussed aging mechanisms.

Acknowledgements
We thank João Pedro de Magalhães and two anonymous reviewers for fruitful comments on the manuscript.

Funding
Not applicable.

Authors' contributions
SB and HB conceived the project. CKMS performed all metabolic measurements with the help of JO who set-up the equipment. JO, SB and YH evaluated the data and conducted the statistical tests. JO prepared the figures. All authors contributed to the draft and improvement of the manuscript and reviewed the final version. All authors read and approved the final manuscript.

Competing interests
The authors declare that they have no competing interests.

Author details
[1]Faculty of Biology, University of Duisburg-Essen, Essen, Germany. [2]Faculty of Forestry and Wood Sciences, Czech University of Life Sciences, Praha, Czech Republic. [3]Faculty of Science, University of South Bohemia, České Budějovice, Czech Republic.

References
1. Höhn A, Weber D, Jung T, Ott C, Hugo M, Kochlik B, Kehm R, Konig J, Grune T, Castro JP. Happily (n)ever after: aging in the context of oxidative stress, proteostasis loss and cellular senescence. Redox Biol. 2017;11:482–501.
2. Kirkwood TBL, Holliday R. The evolution of ageing and longevity. Proc R Soc Lond B Biol Sci. 1979;205:531–46.
3. Deerenberg C, Pen I, Dijkstra C, Arkies B-J, Visser GH, Daan S. Parental energy expenditure in relation to manipulated brood size in the European kestrel Falco tinnunculus. ZACS. 1995;99:39–48.
4. McNab BK. The energetics of reproduction in endotherms and its implication for their conservation. Integr Comp Biol. 2006;46:1159–68.
5. Speakman JR. The physiological costs of reproduction in small mammals. Philos Trans R Soc Lond Ser B Biol Sci. 2008;363:375–98.
6. Heldstab SA, van Schaik CP, Isler K. Getting fat or getting help? How female mammals cope with energetic constraints on reproduction. Front Zool. 2017;14:1–11.
7. Kirkwood TBL. Understanding ageing from an evolutionary perspective. J Intern Med. 2008;263:117–27.
8. Blount JD, Vitikainen EIK, Stott I, Cant MA. Oxidative shielding and the cost of reproduction. Biol Rev. 2016;91:483–97.
9. de Magalhaes JP, Costa J, Church GM. An analysis of the relationship between metabolism, developmental schedules, and longevity using phylogenetic independent contrasts. J Gerontol A Biol Sci Med Sci. 2007;62:149–60.
10. Furness LJ, Speakman JR. Energetics and longevity in birds. Age. 2008;30:75–87.
11. Munshi-South J, Wilkinson GS. Bats and birds: exceptional longevity despite high metabolic rates. Ageing Res Rev. 2010;9:12–9.
12. Selman C, McLaren JS, Collins AR, Duthie GG, Speakman JR. The impact of experimentally elevated energy expenditure on oxidative stress and lifespan in the short-tailed field vole Microtus agrestis. Proc Biol Sci. 2008;275:1907–16.
13. Speakman JR, Talbot DA, Selman C, Snart S, McLaren JS, Redman P, Krol E, Jackson DM, Johnson MS, Brand MD. Uncoupled and surviving: individual mice with high metabolism have greater mitochondrial uncoupling and live longer. Aging Cell. 2004;3:87–95.
14. Speakman JR, Garratt M. Oxidative stress as a cost of reproduction: beyond the simplistic trade-off model. BioEssays. 2014;36:93–106.
15. Speakman JR, Blount JD, Bronikowski AM, Buffenstein R, Isaksson C, Kirkwood TBL, Monaghan P, Ozanne SE, Beaulieu M, Briga M, et al. Oxidative stress and life histories: unresolved issues and current needs. Ecol Evol. 2015;5:5745–57.
16. Burda H, Honeycutt RL, Begall S, Locker-Grütjen O, Scharff A. Are naked and common mole-rats eusocial and if so, why? Behav Ecol Sociobiol. 2000;47:293–303.
17. Bappert M-T, Burda H, Begall S. To mate or not to mate? Mate preference and fidelity in monogamous Ansell's mole-rats, Fukomys anselli, Bathyergidae. Folia Zool. 2012;61:71–83.
18. Dammann P, Burda H. Sexual activity and reproduction delay ageing in a mammal. Curr Biol. 2006;16:R117–8.
19. Burda H. Individual recognition and incest avoidance in eusocial common mole-rats rather than reproductive suppression by parents. Experientia. 1995;51:411–3.
20. Schielke CKM, Begall S, Burda H. Reproductive state does not influence activity budgets of eusocial Ansell's mole-rats, Fukomys anselli (Rodentia, Bathyergidae): a study of locomotor activity by means of RFID. Mammal Biol. 2012;77:1–5.
21. Skliba J, Lovy M, Hrouzkova E, Kott O, Okrouhlik J, Sumbera R. Social and environmental influences on daily activity pattern in free-living subterranean rodents: the case of a eusocial bathyergid. J Biol Rhythm. 2014;29:203–14.
22. Henning Y, Vole C, Begall S, Bens M, Broecker-Preuss M, Sahm A, Szafranski K, Burda H, Dammann P. Unusual ratio between free thyroxine and free triiodothyronine in a long-lived mole-rat species with bimodal ageing. PLoS One. 2014;9:e113698.
23. Begall S, Berendes M, Schielke CK, Henning Y, Laghanke M, Scharff A, van Daele P, Burda H. Temperature preferences of African mole-rats (family Bathyergidae). J Therm Biol. 2015;53:15–22.
24. Lighton J. Measuring metabolic rates - a manual for scientists. New York: Oxford University Press; 2008.
25. Amaratunga D, Ge N. Step-down trend tests for identifying the minimum effective dose. J Biopharm Stat. 1998;8:151–62.
26. Venkatraman E: clinfun: clinical trial design and data analysis functions. R package. R package version 1.0.10 edition; 2015.
27. Bates D, Machler M, Bolker B, Walker S: lme4: linear mixed-effects models using Eigen and S4. R package version 1.1–8 edition; 2015.
28. R Core Team. R: A language and environment for statistical computing. 2015.
29. Zelová J, Sumbera R, Sedlácek F, Burda H. Energetics in a solitary subterranean rodent, the silvery mole-rat, Heliophobius argenteocinereus, and allometry of RMR in African mole-rats (Bathyergidae). Comp Biochem Physiol A Mol Integr Physiol. 2007;147:412–9.
30. Douhard F, Lemaitre JF, Rauw WM, Friggens NC. Allometric scaling of the elevation of maternal energy intake during lactation. Front Zool. 2016;13:32.
31. Zheng G-X, Lin J-T, Zheng W-H, Cao J, Zhao Z-J. Energy intake, oxidative stress and antioxidant in mice during lactation. Zool Res. 2015;36:95–102.
32. Schmidt CM, Jarvis JUM, Bennett NC. The long-lived queen: reproduction and longevity in female eusocial Damaraland mole-rats (Fukomys damarensis). Afr Zool. 2013;48:193–6.
33. Scantlebury M, Speakman JR, Oosthuizen MK, Roper TJ, Bennett NC. Energetics reveals physiologically distinct castes in a eusocial mammal. Nature. 2006;440:795–7.
34. Keil G, Cummings E, de Magalhães JP. Being cool: how body temperature influences ageing and longevity. Biogerontology. 2015;16:383–97.
35. Douhard F, Lemaître J-F, Rauw WM, Friggens NC: Allometric scaling of the elevation of maternal energy intake during lactation. Front Zool. 2016;13. doi:10.1186/s12983-016-0164-y.
36. Yang DB, Xu YC, Wang DH, Speakman JR. Effects of reproduction on immuno-suppression and oxidative damage, and hence support or otherwise for their roles as mechanisms underpinning life history trade-offs, are tissue and assay dependent. J Exp Biol. 2013;216:4242–50.

37. Xu Y-C, Yang D-B, Speakman JR, Wang D-H. Oxidative stress in response to natural and experimentally elevated reproductive effort is tissue dependent. Funct Ecol. 2014;28:402–10.

38. Vaanholt LM, Milne A, Zheng Y, Hambly C, Mitchell SE, Valencak TG, Allison DB, Speakman JR. Oxidative costs of reproduction: Oxidative stress in mice fed standard and low antioxidant diets. Physiol Behav. 2016;154:1–7.

39. Sudyka J, Casasole G, Rutkowska J, Cichoń M. Elevated reproduction does not affect telomere dynamics and oxidative stress. Behav Ecol Sociobiol. 2016;70:2223–33.

40. Oklejewicz M, Daan S. Enhanced longevity in tau mutant Syrian hamsters, *Mesocricetus auratus*. J Biol Rhythm. 2002;17:210–6.

41. Echtay KS, Roussel D, St-Pierre J, Jekabsons MB, Cadenas S, Stuart JA, Harper JA, Roebuck SJ, Morrison A, Pickering S, et al. Superoxide activates mitochondrial uncoupling proteins. Nature. 2002;415:96–9.

42. Keipert S, Voigt A, Klaus S. Dietary effects on body composition, glucose metabolism, and longevity are modulated by skeletal muscle mitochondrial uncoupling in mice. Aging Cell. 2011;10:122–36.

43. Salin K, Auer SK, Rudolf AM, Anderson GJ, Cairns AG, Mullen W, Hartley RC, Selman C, Metcalfe NB. Individuals with higher metabolic rates have lower levels of reactive oxygen species in vivo. Biol Lett. 2015;11:20150538.

44. Begall S, Burda H. Reproductive characteristics and growth in the eusocial Zambian Common mole-rat (Cryptomys sp., Bathyergidae). Mammal Biol - Z Säugetierkunde. 1998;63:297–306.

45. Burda H. Reproductive biology (behaviour, breeding, and postnatal development) in subterranean mole-rats, *Cryptomys hottentotus* (Bathyergidae). Mammal Biol - Z Säugetierkunde. 1989;54:360–76.

46. Brand MD. Uncoupling to survive? The role of mitochondrial inefficiency in ageing. Exp Gerontol. 2000;35:811–20.

47. Busiello RA, Savarese S, Lombardi A. Mitochondrial uncoupling proteins and energy metabolism. Front Physiol. 2015;6:36.

48. Andziak B, O'Connor TP, Qi W, DeWaal EM, Pierce A, Chaudhuri AR, Van Remmen H, Buffenstein R. High oxidative damage levels in the longest-living rodent, the naked mole-rat. Aging Cell. 2006;5:463–71.

49. Detillion CE, Craft TKS, Glasper ER, Prendergast BJ, DeVries AC. Social facilitation of wound healing. Psychoneuroendocrinology. 2004;29:1004–11.

12

Speed dependent phase shifts and gait changes in cockroaches running on substrates of different slipperiness

Tom Weihmann[1*] (iD), Pierre-Guillaume Brun[2] and Emily Pycroft[3]

Abstract

Background: Many legged animals change gaits when increasing speed. In insects, only one gait change has been documented so far, from slow walking to fast running, which is characterised by an alternating tripod. Studies on some fast-running insects suggested a further gait change at higher running speeds. Apart from speed, insect gaits and leg co-ordination have been shown to be influenced by substrate properties, but the detailed effects of speed and substrate on gait changes are still unclear. Here we investigate high-speed locomotion and gait changes of the cockroach *Nauphoeta cinerea*, on two substrates of different slipperiness.

Results: Analyses of leg co-ordination and body oscillations for straight and steady escape runs revealed that at high speeds, blaberid cockroaches changed from an alternating tripod to a rather metachronal gait, which to our knowledge, has not been described before for terrestrial arthropods. Despite low duty factors, this new gait is characterised by low vertical amplitudes of the centre of mass (COM), low vertical accelerations and presumably reduced total vertical peak forces. However, lateral amplitudes and accelerations were higher in the faster gait with reduced leg synchronisation than in the tripod gait with distinct leg synchronisation.

Conclusions: Temporally distributed leg force application as resulting from metachronal leg coordination at high running speeds may be particularly useful in animals with limited capabilities for elastic energy storage within the legs, as energy efficiency can be increased without the need for elasticity in the legs. It may also facilitate locomotion on slippery surfaces, which usually reduce leg force transmission to the ground. Moreover, increased temporal overlap of the stance phases of the legs likely improves locomotion control, which might result in a higher dynamic stability.

Keywords: Leg coordination, Body dynamics, Biomechanics, Poly-pedal locomotion, Insect, Arthropod

Background

Alternating sets of synchronously active diagonally adjacent legs, i.e. tripods in insects and four-leg sets in spiders, are largely regarded as the dominating coordination pattern employed by fast running insects and arachnids [1–3]. The sets either comprise the legs L1, R2, L3 (and R4) or R1, L2, R3 (and L4) with L and R indicating the left and right side legs, respectively, both counted fore to aft. These sets are accounted as the physical basis for spring-mass like dynamics as observed in the locomotion of some species [4–6]. Strictly alternating sets of legs are characterised by anti-cyclic activity of adjacent ipsilateral legs

and the contralateral legs of the pairs of legs, i.e. by phase shifts of 0.5 [7–9]. Deviations from the alternating pattern [10–14], however, have rarely been analysed in the context of physical and biomechanical constraints and were mostly attributed to anatomical differences or inherent variability of poly-pedal locomotor systems. Nevertheless, characteristic changes in the speed-dependent increase of stride frequencies and oxygen consumption with simultaneously lacking aerial phases, as found in some studies [2, 15–17], seem to indicate an additional gait transition for high running speeds in insects and spiders.

In bouncing gaits such as running, trotting and multi-legged equivalents, significant proportions of movement energy can be elastically stored in the initial and recovered in the final part of the stance phase with optimised leg properties [18], which helps to economise locomotion.

---
* Correspondence: tom.weihmann@uni-koeln.de
[1]Department of Animal Physiology, Institute of Zoology, University of Cologne, Zülpicher Strasse 47b, 50674 Cologne, Germany
Full list of author information is available at the end of the article

However, the stiffness of energy-storing elastic components such as sclerites, apodemes or other skeletal structures [19–21] must be matched to each other in all legs involved to provide concerted loading and unloading rates. In cockroaches, hind legs are characterised by distal joints with axes in parallel with the main ground force direction, which facilitates passive elastic energy storage in the hip joint [20].

Bouncing gaits, like trot and, with some restrictions, also gallop, are characterised by rhythmic upwards and downwards movements of the COM with in-phase oscillations of kinetic and potential energy [22, 23]. In running and trotting, the initial downwards movement is reversed by a single impulse of one leg or a set of synchronously active legs [4]. In gallop-like footfall patterns of vertebrates, such concerted stance phases decompose in consecutive ground contacts of the single legs [23], which affects locomotion dynamics and energy efficiency.

The majority of insect species seem to have significantly lower maximum running speeds than specialist runners, which are in the focus of many existing studies [24–28]. One of these specialists is the blattid cockroach *Periplaneta americana* which is characterised by extraordinarily long legs and a linear increase of the stride frequency over a wide range of running speeds [29]. Non-specialist runners, such as the blaberid cockroach species *Blaberus discoidalis* and *Nauphoeta cinerea* seem to be limited in their maximum leg cycle frequency but must still be able to attain high running speeds since successful predator avoidance is also crucial in these species. Just like in bipeds [30] and quadrupeds [31–33], the dependencies of stride frequencies to running speed are curvilinear in these blaberid cockroach species (cp. [2, 15]), which seems to indicate a gait change also occurs for these insects. However, the physical basis for the suspected gait transition has not yet been revealed. In a recent study on arachnid locomotion [34] we were able to show that fast moving mites employ temporally distributed footfall patterns at maximum running speeds and that such patterns may increase locomotion efficiency in arthropods. In nature, insects are commonly faced with slippery or unsteady substrates. Such surfaces do not provide secure footholds and can lead to unpredictable perturbations [35–37]. To our knowledge, no studies considering coordinative adaptations to slippery and unsteady substrates exist for fast moving arthropods. Accordingly, we examine here whether or not the saturating stride frequencies found in blaberid cockroaches are accompanied by changed footfall patterns and how they can affect running efficiency and endurance at high speeds on substrates with different grit sizes and slipperiness.

Methods
Animals
Thirteen adult male *Nauphoeta cinerea* (body length: 27 ± 1.3 mm, mean ± s.d.) were taken from a laboratory colony. This species has been examined with regard to their capabilities in substrate attachment and climbing in a couple of previous studies (e.g. [38]) which makes them a good model to examine the impact of substrate properties on locomotion. Moreover, *N. cinerea* is closely related to *Blaberus discoidalis*, whose locomotion has been extensively examined in the past (e.g. [20, 39–41]). Accordingly, our results can be well compared to those of the anatomically and behaviourally similar species.

The insects were kept in plastic containers at 25 °C and were supplied with dog food and water ad libitum. We used males only to prevent potential bias due to gravid females. When selecting study animals, we excluded cockroaches that were unwilling to run. All wings (representing about 2% of the body mass) were removed and three hollow Styrofoam spheres (2.5 mm diameter), coated in white paint were glued (using a mixture of paraffin and pine resin) onto the pronotum, the second to third thoracic tergum and close to the end of the abdomen, to provide markers for digitization (Fig. 1). The animals were allowed to recover for at least one hour before the trials. The total mass of the three spheres and the wax resin mixture amounted to 6.5 ± 1.2% of the wingless animals' body mass. With markers, the mass of the animals was 450 ± 76 mg.

Running track
Cockroaches were tested in a running track of 50 cm length and 3.5 cm width, with a floor covered with sandpaper of 30 µm particle size (Ultra Tec, Santa Ana, CA, USA). The walls were 5 cm high and covered with slippery 1 µm particle size sandpaper to prevent the animals from escaping. The central hardboard section (12 cm long) of the walkway could be exchanged for another piece. The pieces were either covered with non-slippery 30 µm particle size sandpaper or semi-slippery 12 µm particle size sandpaper. Different asperity sizes can facilitate or interfere with the abilities of claws or tarsal attachment pads to engage with a substrate. Micro-rough substrates within a range from some tenths to a few µm particle size impedes both, the claws and the attachment pads, from achieving good traction, which results in reduced friction and a reduced ability to climb [42–44]. Whether or not a substrate was slippery for *N. cinerea* cockroaches was tested on vertical substrates of 0.1, 1, 3, 12 and 16 µm particle size. While the specimens could not climb on 0.1 to 3 µm sandpaper and had difficulty walking on 12 µm sandpaper, they coped with 16 µm sandpaper. In the central region, the walls of the running track were made of Perspex allowing video recording of the kinematics in the lateral view (see below). All examined runs were escape runs. The insects were spurred by short puffs of air or by touching their feet or cerci with a fine paint brush. Recordings of straight and continuous runs were saved for further analysis.

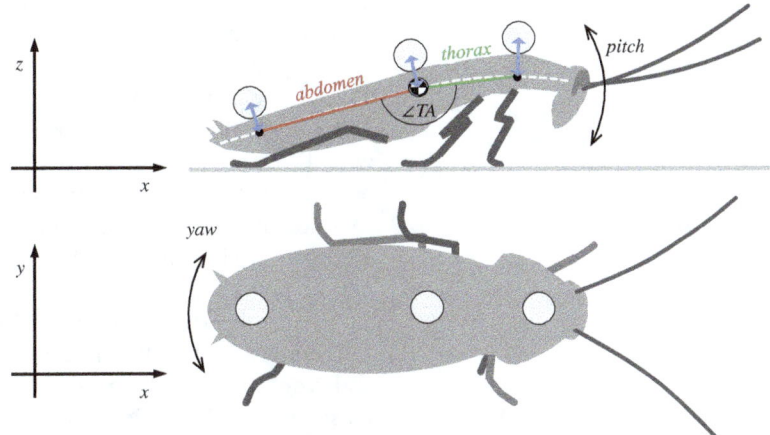

Fig. 1 Sketch of a running cockroach drawn after a single frame of a typical sequence. Upper row: Side view of a specimen with markers on the pronotum, the metathorax and the caudal tip of the abdomen. The lateral edge along the body (white dashed line) was used to define the centre-line onto which the positions of the markers were projected (blue arrows). The projection of the thoracic marker was assumed as the position of the centre of mass. The position of the thorax was defined by the connecting line between the projections of the pronotum and the thorax markers (green line) while the position of the abdomen was defined by the connecting line between the projections of the thorax marker and the abdomen marker onto the mid-line (red line). The angle between thorax and abdomen was ∠TA. For clarity only the legs of one tripod (R1, L2, R3) are shown. Lower row: Top view of a specimen with markers applied to the pronotum, the thorax and the abdomen. Both sets of legs are in contact with the ground, with the tripod made up by R1, L2 and R3 pictured just before taking off and the tripod consisting of L1, R2 and L3 having just touched down

Recording and digitisation

The runs were recorded in top view with a Photron FastCam SA2 (Photron Ltd., Tokyo, Japan, resolution 2048 × 2048 pixels) high-speed video camera which was oriented perpendicularly to the walkway. In the central region of the running track with the exchangeable surface and the Perspex walls, a bracket with a 45° mirror provided a side view of the running insects.

Recordings were made at 250 frames per second and with a shutter time of 2 ms. The central region of the track filled the camera's field of view, resulting in a spatial resolution of approximately 17 pixels per mm. The track was illuminated with two white LED lights (XS40, Spectrum Illumniation Inc., Montague MI, USA) fitted with red filters to reduce their brightness in the spectral range visible to the insects.

The videos were digitised in ProAnalyst Lite Edition (Xcitex Inc., Massachusetts, USA). A 72 × 16 × 9.6 mm Lego™ brick assembly was used for calibration. Top and side views were calibrated separately taking into regard slight aspect variations caused by the mirror. Coordinate axes were adjusted such that the fore-aft axis was aligned to the direction of the walkway. The three marker points on the pronotum, thorax, and abdomen were tracked automatically. Additionally, in the top view, manual tracking was used to digitise the positions of the foot tips (Fig. 2) during contact (cp. [16]). Only straight runs with constant running speed and where animals did not touch the walls were considered. Runs in which the running speed varied by more than 10% were excluded from analyses.

Analyses

In order to assess the relative frequencies of the different leg coordination schemes; initially all recorded sequences were roughly classified. To this end, it was determined whether the ground contacts of front and rear legs of one side were in phase; in this case they were classified as alternating tripod gait (Fig. 3a). Traditionally metachronal leg coordination, has been defined by reference to slow movement sequences or those with high duty factors (e.g. [9, 45, 46]) at which the wave travels down the entire grouping of appendages. Wave propagation appears somewhat less clear when duty factors decrease and relative fluctuations in touch down and take off increase. However, for bouncing gaits, such as trotting and the multi-legged equivalents, synchronised sets of legs, such as tripods, represent the functional unit which becomes inoperative when the contact phases of the single legs dissipate. As a direct consequence, such a temporal dissipation results in a loss of the typical vertical body oscillations (see Fig. 4). Accordingly, in order to simplify the terminology of the new high speed gait and because of the considerable similarity of the gait patterns (Fig. 3b) all runs with low synchronicity between front and rear legs of one side were classified as metachronal.

Ninety runs were recorded on the non-slippery and 81 runs on the slippery substrate. 11 runs were classified as metachronal (12%) on the non-slippery substrate, and eight on the slippery substrate (10%). Metachronal runs seemed to occur predominantly at high running speeds.

In order to increase the sample size of metachronal runs allowing statistical comparisons between the paradigms,

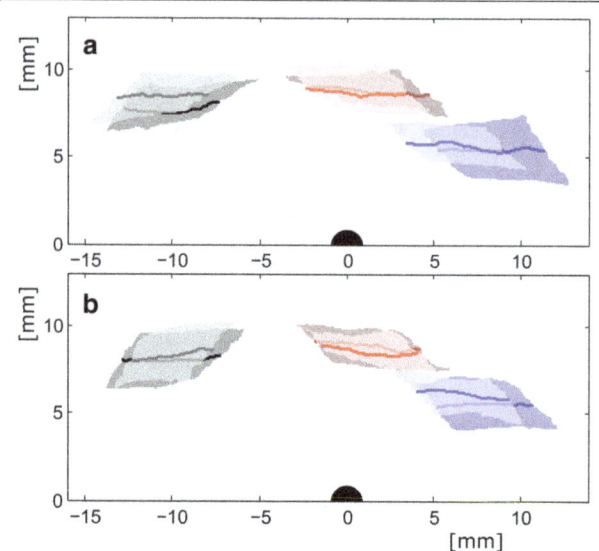

Fig. 2 Contact areas for metachronal (dark sublayer) and alternating (bright upper layer) gait patterns on non-slippery and slippery substrate. **a** Averaged contact positions on the non-slippery substrate with median trajectories of the legs' tarsi (dark lines in centre of each layer) with respect to the COM (black semi-circle at the point of origin). Fore legs are depicted in blue, middle legs in red and rear legs in black. Contact areas are bound in anterior-posterior direction by the medians of the anterior and posterior extreme positions for a given lateral distance to the COM. In the lateral direction, the boundaries are the 25% and the 75% quartiles of the lateral distributions for distinct anterior-posterior positions of the tarsi (see [16]). **b** Averaged contact areas on slippery substrate with median trajectories of the legs' tarsi with respect to the COM

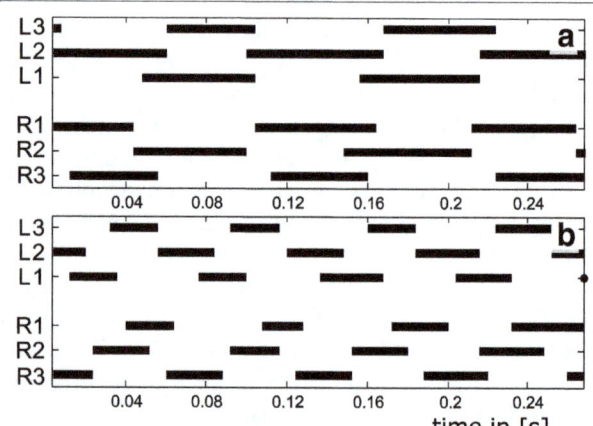

Fig. 3 Example gait patterns; bars depict ground contacts of the legs which are counted fore to aft with L indicating left and R indicating right side legs. **a** Alternating tripodal run at an average speed of 0.1 ms⁻¹ with an average phase shift between ipsilateral fore and rear legs of 0.91 and an average duty factor of 0.52. **b** High speed metachronal run at an average running speed of 0.24 ms⁻¹ with an average phase shift between ipsilateral rear and fore legs of 0.71 and an average duty factor of 0.41

20 additional fast runs were recorded. Finally, 24 of the best runs, spanning the full speed range of the examined runs, were digitized and analysed for the non-slippery condition; 11 of them being alternating tripodal gaits and 13 metachronal. On the slippery substrate 25 runs were analysed with 12 being alternating and 13 metachronal. The ratios between metachronal and tripodal runs implicitly indicate that not all animals could be urged to run quickly. Eventually, the samples cover 6 to 8 individuals which contributed one or two high quality runs. For each run a sequence of 3.6 ± 1 consecutive strides was digitised.

The phase shift between the touch-downs of the ipsilateral front and rear leg was used to quantitatively distinguish alternating tripod gait from metachronal gait patterns. Runs were classified as metachronal if the median phase values of a run differed by more than ± 0.2 from synchrony (indicated by a phase value of 0 or 1) (cp. [8]). Differences between phase relations were tested for significance using the Kuiper two-sample test which is a circular analogue of the Kolmogorov-Smirnov test [47].

Before analysis, a correction was applied to the side view data to account for differences in the heights of the markers (within and between specimens) due to variation in the amount of wax used to apply each marker. The dorsal cuticle of *N. cinerea* narrows towards a visible lateral edge (Fig. 1a). This anatomical structure was easily visible in the lateral view and was taken to be the dorsal-ventral centre line. The distances between the centres of the markers and this midline were determined for each individual. Then, the centre points of the markers were projected orthogonally onto the centre line. Since the COM of the animals was on an anterior-posterior level with the hind limb coxae, the thoracic marker was just in front and very close to the COM of the cockroach body. Therefore it was taken as its representative.

Once the position of the COM was obtained in the side view further parameters could be extracted in a right handed system of coordinates. Extracted parameters included the height of the COM above the substrate, distance covered per stride, pitch, yaw and the angle between the thorax and abdomen. Fore-aft, lateral and vertical velocities as well as accelerations of the body were then calculated using the derivative of the COM position data and by smoothing the primary data by fitting gliding second-order polynomials to the time series, including four adjacent points on both sides at each position [48]. Pitch was calculated in the side view as the angle between substrate and the connecting line between the pronotum marker and the thorax marker. To calculate yaw, the line connecting the pronotum marker and the abdominal marker in the top view was used. The connecting lines between pronotum and thorax marker and between thorax and abdomen marker were used to calculate the ventral angle between thorax and abdomen

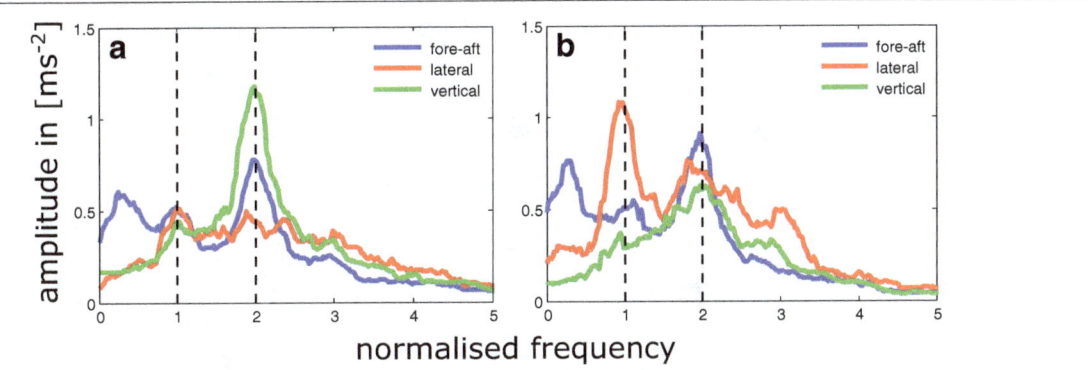

Fig. 4 Median trends for the frequency spectra of COM accelerations in fore-aft (blue), lateral (red) and vertical (green) direction in ms^{-2}. All frequencies were normalised by dividing specific values by the corresponding stride frequency. **a** Runs with alternating tripodal leg coordination **b** Runs with metachronal leg coordination. Substrate slipperiness did not affect the graphs and runs of both conditions were merged

in side view. All body oscillations were referred to the stride periods of the second legs; lateral data was inverted for tripods consisting of L1, R2 and L3 such that all data was normalised onto the reverse set of legs (R1, L2, R3).

Moreover, kinematic parameters of the legs such as stride frequency (f_T), duty factors (β), contact and swing rates (t_C^{-1} and t_S^{-1}) and the positions of the legs in contact with the ground (see [16]) were also examined. Touch-down and take-off positions as well as the lateral distance of the feet to the COM were extracted for all legs while the second legs were used as proxy for the analyses of the distances travelled by the COM during contact with the ground (s_C) and during a whole stride (s_T). Fore-aft and lateral distances of the feet with respect to the COM (Fig. 2) were measured along the x- and y-axis of a body-fixed coordinate system in top view. Data for corresponding left and right legs were merged.

The changes in contact duration (t_C) and swing duration (t_S) with speed are highly non-linear (Additional file 1: Figure S8). Both measures can be linearized by considering the reciprocal values (t_C^{-1} and t_S^{-1}), i.e. contact and swing rate. The dependencies on running speed were then approximated by using linear least squares fits. The contact rate increases linearly with running speed. On the non-slippery substrate the swing rates have sloped dependencies at lower speeds and adopt constant values at high speeds. The point of intersection indicates a transitional velocity (Fig. 5).

Stride frequencies and duty factors derive from these durations and rates. Their velocity dependencies are non-linear as well. Their approximate speed dependencies were calculated by following Weihmann [16]. Thus, stride frequencies (f_T) were calculated as $f_T = t_T^{-1} = (t_C + t_S)^{-1}$. Their slopes are linear and relatively steep at low speeds and decrease in a non-linear manner at speeds above the transition point (Fig. 5). With t_T being the stride duration, i.e. the reciprocal of the stride frequency, duty factors were calculated conventionally ($\beta = t_C/t_T$) for all pairs of legs. Due to the specific characteristics of t_C and t_T, their

velocity dependencies displayed graphs consisting of two adjacent hyperbola for each leg. The curvatures of the graphs were higher at low running speeds and low above the transition speed (Fig. 5).

On the slippery substrate, metachronal runs occurred at much lower running speeds which resulted in a wide overlap region (Additional file 1: Figure S6). Therefore, no saturation of stride frequencies and swing rates could be observed.

All analyses were performed using MatLab scripts (MATLAB 7.10.0; The MathWorks, Natick, MA, USA). All values are provided either as mean ± s.d. or as median and the 25% and 75% percentiles (Q25/Q75). Accordingly for pairwise comparisons, t-tests or non-parametric tests such as the Mann-Whitney U test were used. For multivariate analyses one way analyses of variance (ANOVA) with Tukey-Kramer post-hoc tests were applied. If not specified otherwise, all statistical tests refer to a 5% significance level.

Oscillations of the COM accelerations were analysed in all three dimensions using Matlab's Fast Fourier Transformation (FFT) algorithm. Overall ground reaction forces of the walking legs determine the accelerations of the animal's body. Therefore, body oscillations provide insight into the application of leg forces over time. The frequency spectrums of the acceleration amplitudes were calculated for each running sequence and frequency values were normalised due to division by the individual stride frequency of each run (Fig. 4).

In order to assess the functional basis of the COM oscillations, tripod synchrony factors (TSF) were determined in accordance with the method proposed by Spagna et al. [49], i.e. as a normalised fraction of contact phase overlap between legs in the same set of legs (Fig. 6a). Consequently, the synchrony factor adopts values between 0 and 1, at which 1 corresponds to perfect synchronicity. Additionally, we determined the relative temporal overlap of consecutive stance phases of the sets of legs. In contrast to the "Per cent double support phase per stride"

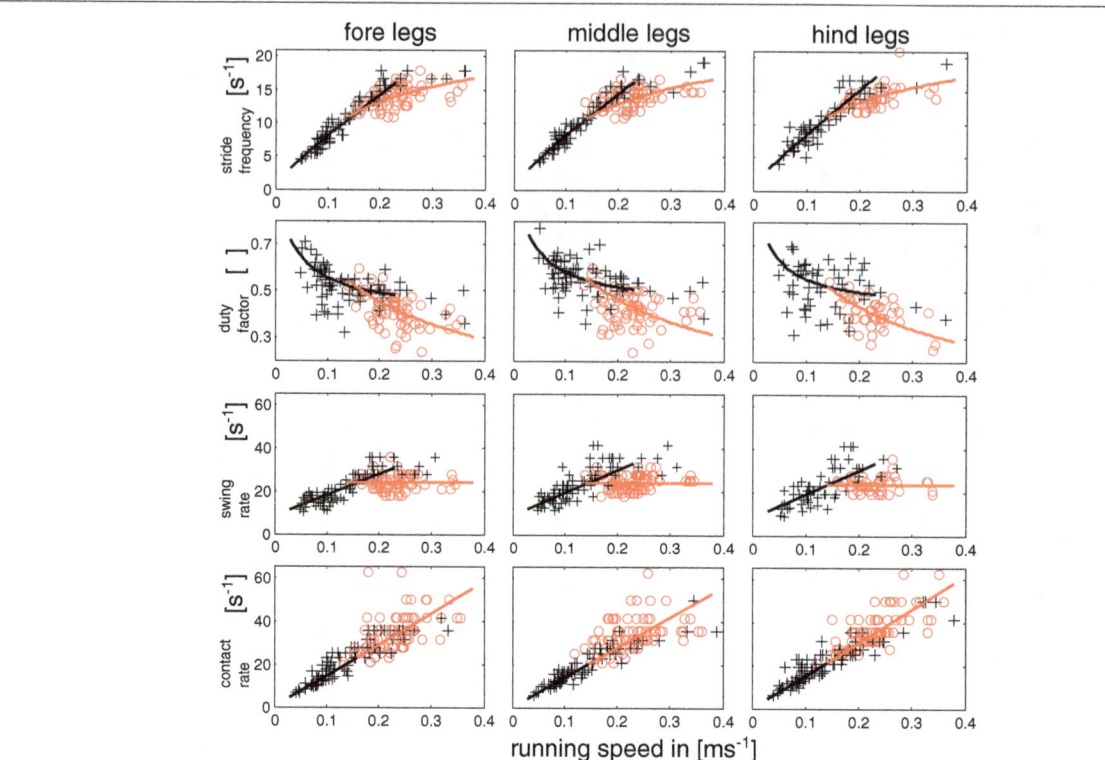

Fig. 5 Stride frequencies, duty factors, swing rates and contact rates plotted against running speed for all walking legs and non-slippery conditions. First row: Stride frequency (f_T); Second row: duty factors (β); Third row: swing rates (t_S^{-1}); Fourth row: contact rates (t_C^{-1}). Red circles are measured values for runs with metachronal leg coordination while black crosses depict values from alternating tripodal runs. Black and red lines are linear least squares regressions in the third and fourth rows while these regressions were used to calculate the approximation lines in the first two rows (further explanations can be found in the methods section). Red approximation curves refer to metachronal and black to alternating runs

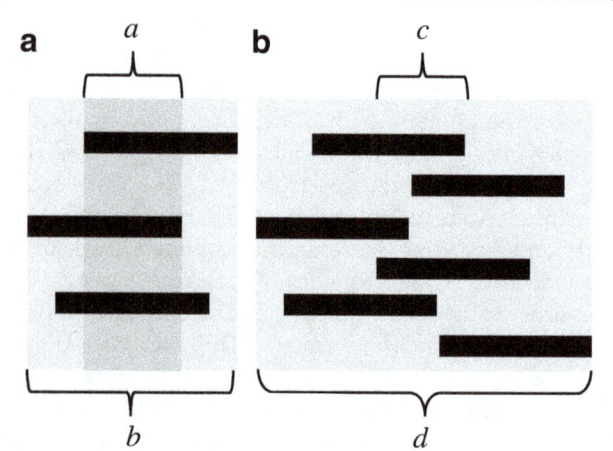

Fig. 6 Sketches explaining the calculation of the tripod synchrony factor (TSF) and the temporal overlap ratio between consecutive sets of legs. **a** The TSF is calculated as: $TSF = a/b$. **b** The temporal overlap between consecutive sets of legs is calculated as c/d. Black bars depict intervals with legs in contact with the ground

introduced by Ting et al. [2] we did not focus on those phases with six legs on the ground but determined the time span of a set's ground contact from the first contact of one of the legs till the last leg of the set has lost ground contact. The fraction with at least one leg of both sets in contact with the ground, was then related to the overall time span of the ground contacts of the two sets of legs (Fig. 6b).

By using the 3D fluctuations of the COM over the strides of the second legs and the average mass, the specimens' fore-aft (E_{kx}), lateral (E_{ky}) and vertical (E_{kz}) kinetic energies as well as the potential energy (E_p) of the COM were calculated. The kinetic energies were calculated as the product $0.5 \cdot m \cdot v^2$ with m being body mass and v being fore-aft, lateral or vertical velocities respectively. The gravitational potential energy was calculated as the product $m \cdot g \cdot h$ where m is the body mass, g represents acceleration due to gravity and h is the vertical distance between COM and the substrate. According to Full and Tu [25], the mass specific external mechanical energy per unit distance (M_{COM}) was then calculated by relating the sum of the changes in these COM energies to the distance travelled during a stride, which is equivalent to the external mechanical work and has units of $J \cdot kg^{-1} \cdot m^{-1}$ [25, 31, 50].

$$M_{COM} = \frac{\Delta E_{kx} + \Delta E_{ky} + \Delta E_{kz} + \Delta E_p}{body\ mass \cdot stride\ length}$$

In order to assess locomotion dynamics, recovery and congruity for E_{kx} and E_p were calculated for the stride periods of the second legs. Recovery provides information about the percentage of exchange between kinetic and potential energy and was calculated in accordance with Ahn et al. [51]. Thus, high percentages would indicate out-of-phase fluctuations of the two energy forms, i.e. COM dynamics similar to walking. Low percentages would indicate in-phase fluctuations as similar to running [31]. Congruity, in turn, compares the curve progression of kinetic and potential energy. Values of 100% occur if both trajectories follow the same trend throughout a whole gait cycle and would indicate a running gait; opposed curve progressions would result in low values, indicating walking gaits [51, 52].

Results
Running speed
Maximum running speed was reduced on the slippery substrate. While values of up to 0.35 ms^{-1} were observed on non-slippery the animals reached only 0.25 ms^{-1} on the slippery substrate (Table 1 in Appendix). Maximal attainable stride frequencies of about 15 s^{-1} did not differ between slippery and non-slippery substrates (Fig. 5, Additional file 1: Figure S6). The values are very similar to those observed in *Blaberus discoidalis* (cp. Fig. 7 in [15]).

On coarse substrate, running speed was significantly higher in metachronal runs (mean: 0.21 ms^{-1}; Q25: 0.17 ms^{-1}/Q75: 0.24 ms^{-1}), while alternating tripodal patterns were used predominantly at lower running speeds (0.12 ms^{-1}; 0.09/0.2 ms^{-1}) (cp. Table 1 in Appendix). Running speeds of metachronal runs on non-slippery substrate were also significantly higher than those of the slippery conditions while the other groups did not differ from each other.

Posture
In the working range of the fore legs, significant shifts in anterior direction occurred during metachronal coordination with respect to alternating tripodal runs (Fig. 2; Additional file 1: Table S1). The working ranges shifted for about 1.3 mm on slippery and 2.4 mm on non-slippery substrate. Positional shifts in the second and third legs were smaller and barely significant. Contact lengths (s_C) were significantly smaller in alternating tripodal runs on slippery substrate. In the other conditions no differences could be observed.

The average height of the COM did not change significantly with substrate properties or leg coordination; it was always about 3.74 ± 0.48 mm (Table 1 in Appendix).

Temporal measures
On the non-slippery substrate, contact rates (t_C^{-1}) increased linearly with speed (Fig. 5); the slopes (145/140/153) were similar for front, middle and rear legs and no change could be observed between alternating and metachronal leg coordination. Swing rates (t_S^{-1}) also had similar slopes (98/104/113) and y-intercepts (8.4/9/7.7) at lower speeds, i.e. with alternating tripodal leg coordination. At high running speeds and metachronal leg coordination, swing rates were constant and very similar among the legs (24 ± 3.5 s^{-1}/ 24.3 ± 3.7 s^{-1}/ 23.7 ± 3.4 s^{-1}).

Phase relations of the legs
As expected, on non-slippery substrate, *N. cinerea* showed ipsilateral phase shifts of about 0.5, i.e. alternating sets of legs at lower and intermediate running speeds. The circular mean of the phase lag of front leg touch-downs in the stride period of middle legs ($_1L_2$ following the nomenclature of Shultz 1987 [53]) was 0.49 with a confidence interval (CI) between 0.47 and 0.51. The value for the middle legs with regard to rear legs ($_2L_3$) was 0.45 (CI: 0.42/0.47) and the contralateral phases of the rear legs were about 0.5 (0.48/0.51) (Additional file 1: Figure S7, Additional files 2 and 3). At high running speeds, however, ipsilateral phase shifts between the front and middle legs were about (0.32; 0.29/0.34) and deviated significantly ($p < 0.001$) from 0.5. Accordingly, front and rear legs did not act synchronously and the legs of each set made ground contact successively rather than synchronously (Fig. 3b). Thus, coordination patterns were metachronal rather than alternating at high running speeds, though contralateral phases of the rear legs were still about 0.5 (0.48/0.51). On slippery substrate phase values for alternating ($_1L_2$: 0.48; 0.46/0.5, $_2L_3$: 0.42; 0.41/0.44) and metachronal ($_1L_2$: 0.33; 0.31/0.36, $_2L_3$: 0.4; 0.38/0.42) runs were similar to those on non-slippery substrate (Additional file 1: Figure S7).

In accordance with the changed phase lag between front and middle legs the tripodal synchrony factor (TSF) also differed between alternating tripodal and metachronal runs. While the TSF adopted values of about 0.61 ± 0.15 ($n = 161$) in alternating runs, the values decreased when the animals used metachronal leg coordination (0.25 ± 0.13; $n = 119$; see Fig. 7), i.e. at higher running speeds ($p < 0.001$).

Duty factors and temporal overlap of leg activity
The speed dependencies of the duty factors were similar in all pairs of legs. Duty factors decreased from about 0.7 at low running speeds to values around and below 0.5 at maximum running speeds (Figs. 3 and 5).

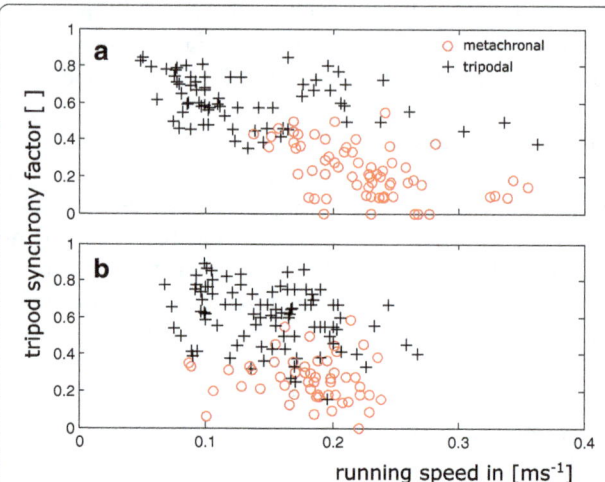

Fig. 7 Tripod synchrony factors (TSF) after Spagna et al. [49] plotted against running speed. Red circles refer to runs with metachronal leg coordination and black crosses to tripodal runs. **a** Non-slippery conditions **b** Slippery conditions, which led to lower maximum speeds and strong overlap in the speed ranges of tripodal and metachronal runs

Although duty factors reached relatively low values at high running speeds on non-slippery substrate and even fell below the critical value of 0.5, the consecutive footfalls of the legs of a set and increased temporal overlap of the consecutive sets of legs, prevented periods without contact with the ground. For runs above 0.15 ms^{-1} running speed, the temporal overlap of the consecutive sets of legs was significantly higher ($p < 0.001$) in runs with metachronal leg coordination (0.31 ± 0.11; $n = 106$) compared to alternating tripodal runs (0.23 ± 0.1; $n = 72$). Below the transitional speed of 0.15 ms^{-1} metachronal leg coordination was effectively lacking; in alternating runs the temporal overlap adopted values of 0.29 ± 0.12 ($n = 87$) which did not differ from the values for metachronal runs at higher running speeds.

Body oscillations
The FFT analyses of the COM accelerations revealed differences between alternating and metachronal runs in vertical and lateral direction, while the frequency spectrum in fore-aft direction differed only little (Fig. 4). On slippery and non-slippery substrate, tripodal runs were characterised by relatively high vertical and low horizontal amplitudes while metachronal runs had lower vertical and higher lateral amplitudes. In tripodal and metachronal runs, the frequency-maxima of the fore-aft and vertical oscillations corresponded to two times the stride frequency while the maxima of the lateral oscillations were equal to stride frequency.

In all examined sequences, recovery was low (0.09; 0.03/ 0.16) and congruity was always quite high (0.7; 0.59/0.81). Running speed and leg coordination did not have any impact.

On both substrates, the external M_{COM} was significantly higher in tripodal than in metachronal runs (1.77 ± 0.78 J·

kg^{-1}·m^{-1} vs. 1.47 ± 0.69 J· kg^{-1}·m^{-1}; Table 1 in Appendix). The differences were about 20% on non-slippery and 10% on slippery substrate. This coincides with differences in the peak-to-peak amplitudes of the vertical oscillations, which were also significantly decreased in metachronal runs (0.51–0.63 mm) compared with tripodal runs (0.94–1.02 mm) (Table 1 in Appendix). The lateral amplitudes were not different on the 5% significance level. However, on slippery substrate, with metachronal leg coordination, lateral peak-to-peak amplitudes (1.42 mm) tended to be smaller than for the other conditions (1.55–1.73 mm; $p = 0.0506$; Table 1 in Appendix).

On average the COM changed its position relative to the running direction once during a stride in metachronal and tripodal runs (Additional file 1: Figures S1-S4). Though lateral oscillation patterns were not particularly consistent, during alternating runs, the position of the COM swayed predominantly to the side of the body with the front and rear legs on the ground while the COM swayed to that side with the second leg in contact with the ground during metachronal runs. Only in metachronal runs on slippery substrate were we able to find distinct lateral oscillation patterns. Concurrently, the average trajectory of the yaw angle was markedly sine shaped here (cp. [54]) while it was rather cosine shaped during alternating runs on non-slippery substrate. Accordingly, at touch down, yaw angles started with low values in metachronal runs on slippery substrate while in alternating runs on non-slippery substrate yaw angles were maximal when the legs touched the ground. The trajectories of the yaw angle seemed generally rather sine shaped on slippery and rather cosine shaped on non-slippery substrate (Additional file 1: Figure S5).

The mean pitch and yaw angles (Fig. 1; Additional file 1: Figure S5) were about zero in all running conditions. However, the pitch amplitudes were significantly lower in metachronal runs (about 4.5°) compared to alternating tripodal runs (6° to 7.5°). This difference was higher on non-slippery substrate (Table 1 in Appendix). The amplitudes of yaw were not significantly different from each other and ranged from 6.8° to 8.1° (Table 1 in Appendix). The mean angles between thorax and abdomen were between 160° and 173° and significantly higher in alternating tripodal runs than in metachronal runs (Table 1 in Appendix). Accordingly, the flexion of the cockroaches' body was more pronounced in metachronal runs and their position was relatively outstretched in tripodal runs. The amplitudes of this angle (4.1° - 4.7°) were not different between the running conditions.

Discussion
In ambulatory locomotor systems, gaits and gait changes have been widely examined in two- and four-legged vertebrates. Arthropods, in turn, are still largely regarded as being restricted to only two different gaits; namely a highly

feedback-controlled slow walking gait with metachronal leg coordination and the rather feedforward controlled running gait which is characterised by alternating sets of diagonally adjacent legs [10, 15, 39, 55]. Indeed, some species such as wood ants or fruit flies seem to use only these gaits [11, 52]. In these species the tripodal synchrony factors (TSF) are low at very low, increase at intermediate and reach a plateau of high values at high running speeds. However, our results clearly show that *N. cinerea* exhibits two fast running gaits which supplement the typical slow and mostly unsteady metachronal gait. Thus, in *N. cinerea* the TSF values decrease again at running speeds higher than 0.15 ms^{-1} (Fig. 7). Reduced synchronicity, in turn, is caused by phase shifts between the ipsilateral legs that deviate significantly from 0.5. Since phase shifts are relatively constant for each co-ordination pattern, gradually decreasing TSF values are caused by speed-dependent decreasing contact durations (Fig. 5). The slopes of t_C^{-1} against running speed and those of the contact lengths (s_C; Fig. 8) are similar in all walking legs. Therefore, the different lengths of fore, middle and rear legs [41] apparently do not affect stride length and are not the reason for changed phase shifts at high running speeds.

Decreasing phase shifts and resulting consecutive ground contacts of the legs of a tripodal set facilitate permanent body support although swing durations increase relative to single leg contact durations, which enables further increasing stride lengths without the need for ballistic phases of the COM (Fig. 8). This is reflected by a significantly higher temporal overlap between consecutive sets of legs in metachronal runs if compared with tripodal gait patterns.

Constant swing durations limit the contraction speeds of the involved muscles and prevent excessive metabolic costs which would occur when the legs would be swung

in decreasingly shorter intervals necessary for running with uniform, short strides and therefore largely increased stride numbers per distance covered [56]. Correspondingly, high speed metachronal leg coordination can limit the metabolic costs necessary for swinging the legs back to touch-down position and reduces the number of strides per distance travelled, which both contribute significantly to total running costs [57–59]. A reduction in stride numbers may be particularly significant in small legged organisms with their increased relative joint stiffnesses. When body size and limb weight decrease, relative cross-section areas of leg muscles and joint membranes increase. Accordingly, damping within these structures increases relative to inertial forces of the legs and several researchers hypothesized that movements of small animals are dominated by passive joint forces rather than inertia [60–62]. Even though, dissipation within the leg joints likely affects leg swing and accompanied metabolic costs, for cockroaches, our results indicate that stance phases seem to be governed by the inertia of the body which enables exploitation of the advantages of synchronized sets of legs such as high static stability and the employment of elastic energy storage at intermediate running speeds.

Since power is the product of force and contraction speed, muscles generate their highest power output at intermediate contraction speeds, which makes them particularly efficient in this range. Eccentric load, as typical for anti-gravity muscles during stance, is limited and quickly leads to overstrain [63, 64]. In spring-mass systems, vertical oscillation amplitudes and maximum vertical ground reaction forces increase with speed [65–67]. Our results in *N. cinerea* do also show pronounced vertical oscillations during alternating tripodal runs (Fig. 4, Additional file 1: Figures S1-S4, Table 1 in Appendix). Accordingly, at high running speed, the interplay of concordant body pitch, high vertical amplitudes and accompanied low reversal points may lead to unfavourable joint angles in the already bent fore and middle legs [41]. Moreover, excessive vertical amplitudes are prevented by the generally low position of the COM in cockroaches. High amplitudes and ground reaction forces would also lead to high contraction speeds in concentrically acting and high stresses in eccentrically loaded muscles. Therefore, further adherence to spring-mass dynamics at very high running speeds would be disadvantageous in particular as blaberid cockroaches use fast sprints only occasionally (cp. [39]), and seem rarely adapted to prolonged fast leg muscle activity.

Energy efficiency in bouncing gaits such as running, trotting and the alternating tripodal gait of insects such as cockroaches mostly relies on spring-mass dynamics. This requires matched stiffnesses of the energy storing components in all walking legs to enable concerted loading and unloading rates [34], particularly in poly-pedal locomotor

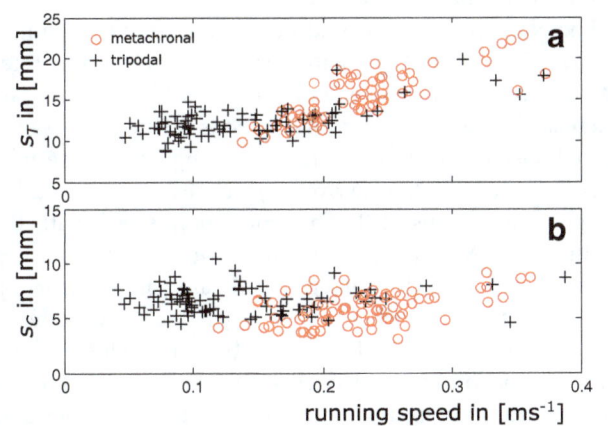

Fig. 8 Distances covered by the COM during stride and contact duration of the second legs. **a** Stride length (s_T) of the second legs. Red circles refer to runs with metachronal leg coordination while black crosses refer to alternating tripodal runs **b** Distances covered by the COM during leg contact (s_C)

systems. In blaberid cockroaches hind leg stiffness is largely determined by the passive mechanical properties of the legs themselves [20, 68]. Except for the coxa, the axes of all involved joints are almost in parallel with the ground force vectors, which makes cockroach hind legs particularly suitable for elastic storage and recovery of movement energy in vertical directions. As a result, the hip joint effectively stores and recycles displacement energy over a wide range of oscillation frequencies [20]. In the front and middle legs, however, the joint axes are perpendicular to the main ground force direction. Accordingly leg stiffness depends strongly on the activity of leg muscles and their contraction properties. Whether or not elastic mechanisms take effect in these legs has not been examined yet. However, to enable efficient use of the hind leg spring, stride frequencies and timing of the anterior legs have to be adjusted to that of the former.

COM accelerations represent causative overall ground forces. The FFT analysis of COM accelerations in *N. cinerea* revealed peaks at two-times stride frequency for vertical and fore-aft fluctuations and one-time stride frequency for lateral oscillations, which is in agreement with spring-mass running dynamics [16]. However, along with the changes in inter-leg coordination as discussed above the FFT analysis also revealed distinct spectral changes. Thus, in alternating tripodal runs the vertical amplitudes were significantly higher than in metachronal runs while lateral amplitudes were higher in the latter (Fig. 4). Accordingly, the major plane of COM oscillations changed from sagittal to horizontal.

Direct measurements also showed significantly higher vertical peak-to-peak amplitudes of the COM in alternating tripodal gaits (Table 1 in Appendix). Accordingly, the mass specific external mechanical energy was also significantly higher in alternating runs. However, bouncing gaits such as the alternating tripodal pattern of insects enable efficient loading of elastic leg structures and benefit from these structures' elastic recoil. This internal energy storage and transmission system is not externally visible and increases the apparent energy fluctuations and total energetic costs while effective metabolic costs may be considerably lower [15, 40, 69].

In alternating runs the spectrum of lateral acceleration amplitudes was blurred and did not show a clear maximum (Fig. 4a). However, such a clear maximum close to one-time stride frequency is required by the model proposed by Schmitt and Holmes [70] to enable dynamic stability in the horizontal plane. Therefore, the dynamics assumed for their lumped two-legged model do better correspond to those observed for metachronal runs with their pronounced lateral oscillations and a clear spectral maximum. Since static stability is reduced with temporally distributed touch-downs within a set of legs, dynamically stabilizing effects imposed by COM dynamics

similar to the lateral leg spring model [6] and distributed mechanical feedback [3] may replace static stability as the major stabilizing mechanism at high running speeds. This change towards passive stabilization and feed forward control can also be related to the reduced ratio between contact duration and the fastest possible reflex response [16, 71]. Thus, in the range of the transitional speed of about 0.15 ms^{-1} contact durations fall below 40 ms (Fig. 5) whereas the shortest reflex responses in cockroaches are about 20 ms [72]. Accordingly, half the stance duration has passed and the legs' ground forces have reached their maximum before a reflex response could at all affect leg activity.

The mean angles between thorax and abdomen were slightly but significantly lower in metachronal runs (Table 1 in Appendix); the total difference was about 5°. Nevertheless, the positional change led to a smaller distance between the caudal abdomen tip and the substrate surface, which might result in intermittent ground contacts as found in wood ants [27] and may have a stabilizing effect for the vertical position of the COM. However, the bent body posture could also simply be the result of anatomical constraints in external leg muscles [73] that emerge at high running speeds.

Comparative

Blaberus discoidalis is another blaberid cockroach species extensively used for experimental examinations. This species shows a similar saturation of the stride frequency with increasing running speed as *N. cinerea* while aerial phases are also lacking [2, 15]. In *N. cinerea*, stride frequency saturation is primarily caused by swing phases with constant and therefore relative to the contact phases increasing durations. Without significant aerial phases, constant swing phases, in turn, are indicative for changed leg coordination. Therefore, *B. discoidalis* seems to pass through the same gait change as *N. cinerea*.

For Namibian tenebrionid desert beetles Bartholomew et al. [17] reported constant oxygen consumption rates at running speeds above about 0.13 ms^{-1}, while this rate increased linearly at running speeds below the transitional speed. Moreover, at high running speeds the beetles had very stable COM trajectories without visible height fluctuations [17]. This corresponds well with the reduced COM fluctuations found for *N. cinerea* when running with metachronal leg coordination, in particular on slippery substrate (see below). However, due to the flattened shape of the beetles' body, lift, which would reduce vertical ground reaction forces, may also contribute to the constant cost of transport at high running speeds [17], while drag should play only a minor role at running speeds significantly below 1 ms^{-1} and given Reynolds numbers [24].

Apart from insects, gait changes from alternating to metachronal leg coordination also seem to occur in fast running arachnids. Thus, a change from linearly increasing to constant swing rates and a concurrently reduced increase of stride frequencies were also found in the vagrant spiders *Ancylometes bogotensis* and *Cupiennius salei* [16, 74]. Recent experiments in tiny mites with body lengths of only about 1 mm also revealed coordinative changes at their highest running speeds [34] which indicate dynamical changes even in some of the smallest terrestrial runners.

However, insects do not always change gaits when attaining high running speeds. The blattid cockroach *Periplaneta americana*, for example, is specialized in extremely fast escape runs. This species increases its stride frequency nearly linearly over a wide range of running speeds [29]. At their highest speeds they even lift the frontal legs off the ground and use only their hind legs for propulsion [25]. Accordingly, they cannot adopt metachronal leg coordination and seem to exploit other mechanisms to limit metabolic expenses. Interestingly, the mass specific external mechanical energy found for running *N. cinerea* (1.47 to 1.77 J·kg^{-1}·m^{-1}) is quite similar to values of about 1.5 J·kg^{-1}·m^{-1} as measured for *P. americana* [25].

Linearly increasing stride frequencies were also found in wood ants and fruit flies [11, 52] while the fast running North African dessert ant *Cataglyphis fortis* reduces the increase of the stride frequency at very high running speeds [26]. In *C. fortis*, however, the synchronisation of the legs within a set is also maintained at high running speeds, and stride frequencies increase slower at high running speeds due to the occurrence of aerial phases.

According to [75] the jumping bristletail *Petrobius* reacts upon disturbances by jumps or a peculiar high speed jumping gait. Though no detailed data are available, this jumping gait indeed might reproduce mammalian gallop dynamics with three consecutively active symmetric pairs of legs instead of four sequenced legs making up a set which replaces itself after a ballistic phase and causes redirection of the COM from downwards to upwards. Leg coordination patterns similar to that of *Petrobius* were described recently for the relatively slow locomotion of some South African species of dung beetles [76]. The COM dynamics of these beetles' "gallop" gait, however, rather seems to employ inverted-pendulum dynamics (e.g. [50, 77]) allowing them to travel at relatively low metabolic costs on deformable granular media [78].

Impact of slipperiness
On slippery substrate the onset of metachronal leg coordination occurs at significantly lower running speeds (Table 1 in Appendix) and maximum running speeds were generally lower (Fig. 7) compared with non-slippery conditions. Accordingly, the slippery sand paper seems to prevent the transmission of high horizontal forces as necessary to propel the animals at velocities above 0.25 ms^{-1}.

In insects using the alternating tripodal gait, the front and middle legs generate lateral ground reaction forces and brace against each other [52, 79]. In *N. cinerea*, desynchronization of the legs within the alternating sets as found in metachronal runs arises primarily from changing phase relations between these legs. Therefore, the early onset of metachronal coordination on slippery substrate may indicate active avoidance of lateral bracing of front and middle legs which might be of significant functional impact on coarse substrates. In principle, due to lateral bracing, elastic structures in the legs can be loaded that recoil in the second half of the contact phases and might contribute to energy recovery during a stride. Moreover, bracing can also be a mechanism to control the lateral dynamics of the COM and therefore dynamic stability (see [54]). However, on slippery and granular substrates lateral forces are difficult to transfer onto the ground which is also reflected by the significantly shorter contact lengths (s_C) found in alternating runs on slippery substrate (Additional file 1: Table S1).

Along with a significant reduction of vertical COM accelerations, metachronal coordination patterns imply more evenly distributed ground forces, reduce required force peaks and can increase energy efficiency if elastic mechanisms are not applicable (cp. [23]). They also prevent lateral bracing between front and middle legs and the risk of outward slipping, which could cause severe disturbance. Accordingly, such patterns seem to be particularly advantageous on slippery substrates.

Additionally, metachronal leg coordination increases the temporal overlap of the consecutive sets of legs. Therefore permanent ground contact of at least some legs is permitted, although stride lengths increase and duty factors decrease significantly (Figs. 4, 5, and 6). A permanent connection between substrate and some walking legs prevents interruption of proprioceptive information about the animal's position with respect to the ground and may increase controllability of basically interference-prone locomotion (cp. [71]).

Conclusion
At high running speeds, when the vertical amplitudes of the body and therefore the use of elasticity are limited, cockroaches avoid the disadvantages of bouncing by temporal dissociation of the alternating sets of legs which might also cause reduced metabolic costs and increased dynamic stability in the horizontal plane. In other words, the change from the alternating tripodal to a metachronal gait pattern at high running speeds can help arthropods to avoid overstraining involved muscles and may facilitate energy efficient high speed locomotion at the same time. Since similar kinematic adaptations were found in a range of unrelated species the high speed metachronal gait is probably widely used by legged terrestrial arthropods and may be a characteristic for fast escape manoeuvres.

Appendix

Table 1 Medians, interquartile ranges and statistical comparisons of running speed, vertical position and amplitude of the COM, lateral amplitude of the COM, amplitudes of pitch and yaw, mean values and amplitudes of the angle between thorax and abdomen and the means of the mass specific external mechanical energy (M_{COM}) as described in the methods section. The values for alternating tripodal (alt) and metachronal (met) runs on slippery (s) and non-slippery (ns) substrates were tested against each other via one-way ANOVA. The sample sizes (n) refer to the numbers of examined strides. Significant differences on the 5% level are indicated by black dots

Parameter	Slipperyness	Pattern	Median (Q25/Q75)	Unit	n	Significance			
						ns	ns	s	s
						alt	met	alt	met
running	ns	alt	0.12 (0.09/0.20)	ms^{-1}	62		•		
speed	ns	met	0.21 (0.17/0.24)		74	•		•	•
	s	alt	0.16 (0.13/0.19)		74		•		
	s	met	0.18 (0.13/0.20)		69		•		
vertical	ns	alt	3.66 (3.36/4.11)	mm	62				
position	ns	met	3.74 (3.44/4.04)		74				
	s	alt	3.82 (3.38/4.22)		74				
	s	met	3.65 (3.32/4.03)		69				
vertical	ns	alt	0.94 (0.54/1.31)	mm	62		•		•
peak-to-peak	ns	met	0.63 (0.37/0.80)		74	•		•	
amplitude	s	alt	1.02 (0.85/1.30)		74		•		•
	s	met	0.51 (0.36/1.02)		69	•		•	
lateral	ns	alt	1.55 (1.03/2.56)	mm	62				
peak-to-peak	ns	met	1.73 (1.05/2.21)		74				
amplitude	s	alt	1.55 (1.17/2.23)		74				
	s	met	1.42 (1.04/1.99)		69				
pitch	ns	alt	7.5 (4.6/10.1)	°	62		•		•
peak-to-peak	ns	met	4.5 (3.1/7.0)		74	•			
amplitude	s	alt	5.9 (3.8/8.0)		74				•
	s	met	4.6 (2.6/7.2)		69	•		•	
yaw	ns	alt	8.0 (4.9/10.9)	°	62				
peak-to-peak	ns	met	6.8 (4.6/10.3)		74				
amplitude	s	alt	8.1 (5.5/10.6)		74				
	s	met	8.0 (4.7/10.6)		69				
∠TA	ns	alt	172.0 (168.2/172.8)	°	62		•		•
	ns	met	167.1 (156.3/168.6)		74	•		•	
	s	alt	168.0 (165.8/172.6)		74		•		•
	s	met	163.6 (159.3/165.5)		69	•		•	
∠TA	ns	alt	4.5 (3.1/6.5)	°	62				
peak-to-peak	ns	met	4.4 (2.8/5.8)		74				
amplitude	s	alt	4.7 (3.3/6.5)		74				
	s	met	4.1 (3.1/5.3)		69				

Table 1 Medians, interquartile ranges and statistical comparisons of running speed, vertical position and amplitude of the COM, lateral amplitude of the COM, amplitudes of pitch and yaw, mean values and amplitudes of the angle between thorax and abdomen and the means of the mass specific external mechanical energy (M_{COM}) as described in the methods section. The values for alternating tripodal (alt) and metachronal (met) runs on slippery (s) and non-slippery (ns) substrates were tested against each other via one-way ANOVA. The sample sizes (n) refer to the numbers of examined strides. Significant differences on the 5% level are indicated by black dots *(Continued)*

Parameter	Slipperyness	Pattern	Median (Q25/Q75)	Unit	n	Significance			
M_{COM}	ns	alt	1.86 (1.29/2.46)	$J \cdot kg^{-1} \cdot m^{-1}$	62		•		•
	ns	met	1.47 (1.07/1.81)		74	•		•	
	s	alt	1.82 (1.57/2.34)		74		•		•
	s	met	1.63 (1.13/2.01)		69	•		•	

Additional files

Additional file 1: Figure S1. COM kinematics over second legs' strides for alternating tripodal runs on non-slippery substrate. A stride consists of a contact phase and the subsequent swing phase. The black solid line shows the median course of a value and the grey shaded area the interquartile range. First column: fore-aft direction (X); Second column: lateral direction (Y); Third column: vertical direction (Z). First row: distance in m; Second row: velocity in ms^{-1}; Third row: acceleration in ms^{-2}. **Figure S2.** COM kinematics over second legs' strides for metachronal runs on non-slippery substrate. A stride consists of a contact phase and the subsequent swing phase. The black solid line shows the median course of a value and the grey shaded area the interquartile range. First column: fore-aft direction (X); Second column: lateral direction (Y); Third column: vertical direction (Z). First row: distance in m; Second row: velocity in ms^{-1}; Third row: acceleration in ms^{-2}. **Figure S3.** COM kinematics over second legs' strides for alternating tripodal runs on slippery substrate. A stride consists of a contact phase and the subsequent swing phase. The black solid line shows the median course of a value and the grey shaded area the interquartile range. First column: fore-aft direction (X); Second column: lateral direction (Y); Third column: vertical direction (Z). First row: distance in m; Second row: velocity in ms^{-1}; Third row: acceleration in ms^{-2}. **Figure S4.** COM kinematics over second legs' strides for metachronal runs on slippery substrate. A stride consists of a contact phase and the subsequent swing phase. The black solid line shows the median course of a value and the grey shaded area the interquartile range. First column: fore-aft direction (X); Second column: lateral direction (Y); Third column: vertical direction (Z). First row: distance in m; Second row: velocity in ms^{-1}; Third row: acceleration in ms^{-2}. **Figure S5.** The courses of pitch, yaw and ∠TA over the second legs' strides. The black solid line shows the median course of an angle and the grey shaded area the interquartile range. Upper row: Alternating tripodal runs on non-slippery substrate; Second row: Metachronal runs on non-slippery substrate; Third row: Alternating tripodal runs on slippery substrate; Fourth row: Metachronal runs on slippery substrate. **Figure S6.** Stride frequencies, duty factors, swing rates and contact rates plotted against running speed for all walking legs and slippery conditions. First row: Stride frequency (f_T); Second row: duty factors (β); Third row: swing rates (t_S^{-1}); Fourth row: contact rates (t_C^{-1}). Red circles are measured values for runs with metachronal leg coordination while black crosses depict values from alternating tripodal runs. **Figure S7.** Phase shifts between ipsilateral legs and between the contralateral rear legs during alternating tripodal (white) and metachronal (red) runs on non-slippery (left) and slippery substrate. A, E) Phase values for the touch-downs of the fore legs in the stride period of the rear legs. B, F) Phase values for the touch-downs of the middle legs in the stride period of the rear legs. C, G) Phase values for the touch-downs of the fore legs in the stride period of the middle legs. D, H) Phase values for the touch-downs of the contralateral rear legs. **Figure S8.** Stride durations, swing durations and contact durations plotted against running speed for all walking legs. Stride duration: row one and four; contact duration: row two and five; swing duration: row three and six. The rows one to three refer to non-slippery conditions whereas the rows four to six refer to slippery conditions. Red circles are measured values for runs with metachronal leg coordination while black crosses depict values from alternating tripodal runs. Black (alternating runs) and red (metachronal) lines in the upper three rows were calculated on the basis of the linear least squares regressions for t_S^{-1} and t_C^{-1} (see Fig. 4). (PDF 2640 kb)

Additional file 2: Typical running sequence with metachronal leg coordination. The sequence is slowed down to 1/10 of the recording speed. The mean running speed was 0.2 ms-1, the mean phase shift between the ipsilateral legs was 0.77 and the mean duty factor was 0.42. (AVI 1512 kb)

Additional file 3: Typical running sequence with tripodal leg coordination (1/10 recording speed). Beginning on frame 80 the animal increases its mean speed within a fraction of a stride from about 0.1 ms-1 towards a mean running speed of 0.16. Before the change the mean phase shift between ipsilateral legs was 0.9 and the mean duty factor was 0.52. With the higher running speed the mean phase shift between the ipsilateral legs decreased to 0.82 and the mean duty factor to 0.45. (AVI 2343 kb)

Abbreviations
∠TA: Ventral angle between thorax and abdomen; COM: Centre of mass; f_T: Stride frequency, i.e. the inverse of the stride duration t_T; M_{COM}: Mass specific external mechanical energy per unit distance; s_C: Distance travelled by the COM during contact; s_T: Distance travelled by the COM during stride; t_C^{-1}: Contact rate, i.e. the inverse of the contact duration t_C; t_S^{-1}: Swing rate, i.e. the inverse of the swing duration t_S; TSF: Tripod synchrony factor

Acknowledgements
We would like to thank Walter Federle who hosted the experiments and subsequent analyses in his lab and Ansgar Büschges who actively supported the work on the present manuscript.

Funding
This work was supported by the German Research Foundation (DFG, We 4664/2–1 and 3–1) to TW.

Authors' contributions
TW conceived, designed and coordinated the study, analysed the data, carried out the statistical analyses and drafted the manuscript; PGB participated in data acquisition and analysis; EP participated in data acquisition and revised the manuscript critically. All authors read and approved the final manuscript.

Competing interests
The authors declare that they have no competing interests.

Author details

[1]Department of Animal Physiology, Institute of Zoology, University of Cologne, Zülpicher Strasse 47b, 50674 Cologne, Germany. [2]Ecole Normale Supérieure de Lyon Département de Biologie, Lyon, France. [3]Department of Zoology, University of Cambridge, Downing Street, Cambridge CB2 3EJ, UK.

References

1. Bowerman RF. The control of walking in the scorpion. J Comp Physiol. 1975;100:197–209.
2. Ting LH, Blickhan R, Full RJ. Dynamic and static stability in hexapedal runners. J Exp Biol. 1994;197:251–69.
3. Spagna JC, Goldman DI, Lin PC, Koditschek DE, Full RJ. Distributed mechanical feedback in arthropods and robots simplifies control of rapid running on challenging terrain. Bioinspir Biomim. 2007;2:9–18.
4. Blickhan R, Full RJ. Similarity in multilegged locomotion: bouncing like a monopode. J Comp Physiol A. 1993;173:509–17.
5. Sensenig AT, Shultz JW. Mechanical energy oscillations during locomotion in the harvestman Leiobunum vittatum (Opiliones). J Arachnol. 2006;34:627–33.
6. Full RJ, Koditschek DE. Templates and anchors: neuromechanical hypotheses of legged locomotion on land. J Exp Biol. 1999;202:3325–32.
7. Wilson DM. Insect walking. Annu Rev Entomol. 1966;11:103–22.
8. Graham D. A behavioural analysis of the temporal organisation of walking movements in the 1st Instar and adult stick insect (Carausius morosus). J Comp Physiol. 1972;81:23–52.
9. Wendler G. Laufen und Stehen der Stabheuschrecke Carausius morosus: Sinnesborstenfelder in den Beingelenken als Glieder von Regelkreisen. Z Vergl Physiol. 1964;48:198–250.
10. Hughes GM. The co-ordination of insect movements. I: the walking movements of insects. J Exp Biol. 1952;29:267–84.
11. Wosnitza A, Bockemühl T, Dübbert M, Scholz H, Büschges A. Inter-leg coordination in the control of walking speed in drosophila. J Exp Biol. 2013;216:480–91.
12. Mendes CS, Bartos I, Akay T, Márka S, Mann RS. Quantification of gait parameters in freely walking wild type and sensory deprived Drosophila melanogaster. elife. 2013;2:e00231.
13. Ward TM, Humphreys WF. Locomotion in burrowing and vagrant wolf spiders (Lycosidae). J Exp Biol. 1981;92:305–21.
14. Biancardi CM, Fabrica CG, Polero P, Loss JF, Minetti AE. Biomechanics of octopedal locomotion: kinematic and kinetic analysis of the spider Grammostola Mollicoma. J Exp Biol. 2011;214:3433–42.
15. Full RJ, Tu MS. Mechanics of six-legged runners. J Exp Biol. 1990;148:129–46.
16. Weihmann T. Crawling at high speeds: steady level locomotion in the spider Cupiennius salei - global kinematics and implications for Centre of Mass Dynamics. PLoS One. 2013;8:e65788.
17. Bartholomew GA, Lighton JRB, Louw GN. Energetics of locomotion and patterns of respiration in tenebrionid beetles from the Namib Desert. J Comp Physiol B. 1985;155:155–62.
18. Alexander RM, Bennet-Clark HC. Storage of elastic strain energy in muscle and other tissues. Nature. 1977;265:114–7.
19. Sensenig AT, Shultz JW. Mechanics of cuticular elastic energy storage in leg joints lacking extensor muscles in arachnids. J Exp Biol. 2003;206:771–84.
20. Dudek DM, Full RJ. Passive mechanical properties of legs from running insects. J Exp Biol. 2006;209:1502–15.
21. Bennet-Clark HC. The energetics of the jump of the locust Schistocerca Gregaria. J Exp Biol. 1975;63:53–83.
22. Minetti AE, Ardig OL, Reinach E, Saibene F. The relationship between mechanical work and energy expenditure of locomotion in horses. J Exp Biol. 1999;202:2329–38.
23. Ruina A, Bertram JE, Srinivasan M. A collisional model of the energetic cost of support work qualitatively explains leg sequencing in walking and galloping, pseudo-elastic leg behavior in running and the walk-to-run transition. J Theor Biol. 2005;237:170–92.
24. Full RJ, MAR K. Drag and lift on running insects. J Exp Biol. 1992;176:89–101.
25. Full RJ, Tu MS. Mechanics of a rapid running insect: two-, four- and six-legged locomotion. J Exp Biol. 1991;156:215–31.
26. Wahl V, Pfeffer SE, Wittlinger M. Walking and running in the desert ant Cataglyphis fortis. J Comp Physiol A Neuroethol Sens Neural Behav Physiol. 2015;201:645–56.
27. Reinhardt L, Weihmann T, Blickhan R. Dynamics and kinematics of ant locomotion: do wood ants climb on level surfaces? J Exp Biol. 2009;212:2426–35.
28. Weihmann T, Blickhan R. Comparing inclined locomotion in a ground-living and a climbing ant species: sagittal plane kinematics. J Comp Physiol A Neuroethol Sens Neural Behav Physiol. 2009;198:1011–20.
29. Delcomyn F. The locomotion of the cockroach Periplaneta americana. J Exp Biol. 1971;54:443–52.
30. Gatesy SM, Biewener A. Bipedal locomotion: effects of speed, size and limb posture in birds and humans. J Zool Lond. 1991;224:127–47.
31. Cavagna GA, Heglund NC, Taylor CR. Mechanical work in terrestrial locomotion: two basic mechanisms for minimizing energy expenditure. Am J Phys. 1977;233:R243–61.
32. Heglund NC, Taylor CR. Speed, stride frequency and energy cost per stride: how do they change with body size and gait? J Exp Biol. 1988;138:301–18.
33. Heglund NC, Taylor CR, McMahon TA. Scaling stride frequency and gait to animal size: mice to horses. Science. 1974;186:1112–3.
34. Weihmann T, Goetzke HH, Günther M. Requirements and limits of anatomy-based predictions of locomotion in terrestrial arthropods with emphasis on arachnids. J Paleontol. 2015;89:980–90.
35. Günther M, Weihmann T. Climbing in hexapods: a plain model for heavy slopes. J Theor Biol. 2012;293:82–6.
36. Le Jeune TM, Willems PA, Heglund NC. Mechanics and energetics of human locomotion on sand. J Exp Biol. 1998;201:2071–80.
37. Bohn HF, Federle W. Insect aquaplaning: nepenthes pitcher plants capture prey with the peristome, a fully wettable water-lubricated anisotropic surface. PNAS. 2004;101:14138–43.
38. Clemente CJ, Federle W. Pushing versus pulling: division of labour between tarsal attachment pads in cockroaches. Proc Biol Sci. 2008;275:1329–36.
39. Bender JA, Simpson EM, Tietz BR, Daltorio KA, Quinn RD, Ritzmann RE. Kinematic and behavioral evidence for a distinction between trotting and ambling gaits in the cockroach Blaberus discoidalis. J Exp Biol. 2011;214:2057–64.
40. Full RJ, Blickhan R, Ting LH. Leg design in hexapedal runners. J Exp Biol. 1991;158:369–90.
41. Kram R, Wong B, Full RJ. Three-dimensional kinematics and limb kinetic energy of running cockroaches. J Exp Biol. 1997;200(Pt 13):1919–29.
42. Bullock JMR, Federle W. The effect of surface roughness on claw and adhesive hair performance in the dock beetle Gastrophysa Viridula. Insect Science. 2011;18:298–304.
43. Voigt D, Schuppert JM, Dattinger S, Gorb SN. Sexual dimorphism in the attachment ability of the Colorado potato beetle Leptinotarsa Decemlineata (Coleoptera : Chrysomelidae) to rough substrates. J Insect Physiol. 2008;54:765–76.
44. Dai Z, Gorb SN, Schwarz U. Roughness-dependent friction force of the tarsal claw system in the beetle Pachnoda marginata (Coleoptera, Scarabaeidae). J Exp Biol. 2002;205:2479–88.
45. Anderson B, Shultz J, Jayne B. Axial kinematics and muscle activity during terrestrial locomotion of the centipede Scolopendra heros. J Exp Biol. 1995;198:1185–95.
46. Jamon M, Clarac F. Locomotor patterns in freely moving crayfish (Procambarus Clarkii). J Exp Biol. 1995;198:683–700.
47. Velasco MJ, Philipp B: Circular statistics toolbox for Matlab. 2009.
48. Weihmann T, Karner M, Full RJ, Blickhan R. Jumping kinematics in the wandering spider Cupiennius salei. J Comp Physiol A Neuroethol Sens Neural Behav Physiol. 2010;196:421–38.
49. Spagna JC, Valdivia EA, Mohan V. Gait characteristics of two fast-running spider species (Hololena adnexa and Hololena curta), including an aerial phase (Araneae: Agelenidae). J Arachnol. 2011;39:84–91.
50. Blickhan R, Full RJ. Locomotion energetics of the ghost crab: II mechanics of the center of mass. J Exp Biol. 1987;130:155–74.
51. Ahn A, Furrow E, Biewener A. Walking and running in the red-legged running frog, Kassina maculata. J Exp Biol. 2004;207:399–410.
52. Reinhardt L, Blickhan R. Level locomotion in wood ants: evidence for grounded running. J Exp Biol. 2014;217:2358–70.
53. Shultz JW. Walking and surface film locomotion in terrestrial and semiaquatic spiders. J Exp Biol. 1987;128:427–44.
54. Schmitt J, Garcia M, Razo RC, Holmes P, Full RJ. Dynamics and stability of legged locomotion in the horizontal plane: a test case using insects. Biol Cybern. 2002;86:343–53.
55. Cruse H, Dürr V, Schmitz J. Insect walking is based on a decentralized architecture revealing a simple and robust controller. Philos Transact A Math Phys Eng Sci. 2007;365:221–50.

56. Nishii J. An analytical estimation of the energy cost for legged locomotion. J Theor Biol. 2006;238:636–45.

57. Marsh RL, Ellerby DJ, Carr JA, Henry HT, Buchanan CI. Partitioning the Energetics of walking and running: swinging the limbs is expensive. Science. 2004;303:80–3.

58. Fedak MA, Heglund NC, Taylor CR. Energetics and mechanics of terrestrial locomotion. II. Kinetic energy changes of the limbs and body as a function of speed and body size in birds and mammals. J Exp Biol. 1982;97:23–40.

59. Farley CT, Taylor CR. A mechanical trigger for the trot-gallop transition in horses. Science. 1991;19:306–8.

60. Hooper SL, Guschlbauer C, Blumel M, Rosenbaum P, Gruhn M, Akay T, Buschges A. Neural control of unloaded leg posture and of leg swing in stick insect, cockroach, and mouse differs from that in larger animals. J Neurosci. 2009;29:4109–19.

61. Ache J, Matheson T. Passive joint forces are tuned to limb use in insects and drive movements without motor activity. Curr Biol. 2013;23:1418–26.

62. Reilly SM, McElroy EJ, Biknevicius AR. Posture, gait and the ecological relevance of locomotor costs and energy-saving mechanisms in tetrapods. Zoology. 2007;110:271–89.

63. Armstrong RB, Ogilvie RW, Schwane JA. Eccentric exercise-induced injury to rat skeletal muscle. J Appl Physiol Respir Environ Exerc Physiol. 1983;54:80–93.

64. Lindstedt SL, LaStayo PC, Reich TE. When active muscles lengthen: properties and consequences of eccentric contractions. Physiology. 2001;16:256–61.

65. Keller TS, Weisbrger AM, Ray JL, Hasan SS, Shiavi RG, Spengler DM. Relationship between vertical ground reaction force and speed during walking, slow jogging, and running. Clin Biomech. 1996;11:253–9.

66. McMahon TA, Cheng GC. The mechanics of running: how does stiffness couple with speed? J Biomech. 1990;23(Suppl 1):65–78.

67. Blickhan R. The spring-mass model for running and hopping. J Biomech. 1989;22:1217–27.

68. Günther M, Weihmann T. The load distribution among three legs on the wall: model predictions for cockroaches. Arch Appl Mech. 2011;81:1269–87.

69. Cavagna GA, Saibene FP, Margaria R. Mechanical work in running. J Appl Physiol. 1964;19:249–56.

70. Schmitt J, Holmes P. Mechanical models for insect locomotion: dynamics and stability in the horizontal plane-II. Application. Biol Cybern. 2000;83:517–27.

71. Sponberg S, Full RJ. Neuromechanical response of musculo-skeletal structures in cockroaches during rapid running on rough terrain. J Exp Biol. 2008;211:433–46.

72. Schaefer PL, Kondagunta GV, Ritzmann RE. Motion analysis of escape movements evoked by tactile stimulation in the cockroach Periplaneta Americana. J Exp Biol. 1994;190:287–94.

73. Carbonell CS. The thoracic muscles of the cockroach Periplaneta americana (L.). Washington: Smithsonian Institution; 1947.

74. Weihmann T. Biomechanische Analyse der ebenen Lokomotion von Ancylometes bogotensis (Keyserling, 1877) (Chelicerata, Arachnida, Lycosoidea). doctoral thesis. Friedrich Schiller Universität. Jena: Biologisch-Pharmazeutische Fakultät; 2007.

75. Manton SM. The evolution of arthropodan locomotiory mechanisms part 10. Locomotory habits, morphology and evolution of the hexapod classes. Zool J Linnean Soc. 1972;51:203–400.

76. Smolka J, Byrne MJ, Scholtz CH, Dacke M. A new galloping gait in an insect. Curr Biol. 2013;23:R913–5.

77. Alexander R. Energy-saving mechanisms in walking and running. J Exp Biol. 1991;160:55–69.

78. Li C, Zhang T, Goldman DI. A terradynamics of legged locomotion on granular media. Science. 2013;339:1408–12.

79. Dickinson MH, Farley CT, Full RJ, Koehl MAR, Kram R, Lehman S. How animals move: an integrative view. Science. 2000;288:100–6.

13

Trade-offs in the production of animal vocal sequences: insights from the structure of wild chimpanzee pant hoots

Pawel Fedurek[1], Klaus Zuberbühler[2,3]* and Stuart Semple[4]

Abstract

Background: Vocal sequences - utterances consisting of calls produced in close succession - are common phenomena in animal communication. While many studies have explored the adaptive benefits of producing such sequences, very little is known about how the costs and constraints involved in their production affect their form. Here, we investigated this issue in the chimpanzee (*Pan troglodytes schweinfurthii*) pant hoot, a long and structurally complex vocal sequence comprising four acoustically distinct phases – introduction, build-up, climax and let-down.

Results: We found that in each of these phases, and for the sequence as a whole, there was a negative relationship between the number of calls produced and their average duration. There was also a negative relationship between the total duration of some adjacent phases. Significant relationships between the fundamental frequency of calls and their number or duration were found for some phases of the sequence, but the direction of these relationships differed between particular phases.

Conclusions: These results indicate that there are trade-offs in terms of signal duration at two levels in pant-hoot production: between call number and duration, and between the relative durations of successive phases. These trade-offs are likely to reflect biomechanical constraints on vocal sequence production. Phase-specific trade-offs also appear to occur between fundamental frequency and call number or duration, potentially reflecting that different phases of the sequence are associated with distinct types of information, linked in different ways to call pitch. Overall, this study highlights the important role of costs and constraints in shaping the temporal and acoustic structure of animal vocal sequences.

Keywords: Acoustic trade-offs, Call sequences, Chimpanzee, Compression, Menzerath's law, Pant hoot

Background

Vocal signals are an integral part of animal communication and have important functions, ranging from attracting mating partners to coordinating activities between group members [1, 2]. Vocal sequences, utterances consisting of a series of calls produced in close succession, are common phenomena and found across a wide range of animal taxa [3]. The adaptive benefits of such signals have been widely researched. For example, repeated production of the same call type has been found to reduce the probability of signal misinterpretation by the

receiver [4], while production of vocal sequences composed of different call types can enhance the communicative potential of individual calls or different combinations of calls [5–7], facilitate individual recognition [8], or play a role in attracting mates [9, 10] or repelling sexual rivals [11, 12].

While a range of adaptive benefits of vocal sequences have been demonstrated, much less attention has been paid to the potential costs and constraints involved in producing such signals. Although vocalising in itself has a metabolic cost, this appears to be relatively low [13–15]; however, the production of long vocal sequences may involve further energetic costs linked to the fine muscle control that is needed - over several levels of vocal production - to generate these complex utterances. Specifically, vocal sequence production may be affected by

* Correspondence: kz3@st-and.ac.uk
[2]Institute of Biology, University of Neuchâtel, Neuchâtel, Switzerland
[3]School of Psychology and Neuroscience, University of St Andrews, St Andrews, Scotland, UK
Full list of author information is available at the end of the article

biomechanical constraints related to lung capacity, breathing control [16], airflow control at the source, and movements of the vocal tract [17, 18]. Additionally, a potential constraint on vocal sequence utterance is related to the risk of hyperventilation, which may occur if vocalisations are produced in too rapid succession [19]. These costs and constraints could lead to significant trade-offs in how vocal sequences are constructed.

A recent study of male gelada (*Theropithecus gelada*) vocal sequences provided evidence for just such a trade-off: a negative correlation was found between the number of calls in a sequence and the average duration of these constituent calls [20]. The production of sequences with a greater number of calls thus only appears possible if shorter calls are used within them, which may reflect energetic or breathing constraints on vocal production [20]. This pattern is consistent with Menzerath's law, a linguistic law which states that the larger the construct, the smaller is the size of its constituents [20–23]. This law has been linked mathematically to compression - the information-theoretic principle of minimising code length - and it has been argued that this is a universal principle not only of animal behaviour [24], but also of biological information systems in the broadest sense [20].

In vocal sequences with distinct phases - such as orlotan bunting (*Emberiza hortulana*) song [25], rock hyrax (*Procavia capensis*) calls [11] or chimpanzee pant hoots [26] - another potential trade-off is in the overall investment of effort between phases. There is evidence that different phases in such sequences can be associated with different types of information, and be relevant to different potential receivers [11, 12, 27, 28]. Consequently, social factors such as audience composition may affect how signallers potentially benefit from allocating more effort to one phase or another. If energetic or breathing-related constraints apply to the whole sequence, individuals may benefit by allotting more to one phase at the cost of what is possible to allot to another, depending on their specific circumstances. While it has been shown that callers can modify the duration of specific phases or notes within a sequence [11, 29], it is unclear whether such adjustments at the level of whole phases affect the duration of other phases.

Duration - of calls or sequence phases - is, however, only one measure of cost, and constraints may apply to other, not necessarily temporal, acoustic features of vocal sequences. One spectral acoustic feature of calls that has been associated with energetic costs is fundamental frequency (F0) [30]. In a number of animals, including Japanese quails (*Coturnix japonica*) [31], Alston's singing mice (*Scotinomys teguina*) [32] and humans (*Homo sapiens*) [33], low frequency of calls reflects good health or condition of the caller, partially because such calls are

energetically costly to produce. However, in other animals, such as red deer (*Cervus elaphus*) [34], chacma baboons (*Papio cynocephalus ursinus*) [35] and white-handed gibbons (*Hylobates lar*) [36] producing high rather than low-frequency calls is associated with good quality among males. This could be because high-frequency calling requires a high sub-glottal pressure and elevated muscular effort, and therefore incurs metabolic costs, but more likely is because calling at high frequencies requires significant motor control of the larynx [17, 18, 37].

It is possible, therefore, that there is a trade-off between F0 on the one hand, and call duration or number on the other hand, with the nature of this trade-off depending on whether high or low frequency calls are more energetically costly. For example, if it is particularly costly to produce low frequency calls, it would be expected that the longer or more numerous are the calls in a sequence, the higher would be their F0. If producing calls of high frequency is especially costly, the opposite relationship should be expected. To our knowledge there have been no studies examining directly the possibility that there is a trade-off in vocal sequences between call pitch and either call duration or call number.

In this study, we tested for evidence of trade-offs in chimpanzee pant hoots. This complex vocal sequence consists of four distinct phases ([38]; Fig.1; see Additional file 1 for an example of a recording). Pant-hooting usually starts with the introduction phase, consisting of low-frequency and low-amplitude calls, which then grade into the build-up phase, consisting of a series of short, low-frequency calls [26]. The build-up, in turn, grades into the climax phase, the loudest part of the sequence that can include one or several 'screams' (i.e. climax calls). This is often followed by the let-down phase, which has similar acoustic features to the build-up phase [26]. There is considerable within-[29] and between- [38] individual variation in terms of the number of calls within all four phases of the sequence. Pant hoots have multiple social functions, ranging from signalling social status and bonds, to coordinating grouping and proximity [39–42], and recent evidence indicates that different phases fulfil different communicative functions [43].

To explore potential trade-offs in construction of this complex vocal sequence, we tested first for a negative relationship between call number and duration in each phase, and for the overall pant hoot. Next, we tested whether the durations of adjacent phases in the sequence are negatively related. Finally, we tested whether in each phase, F0 is related to call number or duration; for this analysis there was no clear expectation as to the direction of relationship, as it is unclear whether low- or high-frequency calling is particularly costly for male chimpanzees [44, 45].

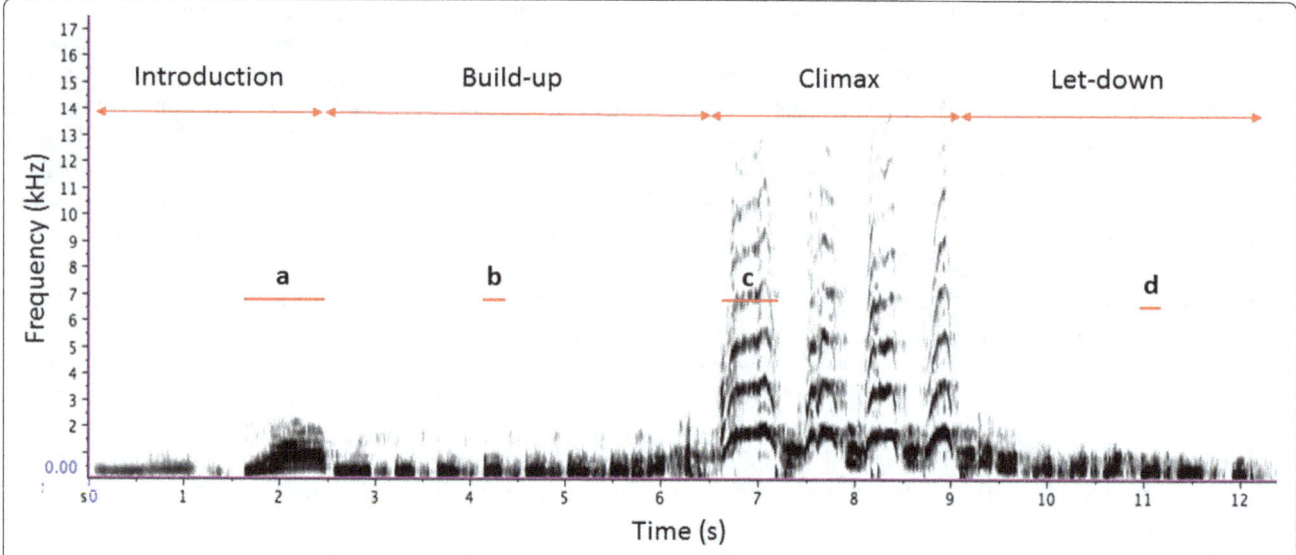

Fig. 1 Spectrographic representation of a pant hoot, with the four phases and their calls. **a** – an introduction call, **b** – a build-up call, **c** – a climax call, **d** – a let-down call. In this example, the introduction consists of two calls, the build-up of nine calls, the climax of four calls, and the let-down of eight calls. Red lines below "**a**", "**b**", "**c**" and "**d**" represent durations of calls within the four phases

Methods

Study site and study subjects

The study was carried out on the Sonso chimpanzee community of Budongo Forest, Uganda. The group has been continuously observed since 1990 and is well habituated to the presence of human observers [46]. At the time of the study, the community contained 75 individuals, with a core home range of around 15 km^2. Study subjects were adult ($N = 11$: $\geq$ 16 years) and late adolescent ($N = 2$: $\geq$ 13–15 years; [47]) males. See Additional files 2, 3, 4 and 5 for information on study males' age, their dominance rank, and the number of pant hoot recordings per individual.

Sampling methods

Fieldwork was conducted between June and October 2013, February and September 2014 and January and December 2015. Data were collected between 0700 and 1630 h local time. Data collection methods for this study were entirely non-invasive.

Each day, an arbitrarily chosen male was followed for the whole day. Pant hoots were audio-recorded from the focal male and, if possible, all other males present in his party, using a Marantz Professional PMD661 solid-state recorder and a Sennheiser ME67 directional microphone. In addition, the context of pant hoot production (travelling or feeding) was noted.

Data collected and definitions

Context. Pant hoots are usually produced in travelling and feeding contexts [42]. Pant hoots given when arriving at a feeding site (e.g. approaching or climbing a feeding tree), or during feeding, were classified as 'feeding' pant hoots. We classified pant hoots produced when moving on the ground (as opposed to arriving at a feeding site or feeding) as 'travel' pant hoots [42].

Dominance rank. This was calculated using the Elo-rating procedure, which is based on sequences of agonistic interactions between individuals [48]; see Additional file 3).

Selection of recordings and acoustic features

An utterance was defined as a "pant hoot" only if it contained the climax phase [26, 42]. We only considered recordings for analyses if they were of high quality without background noise. As well as the number of calls in each phase and the whole sequence, and the duration of calls, we assessed the F0 of calls (peak frequency in Hz of the F0 at the middle of a call) and phase duration (time in seconds between the start of the first call and the end of the last call of a phase).

Statistical analyses

We used linear mixed-effect models (LMM) with maximum likelihood estimates using R, version 3.1.2 [49] and the lme 4 package, version 1.0–7 [50]. In models testing for a negative relationship between call duration and number, call duration was the dependent variable, and the number of calls (per phase or in the entire pant hoot utterance) was the test fixed variable. Since behavioural state might affect the acoustic structure of pant-hooting [51], the context of call production (travelling vs. feeding) was included as a control fixed variable. In models testing for a negative relationship between the durations of adjacent phases, the dependent variable was the duration of build-up, climax, or let-down, respectively, and the fixed variable was the duration of the preceding phase (i.e. introduction, build-up, or climax, respectively). The context of call production was entered as a fixed control variable. In this particular analysis we excluded all pant hoots with missing build-up ($N = 47$) or let-down ($N = 55$) phases. In models testing whether, within a phase, call F0 was related to call number or duration, call F0 was the dependent variable, and both call duration and the number of calls in a phase were fixed test variables. In addition to context of call production, age and dominance rank of the caller were entered as control fixed variables, since these two attributes correlate with F0 of pant-hooting. In all our models we entered as random intercept caller ID, together with random slopes for all the fixed variables within individuals. We entered pant hoot ID as another random intercept since we measured multiple calls from the same pant hoot. Recordings with incomplete introduction phases ($N = 50$) were not incorporated in the analyses concerning the introduction and the entire pant hoot.

We used a likelihood ratio test (LRT) to test the full model against a null model (comprising the intercept and random effects) and to test the significance of individual independent variables [52, 53]. There was no collinearity between the examined independent variables (variance inflation factors of the independent variables were below the value of 2). Prior to the analyses, if necessary, variables were transformed to achieve more symmetrical distributions (see Additional files 4 and 5 for details on which transformation type was used for each variable), and values of all quantitative variables were scaled to a mean of 0 and standard deviation of 1. We ran bootstraps to estimate 95% confidence intervals around the estimates of each fixed effect.

Since data from each call within a sequence were used in three different models (two on the phase level and one on the entire pant hoot level), we controlled the Type I error rate by the sequential Bonferroni technique [54, 55], using a Bonferroni adjustment (k) equal to 3. Since in the analyses with phase duration data from the build-up and the climax were used twice, we applied a Bonferroni adjustment equal to 2.

Results

Descriptive statistics for duration, number and F0 of calls in each phase and the entire pant hoot, and for the duration of the phases and overall sequence, are shown in Table 1.

Is there a negative relationship between call duration and number?

There were significant negative relationships between call duration and the number of calls in all four phases - introduction (Fig. 2a), build up (Fig. 2b), climax (Fig. 2c), let-down (Fig. 2d) - and for the entire pant hoot (Fig. 2e) (Table 2).

Is there a negative relationship between durations of adjacent phases?

There was a significant negative relationship between the duration of the introduction and build-up (estimate ± SE = −0.11 ± 0.04, χ^2 = 5.53, p = 0.019, 95% CI = −0.21 to −0.02; Fig. 3a), and of the build-up and climax phases (estimate ± SE = −0.09±0.04, χ^2 = 5.93, p = 0.015, 95% CI = −0.18 to −0.02; Fig. 3b). The durations of the climax and the let-down phases were not related (estimate ± SE = −0.08±0.11, χ^2 = 0.52, p = 0.469, 95% CI = −0.32 to 0.14; Fig. 3c).

Is there a relationship between call F0 and call duration?

There was a significant positive relationship between call F0 and duration in the climax (Table 3; Fig. 4c) and a significant negative relationship between these two variables in the build-up (Table 3; Fig. 4b). There was no relationship between call F0 and duration in the introduction or let-down phases (Table 3; Fig. 4a and d).

Is there a relationship between call F0 and call number?

There was a positive relationship between call F0 and the number of calls in the climax and let-down (Table 3; Fig. 5c and d). There was no relationship between these variables in the introduction or build-up phases (Table 3; Fig. 5a and b).

Discussion

In this study of wild chimpanzee pant hoots, we found negative relationships between the number and duration of calls, both at the level of phases within the pant hoot,

Table 1 Mean (±SD) values of call duration, number of calls and call F0, per phase and in the whole pant hoot, and the duration of each phase and the entire sequence

	Introduction	Build-up	Climax	Let-down	Entire pant hoot
Call duration (s)	0.48±0.31	0.21±0.07	0.57±0.24	0.20±0.04	0.37±0.27
Phase duration (s)	5.07±2.10	2.47±1.12	1.20±0.60	1.11±0.82	8.05±3.05
N calls	6.68±3.01	5.78±2.48	2.25±1.00	4.27±2.44	14.61±4.04
Call F0 (Hz)	400.04±180.30	302.17±92.81	1182.67±265.24	339.43±82.28	473.74±340.6

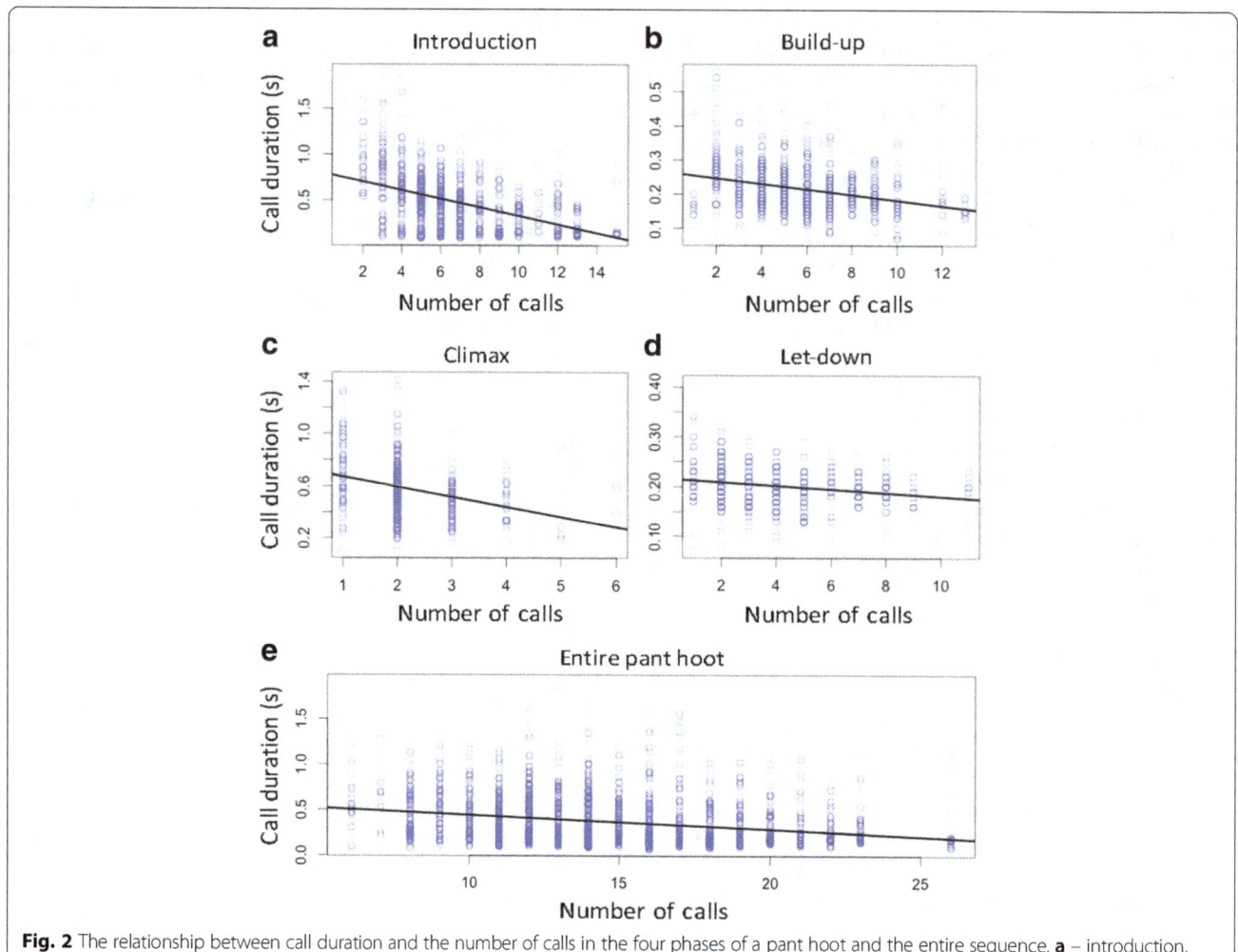

Fig. 2 The relationship between call duration and the number of calls in the four phases of a pant hoot and the entire sequence. **a** – introduction, **b** – build-up, **c** – climax, **d** – let-down, **e** – entire pant hoot. Black line represents regression line; circles represent data points

and for the entire vocal sequence. Negative relationships were also found between the durations of some adjacent phases, namely introduction and build-up, and build-up and climax. While relationships were found in some phases between call F0 and either the number of calls or their durations, the direction of these associations varied between phases. These results imply that there are trade-offs in terms of duration at two levels in pant hoot production - between call number and duration, and between relative duration of successive phases - and that trade-offs between fundamental frequency and call number or duration also occur, with the nature of these being phase-specific.

Our finding of strong, negative relationships between the number of calls and their durations provides further evidence that Menzerath's linguistic law, which reflects the principle of compression, holds in the vocal communication of non-human animals, adding to similar recent evidence from a study of male gelada call sequences [20]. Importantly, agreement with Menzerath's law here

was seen both in phases with relatively long constituent calls (introduction and climax), and in those with shorter constituent calls (build-up and let-down), implying that compression acts similarly across the distinct parts of pant hoots, regardless of the relative length of constituent calls.

Previous studies have proposed that patterns consistent with compression may be less likely to emerge in situations where vocal signals are directed at distant audiences [20, 24]. For example, in female Barbary macaques, copulation call sequences given around the time of ovulation contain more calls than sequences given early in the cycle, but these calls are longer - not shorter - in duration than those in early cycle sequences [56]. It has been proposed that this pattern may be due to the fact that in this type of long-range communication (female copulation calls appear to function to attract males from large distances), there is a conflict between compression and transmission success, with pressure for the latter being more important [20, 24].

Table 2 The relationship between call duration and the investigated (fixed) variables in the introduction, build-up, climax, let-down, and entire pant hoot

Introduction

Independent variable	Estimate ± SE	x^2	p value	95% confidence interval
Number of calls	**−0.45±0.04**	**19.56**	**<0.001**	**−0.52 to − 0.35**
Context	0.25±0.08	7.03	0.008	0.08 to 0.41

Build-up

Independent variable	Estimate ± SE	x^2	p value	95% confidence interval
Number of calls	**−0.15±0.06**	**4.41**	**0.036**	**−0.32 to − 0.01**
Context	−0.03±0.18	0.03	0.857	−0.47 to 0.37

Climax

Independent variable	Estimate ± SE	x^2	p value	95% confidence interval
Number of calls	**−0.32±0.07**	**12.54**	**<0.001**	**−0.49 to − 0.14**
Context	0.35±0.11	5.99	0.014	0.09 to 0.57

Let-down

Independent variable	Estimate ± SE	x^2	p value	95% confidence interval
Number of calls	**−0.14±0.06**	**10.94**	**<0.001**	**−0.37 to − 0.11**
Context	0.17±0.15	1.11	0.291	−0.14 to 0.55

Entire pant hoot

Independent variable	Estimate ± SE	x^2	p value	95% confidence interval
Number of calls	**−0.25±0.03**	**23.16**	**<0.001**	**−0.31 to − 0.19**
Context	0.09±0.06	1.55	0.213	−0.06 to 0.20

Test variables are in bold. (LMM; dependent variable: call duration; random intercepts: pant hoot ID and caller ID)

Our results, however, indicate that compression can play an important role in shaping long-distance vocal communication. In pant hoots, the negative relationship between the number and duration of calls was present both in high-amplitude phases, such as the climax (directed, at least in part, at distant receivers) and in low-amplitude phases, such as the introduction (directed primarily at nearby individuals).

In addition to a negative relationship between call number and duration in pant hoots, we found evidence that the durations of particular phases within this vocal sequence depend on the duration of the adjacent phases. Specifically, there was a negative correlation between the duration of the introduction and the build-up, and between the duration of the build-up and the climax. These results imply trade-offs in investment into different phases. Previous analyses of pant hoots suggest that prolonging the duration of particular phases, such as the build-up or the climax, may be used as effective territorial displays or to coordinate chorusing [29, 43]. However, it appears that, in some cases, if one phase is longer in total duration, the subsequent one tends to be shorter; thus, plasticity in phase duration appears somewhat constrained at a broader level. A lack of significant relationship between the durations of the two last phases in the pant hoot - climax and let-down – may be due to the let-down not having a following phase, such that constraints on its duration are relaxed. Many vocal sequences, across a wide range of taxa, are comprised of specific phases or notes produced in a conservative order [11, 12, 25, 57–59]; these provide the opportunity to test the generality of trade-offs in investment between different parts of the sequence.

Together, the results of analyses of call and phase duration indicate that there are trade-offs at two levels in pant hoot production: between call number and call length (if more calls are given, these tend to be shorter in length; or, if longer calls are given, these tend to be fewer in number), and between relative allocation of acoustic activity into subsequent phases (if one phase is longer, the subsequent one tends to be shorter). Theoretical analyses of communication indicate that reducing signal duration decreases transmission fidelity [60], so it

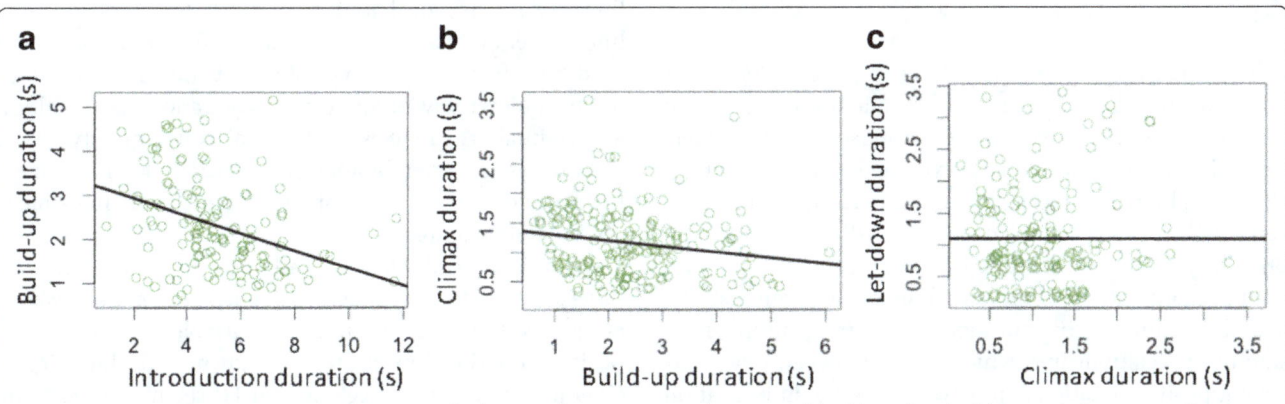

Fig. 3 The relationship between the durations of adjacent phases. **a** – introduction and build up, **b** – build-up and climax, **c** – climax and let-down. Black line represents regression line; circles represent data points

Table 3 The relationship between call F0 and the investigated (fixed) variables in the introduction, build-up, climax, and let-down

Introduction

Independent variable	Estimate ± SE	χ^2	p value	95% confidence interval
Number of calls	**0.07±0.06**	**1.16**	**0.282**	**−0.06 to 0.23**
Call duration	**0.06±0.12**	**0.28**	**0.595**	**−0.19 to 0.32**
Context	0.23±0.10	3.62	0.057	−0.01 to 0.47
Age	0.18±0.37	0.21	0.647	−0.65 to 1.18
Dominance rank	0.15±0.18	0.64	0.424	−0.27 to 0.58

Build-up

Independent variable	Estimate ± SE	χ^2	p value	95% confidence interval
Number of calls	**−0.06±0.04**	**2.60**	**0.106**	**−0.16 to 0.02**
Call duration	**−0.23±0.07**	**6.32**	**0.012**	**−0.38 to − 0.06**
Context	−0.17±0.11	1.72	0.190	−0.45 to 0.12
Age	−0.04±0.10	0.18	0.672	−0.24 to 0.24
Dominance rank	0.06±0.10	0.39	0.533	−0.23 to 0.27

Climax

Independent variable	Estimate ± SE	χ^2	p value	95% confidence interval
Number of calls	**0.14±0.05**	**6.01**	**0.014**	**0.03 to 0.24**
Call duration	**0.31±0.07**	**8.61**	**<0.001**	**0.15 to 0.46**
Context	−0.38±0.13	7.53	0.006	−0.73 to −0.12
Age	0.04±0.06	0.43	0.513	−0.09 to 0.31
Dominance rank	0.04±0.13	0.09	0.758	−0.26 to 0.32

Let-down

Independent variable	Estimate ± SE	χ^2	p value	95% confidence interval
Number of calls	**0.18±0.05**	**9.31**	**0.002**	**0.07 to 0.28**
Call duration	**0.04±0.07**	**0.27**	**0.600**	**−0.12 to 0.19**
Context	−0.28±0.12	5.45	0.019	−0.57 to −0.05
Age	−0.03±0.12	0.08	0.772	−0.35 to 0.23
Dominance rank	0.02±0.06	0.12	0.726	−0.17 to 0.17

Test variables are in bold. (LMM; dependent variable: fundamental frequency; random intercepts: pant hoot ID and caller ID)

is likely that the patterns seen here in pant hoots reflect a compromise between pressure to maximise efficacy of communication and constraints imposed by the energetic demands of producing extended vocal sequences [13–15], biomechanical constraints relating to lung capacity and airflow control [17, 18], or associated breathing-related limitations [16, 19, 61].

Our examination of potential links between call F0 and call number or duration revealed a number of significant relationships, which varied between phases. A strong positive relationship between call F0 and duration was seen in the climax, and a strong negative relationship was seen in the build-up, while no relationship was seen in the introduction or let-down. These findings suggest that, across these different phases, separate trade-offs are (or are not) occurring between pitch and calling effort. For example, the positive relationship between call duration and F0 in the climax indicates that individual calls can either be short and low-pitched or long and high-pitched. In mammals, F0 is mediated by sub-glottal air pressure generated in the lungs, with higher air pressure generating higher F0 as a result of an increased rate of vocal fold vibrations [37]. Our result, therefore, might be a by-product of differences in sub-glottal air pressure, with higher air pressures generating calls that are both longer and higher-pitched. This would indicate that chimpanzees have limited active control over the movement of their larynx, very much in contrast to humans who are able to produce a stable F0 during speech production, more or less independent of sub-glottal air pressure [37, 62]. The negative relationship between call duration and F0 in the build-up may be due to the fact that calls in this phase are much shorter than in the climax; it is possible that there is a critical threshold of call length, above which pitch inevitably rises due to the link with sub-glottal air pressure, but that this threshold is not reached in the build-up phase.

At a functional level, the different relationships between call F0 and duration found in different phases suggest that specific phases within a pant hoot have distinct functions modulated by their pitch [43]. For example, the low-frequency build-up phase seems to be directed (at least in part) to the nearby individuals, since callers adjust its duration depending on the vocal response of the nearby males [29]. The high-frequency high-amplitude climax, on the other hand, seems to be directed at distant receivers [63] and may be an honest signal of individual quality [44]. According to the "calling at the edge" hypothesis [45], mammals calling at near maximum F0 struggle to maintain a harmonic F0, since calling at such extreme frequencies distorts F0 harmonics, resulting in non-linear phenomena (i.e. non-linearity in the vocal fold dynamics) [64]. Indeed, non-linear phenomena are considerably more common in the loud high-frequency climax phase of the pant hoot than in the quieter low-frequency introduction [64]. Calling at maximal frequencies may signal caller quality, since individuals in better biological condition are more likely to produce climaxes that are free from non-linear phenomena (e.g. [45]).

Analysis of call F0 and call number again revealed differences between phases: in only two phases was a clear link found between these variables– a significant positive relationship in the let-down and the climax. Overall, our results in relation to F0 seem to reflect the literature showing inconsistent relationships between call F0 and temporal features. For example, a positive

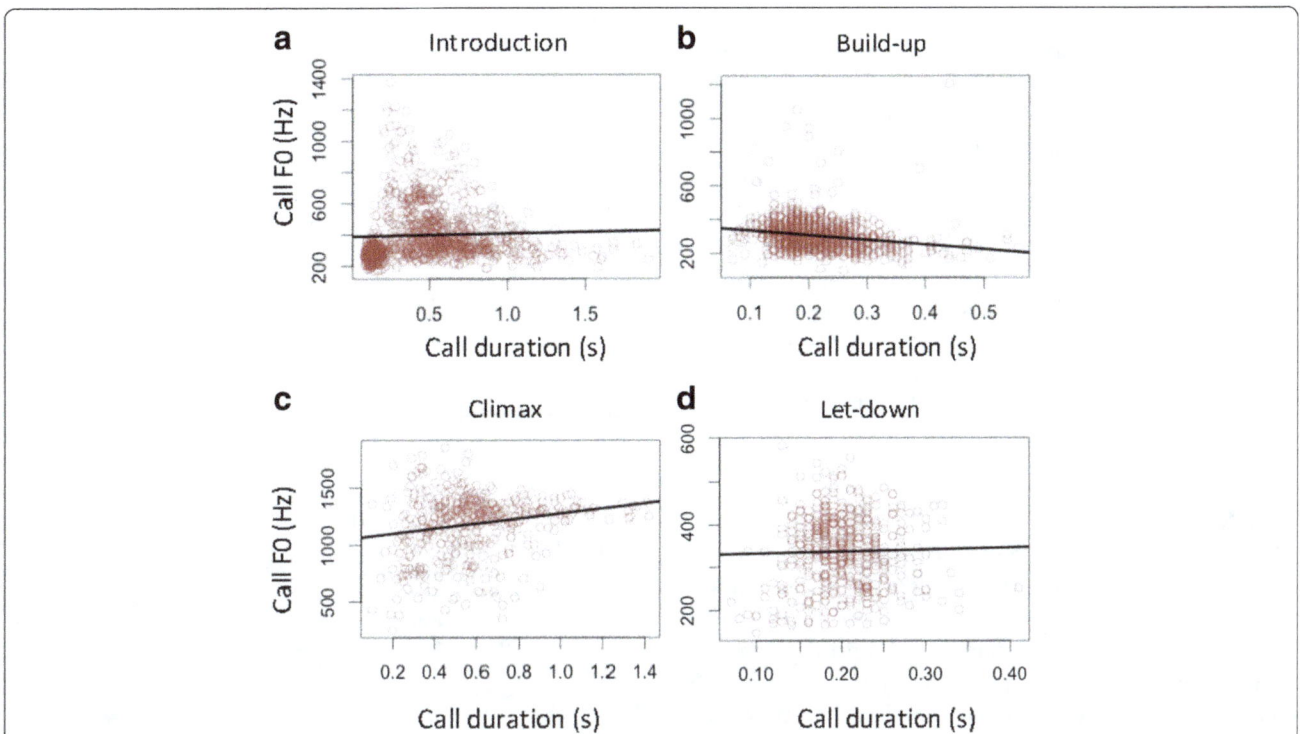

Fig. 4 The relationship between call F0 and duration in the four phases of a pant hoot. **a** – introduction, **b** – build-up, **c** – climax, **d** – let-down. Black line represents regression line; circles represent data points

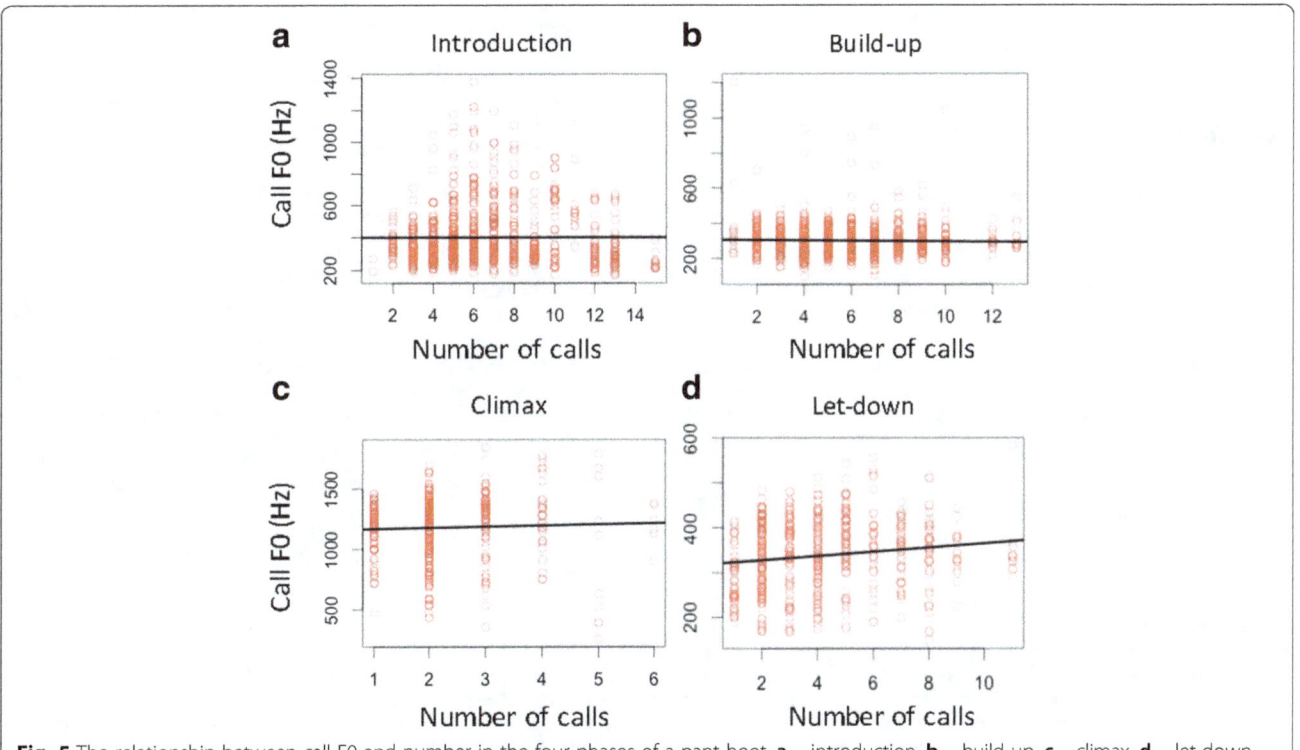

Fig. 5 The relationship between call F0 and number in the four phases of a pant hoot. **a** – introduction, **b** – build-up, **c** – climax, **d** – let-down. Black line represents regression line; circles represent data points

relationship between F0 and both call duration and sequence length was found in chimpanzee victim screams [65]. Similarly, baboon grunts produced in strongly affective situations are both longer and higher frequency than grunts produced in more relaxed situations [66]. In contrast, calls with lower F0 tend to be also longer in Japanese quails [31], while in domestic dogs (*Canis familaris*) [67] F0 and duration of aggressive barks are not correlated. Data from a range of animals, therefore, indicate that there is no elemental, overarching trade-off between temporal features of call sequences and F0 of the constituent calls; that diverse, context-specific trade-offs may be important merits future research.

Conclusions

Identifying the basic patterns of organisation of animal signals can provide important insights into the relationship between their structure and function [2, 68, 69] and can also shed light on the fundamental principles underpinning signal evolution [20, 24]. In this study we focussed on the relationship between temporal and spectral variables of wild chimpanzee pant hoots. Our results suggest that costs and constraints involved in vocal production, balanced against the potential benefits to signallers accrued from variation in signal form, lead to trade-offs of multiple kinds. This study highlights the key role that such costs and constraints can play in shaping the temporal and acoustic structure of animal vocal sequences.

Additional files

Additional file 1: An audio recording of a pant hoot given by an adult male. (WAV 1077 kb)

Additional file 2: Table with estimated age and the dominance rank of the study males, and the number of recordings per male. (DOCX 52 kb)

Additional file 3: Calculation of the dominance status of the study males. (DOCX 107 kb)

Additional file 4: Table with the type of variable transformation in models concerning the relationship between call duration and the investigated (fixed) variables in the introduction, build-up, climax, let-down, and the entire sequence. (DOCX 47 kb)

Additional file 5: Table with the type of variable transformation in models concerning the relationship between call F0 and the investigated (fixed) variables in the introduction, build-up, climax, and let-down. (DOCX 49 kb)

Acknowledgements

We are grateful to the management and staff of the Budongo Conservation Field Station for their support and assistance. We thank the Uganda Wildlife Authority and the Uganda National Council for Science and Technology for permission to conduct the study. Comments provided by the editor Denis Réale and an anonymous reviewer considerably improved the paper. We also thank Roger Mundry, Christof Neumann, and Liz Campbell for statistical advice.

Funding

The study was funded by Swiss National Science Foundation (310030_143359) and European Research Council project grants awarded to KZ (PRILANG 283871).

Authors' contributions

PF: study design, data collection, analysis and interpretation, drafting the article; KZ: provision of necessary tools and resources, data interpretation, drafting the article; SS: study design, data interpretation, drafting the article. All authors read and approved the final manuscript.

Competing interests

The authors declare that they have no competing interests.

Author details

[1]Department of Primatology, Max Planck Institute for Evolutionary Anthropology, Leipzig, Germany. [2]Institute of Biology, University of Neuchâtel, Neuchâtel, Switzerland. [3]School of Psychology and Neuroscience, University of St Andrews, St Andrews, Scotland, UK. [4]Centre for Research in Evolutionary, Social and Interdisciplinary Anthropology, University of Roehampton, London, UK.

References

1. Smith WJ. The behavior of communicating: an ethological approach. Cambridge: Harvard University Press; 1977.
2. Owings DH, Morton ES. Animal vocal communication: a new approach. Cambridge: Cambridge University Press; 1998.
3. Kershenbaum A, Blumstein DT, Roch MA, Akçay Ç, Backus G, Bee MA, et al. Acoustic sequences in non-human animals: a tutorial review and prospectus. Biol Rev. 2014;91:13–52.
4. Wiley RH. Errors, exaggeration, and deception in animal communication. In: Behavioral mechanisms in evolutionary ecology. Edited by Real L. Chicago: University of Chicago Press; 1994. p. 157–189.
5. Schlenker P, Chemla E, Zuberbühler K. What do monkey calls mean? Trends Cogn Sci. 2016;20:894–904.
6. Arnold K, Zuberbühler K. Language evolution: semantic combinations in primate calls. Nature. 2006;441:303.
7. Engesser S, Crane JMS, Savage JL, Russell AF, Townsend SW. Experimental evidence for phonemic contrasts in a nonhuman vocal system. PLoS Biol. 2015;13:e1002171.
8. Pollard KA, Blumstein DT. Evolving communicative complexity: insights from rodents and beyond. Phil Trans R Soc B. 2012;367:1869–78.
9. Catchpole CK, Slater PJB. Bird song: biological themes and variations. Cambridge: Cambridge University Press; 2003.
10. Rand AS, Ryan MJ. The adaptive significance of a complex vocal repertoire in a neotropical frog. Ethology. 1981;57:209–14.
11. Koren L, Geffen E. Complex call in male rock hyrax (Procavia capensis): a multi-information distributing channel. Behav Ecol and Sociobiol 2009;63:581-590.
12. Rehsteiner U, Geisser H, Reyer HU. Singing and mating success in water pipits: one specific song element makes all the difference. Anim Behav. 1998;55:1471–81.
13. Noren DP, Holt MM, Dunkin RC, Williams TM. The metabolic cost of communicative sound production in bottlenose dolphins (Tursiops truncatus). J Exp Biol 2013;216:1624-1629.
14. Oberweger K, Goller F. The metabolic cost of birdsong production. J Exp Biol. 2001;204:3379–88.
15. Franz M, Goller F. Respiratory patterns and oxygen consumption in singing zebra finches. J Exp Biol. 2003;206:967–78.
16. MacLarnon AM, Hewitt GP. The evolution of human speech: the role of enhanced breathing control. Am J Phys Anthropol. 1999;109:341–63.

17. Riede T, Brown C. Body size, vocal fold length and fundamental frequency-implications for mammal vocal communication. In: Wessel A, Menzel R, Tembrock G, editors. Quo Vadis, Behavioural biology? Past, present and future of an evolving science, vol. 380. Halle: Nova Acta Leopoldina; 2013. p. 295–314.

18. Titze IR, Riede T. A cervid vocal fold model suggests greater glottal efficiency in calling at high frequencies. PLoS Comput Biol. 2010;6:e1000897.

19. Hewitt G, MacLarnon A, Jones KE. The functions of laryngeal air sacs in primates: a new hypothesis. Folia Primatol. 2002;73:70–94.

20. Gustison ML, Semple S, Ferrer-i-Cancho R, Bergman TJ. Gelada vocal sequences follow Menzerath's linguistic law. Proc Natl Acad Sci. 2016;20:1522072.

21. Fenk A, Fenk-Oczlon G. Menzerath's law and the constant flow of linguistic information. In: Contributions to quantitative linguistics. Edited by Köhler R and Rieger BB. Trier: Springer; 1991. p. 11–31.

22. Menzerath P. Die Architektonik des deutschen Wortschatzes. F. Dümmler; 1954.

23. Altmann G. Prolegomena to Menzerath's law. Glottometrika. 1980;2:1–10.

24. Ferrer-i-Cancho R, Hernández-Fernández A, Lusseau D, Agoramoorthy G, Hsu MJ, Semple S. Compression as a universal principle of animal behavior. Cogn Sci. 2013;37:1565–78.

25. Osiejuk TS, Ratyńska K, Dale S. What makes a 'local song'in a population of ortolan buntings without a common dialect? Anim Behav. 2007;74:121–30.

26. Mitani JC, Gros-Louis J. Chorusing and call convergence in chimpanzees: tests of three hypotheses. Behaviour. 1998;135:1041–64.

27. Galeotti P, Saino N, Sacchi R, Moller AP. Song correlates with social context, testosterone and body condition in male barn swallows. Anim Behav. 1997;53:687–700.

28. Fedurek P, Slocombe KE, Zuberbühler K. Chimpanzees communicate to two different audiences during aggressive interactions. Anim Behav. 2015;110:21–8.

29. Fedurek P, Schel AM, Slocombe KE. The acoustic structure of chimpanzee pant-hooting facilitates chorusing. Behav Ecol Sociobiol. 2013;67:1781–9.

30. Taylor AM, Reby D. The contribution of source–filter theory to mammal vocal communication research. J Zool. 2010;280:221–36.

31. Beani L, Briganti F, Campanella G, Lupo C, Dessi-Fulgheri F. Effect of androgens on structure and rate of crowing in the Japanese quail (Coturnix japonica). Behaviour 2000;137:417-435.

32. Pasch B, George AS, Hamlin HJ, Guillette LJ, Phelps SM. Androgens modulate song effort and aggression in Neotropical singing mice. Horm Behav. 2011;59:90–7.

33. Puts DA, Doll LM, Hill AK. Sexual selection on human voices. In: Evolutionary perspectives on human sexual psychology and behavior. Edited by Weekes-Shackelford VA and Shackelford TK. New York: Springer; 2014. p. 69–86.

34. Reby D, Charlton BD, Locatelli Y, McComb K. Oestrous red deer hinds prefer male roars with higher fundamental frequencies. Proc R Soc Lond B Biol Sci. 2010;277:2747–53.

35. Fischer J, Kitchen DM, Seyfarth RM, Cheney DL. Baboon loud calls advertise male quality: acoustic features and their relation to rank, age, and exhaustion. Behav Ecol Sociobiol. 2004;56:140–8.

36. Barelli C, Mundry R, Heistermann M, Hammerschmidt K. Cues to androgens and quality in male gibbon songs. PLoS One. 2013;8:e82748.

37. Titze IR. On the relation between subglottal pressure and fundamental frequency in phonation. J Acoust Soc Am. 1989;85:901–6.

38. Marler P, Hobbett L. Individuality in a long-range vocalization of wild chimpanzees. Z Für Tierpsychol. 1975;38:97–109. 39

39. Aureli F, Schaffner CM, Boesch C, Bearder SK, Call J, Chapman CA, et al. Fission-fusion dynamics: new research frameworks. Curr Anthropol. 2008;49:627–54.

40. Mitani JC, Nishida T. Contexts and social correlates of long-distance calling by male chimpanzees. Anim Behav. 1993;45:735–46.

41. Fedurek P, Machanda ZP, Schel AM, Slocombe KE. Pant hoot chorusing and social bonds in male chimpanzees. Anim Behav. 2013;86:189–96.

42. Fedurek P, Donnellan E, Slocombe KE. Social and ecological correlates of long-distance pant hoot calls in male chimpanzees. Behav Ecol Sociobiol. 2014;68:1345–55.

43. Fedurek P, Zuberbühler K, Dahl CD. Sequential information in a great ape utterance. Sci Rep. 2016;6:38226.

44. Fedurek P, Slocombe KE, Enigk DK, Emery Thompson M, Wrangham RW, Muller MN. The relationship between testosterone and long-distance calling in wild male chimpanzees. Behav Ecol Sociobiol. 2016;70:659–72.

45. Riede T, Arcadi AC, Owren MJ. Nonlinear acoustics in the pant hoots of common chimpanzees (Pan troglodytes): vocalizing at the edge. J Acoust Soc Am. 2007;121:1758-1767.

46. Reynolds V. The chimpanzees of the Budongo forest: ecology, behaviour and conservation. Oxford: Oxford University Press; 2005.

47. Goodall J. The chimpanzees of Gombe: patterns of behavior. Cambridge: Harvard University Press; 1986.

48. Neumann C, Duboscq J, Dubuc C, Ginting A, Irwan AM, Agil M, et al. Assessing dominance hierarchies: validation and advantages of progressive evaluation with Elo-rating. Anim Behav. 2011;82:911–21.

49. R Core Team. R: A language and environment for statistical computing. In: R foundation for statistical computing. 2014. http://www.Rproject.org/. Accessed 13 May 2017.

50. Bates D, Maechler M, Bolker B, Walker S. lme4: linear mixed-effects models using Eigen and S4. R package version. 2014;1:1–23.

51. Notman H, Rendall D. Contextual variation in chimpanzee pant hoots and its implications for referential communication. Anim Behav. 2005;70:177–90.

52. Forstmeier W, Schielzeth H. Cryptic multiple hypotheses testing in linear models: overestimated effect sizes and the winner's curse. Behav Ecol Sociobiol. 2011;65:47–55.

53. Barr DJ, Levy R, Scheepers C, Tily HJ. Random effects structure for confirmatory hypothesis testing: keep it maximal. J Mem Lang. 2013;68:255–78.

54. Holm SA. Simple sequentially rejective multiple test procedure. Scand J Stat. 1979;6:65–70.

55. Rice WR. Analyzing tables of statistical tests. Evolution. 1989;43:223–5.

56. Semple S, McComb K. Perception of female reproductive state from vocal cues in a mammal species. Proc R Soc Lond B Biol Sci. 2000;267:707–12.

57. Haimoff EH. Convergence in the duetting of monogamous old world primates. J Hum Evol. 1986;15:51–9.

58. Mitani JC. Sexual selection and adult male orangutan long calls. Anim Behav. 1985;33:272–83.

59. Nelson DA, Poesel A. Segregation of information in a complex acoustic signal: individual and dialect identity in white-crowned sparrow song. Anim Behav. 2007;74:1073–84.

60. Shannon CEA. Mathematical theory of communication. ACM SIGMOBILE mob Comput. Commun Rev. 1948;5:3–55.

61. Fattu JM, Suthers RA. Subglottic pressure and the control of phonation by the echolocating bat, Eptesicus. J Comp Physiol A Neuroethol Sens Neural Behav Physiol. 1981;143:465–75.

62. Lieberman P, Knudson R, Mead J. Determination of the rate of change of fundamental frequency with respect to subglottal air pressure during sustained phonation. J Acoust Soc Am. 1969;45:1537–43.

63. Mitani JC, Gros-Louis J, Macedonia JM. Selection for acoustic individuality within the vocal repertoire of wild chimpanzees. Int J Primatol. 1996;17:569–83.

64. Riede T, Owren MJ, Arcadi AC. Nonlinear acoustics in pant hoots of common chimpanzees (Pan troglodytes): frequency jumps, subharmonics, biphonation, and deterministic chaos. Am J Primatol 2004;64:277–229.

65. Slocombe KE, Zuberbühler K. Chimpanzees modify recruitment screams as a function of audience composition. Proc Natl Acad Sci. 2007;104:17228–33.

66. Rendall D, Seyfarth RM, Cheney DL, Owren MJ. The meaning and function of grunt variants in baboons. Anim Behav. 1999;57:583–92.

67. Taylor AM, Reby D, McComb K. Context-related variation in the vocal growling behaviour of the domestic dog (Canis familiaris). Ethology. 2009;115:905–15.

68. Wiley RH, Richards DG. Physical constraints on acoustic communication in the atmosphere: implications for the evolution of animal vocalizations. Behav Ecol Sociobiol. 1978;3:69–94.

69. Waser PM, Waser MS. Experimental studies of primate vocalization: specializations for long-distance propagation. Ethology. 1977;43:239–63.

14

Sequential social experiences interact to modulate aggression but not brain gene expression in the honey bee (*Apis mellifera*)

Clare C. Rittschof

Abstract

Background: In highly structured societies, individuals behave flexibly and cooperatively in order to achieve a particular group-level outcome. However, even in social species, environmental inputs can have long lasting effects on individual behavior, and variable experiences can even result in consistent individual differences and constrained behavioral flexibility. Despite the fact that such constraints on behavior could have implications for behavioral optimization at the social group level, few studies have explored how social experiences accumulate over time, and the mechanistic basis of these effects. In the current study, I evaluate how sequential social experiences affect individual and group level aggressive phenotypes, and individual brain gene expression, in the highly social honey bee (*Apis mellifera*). To do this, I combine a whole colony chronic predator disturbance treatment with a lab-based manipulation of social group composition.

Results: Compared to the undisturbed control, chronically disturbed individuals show lower aggression levels overall, but also enhanced behavioral flexibility in the second, lab-based social context. Disturbed bees display aggression levels that decline with increasing numbers of more aggressive, undisturbed group members. However, group level aggressive phenotypes are similar regardless of the behavioral tendencies of the individuals that make up the group, suggesting a combination of underlying behavioral tendency and negative social feedback influences the aggressive behaviors displayed, particularly in the case of disturbed individuals. An analysis of brain gene expression showed that aggression related biomarker genes reflect an individual's disturbance history, but not subsequent social group experience or behavioral outcomes.

Conclusions: In highly social animals with collective behavioral phenotypes, social context may mask underlying variation in individual behavioral tendencies. Moreover, gene expression patterns may reflect behavioral tendency, while behavioral outcomes are further regulated by social cues perceived in real-time.

Keywords: Behavioral genomics, Personality, Social insects, Plasticity, Aggression, Collective behavior, Timescale

Background

For social animals, behavioral phenotypes exist at both the individual and group levels [1–8]. Understanding the mechanistic and social factors that shape phenotypes at these two levels remains a fundamental challenge in social behavior research. In some cases, the behavioral tendency of the most extreme group member, or the average tendencies across group members, are good predictors of group-level behavioral phenotypes [9–11].

In highly structured societies however, individuals continuously modulate their behavior in response to social cues from colony mates, a process that optimizes group level phenotypes depending on environmental conditions and pre-set heritable rules [2, 12–17]. As a result, individual behavior varies across social contexts, a phenomenon known as behavioral flexibility [18–20].

Despite the highly flexible nature of individual behavior in complex societies, some social inputs can have long lasting effects on both behavioral tendencies and behavioral flexibility [18, 21], influencing individual behavioral phenotype in novel social scenarios encountered later in life. One well-known context for this phenomenon is that

Correspondence: clare.rittschof@uky.edu
Department of Entomology, University of Kentucky, S-225 Ag. Science Center North, Lexington, KY 40546, USA

of early-life social experiences, which can have persistent effects throughout life, and may even be robust to additional social inputs [22]. In the field of developmental plasticity, a growing body of literature attempts to predict how individuals weigh information from their past and current environments to optimize their phenotypes [23]. Empirical studies that assess how social experiences accumulate over time to affect individual behavioral outcome and flexibility are also necessary to interpret behavioral optimization at the group level.

In the honey bee (*Apis mellifera*), worker bees perform aggressive behaviors in the context of nest defense, which is a collective activity that is modulated at the colony level by ecological conditions [24–26]. For individuals, responsiveness to aggression inducing cues and the decision to engage in tasks associated with nest defense are influenced both directly by ecological cues and indirectly by social interactions with nestmates; these interactions occur in a variety of contexts throughout both the pre-adult and adult life stages [25, 27–31]. Despite the large degree of social sensitivity inherent to aggressive behaviors, it is unknown whether or how an individual's sequential social experiences cumulatively influence aggression levels or behavioral flexibility in defensive social contexts. If social information accumulated over time influences behavioral outcome, there may be constraints to individual and group level behavioral plasticity that prevent an optimal response to given ecological conditions [32]. Moreover, the group-level impacts of these cumulative individual effects are both unknown and difficult to measure due to the fact that honey bees live in large, complex societies composed of about 20,000- 40,000+ individual workers [1, 33]. The first goal of the current study is to evaluate whether social experiences early in adult life influence individual behavioral outcome, flexibility, and group level aggressive response in subsequent social contexts. To do this in the highly social honey bee, I combine small scale field and lab based social manipulations and behavioral assays of aggression.

The cumulative effects of social experiences on behavior may depend on the nature of the underlying mechanisms that entrain previous experiences relative to those that regulate behavioral outcome on a more proximal timescale. In some cases, these mechanisms operate at different levels of biological organization; for example, early-life social experiences may affect brain structure, while subsequent experiences modulate brain biochemistry [34]. In the context of honey bee aggression, genomics studies demonstrate that brain gene expression patterns track socially-induced behavioral variation, not only for stable shifts in aggression, but also for more rapid and transient changes in phenotype that occur on the order of minutes [24, 25, 28]. Moreover, shifts in aggression across very different timescales and contexts are associated with transcriptional variation in overlapping sets of genes [25, 28, 35]. These findings predict that transcriptomic patterns will track behavioral outcome associated with sequential social experiences, for example, resulting in an interaction effect of multiple experiences on gene expression. In addition to the behavioral analyses described above, the second goal of the current study is to evaluate how transcriptomic patterns reflect cumulative social experience and whether they parallel behavioral effects. To do this, I analyze a small set of previously published honey bee aggression biomarker genes.

Methods

I manipulated early adult social experience by implementing a full colony chronic disturbance paradigm following Rittschof and Robinson [25]. Briefly, I constructed two pairs of small colonies made up of about 4000 one-day-old adult bees collected from 8 to 10 source colonies headed by naturally mated queens. One-day-old bees were combined and then assigned randomly across colonies of each pair, such that a wide array of genotypes of European descent were evenly represented across each pair. I marked each bee on the thorax with paint to precisely control colony size, and then introduced a naturally mated queen to each colony. I provisioned hives with *ad libitum* food, including a partial frame of pollen and a full frame of honey. Colonies were provisioned with food because young bees do not begin to forage in strong numbers until 6–7 days of age. However, colonies were allowed to forage freely throughout the experiment (following [25]). I established the hives in an apiary and commenced the chronic disturbance paradigm [25]: one colony of the pair (selected at random) was left undisturbed as a control, while the other colony was exposed to a combination of artificial alarm pheromone (an aggression-inducing social cue) and physical agitation (opening the colony and lifting and dropping frames in a controlled manner) on a chronic basis (twice a day, once in the morning between 08:00 and 10:00 and once in the afternoon between 13:00 and 15:00) over the course of the first 8 days of adult life. Relative to the control, this treatment results in a highly robust and significant decrease in aggressive behavior measured at the colony level. This effect persists for at least 24 h following the final disturbance [36]. Brain biomarker gene expression patterns for chronically disturbed bees are also consistent with low aggression [25, 36]. A range of behavioral groups, including foragers, soldiers, and bees collected from inside the hive show these brain gene expression effects [25], which persist for 48–72 h following the final disturbance treatment [25, 36]. Thus, this artificial manipulation of the social environment results in stable changes in both behavior and gene expression.

Following this early-adulthood disturbance manipulation, I collected bees for a second, laboratory-based manipulation of social context. This experiment involved combining bees originating from the disturbed and undisturbed colonies together into new social groups, and assaying the aggressive behaviors of these groups. This laboratory-based manipulation was derived from the nestmate recognition assay [37]. In previous work we showed that when kept in small groups over a relatively brief time period (overnight), bees originating from different colonies can discriminate their new groupmates from foreign bees and respond aggressively towards an intruder [36]. Thus, bees kept in a small group in the lab develop a social identity that can be used to investigate how individuals behave under different social conditions.

To manipulate the social conditions in the lab, I collected bees from disturbed and undisturbed colonies and combined individuals into small social groups (8 bees per group) that differed in the ratio of disturbed to undisturbed group members, a design analogous to forming groups composed of different numbers of high and low aggression personality individuals ([9], Fig. 1). Groups consisted of 8 undisturbed, 6 undisturbed and 2 disturbed, 4 undisturbed and 4 disturbed, 2 undisturbed and 6 disturbed, or 8 disturbed individuals. I performed these collections on the evening of the 8th day of colony life, 5 h following the final disturbance treatment (prior to collection I first performed a short ~30 s field assay to confirm that disturbed colonies showed the predicted decreased aggressive response compared to undisturbed colonies [25, 36]). To collect enough bees for the groups, I opened each colony and vacuumed bees from the frame containing capped honey. Collecting from the honey frame maximized the chances that I collected roughly the same distribution of bee task groups from both colonies. By 8 days of age, small colonies composed of single-aged bees stratify into a range of behavioral groups (e.g., [38]) including nurses, foragers, and guards. All of these castes could be present on the honey frame and represented in this study. Bees were transferred to plastic bags and anesthetized on ice for ~5 min until sedated. Sedated bees were then transferred into petri dishes in different ratios of disturbed and undisturbed individuals. Sedation is required to eliminate conflict among group members originating from two different colonies (Rittschof, personal observation). I monitored groups until all bees recovered from anesthesia, replacing dead bees as needed. During this monitoring period, I confirmed that there were no aggressive interactions between bees as a result of combining individuals from two different colonies into a single dish. I repeated this entire experiment, including colony construction, chronic disturbance, and the group collections and behavioral across two pairs of colonies. During the second replicate, I

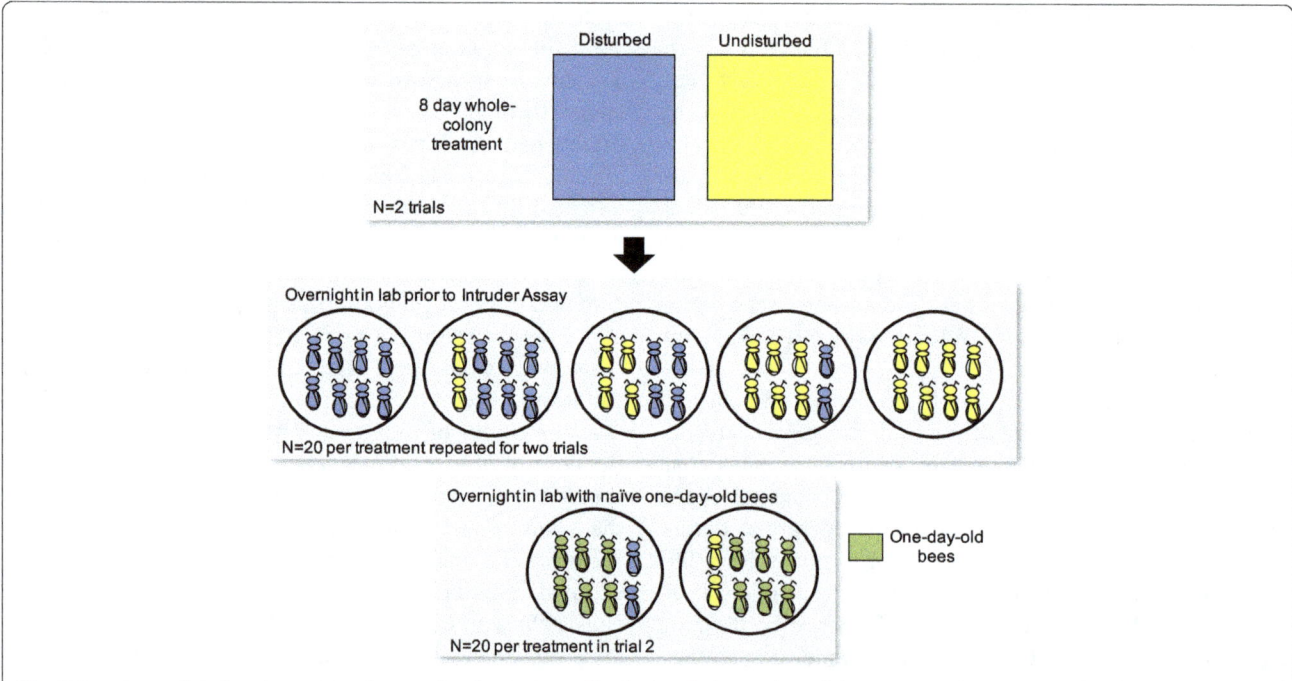

Fig. 1 Experimental design. I constructed two pairs of experimental colonies. Each member of the pair was identical in terms of number of individuals, age, and genetic background. One of each pair was disturbed on a chronic basis for 8 days while the other was left undisturbed to generate low (disturbed, *blue*) and high (undisturbed, *yellow*) aggression individuals. We collected and anesthetized individuals, and then combined them into groups of eight for the lab-based assay of aggression. In the second replicate (the second pair of colonies), we added two treatment groups with docile one-day-old adult bees (*green*)

added two treatment groups to the 5 listed above. These groups were composed of 6 highly docile one-day-old bees (originating from a single, naturally-mated colony) and either two undisturbed or two disturbed bees. This additional treatment allowed me to investigate how disturbed and undisturbed individuals further alter their behavior in the presence of individuals that are highly docile and largely unresponsive to aggressive cues. In this case, the increased docility is a function of age and not social experience, but both factors contribute to variation in individual aggression in a natural colony context [28], and so it is likely that individual behaviors influence group members in comparable ways. For all groups, bees were provisioned with *ad libitum* with 50% sucrose, and petri dishes were transferred to a dark 34 °C incubator overnight for behavioral assessments beginning the following morning.

Behavioral assessments were performed between 08:00 and 15:00 the following day by two observers. Because bees were paint-marked according to their colony of origin I was unable to blind the experiment. Petri dishes were transferred from the incubator to a temperature-controlled room (25-30 °C) for behavioral analysis. Dishes were arranged in random order and left undisturbed for one hour prior to the initiation of behavioral observations. I assayed individual and group level aggressive behaviors using the Intruder Assay as described in [36]. Briefly, I collected an intruder bee from the entrance of a randomly selected colony, introduced this bee through a small hole into the petri dish containing the eight focal bees, and tallied aggressive behaviors displayed towards the intruder bee. Because all bees were paint-marked, I could assign tallies to either undisturbed or disturbed individuals. Aggressive behaviors include antennation, antennation with mandibles opened, biting, mounting the intruder and flexing the abdomen, and stinging. From these tallies I calculated an aggression index (a tally of aggressive behaviors weighted for severity of behavior, [36, 39]) on a per bee basis, as well as the total level of aggression displayed by the group (on a per bee basis). Following behavioral observations, bees were immediately flash-frozen in liquid nitrogen for later gene expression analysis. Bees were stored separately as a function of social group (the ratio of disturbed to undisturbed bees), but multiple groups of 8 bees were mixed into a single container for storage.

Following protocols described in [25], I used quantitative PCR to evaluate brain expression levels for four biomarker genes. These genes were selected based on previous microarray studies that showed a robust association between brain expression levels and variation in aggressive behavior across social, developmental, and evolutionary contexts [28]. We further validated that these genes are differentially expressed in the brain specifically as a function of chronic disturbance [25]. These four genes are involved in a range of pathways including stress response and alcohol metabolism [25]. Though I have not demonstrated a causal relationship between these genes and aggression level, they are predictive of aggression across many timescales for behavioral variation, and so provide a means to compare the effects of cumulative social experiences on the molecular state of the brain versus behavior.

I dissected brains following Schulz and Robinson [40] and extracted nucleic acids using RNeasy kits including an on-column treatment to remove genomic DNA (Qiagen, Valencia CA, USA). I synthesized cDNA from 200 ng RNA using ArrayScript (Ambion, Life Technologies, Grand Island, NY, USA) reverse transcriptase and a spiked-in internal control to estimate the quality of the synthesis. I performed qPCR on an ABI Prism 7900 in triplicate 10 uL reactions in 384-well plates using PerfeCTa SYBR Green Fastmix (Quanta Biosystems, Gaithersburg, MD, USA). I normalized biomarker genes to the geometric mean of two constitutively expressed control genes, *Actin-1* (*GB44311*) and *Gapdh* (*GB50902*). I verified that control gene expression showed low variance ([41], with Ct standard deviation = 0.18 (*Actin-1*) and 0.20 (*Gapdh*)), and I used two-tailed *t*-tests to verify that expression values did not differ across treatment groups. A stability analysis using GeNorm recommended using the geometric mean of both genes as the endogenous control [42].

All data was analyzed using JMP Pro 12.1. Behavioral data were analyzed using non-parametric statistics because assumptions of normality and equal variance were not met in all comparisons. Observers showed some variation in behavioral scoring, but there were no significant effects of observer on the outcome of any reported results. Gene expression data were analyzed using a relative standard curve method (e.g., [43]), and assessed for normality on a gene by gene basis. All genes except *GB53860* met assumptions for parametric statistical analyses. For *GB53860* I implemented non-parametric tests and generalized linear models (noted in text).

Results

I first compared per-bee aggression scores for each bee type (undisturbed, disturbed, one-day-old) across all small group laboratory aggression assays, regardless of group treatment. As predicted based on previous studies, disturbance history and age significantly predicted aggression scores in this overall analysis (Kruskal-Wallis Test, $X^2_2 = 12.72$, $P < 0.0017$), with undisturbed bees showing the highest average aggression, and one-day-olds showing the lowest. However, a subsequent analysis of aggression score as a function of lab social group and disturbance history

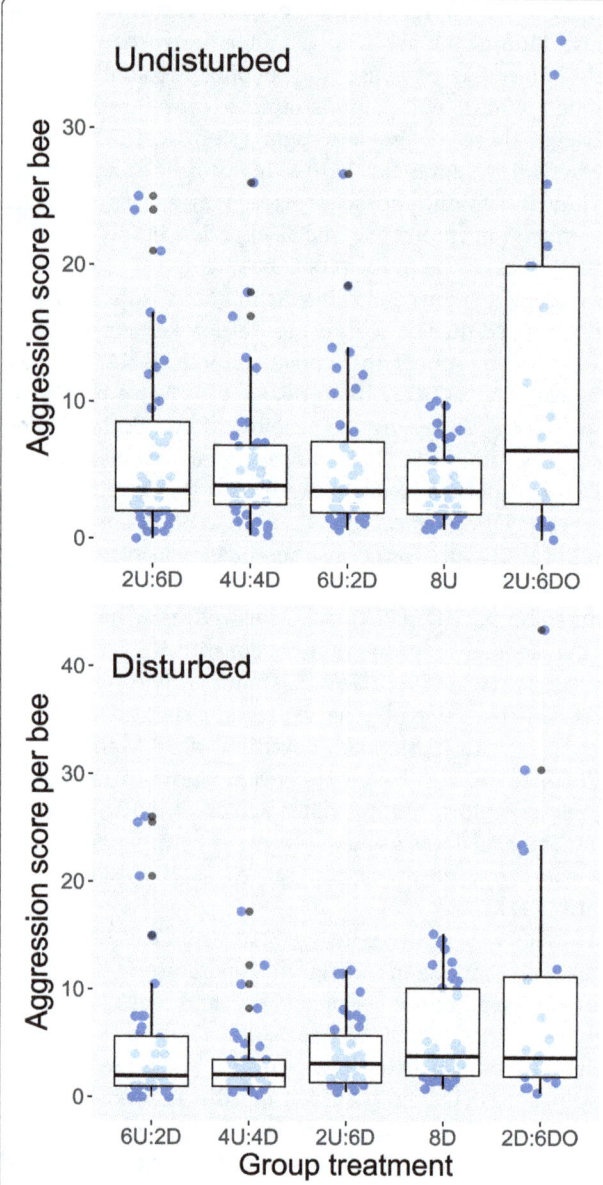

Fig. 2 Analysis of aggression as a function of social group. Undisturbed bees showed relatively consistent aggression levels regardless of social group composition while disturbed individuals significantly modulated their aggression in response to social group composition (*top* and *bottom* panels, respectively). "U", "D", and "DO" indicate the number of undisturbed, disturbed, and one-day-old bees in each group, respectively. A post-hoc analysis of disturbed bee behavior, using a Wilcoxon Test for each pair, significantly distinguished three treatment categories, 6U:2D and 4U:4D, 2U:6D, and 8D and 2D:6DO. Box hinges show the 1st and 3rd quartiles, whiskers indicate 1.5*IQR from the hinge, and the central tendency line indicates the median. Data points represent scores for individual replicates

greater behavioral flexibility as a function of social context compared to undisturbed bees. Disturbed bee aggression increased with decreasing numbers of undisturbed bees in the group, and was highest in groups that contained extremely docile one-day-old bees (Fig. 2). This suggests that disturbed bees increase their aggression levels to compensate for a shortage of high-aggression individuals. Undisturbed individuals show relatively invariant aggression scores as a function of social group treatment, with the exception of groups containing extremely docile one-day-old bees, in which they increase their aggression effort to some degree (Fig. 2). When I compared aggression for disturbed and undisturbed bees kept together in the same social group, disturbed bees were significantly less aggressive in most cases (Wilcoxon Test blocked for group, Fig. 3). As expected, one-day-old bees were significantly less aggressive than their older group counterparts regardless of disturbance experience (Fig. 3).

Total group aggression scores differed slightly but not significantly as a function of social group composition (Wilcoxon Test, $X_6^2 = 3.99$, $P = 0.68$, Fig. 4). There was no simple relationship between total group score and the prior disturbance experience of individuals within the groups. Notably, the groups that successfully killed the intruder bee had significantly higher aggression scores (Wilcoxon Test, $X_1^2 = 21.56$, $P < 0.0001$), suggesting a strong relationship between scoring methods and a biologically relevant aggression outcome; this outcome, however, also did not vary as a function of group composition (Chi-squared Test, $X_6^2 = 5.59$, $P = 0.47$).

To determine whether sequential social experiences influence brain gene expression patterns, I used quantitative PCR to evaluate mRNA levels for four genes identified as aggression biomarkers in a previous study [25]. I evaluated gene expression for disturbed and undisturbed individuals from two group composition treatments, those kept in mixed (4 undisturbed and 4 disturbed) versus uniform (all disturbed or undisturbed) groups. For disturbed bees, individuals across these two social group types showed significant variation in behavior, with higher aggression levels in the uniform groups (Fig. 2). I first assessed whether biomarker expression predicted disturbance history, and found significant effects in the predicted direction for 3 of 4 genes (Table 1, Fig. 5) [25]. However, a subsequent comparison of gene expression level comparing bees kept in mixed versus uniform groups showed that gene expression patterns did not differ as a function of social group treatment even for disturbed individuals (Fig. 5, Table 2). One gene, *Inos*, showed a trend towards significance in the predicted direction for disturbed individuals ($P < 0.109$). A regression analysis for this gene, including disturbance history, lab social group treatment, and their interaction, showed a significant effect of social group across both undisturbed and disturbed individuals,

showed significant variation in behavior as a function of social group for disturbed bees only (Kruskal-Wallis Test, undisturbed: $X_4^2 = 4.33$, $P < 0.36$, disturbed: $X_4^2 = 12.86$, $P < 0.012$, Fig. 2). Thus, disturbed bees show

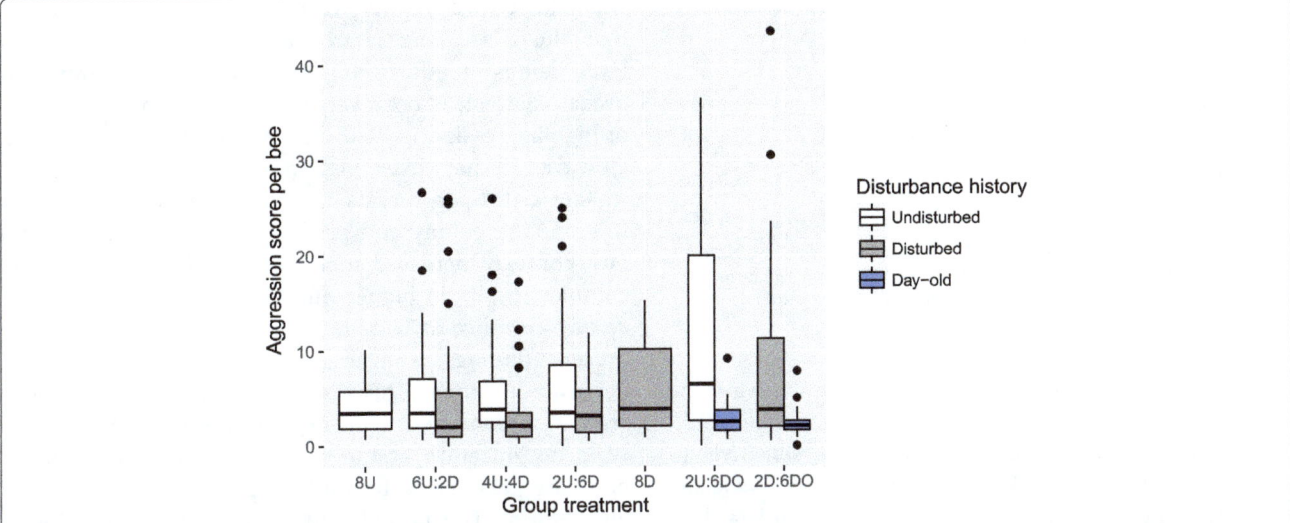

Fig. 3 Comparison of aggression scores for undisturbed, disturbed, and one-day-old bees kept in the same social group. A comparison of aggression scores for bees kept together in mixed groups showed that disturbed bees are typically less aggressive than undisturbed bees when kept together (Wilcoxon Exact Test (one-tailed) 6U:2D $X_1^2 = 4.01$, $P < 0.045$, 4U:4D $X_1^2 = 13.5$, $P < 0.0002$, 2U:6D $X_2^2 = 1.79$, $P < 0.18$). Similarly, and as predicted, one-day-old bees were less aggressive than nine-day-old bees, regardless of disturbance history (Wilcoxon Exact Test (one-tailed) 2U:6DO $X_1^2 = 5.56$, $P < 0.0184$, 2D:6DO $X_1^2 = 4.35$, $P < 0.037$). Aggression scores for bees kept in uniform groups (8U, 8D) are shown for comparison. Box hinges show the 1st and 3rd quartiles, whiskers indicate 1.5*IQR from the hinge, and the central tendency line indicates the median

suggesting some effect of social group composition on brain genomic state, but no interaction between the two terms (whole model: $F_{3,43} = 4.69$, $P < 0.0064$, Disturbance history: $F = 9.42$, $P < 0.0037$, Lab social group composition: $F = 4.16$, $P < 0.048$, interaction: $F = 0.224$, $P = 0.64$). A significant interaction term would parallel the behavioral results, indicating that chronic disturbance history and social group treatment have synergistic effects on the brain molecular state, similar to their effects on behavior. However, we found no evidence for such a pattern, and little evidence of an effect of social group composition on

brain gene expression overall. A similar analysis for the other three genes (but using a generalized linear model with a log link function for *GB53860*) showed no effect of lab social group on gene expression, nor any interaction effects for the sequential social experiences.

Discussion

Results presented here suggest that an individual's disturbance history and subsequent social context interact to influence aggressive behaviors. Overall, disturbed bees tended to show lower aggression scores relative to their undisturbed counterparts, which suggests that chronic disturbance results in a lower aggression behavioral tendency. However, disturbed bees also showed a high degree of behavioral flexibility, modifying their aggression depending on their social group in the lab-based assay. Conversely, undisturbed bees exhibited similar aggression levels regardless of their social group. As a result of these patterns of behavioral flexibility, lab-based social groups showed similar levels of total aggression, with no clear relationship between group composition and group aggression score. This finding suggests the hypothesis that individual honey bees use negative social feedback to modify their behavior in order to stabilize the group aggression effort. In support of this interpretation, undisturbed bees modulated their aggression effort only when kept in groups composed of highly docile one-day-old bees.

Organized societies are unique in that individuals often modulate their behavior to achieve an optimal group-level phenotype that can be set by ecological

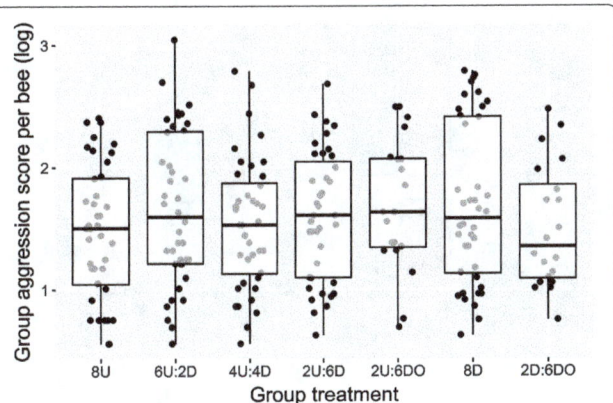

Fig. 4 Group aggression score as a function of social group. Total group aggression score did not significantly vary as a function of social group composition. Box hinges show the 1st and 3rd quartiles, whiskers indicate 1.5*IQR from the hinge, and the central tendency line indicates the median. Data points represent scores for individual replicates

Table 1 Overall, chronic disturbance had significant effects on brain expression for 3 of 4 aggression biomarker genes

Gene (sample size)	t/S	p
Inos (N = 47)	−3.02	**0.002**
Drat (N = 46)	−2.11	0.417
GB53860 (N = 47)	3.75	**0.0003**
Cyp6g1/2 (N = 47)	−1.70	**0.0483**

Values are the result of one-tailed *t*-tests (or a Wilcoxon Exact Test for *GB53860*) to account for the hypothesized direction of change based on previous studies [25]. Sample sizes represent the total number of individuals compared per test. Significant differences are indicated in bold type face

conditions or heritable rules [2], which is not always the case for other types of social groups [9]. As a result, individual behavioral outcome is highly sensitive to social context. Honey bee aggression is strongly socially regulated, and the rules individuals use to modulate their behavior seem to depend on the context for nest defense. Previous studies evaluating the effects of social group composition on individual and group level behavior used colony-level manipulations varying ratios of Africanized and European genotypes, which naturally differ in aggression (Africanized bees are more aggressive [29, 31]). When evaluating aggression displayed in the context of mammalian predator defense, studies generally find that individuals adjust their aggression level to that exhibited by the predominant genotype in the nest; European bees express higher aggression levels when kept in a colony with a majority of Africanized individuals, and *vice versa* [28, 31]. As a result, group level defensive effort in the context of a mammalian predator attack is positively correlated with the number of high aggression individuals present in the colony [31]. However, honey bees also defend their nest against smaller arthropod threats, including conspecifics that try to enter the colony and steal honey [44]. In the natural nest context, increased robbing threat from neighboring colonies results in higher numbers of bees guarding the colony entrance at least temporarily [45]. Similar to my present findings, and in contrast to the mammalian defensive response, guarding shows a pattern of social regulation consistent with negative feedback and phenotypic optimization at the group level; European honey bees are less likely to initiate guarding behavior and guard for shorter periods of time when living in colonies composed primarily of the more aggressive Africanized honey bee [29]. Thus, both negative and positive social feedback could play a role in modulating individual aggressive behaviors depending on context. In a previous study we showed that disturbed colonies have a low overall aggression level in response to a simulated large predator attack [25]. However, in the current study I find that disturbed bees are capable of reaching the levels of aggression exhibited by undisturbed bees. This could be because disturbed bees are less likely to instigate an attack in a large-predator context, and

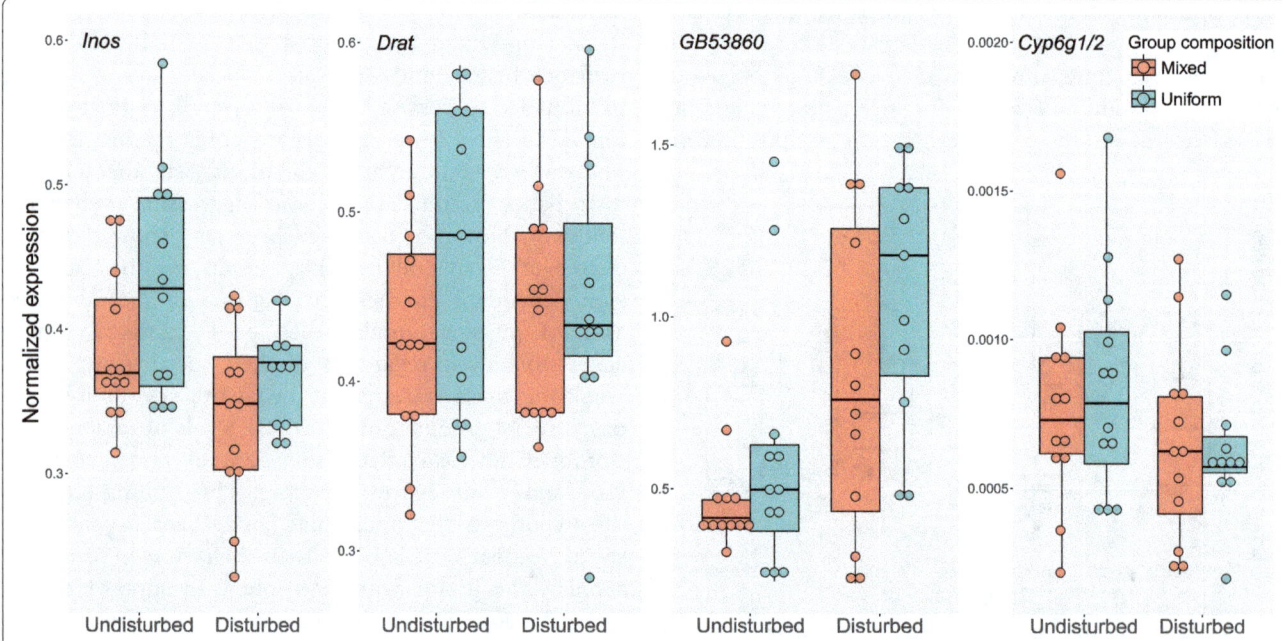

Fig. 5 Gene expression data as a function of disturbance history and social group treatment. *Inos*, *GB53860*, and *Cyp6g1/2* showed significant differences in expression in the predicted direction as a function of disturbance history (Table 1, [25]). A pairwise analysis of the effect of social group treatment within each disturbance history treatment showed no significant effects (Table 2), and a series of linear models incorporating disturbance history, social group treatment, and their interaction, showed significant main effects for *Inos*, with no interaction, and no other significant social group treatment or interaction effects for any other genes

Table 2 Analyzed on a pairwise basis, lab-based social group (mixed, 4U:4D, versus uniform, 8U or 8D) had no effect on gene expression, despite behavioral differences for disturbed bees as a function of group composition (Fig. 2)

Gene	Undisturbed (sample size) t/S	Disturbed (sample size)	Undisturbed p	Disturbed
Inos	1.61 (N = 24)	1.27 (N = 23)	0.121	0.109
Drat	1.43 (N = 23)	0.211 (N = 23)	0.166	0.417
GB53860	162 (N = 24)	154 (N = 23)	0.27	0.10
Cyp6g1/2	0.54 (N = 24)	−0.080 (N = 23)	0.598	0.532

P values represent the outcomes of two-tailed tests for undisturbed bees (where there was no difference in behavioral expression as a function of social group composition) and one-tailed tests for disturbed bees (following predictions based on the finding that individuals displayed higher aggression levels when kept in uniform groups). Parametric tests were performed for all genes except GB53860, which was analyzed using a Wilcoxon Exact Test. Sample sizes represent the total number of individuals compared per test

because response to large predators is organized by positive instead of negative social feedback [44], the colony mounts a very weak total response. In contrast, in a negative feedback context, disturbed bees are driven to exhibit higher levels of aggression when their group mates are docile. The lab-based assay, which quantifies response to an intruder bee [37] appears to resemble the guarding context in terms of both the type of aggression stimulus and the apparent rules of social regulation. However, it remains unclear how the target optimum for group aggression level is set and maintained in the field or the lab.

Negative social feedback can enable social species to redirect task effort, e.g., in the context of optimal foraging [46], but more work is needed to understand how the feedback threshold, and thus the group level phenotypic homeostasis, is set in the context of guarding behavior. I found that group aggression effort was consistent regardless of the sum of individual disturbance histories in a social group. One interpretation of this finding is that a history of chronic predator disturbance does not readily shift the target optimum for group guarding effort. This could be because total group guarding activity is more sensitive to an experience of intruder threat [45], and not the large predator threats simulated by the chronic disturbance (though aggression in these two contexts is correlated to a degree [47]). The optimal guarding effort at the group level is at least somewhat genotype-dependent, as is the social responsiveness of guard bees [29, 47].

My results suggest that disturbed bees, which generally show relatively low aggression levels, also show greater behavioral flexibility in response to social group composition compared to undisturbed bees. Studies in species across the sociality spectrum have shown that the degree of behavioral flexibility exhibited by a particular individual can vary as a function of personality. Moreover, in some species, "proactive" individuals, which are often more aggressive, tend to be less responsive to environmental variation compared to low aggression "reactive" individuals [20, 48], consistent with the current results. However,

it is difficult to determine whether variation in flexibility is truly an inherent property of disturbed, low-aggression individuals, or rather simply a reflection of the pattern of negative social feedback that regulates guarding behavior [29]. This type of social feedback may cause the appearance of increased behavioral flexibility for low aggression individuals who are less responsive to aggression-inducing social cues and therefore more likely to be socially inhibited during the assay. Very few studies in any species have determine whether individual variation in behavioral flexibility is generalizable or behavioral context dependent [19]. In the honey bee however, there are many established contexts for manipulating the colony environment and evaluating behavioral response [49–51]. Future behavioral studies will address whether patterns of behavioral flexibility are consistent across behavioral contexts or are more easily explained in terms of positive or negative social regulatory paradigms.

I evaluated the effects of sequential social experience on the molecular state of the brain, examining a small set of genes associated with aggression in previous studies [25, 28]. If gene expression patterns reflect sequential social experiences, I predicted a disturbance history by social group interaction on gene expression levels, but no such significant interactions were identified. In general agreement with previous studies, chronic disturbance induced a pattern of brain gene expression consistent with low aggression, but there was little evidence that short-term modulation of social context (i.e., in the lab-based manipulations of small groups) further influenced gene expression. This result stands in contrast to the observed social context-dependent shifts in aggressive behavior for disturbed individuals.

This mismatch between aggression and aggression-relevant brain gene expression, but only in acute social contexts, is intriguing. It could reflect the fact that these selected genes are not causally associated with behavioral change. However, the fact that they are context-dependent predictors of aggression suggests there may be more complex processes involved [52]. For instance, it is possible that these genes are associated with flexibility in aggression rather than aggression level *per se*. However, a previous study [28] found that these same genes predict high aggression in older bees that show a relatively high degree of variation in aggression over time and across individuals (in contrast to the high aggression undisturbed bees in the current study). Though many studies associate gene expression with behavioral variation, the connection between these two distant phenotypic levels is still poorly understood. There is ample evidence that an ephemeral or acute social experience can induce widespread changes in gene expression [35, 53, 54], but the immediate response to a social situation is likely mediated by neural electrical signaling [55], perhaps even despite existing differences at the molecular level. For example, in migratory locusts, social context instigates a transition from a solitary to a gregarious phenotype associated with swarming [56]. Gene expression changes accompany this transition, but the initial shift in behavior that is required to stimulate changes in other aspects of phenotype (increased or decreased association with conspecifics), occurs rapidly and precedes gene expression changes in some contexts but not others [57]. Similarly, in an ant, the *foraging* gene varies as a function of age but is not a direct predictor of foraging behavior *per se*, despite the fact that foraging activity in general increases with age [58]. Understanding how behavior retains flexibility in some contexts in spite of variation in the molecular or even structural state of the brain presents a challenging area of future work. This work is relevant particularly in light of the fact that an increasing number of studies use gene expression patterns as markers or predictors of consistent individual differences in behavior [59, 60].

In the honey bee, variation in aggression tendency (reflected in gene expression patterns) only becomes obvious if high levels of inhibitory cues are available to disturbed individuals in real time during the behavioral assay. Thus, here I show that a combination of behavioral and gene expression analyses provides an opportunity to identify cryptic variation in personality under conditions in which behavior may appear invariant across individuals. Conversely, these results also emphasize that the existence of consistent individual differences in behavior and behavioral plasticity within individuals are not mutually exclusive alternatives. Finally, gene expression data provide a unique tool not only to explore the neural basis of personality and behavioral flexibility, but also to differentiate transient versus lasting shifts in environmentally responsive behavioral phenotypes [28, 61], particularly in circumstances where personality variation and behavioral flexibility are linked. Here gene expression patterns reflect the more stable behavioral state that results from chronic disturbance, but not behavioral outcomes in real-time due to variation in the relatively short-lived lab based social group context.

Conclusions

Taken together, the results presented here show that individual behavior is a function of behavioral tendency, social cues experienced in real time, and underlying rules for social modulation of behavior in response to group-level effort. The molecular state of the brain can reveal underlying variation in behavioral tendency and may predict social response, but it does not always match behavioral outcome. In a species with highly socially regulated aggressive behavior, a mechanistic link between aggressive tendency and response to social context provides even greater ability for individuals to fine tune group level behaviors.

Acknowledgements

I want to thank Paul Schreiber for assistance with setting up experiments and collecting behavioral data, Mario Padilla and Bernardo Niño for help setting up experiments, and Christina Grozinger and Gene Robinson for comments on this study.

Funding

Work was funded by the National Science Foundation IOS-1256705 (Gene Robinson and Nathan Price).

Competing interests

The author declares that they have no competing interests.

References

1. Pruitt JN, Riechert SE. How within-group behavioural variation and task efficiency enhance fitness in a social group. Proc Biol Sci. 2011;278:1209–15.
2. Seeley TD. Honey bee colonies are group-level adaptive units. Am Nat. 1997;150:S22–41.
3. Jandt JM, Bengston S, Pinter-Wollman N, Pruitt JN, Raine NE, Dornhaus A, Sih A. Behavioural syndromes and social insects: personality at multiple levels. Biol Rev Camb Philos Soc. 2014;89:48–67.
4. Bonabeau E. Social insect colonies as complex adaptive systems. Ecosystems. 1998;1:437–43.
5. Duarte A, Weissing FJ, Pen I, Keller L. An evolutionary perspective on self-organized division of labor in social insects. Annu Rev Ecol Evol Syst. 2011;42:91–110.
6. Linksvayer TA, Fewell JH, Gadau J, Laubichler MD. Developmental evolution in social insects: regulatory networks from genes to societies. J Exp Zool B Mol Dev Evol. 2012;318:159–69.
7. Pinter-Wollman N. Personality in social insects: How does worker personality determine colony personality. Curr Zool. 2012;58:579–87.
8. Jeanson R, Weidenmuller A. Interindividual variability in social insects - proximate causes and ultimate consequences. Biol Rev Camb Philos Soc. 2014;89:671–87.
9. Pruitt JN, Grinsted L, Settepani V. Linking levels of personality: personalities of the 'average' and 'most extreme' group members predict colony-level personality. Anim Behav. 2013;86:391–9.
10. Calderone NW, Page RE. Genotypic variability in age polyethism and task specialization in the honey bee *Apis mellifera* (Hymenoptera: Apidae). Behav Ecol Sociobiol. 1988;22:17–25.

11. Laskowski KL, Bell AM. Strong personalities, not social niches, drive individual differences in social behaviours in sticklebacks. Anim Behav. 2014;90:287–95.

12. Seeley TD. Social foraging in honey bees: how nectar foragers assess their colony's nutritional status. Behav Ecol Sociobiol. 1989;24:181–99.

13. Fewell JH, Winston ML. Colony state and regulation of pollen foraging in the honey bee, Apis mellifera L. Behav Ecol Sociobiol. 1992;30:387–93.

14. Camazine S. The regulation of pollen foraging by honey bees: how foragers assess the colony's need for pollen. Behav Ecol Sociobiol. 1993;32:265–72.

15. Schmickl T, Crailsheim K. Inner nest homeostasis in a changing environment with special emphasis on honey bee brood nursing and pollen supply. Apidologie. 2004;35:249–63.

16. Lichocki P, Tarapore D, Keller L, Floreano D. Neural networks as mechanisms to regulate division of labor. Am Nat. 2012;179:391–400.

17. Dreller C, Page RE, Fondrk MK. Regulation of pollen foraging in honeybee colonies: effects of young brood, stored pollen, and empty space. Behav Ecol Sociobiol. 1999;45:227–33.

18. Coppens CM, de Boer SF, Koolhaas JM. Coping styles and behavioural flexibility: towards underlying mechanisms. Philos Trans R Soc Lond B Biol Sci. 2010;365:4021–8.

19. Dingemanse NJ, Wolf M. Between-individual differences in behavioural plasticity within populations: causes and consequences. Anim Behav. 2013;85:1031–9.

20. Koolhaas JM, Korte SM, De Boer SF, Van Der Vegt BJ, Van Reenen CG, Hopster H, De Jong IC, Ruis MAW, Blokhuis HJ. Coping styles in animals: current status in behavior and stress-physiology. Neurosci Biobehav Rev. 1999;23:925–35.

21. Bergmuller R, Taborsky M. Animal personality due to social niche specialisation. Trends Ecol Evol. 2010;25:504–11.

22. Weaver IC. Integrating early life experience, gene expression, brain development, and emergent phenotypes: unraveling the thread of nature via nurture. Adv Genet. 2014;86:277–307.

23. Stamps JA, Frankenhuis WE. Bayesian models of development. Trends Ecol Evol. 2016;31:260–8.

24. Alaux C, Robinson GE. Alarm pheromone induces immediate-early gene expression and slow behavioral response in honey bees. J Chem Ecol. 2007;33:1346–50.

25. Rittschof CC, Robinson GE. Manipulation of colony environment modulates honey bee aggression and brain gene expression. Genes Brain Behav. 2013;12:802–11.

26. Kastberger G, Thenius R, Stabentheiner A, Hepburn R. Aggressive and docile colony defence patterns in Apis mellifera: a retreater–releaser concept. J Insect Behav. 2008;22:65–85.

27. Rittschof CC, Coombs CB, Frazier M, Grozinger CM, Robinson GE. Early-life experience affects honey bee aggression and resilience to immune challenge. Sci Rep. 2015;5:15572.

28. Alaux C, Sinha S, Hasadsri L, Hunt GJ, Guzman-Novoa E, Degrandi-Hoffman G, Uribe-Rubio JL, Southey BR, Rodriguez-Zas S, Robinson GE. Honey bee aggression supports a link between gene regulation and behavioral evolution. Proc Natl Acad Sci U S A. 2009;106:15400–5.

29. Hunt GJ, Guzmán-Novoa E, Uribe-Rubio JL, Prieto-Merlos D. Genotype-environment interactions in honeybee guarding behaviour. Anim Behav. 2003;66:459–67.

30. Uribe-Rubio JL, Guzman-Novoa E, Vazquez-Pelaez CG, Hunt GJ. Genotype, task specialization, and nest environment influence the stinging response thresholds of individual Africanized and European honeybees to electrical stimulation. Behav Genet. 2008;38:93–100.

31. Guzmán-Novoa E, Page RE. Genetic dominance and worker interactions affect honeybee colony defense. Behav Ecol. 1994;5:91–7.

32. Auld JR, Agrawal AA, Relyea RA. Re-evaluating the costs and limits of adaptive phenotypic plasticity. Proc Biol Sci. 2010;277:503–11.

33. Winston ML. The Biology of the Honey Bee. Cambridge: Harvard University Press; 1987.

34. Cardoso SD, Teles MC, Oliveira RF. Neurogenomic mechanisms of social plasticity. J Exp Biol. 2015;218:140–9.

35. Rittschof CC, Bukhari SA, Sloofman LG, Troy JM, Caetano-Anolles D, Cash-Ahmed A, Kent M, Lu X, Sanogo YO, Weisner PA, Zhang H, Bell AM, Ma J, Sinha S, Robinson GE, Stubbs L. Neuromolecular responses to social challenge: common mechanisms across mouse, stickleback fish, and honey bee. Proc Natl Acad Sci U S A. 2014;111:17929–34.

36. Li-Byarlay H, Rittschof CC, Massey JH, Pittendrigh BR, Robinson GE. Socially responsive effects of brain oxidative metabolism on aggression. Proc Natl Acad Sci U S A. 2014;111:12533–7.

37. Breed MD. Nestmate recognition in honey bees. Anim Behav. 1983;31:86–91.

38. Whitfield CW, Cziko AM, Robinson GE. Gene expression profiles in the brain predict behavior in individual honey bees. Science. 2003;302:296–9.

39. Richard FJ, Holt HL, Grozinger CM. Effects of immunostimulation on social behavior, chemical communication and genome-wide gene expression in honey bee workers (Apis mellifera). BMC Genomics. 2012;13:558.

40. Schulz DJ, Robinson GE. Biogenic amines and division of labor in honey bee colonies: behaviorally related changes in the antennal lobes and age-related changes in the mushroom bodies. J Comp Physiol A-Sens Neural Behav Physiol. 1999;184:481–8.

41. Pfaffl MW, Tichopad A, Prgomet C, Neuvians TP. Determination of stable housekeeping genes, differentially regulated target genes and sample integrity: BestKeeper - excel-based tool using pairwise correlations. Biotechnol Lett. 2004;26:509–15.

42. Xie F, Xiao P, Chen D, Xu L, Zhang B. miRDeepFinder: a miRNA analysis tool for deep sequencing of plant small RNAs. Plant Mol Biol. 2012;80:75–84.

43. Larionov A, Krause A, Miller W. A standard curve based method for relative real time PCR data processing. BMC Bioinformatics. 2005;6:62.

44. Breed MD, Guzman-Novoa E, Hunt GJ. Defensive behavior of honey bees: organization, genetics, and comparisons with other bees. Annu Rev Entomol. 2004;49:271–98.

45. Couvillon MJ, Robinson EJH, Atkinson B, Child L, Dent KR, Ratnieks FLW. En garde: rapid shifts in honeybee, Apis mellifera, guarding behaviour are triggered by onslaught of conspecific intruders. Anim Behav. 2008;76:1653–8.

46. Czaczkes TJ. How to not get stuck-negative feedback due to crowding maintains flexibility in ant foraging. J Theor Biol. 2014;360:172–80.

47. Collins AM, Rinderer TE, Harbo JR, Bolten AB. Colony defense by Africanized and European honey bees. Science. 1982;218:72–4.

48. Holbrook CT, Wright CM, Pruitt JN. Individual differences in personality and behavioural plasticity facilitate division of labour in social spider colonies. Anim Behav. 2014;97:177–83.

49. Beshers SN, Huang ZY, Oono Y, Robinson GE. Social inhibition and the regulation of temporal polyethism in honey bees. J Theor Biol. 2001;213:461–79.

50. Schulz DJ, Huang Z-Y, Robinson GE. Effects of colony food shortage on behavioral development in honey bees. Behav Ecol Sociobiol. 1998;42:295–303.

51. Huang ZY, Robinson GE. Honey bee colony integration: worker-worker interactions mediate hormononally regulated plasticity in division of labor. Proc Natl Acad Sci U S A. 1992;89:11726–9.

52. Aubin-Horth N, Renn SC. Genomic reaction norms: using integrative biology to understand molecular mechanisms of phenotypic plasticity. Mol Ecol. 2009;18:3763–80.

53. Zayed A, Robinson GE. Understanding the relationship between brain gene expression and social behavior: lessons from the honey bee. Annu Rev Genet. 2012;46:591–615.

54. Sanogo YO, Band M, Blatti C, Sinha S, Bell AM. Transcriptional regulation of brain gene expression in response to a territorial intrusion. Proc Biol Sci. 2012;279:4929–38.

55. Clayton DF. The genomic action potential. Neurobiol Learn Mem. 2000;74:185–216.

56. Wang X, Kang L. Molecular mechanisms of phase change in locusts. Annu Rev Entomol. 2014;59:225–44.

57. Guo W, Wang X, Ma Z, Xue L, Han J, Yu D, Kang L. CSP and takeout genes modulate the switch between attraction and repulsion during behavioral phase change in the migratory locust. PLoS Genet. 2011;7:e1001291.

58. Oettler J, Nachtigal AL, Schrader L. Expression of the foraging gene is associated with age polyethism, not task preference, in the ant Cardiocondyla obscurior. PLoS One. 2015;10:e0144699.

59. Rey S, Boltana S, Vargas R, Roher N, Mackenzie S. Combining animal personalities with transcriptomics resolves individual variation within a wild-type zebrafish population and identifies underpinning molecular differences in brain function. Mol Ecol. 2013;22:6100–15.

60. Bell AM, Aubin-Horth N. What can whole genome expression data tell us about the ecology and evolution of personality? Philos Trans R Soc Lond B Biol Sci. 2010;365:4001–12.

61. Rittschof CC, Robinson GE. Genomics: moving behavioural ecology beyond the phenotypic gambit. Anim Behav. 2014;92:263–70.

15

Asian house rats may facilitate their invasive success through suppressing brown rats in chronic interaction

Hong-Ling Guo[1,2], Hua-Jing Teng[1,2], Jin-Hua Zhang[1], Jian-Xu Zhang[1*] and Yao-Hua Zhang[1*] ⓘ

Abstract

Background: The Asian house rat (*Rattus tanezumi*) and the brown rat (*Rattus norvegicus*) are closely related species and are partially sympatric in southern China. Over the past 20 years, *R. tanezumi* has significantly expanded northward in China and partially replaced the native brown rat subspecies, *R. n. humiliatus*. Although invasive species are often more aggressive than native species, we did not observe interspecific physical aggression between *R. tanezumi* and *R. n. humiliatus*. Here, we focused on whether or not *R. tanezumi* was superior to *R. n. humiliatus* in terms of nonphysical competition, which is primarily mediated by chemical signals.

Results: We performed two laboratory experiments to test different paradigms in domesticated *R. tanezumi* and *R. n. humiliatus*. In Experiment 1, we caged adult male rats of each species for 2 months in heterospecific or conspecific pairs, partitioned by perforated galvanized iron sheets, allowing exchange of chemical stimuli and ultrasonic vocalization. The sexual attractiveness of male urine odor showed a tendency (marginal significance) to increase in *R. tanezumi* caged with *R. n. humiliatus*, compared with those in conspecific pairs. Hippocampal glucocorticoid receptor (*GR*) and brain-derived nutrition factor (*BDNF*) mRNA were upregulated in *R. n. humiliatus* and *R. tanezumi*, respectively, when the rats were caged in heterospecific pairs. In Experiment 2, we kept juvenile male rats in individual cages in rooms with either the same or the different species for 2 months, allowing chemical interaction. The sexual attractiveness of male urine was significantly enhanced in *R. tanezumi*, but reduced in *R. n. humiliatus* by heterospecific cues and mRNA expression of hippocampal *GR* and *BDNF* were upregulated by heterospecific cues in *R. n. humiliatus* and *R. tanezumi*, respectively. Although not identical, the results from Experiments 1 and 2 were generally consistent.

Conclusions: The results of both experiments indicate that nonphysical/chronic interspecific stimuli, particularly scent signals, between *R. n. humiliatus* and *R. tanezumi* may negatively affect *R. n. humiliatus* and positively affect *R. tanezumi*. We infer that chronic interspecific interactions may have contributed to the invasion of *R. tanezumi* into the range of *R. n. humiliatus* in natural habitats.

Keywords: Closely related species, Invasive mechanism, Sexual attractiveness, Neuroendocrine molecules, Chronic stress

Background

The brown rat (*Rattus norvegicus*), the Asian house rat (*R. tanezumi*), and the black rat (*R. rattus*) are three closely related commensal pests [1]. *R. norvegicus* has now spread from northern Asia to all continents except Antarctica, whereas *R. tanezumi* is mainly distributed in eastern, southern, and south-eastern Asia [1–3]. In China, *R. norvegicus* is widespread and has differentiated into four subspecies, including *R. n. norvegicus*, *R. n. soccer*, *R. n. humiliatus*, and *R. n. caraco*, whereas *R. tanezumi* typically lives south of the Yellow River and is sympatric with *R. n. norvegicus* and *R. n. soccer* [1, 4, 5]. *R. tanezumi* is also sympatric with *R. n. caraco* in the Korean peninsula [6]. *R. n. humiliatus*, which is the smallest of the four subspecies and lives mainly in Hebei Province in central North China, is the only subspecies geographically isolated from *R. tanezumi* [4, 7]; however,

* Correspondence: zhangjx@ioz.ac.cn; zhangyh@ioz.ac.cn
[1]State Key Laboratory of Integrated Management of Pest Insects and Rodents in Agriculture, Institute of Zoology, Chinese Academy of Sciences, No.1-5 Beichen West Road, Chaoyang District, Beijing 100101, China
Full list of author information is available at the end of the article

R. tanezumi has recently expanded its range to north of the Yellow River in the south of Hebei Province and partially replaced the native *R. norvegicus* subspecies (i.e., *R. n. humiliatus*) [8–10]. Several factors, including global warming and higher resistance to common rodenticides compared with brown rats, are believed to likely contribute to the invasive success of *R. tanezumi* [5, 11–15].

Exotic invasive species that successfully expand their range and displace native species appear to exhibit superiority in interspecific competition [16]. As closely related species are more likely to compete than those that are more distantly related, their competition, coexistence, and invasion have been extensively studied [16–20]. There are several types of interspecific competition, including indirect resource competition, interspecific aggression, interspecific territoriality, overgrowth, and chemical competition [16, 21]. In the case of rodents, invasive species are often more aggressive than native species [22]. While spreading from Asia to Europe and America during the Middle Ages, *R. norvegicus* generally displaced *R. rattus* in human settlements, where they out-competed *R. rattus* via physical interspecific interactions [11, 23]. It is logical that *R. norvegicus* is superior to *R. rattus* in terms of interspecific aggression, since the former is larger than the latter [1]; however, invasive *R. tanezumi* is generally smaller than native *R. n. humiliatus*. We did not observe that male *R. n. humiliatus* and *R. tanezumi* displayed interspecific aggression in dyadic encounters in a neutral arena (a common laboratory method to investigate aggressive behavior), and they can even live peacefully together for long periods of time when caged in interspecific male–male pairs [1, 19] (unpublished data). Thus, it is necessary to explore whether other types of interspecific competition (e.g., nonphysical interaction) contribute to the successful invasion of *R. tanezumi* into areas containing *R. n. humiliatus*.

Interspecific chemical interactions, including communication signaling and allelopathy, are widespread among prokaryotes, plants, and invertebrates, and are important in the invasion of exotic species [12, 14, 24]. For mammals, odor-mediated communication between different species (e.g., during predator and prey interactions) is also important, and the general laws of chemical ecology apply [25, 26]. Rodents use scent signals extensively in species recognition and interspecific competition [25, 27–30]. In rodents, scent signals, including urine volatile compounds and major urine proteins, are distinctive between closely related species and even between subspecies, including *R. n. humiliatus* and *R. tanezumi* [31–38] (unpublished data). The scent signals released by animals can function in heterospecific, as well as conspecific interactions, without the physical presence of the donor [37, 39–41]. Therefore, in rodents,

where physical antagonism does not exist, interspecific competition may rely partially or completely on interspecific odor-based effects.

Interspecific competition can induce physiological stress responses and inhibitory effects on some phenotypic traits of competitors [20, 42]. In rodents, male scent signals have crucial roles in mediating sexual behavior and are often correlated with reproductive success [43–46]. If interspecific interactions induce strong chronic social stress, they can impair the attractiveness of the odor of male urine to females, and the production of urinary sex pheromones, and consequently disturb rodent reproductive behavior [47–50]. Conversely, changes in the sexual attractiveness of males subjected to interspecific competition may indicate stressful states in their competitors.

Competition-induced stress can activate the hypothalamic–pituitary–adrenal (HPA) axis to release endocrine hormones and neurotransmitters and alter gene expression in some regions of the rodent brain [25, 51–54]. After chronic competition, shifts in glucocorticoid levels are not always detectable in rodents [20]. The hippocampus is exquisitely sensitive to stressors, due to direct emotional input from the basolateral amygdala (BLA) and glucocorticoids (GCs), and because of its high density of GC receptors [55, 56]. The mRNA expression of hippocampal glucocorticoid receptor (*GR*), brain-derived neurotrophic factor (*BDNF*), and the BDNF receptor (*TrkB*) can be affected by stressors through modulation of the HPA axis and emotion-related input from the BLA [52–54]. Levels of *GR*, *BDNF*, and *TrkB* are closely related to emotional behavior, neuronal development, and plasticity, and can thus reflect emotional states in rodents [55, 57–59]. Stressors often impart different effects on adult and juvenile animals [60]. Early exposure to aversive stimuli often causes long-term alterations in many aspects of behavior, such as behavioral regulation, neuroendocrine responsiveness to stress, and mRNA expression of central nervous system genes related to behavioral change in rodents [61].

In the current study, we aimed to explore the potential roles of chronic nonphysical/chemical competition in the natural invasion process of *R. tanezumi* replacing *R. n. humiliatus* using laboratory experiments. Therefore, we performed two long-term experiments to test different paradigms in our laboratory. In Experiment 1, two adult male rats of the same or different species were caged together, partitioned by a perforated galvanized iron sheet; in Experiment 2, juvenile males of these two rat species were exposed to heterospecific or conspecific odors. To examine the effects of these exposures, we then evaluated changes in the sexual attractiveness of male urine odor and determined the mRNA levels of hippocampal *GR*, *BDNF*, and *TrkB*, as well as serum cortisol concentrations.

Results

Effects of chronic interspecific interaction on body weight

Experiment 1: After two months of interaction, the body weight was not significantly different between the control and treatment groups of either *R. n. humiliatus* or *R. tanezumi* (Fig. 1a).

Experiment 2: After two months of interaction, the body weight of immature males exposed to a heterospecific odor was higher than that of the control group ($t = 2.232$, $n = 6$ for each group, $p = 0.050$) in *R. n. humiliatus*, whereas body weight did not differ between the control and treatment groups in *R. tanezumi* (Fig. 1b).

Sexual attractiveness of male urine odor

Experiment 1: In *R. n. humiliatus*, female attraction to male urine did not differ between the control and treatment groups (Fig. 2a), while in *R. tanezumi*, females exhibited a trend towards preferring males caged with *R. n. humiliatus* over those caged with their own species ($z = 1.726$, $n = 12$, $p = 0.080$, marginal significance) (Fig. 2a).

Experiment 2: In *R. n. humiliatus*, heterospecific odor stimulation significantly suppressed the sexual attractiveness of male urine to conspecific females compared with conspecific odor stimulation ($z = 2.884$, $n = 16$, $p = 0.004$) (Fig. 2b). Conversely, the sexual attractiveness of male urine was significantly greater after heterospecific odor stimulation in *R. tanezumi* ($z = 2.373$, $n = 18$, $p = 0.018$) (Fig. 2b).

Serum cortisol and testosterone levels

Experiment 1: A radioimmunoassay demonstrated that serum testosterone and cortisol levels were not significantly different between the control and treatment groups for either *R. n. humiliatus* or *R. tanezumi* (3a

and b). However, serum cortisol concentration was higher in both the control and treatment groups of *R. tanezumi* than those of *R. n. humiliatus* (control group: $t = 2.938$, $n = 12$ for each species, $p = 0.008$; treatment group: $t = 3.395$, $n = 9$-10 for each species, $p = 0.004$) (Fig. 3a).

Experiment 2: Similarly, interspecific odor stimulation had no apparent effect on serum testosterone or cortisol levels within species for either *R. n. humiliatus* or *R. tanezumi* (Fig. 3c and d). *R. tanezumi* also had a higher serum cortisol concentration than *R. n. humiliatus* in both the control ($t = 3.173$, $n = 6$-7 for each species, $p = 0.009$) and treatment ($t = 3.445$, $n = 6$-8 for each species, $p = 0.005$) groups (Fig. 3c).

Gene expression of *GR* and *BDNF* in the hippocampus

Experiment 1: Quantitative real-time PCR demonstrated that *GR* mRNA expression was slightly upregulated in the hippocampus of *R. n. humiliatus* rats exposed to heterospecific stimuli ($t = 1.901$, $n = 8$ for each group, $p = 0.078$, marginal significance), whereas hippocampal *BDNF* was significantly upregulated in *R. tanezumi* exposed to heterospecific stimuli ($z = 3.361$, $n = 8$ for each group, $p = 0.001$) (Fig. 4a and b). Hippocampal *TrkB* mRNA expression exhibited no significant changes within either rat species (Fig. 4a and b).

Experiment 2: Similar to the results of Experiment 1, hippocampal *GR* mRNA expression ($t = 4.125$, $n = 6$ for each group, $p = 0.009$) in *R. n. humiliatus* and *BDNF* mRNA expression ($t = 1.911$, $n = 7$-8 for each group, $p = 0.078$, marginal significance) in *R. tanezumi* were upregulated (Fig. 4c and d). Hippocampal *TrkB* mRNA expression showed no change within either rat species.

Discussion

Our results from binary choice tests suggest that chronic nonphysical competition between *R. tanezumi* and *R. n.*

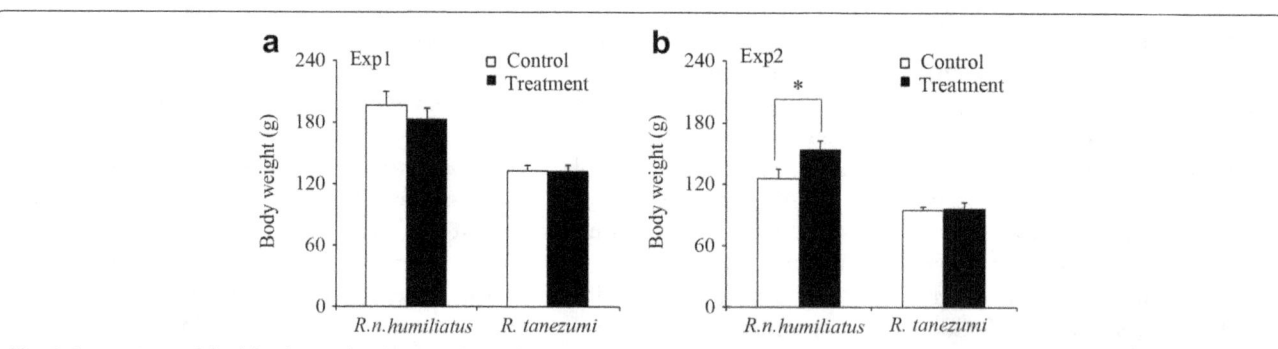

Fig. 1 Comparison of final body weights (mean ± SE, g) between control and treatment groups of *R. norvegicus humiliatus* or *R. tanezumi* in Experiment 1 (Exp1) (**a**) ($n_{control} = n_{treatment} = 9$ for *R. n. humiliatus* and $n_{control} = 9$, $n_{treatment} = 10$ for *R. tanezumi*.) and Experiment 2 (Exp2) (**b**) ($n_{control} = n_{treatment} = 6$ for *R. n. humiliatus*; $n_{control} = 7$ and $n_{treatment} = 8$ for *R. tanezumi*). Independent samples *t*-test or Mann–Whitney *U*-test; *$p \leq 0.05$

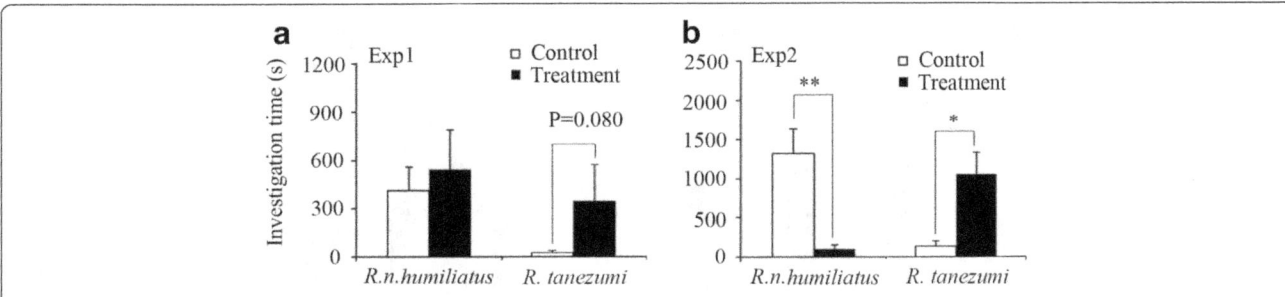

Fig. 2 Investigation time (mean ± SE, sec) spent by female *R. norvegicus humiliatus* or *R. tanezumi* on conspecific male urine samples between control and treatment groups in Experiment 1 (Exp1) (**a**) ($n = 15$ for test female rats of each species), and Experiment 2 (Exp2) (**b**) ($n = 18$ for test females of each species). Paired *t*-test or Wilcoxon signed-rank test; *$p < 0.05$, **$p < 0.01$

humiliatus exerts completely opposite effects on their male scent signals, which were enhanced in *R. tanezumi*, but inhibited in *R. n. humiliatus*. The results from Experiments 1 and 2 were not identical; however, they were generally consistent with one another. As sexual attractiveness is often correlated with reproductive success and fitness in male animals, the reproductive success of *R. n. humiliatus* may be suppressed due to reduced sexual attractiveness, and augmented in *R. tanezumi* due to increased sexual attractiveness resulting from chronic interspecific competition. Such asymmetric competition effects on scent signals and reproductive behavior may have facilitated the invasive *R. tanezumi* population to replace that of the native *R. n. humiliatus* [41, 43, 62–65]. Our results, therefore, warrant further

investigation in the laboratory and the field [46, 66]. Although rodents can emit species-specific ultrasonic vocalizations that are behaviorally important for mating, nursing, aggression, defense, and emotion within species, the effects of ultrasound-mediated chronic competition between closely related species appear to be very weak [67–69]. Therefore, we believe that nonphysical chronic interspecific interactions between *R. tanezumi* and *R. n. humiliatus* are likely to have been primarily mediated by chemical signals in our experiments.

In addition, chronic stress often impairs the sexual attractiveness of urine odor and decreases the levels of volatile pheromones in male mouse urine [47–49]. For example, both the presence of a predator and its scent can inhibit the sexual attractiveness of male mouse urine

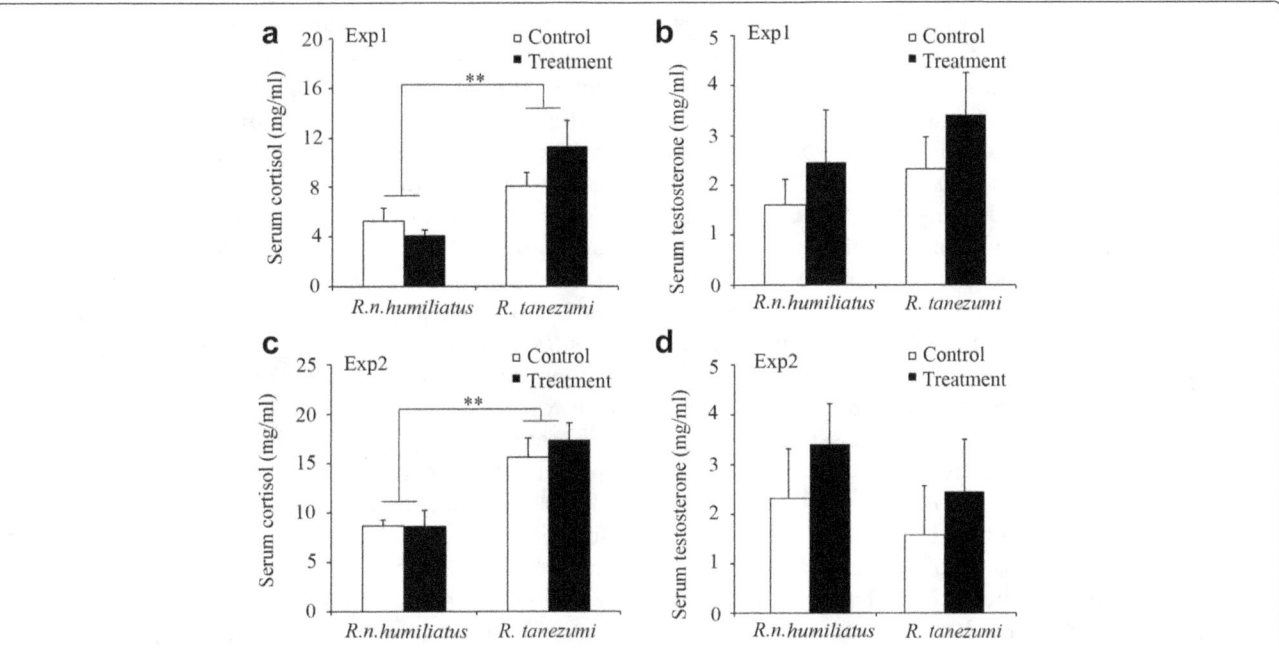

Fig. 3 Serum cortisol and testosterone concentration (mean ± SE) in *R. norvegicus humiliatus* or *R. tanezumi* in Experiment 1 (Exp1) ($n_{control} = 12$, $n_{treatment} = 10$ for *R. n. humiliatus*; $n_{control} = 12$ and $n_{treatment} = 9$ for *R. tanezumi* (**a** and **b**) and Experiment 2 (Exp2) ($n_{control} = n_{treatment} = 6$ for *R. n. humiliatus*; $n_{control} = 7$ and $n_{treatment} = 8$ for *R. tanezumi*) (**c** and **d**). Independent samples *t*-test or Mann-Whitney *U*-test; *$p < 0.05$, **$p < 0.01$

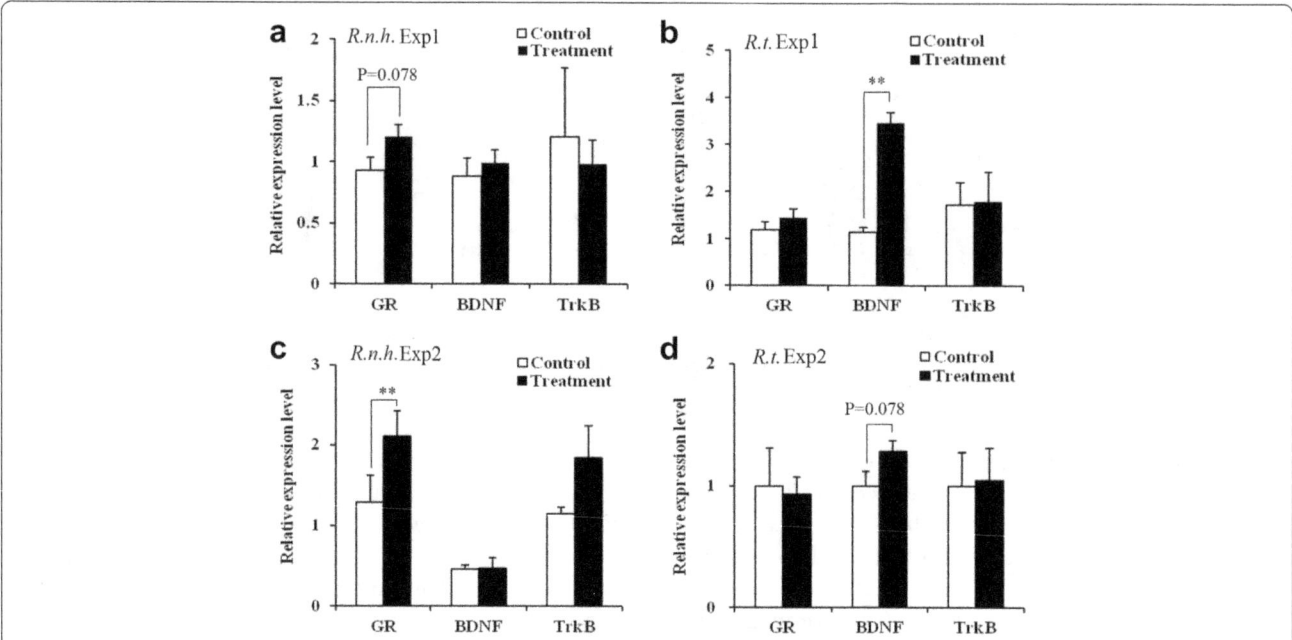

Fig. 4 Comparison of hippocampal expression of *GR*, *BDNF*, and *TrkB* mRNA (mean ± SE) between control and treatment groups of *R. norvegicus humiliatus* (*R.n.h.*) (**a** and **c**) or *R. tanezumi* (*R.t.*) (**b** and **d**) in Experiment 1 (Exp1) ($n_{control} = n_{treatment} = 8$ for *R.n.h.* and $n_{control} = n_{treatment} = 8$ for *R.t.*) and Experiment 2 (Exp2) ($n_{control} = n_{treatment} = 6$ for *R.n.h.*; $n_{control} = 7$ and $n_{treatment} = 8$ for *R.t.*). Independent samples *t*-test or Mann-Whitney *U*-test; *$p < 0.05$, **$p < 0.01$

[48]. However, low predation risk, reflected as a low dose of predator scent, has a positive effect, boosting the sexual attractiveness of male mouse urine [36, 47, 70]. In intraspecific male–male competition, two opponents can form a stable dominance–submission relationship, in which the dominant partner has more volatile phero- mones in the urine and greater sexual attractiveness compared with the submissive partner [49, 71]. Here, particularly in Experiment 2, the sexual attractiveness of male urine was augmented in *R. tanezumi*, but sup- pressed in *R. n. humiliatus*, as a result of exposure to heterospecific stimuli, indicating that *R. n. humiliatus* may be stressed by *R. tanezumi*. Coincidentally, juvenile *R. n. humiliatus* rats exposed to *R. tanezumi* odor gained more body weight than those exposed to their own spe- cies in Experiment 2, possibly reflecting the influence of a stressor [44, 72].

The influence of nonphysical competition-induced stress was further confirmed by the observation of up- regulated hippocampal *GR* mRNA levels in *R. n. humi- liatus* in response to *R. n. humiliatus* cues. As reported in previous studies of competition stress, we detected alterations in the expression of *GR* and *BDNF* mRNA in the hippocampus, but no differences in blood cortisol and testosterone levels in *R. n. humiliatus* experiencing chronic stress due to the presence of *R. tanezumi* [20, 55, 56]. *GRs* are key mediators of the neuroendocrine response to stress, and stressor- specific alterations in *GR* mRNA levels are more

pronounced in male than in female rodents [73]. Hip- pocampal *GR* has been implicated in negative feed- back inhibition of the HPA axis and can mediate the deleterious effects of blood glucocorticoids on hippo- campal neuron survival and function [52, 59]. The ex- pression of hippocampal *GR* can be affected by stressful stimuli and is adaptively downregulated to protect the hippocampus from glucocorticoid hyperse- cretion in acute stress and in the early stages of chronic stress [52, 59]. In the late stages of chronic stress, blood glucocorticoid may return to control levels and hippocampal *GR* expression may be upreg- ulated due to the buffering effect of repeated stress and habituation effects [74]. Therefore, the upregula- tion of hippocampal *GR* mRNA observed in *R. n. humiliatus* in our study may indicate that this species was physiologically impaired by long-term interaction with *R. tanezumi*.

In contrast, hippocampal *BDNF* mRNA levels were upregulated in *R. tanezumi*, particularly in Experiment 1, after exposure to *R. n. humiliatus*. The neurotrophic factor, *BDNF*, is associated with neuronal development, survival, and plasticity, and has been found to decrease in response to acute and mild chronic stress, which may contribute to the neuronal atrophy/death observed in rodents suffering from chronic stress [57, 58]. Con- versely, an enriched environment (EE) can ameliorate stress-induced symptoms, such as anxious behavior and increase hippocampal neurogenesis and *BDNF* protein levels in mice and rats [36, 70]. In this study, heterospecific cues from *R. n. humiliatus* enhanced hippocampal *BDNF* gene expression in *R. tanezumi*, and an EE may therefore improve cognitive function, spatial memory, and local behavioral adaptation of invasive *R. tanezumi*. For example, it is conceivable that, if immature *R. tanezumi* rats of pioneer populations disperse into the range of *R. n. humiliatus*, the resulting heterospecific cues will promote the development and survival of *R. tanezumi*. Moreover, repeated encounters between adult rats of these two species during the invasion process may improve the neuroendocrine state and adaptive behavior of *R. tanezumi*.

Conclusions

The results of both experiments conducted in this study imply that chronic nonphysical interspecific stimuli, particularly scent signals, can have asymmetric effects on *R. n. humiliatus* and *R. tanezumi*, leading to detrimental effects on the sexual attractiveness and neuroendocrine system of the former, and favorable effects on the same factors in the latter. Thus, we infer that chronic interspecific interactions may contribute to the invasive success and northward expansion of *R. tanezumi* and the decline of native *R. n. humiliatus* populations in natural habitats. These results warrant further investigation in the field.

Methods
Animals

Wild *R. tanezumi* and *R. n. humiliatus* were captured from Shanxi Province and Beijing (China), respectively. Each rat species was maintained as an outbred colony of 300–400 rats in our laboratory. The rats used were of the third generations, weaned at 4 weeks of age, and caged in groups of same sex siblings prior to use. All animals were kept in plastic rat cages ($37 \times 26 \times 17$ cm), in two separate rooms (14:10 h light: dark photoperiod, lights on at 5:00 am) and were maintained at 25 °C ± 2 °C. Food (standard rat chow) and water were provided *ad libitum*.

Experiment 1 (assessment of chemical signals and ultrasonic vocalization stimuli in adult animals)

Twenty-two adult male rats of each rat species were randomly selected from the colonies at 14–18 years of age, and 10 paired with the other species and the others paired with their own species as a control. All pairs of rats were housed in the same cage for 2 months, and all cages were partitioned with perforated galvanized iron sheets containing one 0.3 cm diameter hole per cm^2, to allow chemical and ultrasonic interactions (Fig. 5). Body weights were not significantly different between control and treatment groups of the same species (*R. n.*

humiliatus: 169.4 ± 9.069 g vs. 172.5 ± 8.303 g, $t = 0.014$, $p = 0.989$; *R. tanezumi*: 135.4 ± 7.248 g vs. 127.2 ± 4.397 g, $t = 1.281$, $p = 0.218$).

The sexual attractiveness of the odor of the urine from heterospecific or conspecific caged males was assessed using twelve estrous female rats of each species. All female subjects were between 14 and 18 weeks of age and had estrous cycles of 4–5 days, as determined by a vaginal smear examination. Females were used in experiments on the days that they came into estrus. Sixteen female rats of each species were used as urine recipients in Experiment 1.

Experiment 2 (assessment of the effects of odor stimulus on young animals)

Fifteen immature males (4 weeks old) of each species were individually caged and randomly assigned into two groups for 2 months. One group ($n = 8$) was kept in a room with the other species, while the other group ($n = 7$) remained in a room with its own species, as a control. Each rat room was 15 m^2 and contained approximately 120 rats (Fig. 5). Body weights were not significantly different between the two groups of the same species (*R. n. humiliatus*: 50.66 ± 3.972 g vs. 49.31 ± 4.533 g, $t = 0.225$, $p = 0.826$; *R. tanezumi*: 47.36 ± 6.177 g vs. 46.03 ± 9.010 g, $t = 0.338$, $p = 0.740$). Eighteen female rats of each species in estrus were used as urine recipients in Experiment 2.

Urine collection

Within 3 days after the chronic interspecific interaction experiments, we individually collected rat urine using clean metabolic rat cages during the dark phase of the light cycle. Urine collection continued for 8 h daily. The urine from the metabolic cage was collected in a tube immersed in an ice box. Standard rat chow and water were freely available. Urine samples were stored at -20 °C until use. Metabolic cages were washed thoroughly with water between urine collections.

Behavioral tests of sexual attractiveness

Olfactory preference tests were conducted in a two-choice box that consisted of a plastic rat cage that served as a start box and two Plexiglas choice tubes (internal diameter, 7.5 cm; length, 50 cm). The two choice tubes were symmetrically connected to the long side of the start box and each tube had a removable perforated galvanized iron sheet partition 5 cm away from the box to control rat access. An odorant presentation compartment partitioned by a perforated galvanized iron sheet from the other part of the tube was at the distal end of each 10 cm tube (Fig. 5).

Female rats were first test acclimated in the start box for 30 min, then a microscope slide with a urine

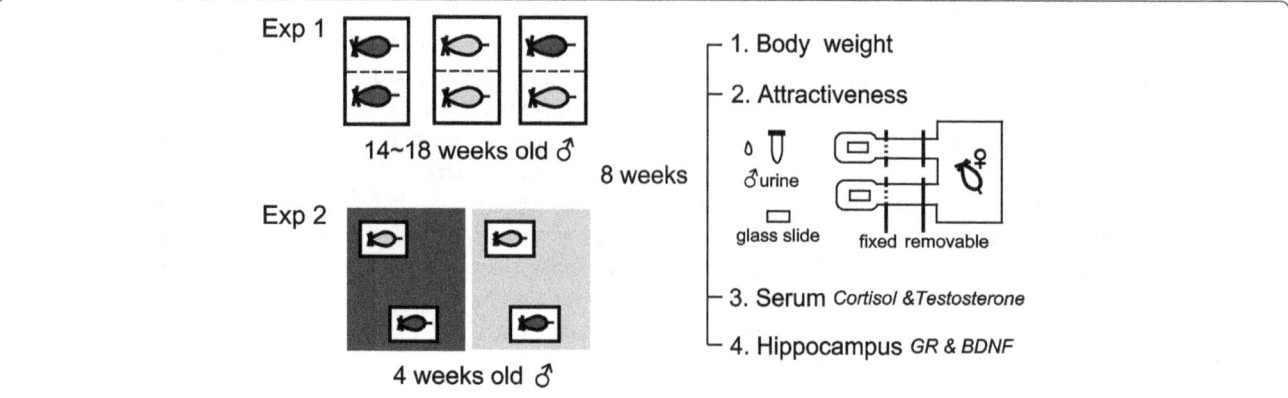

Fig. 5 Experimental set up. In Experiment 1, from 22 adult male rats of each rat species aged from 14 to 18 weeks, 10 were paired with the other species and the others were paired with their own species as a control. All pairs of rats were housed in the same cage for 2 months, and partitioned by perforated galvanized iron sheets. In Experiment 2, one group (n = 8) was kept in a room with the other species, and the other group (n = 7) remained in a room with its own species as a control. Each rat room was 15 m² and contained approximately 120 rats of the same species. Eight weeks later, body weight, attractiveness of urine, serum cortisol and testosterone, and hippocampal *GR* and *BDNF* expression were measured. The attractiveness of male urine was determined using a two-choice box, consisting of a plastic (start box) and two Plexiglas choice tubes (internal diameter, 7.5 cm; length, 50 cm). The choice tubes were symmetrically connected to the start box and each tube had a removable perforated galvanized iron sheet partition 5 cm away from the box to control rat access. An odorant presentation compartment partitioned by a perforated galvanized iron sheet from the other part of the tube was at the distal end of each 10 cm tube. We painted the male urine on a glass slide, and then placed the slide in the presentation compartment

sample (20 µL) was placed into the odorant presentation compartment of each tube. Female rats were simultaneously exposed to two urine samples from males conspecific to them, where one of the samples was from a male that had been housed with a conspecific male, and the other with a heterospecific male. We immediately opened the door to allow the females to freely respond to the urine. Between trials, the start box and choice tubes were cleaned thoroughly with water and 75% ethanol. All tests were recorded on video. Investigation times (i.e., the time each female spent in a choice tube) were determined from video replay using a Noldus ethovision XT system (Noldus, Wageningen, The Netherlands). We recorded the investigation time for 30 min in Experiment 1 and 1 h in Experiment 2 after the female initially entered either of the choice arms. If a test female did not enter either choice tube within 30 min (i.e., investigation time = 0), we did not use the data.

Blood and tissue sampling

Two days after urine collection, all rats were decapitated (within 3 min) and blood samples immediately collected. The hippocampus was immediately dissected, rapidly frozen in liquid nitrogen, and stored at -80 °C until use. Blood samples were incubated at 4 °C in a refrigerator for 12 h and then centrifuged at 3000 rpm for 15 min for serum collection. Serum samples were stored at -80 °C until use. In Experiment 1, individual hippocampi from

eight males were used for RNA isolation and quantitative real-time PCR; in Experiment 2, all of the hippocampi were used.

Cortisol and testosterone analysis

Serum samples were analyzed in duplicate for cortisol using an Iodine[125I] cortisol RIA kit and for testosterone using an Iodine[125I] Testosterone RIA Kit (Beijing North Institute of Biological Technology, China). In detail, 100 µL of iodine-125 labeled cortisol (or testosterone) was incubated with 50 µL of serum at 37 °C for 1 h in a water bath. Then, 500 µL of immune separating agent was added to each sample tube and samples incubated at room temperature for 15 min. Sample tubes were then centrifuged at 3800 rpm for 15 min and the supernatant discarded. The remaining radioactivity bound to the tube was measured using a gamma scintillation counter calibrated for iodine-125 using a radioimmunoassay system (XH6080, Xi'an Nuclear Instrument Factory, Xi'an, China). For both cortisol and testosterone, the intra-assay coefficients of variation were less than 10%, and the inter-assay coefficients of variation were less than 15%.

Quantitative real-time PCR

Isolation of total hippocampal RNA was performed using Trizol reagent (Invitrogen, Life Technologies, Grand Island, NY, USA) according to the manufacturer's instructions. Total RNA concentration was determined using a NanoDrop spectrophotometer (Thermo Fisher Scientific Inc., Waltham, USA). Reverse-transcription

Table 1 Primer sequences for real-time PCR

Gene	Forward Primer 5'–3'	Reverse Primer 5'–3'
GAPDH	GACAATGAATATGGCTAC AGCAAC	TTTATTGATGGTATTCGAGA GAAGG
GR	AGGCAGTGTGAAATTGTA TCCCAC	GAGGCTTACAATCCTCATTC GTGT
BDNF	GAAGGGCCAGGTCGATT AGGTG	GACGGAAACAGAACGAAC AGAA
TrkB	GAGACGAAATCCAGCCC CGACAC	CACAGACTTCCCTTCCTCCA CCG

was performed using a PrimeScript® RT reagent Kit With gDNA Eraser (Perfect Real Time) (Takara Bio Inc., Dalian, China), following the manufacturer's protocol. The resulting cDNA was amplified using an Mx3005P quantitative PCR system (Stratagene, La Jolla, CA, USA) and the relative abundance of the mRNA of the target genes determined using a SYBR Green RealMasterMix Kit (Tiangen, Beijing, China) according to the manufacturer's instructions. PCR primers were designed using NCBI Primer Blast (http://www.ncbi.nlm.nih.gov/tools/primer-blast) and the sequences are listed in Table 1. The housekeeping gene, GAPDH, was used as a control to normalize the relative mRNA levels. Data were analyzed as previously described [35].

Statistical analyses

The distributions of raw data were examined using Kolmogorov–Smirnov tests. If data were normally distributed, t-tests for paired-samples were used for the behavioral data and independent t-tests were used for the body weight, serum hormone and mRNA expression data. If the data were not normally distributed, Wilcoxon signed-rank and Mann–Whitney U tests were used. All statistical analyses were conducted using the SPSS software package (v15.0, SPSS Inc., Chicago, IL, USA). Alpha was set at $P \leq 0.05$.

Funding
This work was supported by grants from the National Natural Science Foundation of China (31572277), the Strategic Priority Research Program of the Chinese Academy of Sciences (No. XDB11010400), and the State Key Laboratory of Integrated Management of Pest Insects and Rodents (ChineseIPM1514).

Authors' contributions
JXZ and ZYH conceived and designed the experiments. HLG, ZYH, HJT, and JHZ performed the experiments YHZ, HLG, and JXZ wrote and revised the manuscript. All authors read and approved the final manuscript.

Competing interests
The authors declare that they have no competing interests.

Author details
[1]State Key Laboratory of Integrated Management of Pest Insects and Rodents in Agriculture, Institute of Zoology, Chinese Academy of Sciences, No.1-5 Beichen West Road, Chaoyang District, Beijing 100101, China. [2]College of Life Sciences, University of Chinese Academy of Sciences, Beijing 100049, China.

References
1. Musser GG, Carleton MD. Superfamily Muroidea. In: Wilson DE, Reeder DM, editors. Mammal species of the world a taxonomic and geographic reference. Baltimore: Johns Hopkins University Press; 2005. p. 894–1531.
2. Robins JH, McLenachan PA, Phillips MJ, Craig L, Ross HA, Matisoo-Smith E. Dating of divergences within the Rattus genus phylogeny using whole mitochondrial genomes. Mol Phylogen Evol. 2008;49:460–6.
3. Deinum EE, Halligan DL, Ness RW, Zhang Y-H, Cong L, Zhang J-X, Keightley PD. Recent evolution in rattus norvegicus is shaped by declining effective population size. Mol Biol Evol. 2015;32:2547–58.
4. Wu DL. Subspecies of the brown rat (Rattus norvegicus Berkenhout) in China. Acta Theriol Sin. 1982;2:107–12.
5. Zhu LB, Qian GZ. The relation between energetic regulation of two rats and their geographical distribution. Acta Theriol Sin. 1985;5:182.
6. Koh HS. Systematic studies on Korean rodents: VI. Analyses of morphometric characters, chromosomal karyotypes and mitochondrial DNA in two species of genus Rattus. Korean J Syst Zool. 1992;8:231–42.
7. Zhang MW, Guo C, Wang Y, Hu ZJ, Chen AG. A review of the studies on the Asian house rat (Rattus tanezumi) in China. Zool Res. 2000;21:487–97.
8. Zou B, Wang TL, Ning ZD, Liu S. The population of the Asian house rat (Rattus tanezumi) has been established in the Linfen District of Shanxi Province. Plant Prot. 1992;18:51.
9. Hou XF, Jiang XC. A survey of the Asian house rat (Rattus tanezumi) in Shijiazhuang District, Hebei Province from 2004 to 2007. Chin J Vector Biol Control. 2008;19:125.
10. Guo S. Population genetics of the Asian house rat in some areas of China. MSc thesis. Beijing: Chinese Center for Disease Control and Prevention; 2012.
11. Lack JB, Hamilton MJ, Braun JK, Mares MA, Van den Bussche RA. Comparative phylogeography of invasive Rattus rattus and Rattus norvegicus in the US reveals distinct colonization histories and dispersal. Biol Invasions. 2013;15:1067–87.
12. Bennett AE, Thomsen M, Strauss SY. Multiple mechanisms enable invasive species to suppress native species. Am J Bot. 2011;98:1086–94.
13. Huang HW, Zou ZT, Tian ZZ, Hu XY, Zhao Y. Test of resistance of rattus tanezumi to warfarin in Xingyi City, Guizhou Province, China. Chin J Vec Biol Contr. 2012;23:554–5.
14. Nielsen BL, Rampin O, Meunier N, Bombail V. Behavioral responses to odors from other species: introducing a complementary model of allelochemics involving vertebrates. Front Neurosci. 2015;9:226.
15. Teng HJ, Zhang YH, Shi CM, Zhang JX. Whole-genome sequencing reveals genetic variations of the Asian house rat. G3-Genes Genom Genet. 2016;6: 1969–77.
16. Amarasekare P. Interference competition and species coexistence. Proc R Soc B. 2002;269:2541–50.
17. Yom-Tov Y, Yom-Tov S, Moller H. Competition, coexistence, and adaptation amongst rodent invaders to Pacific and New Zealand islands. J Biogeogr. 1999;26:947–58.
18. Stuart YE, Campbell TS, Hohenlohe PA, Reynolds RG, Revell LJ, Losos JB. Rapid evolution of a native species following invasion by a congener. Science. 2014;346:463–6.
19. Simeonovska-Nikolova DM. Interspecific social interactions and behavioral responses of Apodemus agrarius and Apodemus flavicollis to conspecific and heterospecific odors. J Ethol. 2007;25:41–8.
20. Liesenjohann M, Liesenjohann T, Palme R, Eccard JA. Differential behavioural and endocrine responses of common voles (Microtus arvalis) to nest predators and resource competitors. BMC Ecol. 2013;13:33.
21. Schoener TW. Field experiments on interspecific competition. Am Nat. 1983; 122:240–85.
22. Harris MR, Siefferman L. Interspecific competition influences fitness benefits of assortative mating for territorial aggression in eastern bluebirds (sialia sialis). PLoS One. 2014;9:e88668.

23. Foster SP. Interspecific competitive interactions between *Rattus norvegicus* and *R. rattus*. *MSc thesis*. Hamilton: University of Waikato; 2010.

24. Vivanco JM, Bais HP, Stermitz FR, Thelen GC, Callaway RM. Biogeographical variation in community response to root allelochemistry: novel weapons and exotic invasion. Ecol Lett. 2004;7:285–92.

25. Apfelbach R, Blanchard CD, Blanchard RJ, Hayes RA, McGregor IS. The effects of predator odors in mammalian prey species: A review of field and laboratory studies. Neurosci Biobehav Rev. 2005;29:1123–44.

26. Sbarbati A, Osculati F. Allelochemical communication in vertebrates: kairomones, allomones and synomones. Cells Tissues Organs. 2006;183:206–19.

27. Hurst JL. The functions of urine marking in a free-living population of house mice, *Mus-domesticus rutty*. Anim Behav. 1987;35:1433–42.

28. Hurst JL. Female recognition and assessment of males through scent. Behav Brain Res. 2009;200:295–303.

29. Brennan PA, Zufall F. Pheromonal communication in vertebrates. Nature. 2006;444:308–15.

30. Johansson BG, Jones TM. The role of chemical communication in mate choice. Biol Rev. 2007;82:265–89.

31. Smadja C, Ganem G. Subspecies recognition in the house mouse: a study of two populations from the border of a hybrid zone. Behav Ecol. 2002;13:312–20.

32. Robertson DHL, Hurst JL, Searle JB, Gunduz I, Beynon RJ. Characterization and comparison of major urinary proteins from the house mouse, *Mus musculus domesticus*, and the aboriginal mouse, *Mus macedonicus*. J Chem Ecol. 2007;33:613–30.

33. Stopkova R, Stopka P, Janotova K, Jedelsky PL. Species-specific expression of major urinary proteins in the house mice (*Mus musculus musculus* and *Mus musculus domesticus*). J Chem Ecol. 2007;33:861–9.

34. Soini HA, Wiesler D, Koyama S, Feron C, Baudoin C, Novotny MV. Comparison of urinary scents of two related mouse species, *Mus spicilegus* and *Mus domesticus*. J Chem Ecol. 2009;35:580–9.

35. Mucignat-Caretta C, Redaelli M, Orsetti A, Perriat-Sanguinet M, Zagotto G, Ganem G. Urinary volatile molecules vary in males of the 2 European subspecies of the house mouse and their hybrids. Chem Senses. 2010;35:647–54.

36. Zhang YH, Zhang JX. Urine-derived key volatiles may signal genetic relatedness in male rats. Chem Senses. 2011;36:125–35.

37. Zhang YH, Zhang JX. A male pheromone-mediated trade-off between female preferences for genetic compatibility and sexual attractiveness in rats. Front Zool. 2014;11:73.

38. Zhang YH, Du YF, Zhang JX. Uropygial gland volatiles facilitate species recognition between two sympatric sibling bird species. Behav Ecol. 2013; 24:1271–8.

39. Roberts SC, Gosling LM. Genetic similarity and quality interact in mate choice decisions by female mice. Nat Genet. 2003;35:103–6.

40. Roberts SA, Simpson DM, Armstrong SD, Davidson AJ, Robertson DH, McLean L, Beynon RJ, Hurst JL. Darcin: a male pheromone that stimulates female memory and sexual attraction to an individual male's odour. BMC Biol. 2010;8:75.

41. Kumar V, Vasudevan A, Soh LJT, Le Min C, Vyas A, Zewail-Foote M, Guarraci FA. Sexual attractiveness in male rats is associated with greater concentration of major urinary proteins. Biol Reprod. 2014;91:150.

42. Dijkstra PD, Verzijden MN, Groothuis TGG, Hofmann HA. Divergent hormonal responses to social competition in closely related species of haplochromine cichlid fish. Horm Behav. 2012;61:518–26.

43. Zhang YH, Liang HC, Guo HL, Zhang JX. Exaggerated male pheromones in rats may increase predation cost. Curr Zool. 2015;62:431–37.

44. Zhang JX, Rao XP, Sun LX, Wang DW, Liu DZ, Zhao CH. Cohabitation impaired physiology, fitness and sex-related chemosignals in golden hamsters. Physiol Behav. 2008;93:1071–7.

45. Andersson M, Simmons LW. Sexual selection and mate choice. Trends Ecol Evol. 2006;21:296–302.

46. Drickamer LC, Gowaty PA, Holmes CM. Free female mate choice in house mice affects reproductive success and offspring viability and performance. Anim Behav. 2000;59:371–8.

47. Zhang JX, Sun LX, Bruce KE, Novotny MV. Chronic exposure of cat odor enhances aggression, urinary attractiveness and sex pheromones of mice. J Ethol. 2008;26:279–86.

48. Huo Y, Fang Q, Shi YL, Zhang YH, Zhang JX. Chronic exposure to a predator or its scent does not inhibit male-male competition in male mice lacking brain serotonin. Front Behav Neurosci. 2014;8:116.

49. Fang Q, Zhang YH, Shi YL, Zhang JH, Zhang JX. Individuality and transgenerational inheritance of social dominance and sex pheromones in isogenic male mice. J Exp Zool Part B. 2016;326:225–36.

50. Liu YJ, Zhang YH, Li LF, Du RQ, Zhang JH, Zhang JX. Cross-fostering of male mice subtly affects female olfactory preferences. PLoS One. 2016;11: e0146662.

51. Stankiewicz AM, Goscik J, Majewska A, Swiergiel AH, Juszczak GR. The effect of acute and chronic social stress on the hippocampal transcriptome in mice. PLoS One. 2015;10:e0142195.

52. Herman JP, Spencer R. Regulation of hippocampal glucocorticoid receptor gene transcription and protein expression in vivo. J Neurosci. 1998;18:7462–73.

53. Nibuya M, Takahashi M, Russell DS, Duman RS. Repeated stress increases catalytic TrkB mRNA in rat hippocampus. Neurosci Lett. 1999;267:81–4.

54. Wosiski-Kuhn M, Erion JR, Gomez-Sanchez EP, Gomez-Sanchez CE, Stranahan AM. Glucocorticoid receptor activation impairs hippocampal plasticity by suppressing BDNF expression in obese mice. Psychoneuroendocrinology. 2014;42:165–77.

55. Kirby ED, Muroy SE, Sun WG, Covarrubias D, Leong MJ, Barchas LA, Kaufer D. Acute stress enhances adult rat hippocampal neurogenesis and activation of newborn neurons via secreted astrocytic FGF2. Elife. 2013;2:e00362.

56. Saaltink DJ, Vreugdenhil E. Stress, glucocorticoid receptors, and adult neurogenesis: a balance between excitation and inhibition? Cell Mol Life Sci. 2014;71:2499–515.

57. Pizarro JM, Lumley LA, Medina W, Robison CL, Chang WLE, Alagappan A, Bah MJ, Dawood MY, Shah JD, Mark B, et al. Acute social defeat reduces neurotrophin expression in brain cortical and subcortical areas in mice. Brain Res. 2004;1025:10–20.

58. Murakami S, Imbe H, Morikawa Y, Kubo C, Senba E. Chronic stress, as well as acute stress, reduces BDNF mRNA expression in the rat hippocampus but less robustly. Neurosci Res. 2005;53:129–39.

59. Kitraki E, Karandrea D, Kittas C. Long-lasting effects of stress on glucocorticoid receptor gene expression in the rat brain. Neuroendocrinology. 1999;69:331–8.

60. Chen LJ, Shen BQ, Liu DD, Li ST. The effects of early-life predator stress on anxiety- and depression-like behaviors of adult rats. Neural Plas. 2014;2014: 163908.

61. Sanchez MM, Ladd CO, Plotsky PM. Early adverse experience as a developmental risk factor for later psychopathology: Evidence from rodent and primate models. Dev Psychopathol. 2001;13:419–49.

62. Wedell N, Tregenza T. Successful fathers sire successful sons. Evolution. 1999;53:620–5.

63. Kortet R, Hedrick A. The scent of dominance: female field crickets use odour to predict the outcome of male competition. Behav Ecol Sociobiol. 2005;59:77–83.

64. Taylor ML, Wedell N, Hosken DJ. The heritability of attractiveness. Curr Biol. 2007;17:R959–60.

65. Hosken DJ, Taylor ML, Hoyle K, Higgins S, Wedell N. Attractive males have greater success in sperm competition. Curr Biol. 2008;18:R553–4.

66. Guillaumet A, Leotard G. Annoying neighbors: multi-scale distribution determinants of two sympatric sibling species of birds. Curr Zool. 2015;61:10–22.

67. Kapusta J, Sales GD. Male-female interactions and ultrasonic vocalization in three sympatric species of voles during conspecific and heterospecific encounters. Behaviour. 2009;146:939–62.

68. Wohr M, Schwarting RKW. Affective communication in rodents: ultrasonic vocalizations as a tool for research on emotion and motivation. Cell Tissue Res. 2013;354:81–97.

69. Musolf K, Meindl S, Larsen AL, Kalcounis-Rueppell MC, Penn DJ. Ultrasonic vocalizations of male mice differ among species and females show assortative preferences for male calls. PLoS One. 2015;10:e0134123.

70. Cao WY, Duan J, Wang XQ, Zhong XL, Hu ZL, Huang FL, Wang HT, Zhang J, Li F, Zhang JY, et al. Early enriched environment induces an increased conversion of proBDNF to BDNF in the adult rat's hippocampus. Behav Brain Res. 2014;265:76–83.

71. Guo H, Fang Q, Huo Y, Zhang Y, Zhang J. Social dominance-related major urinary proteins and the regulatory mechanism in mice. Integr Zool. 2015; 10:543–54.

72. Afolabi AO, Alagbonsi IA, Oke OD. Early prenatal stress increases body weight and reduces nociception in adult male rats. Annu Res Rev Biol. 2014; 4:1431–8.

73. Karandrea D, Kittas C, Kitraki E. Forced swimming differentially affects male and female brain corticosteroid receptors. Neuroendocrinology. 2002;75: 217–26.

16

Edible dormice (*Glis glis*) avoid areas with a high density of their preferred food plant - the European beech

Jessica S. Cornils*[ID], Franz Hoelzl, Birgit Rotter, Claudia Bieber and Thomas Ruf

Abstract

Background: Numerous species, especially among rodents, are strongly affected by the availability of pulsed resources. The intermittent production of large seed crops in northern hemisphere tree species (e.g., beech *Fagus spec.*, oak *Quercus spec.*, pine trees *Pinus spec.*) are prime examples of these resource pulses. Adult edible dormice are highly dependent on high energy seeds to maximize their reproductive output. For juvenile dormice the energy rich food is important to grow and fatten in a very short time period prior to hibernation. While these erratic, often large-scale synchronized mast events provide overabundant seed availability, a total lack of seed production can be observed in so-called mast failure years. We hypothesized that dormice either switch territories between mast and non-mast years, to maximize energy availability or select habitats in which alternative food sources are also available (e.g., fleshy fruits, cones). To analyze the habitat preferences of edible dormice we performed environmental niche factor analyses (ENFA) for 9 years of capture-recapture data.

Results: As expected, the animals mainly used areas with high canopy closure and vertical stratification, probably to avoid predation. Surprisingly, we found that dormice avoided areas with high beech tree density, but in contrast preferred areas with a relatively high proportion of coniferous trees. Conifer cones and leaves can be an alternative food source for edible dormice and are less variable in availability.

Conclusion: Therefore, we conclude that edible dormice try to avoid areas with large fluctuations in food availability to be able to survive years without mast in their territory.

Keywords: Habitat preference, ENFA, Foraging, Rodent, Pulsed resources

Background

Pulsed resources, i.e., large-magnitude, low frequency, and short duration events of increased resource availability, have a huge effect on life-history traits of an individual (e.g., survival and reproduction) as well as on the dynamics of, populations and even whole ecosystems [1–3]. The intermittent production of large seed crops in northern hemisphere tree species (e.g., beech *Fagus spec.*, oak *Quercus spec.*, pine trees *Pinus spec.*) are prime examples of resource pulses. While these erratic, often large-scale synchronized mast events provide overabundant seed availability, a total lack of seed production can be observed in so-called mast failure years. Since these

tree species show unpredictable masting patterns and are unable to yield seeds in two consecutive years, the differences in food availability are extreme, changing from overabundant to completely absent especially in years following a full mast event [4, 5]. The phenomenon of mast synchrony can be explained by three widely tested and supported mechanisms. Firstly, trees may swamp seed predators with as many seeds as possible to enhance the chances of seedling survival (the predator satiation hypothesis, [6–9]). Secondly, seed predators that are swamped in mast years may actually cache more seeds than they are able to retrieve, which would benefit seed dispersal and germination [10]. The third explanation involves weather conditions, which may either enhance or impair pollination success directly, or may affect flowering because selection has favored plants that all respond to weather characteristics in the same way, resulting in high

* Correspondence: jessica.cornils@vetmeduni.ac.at
Department of Integrative Biology and Evolution, University of Veterinary Medicine, Savoyenstraße 1, 1160 Vienna, Austria

synchrony [11]. Under all these scenarios synchrony between individual plants maybe further enhanced by long-term effects of the depletion of resources in masting years.

Numerous species, especially among rodents, are strongly affected by the availability of pulsed resources. In mice (e.g., *Apodemus flavicollis*, *Apodemus sylvaticus*, *Peromyscus leucopus*, *Peromyscus maniculatus*) and the bank vole (*Myodes glareolus*), for example, mast events of beech and oak can cause a rapid population growth and an increased overwinter survival, while abundances are declining when the resource is depleting (e.g., [12–16]).

The impact of resource pulses on reproduction, survival, or hibernation patterns could be shown in several studies (e.g., [13, 17–20]). Eastern chipmunks (*Tamias striatus*) feast on seeds in autumn and store more nuts over the winter in mast years. Thus, this species can manage to raise two litters, with even higher juvenile survival in years with seed masting [21]. Interestingly, tree squirrels (both *Tamiasciurus hudsonicus* and *Sciurus vulgaris*) as well as the arboreal edible dormouse (*Glis glis*) are capable to anticipate future mast events; all three species increase their reproductive investment prior to the actual mast [17, 18, 22]. However, our knowledge on how the habitat choice in a species adapted to pulsed resources and is affected by strong fluctuations in food availability is very limited (but see [13, 23]). For consumers of seeds that show strong year-to-year variation in abundance, at least two different scenarios seem possible: (1) Switching territories between mast years and non-mast years. This "mast-tracking" option should be mostly available to species that are capable of travelling large distances, such as large mammals or birds [24, 25]. (2) Finding a habitat that, in addition to fluctuating seed resources, also provides alternative food sources (e.g., fleshy fruits, cones) in non-mast years. These alternative food sources should be most important for small, non-volant mammals, such as rodents. Year-to-year changes in the composition of food resources will be most relevant, however, for species that are long-lived enough to actually experience both mast seeding and mast failure years. Further, long-term effects of habitat characteristics evidently require a certain degree of site fidelity.

Among the seed-predating rodents, one species that appears to fulfill both of these criteria is the edible dormouse (*Glis glis*). Despite their small size (~100 g), these arboreal hibernators have a maximum longevity of 13 years [26], and hence may be exposed to varying masting situations. Further, previous studies have pointed to a high site-fidelity in edible dormice [18, 23, 27]. Especially for adult females it is known that they only travel as far as necessary from their nesting site to find suitable food [28].

In edible dormice mating occurs after hibernation only in mast years and juveniles are born very late in the active season (end of July to August; only one litter per year in central Europe), just in time with the ripening of beech seeds. Energy rich seeds are crucial for juveniles to grow and gain sufficient body fat stores before their first hibernation season [18, 23, 29]. Since these high caloric seeds are so important for juvenile survival, the optimal habitat for a reproductive female should include masting beech trees to maximize energy availability already during lactation. On the other hand, dormice have to cope in these habitats with low seed availabilities in mast failure years. One adaptation to this extremely reduced food availability is that they entirely skip reproduction in years with mast failures [18, 29]. Interestingly, however, survival rates in adult dormice are even higher in mast failure years than in reproductive years [18, 19, 23, 29]. While this is partly explained by extremely long hibernation seasons, during which the animals escape from predation by remaining hidden in their underground hibernacula [20, 23, 30], not all individuals can afford this strategy. Especially dormice with low body fat reserves have to stay active in mast failure years [20]. Indeed, the daily amount of time spent foraging did not differ in fall between a mast year and a mast failure year (Bieber et al. submitted).

For our analysis we used extreme situations, either non-mast or mast failure years with almost no trees producing seed buds/seeds at the whole study site, or full mast years, with almost all trees flowering. We used pollen densities as an indicator, but since pollen densities may not reflect seed densities in all species (e.g. [11]), we additionally confirmed seed densities visually in the field. However, edible dormice already have to know in spring whether a year will be a full or non-mast year, because growth of gonads and mating has to occur right away, to make sure ripe beech seeds are available to juveniles for prehibernation fattening [31]. Therefore the vast majority (~90%) of the adult dormice establish their territory already before July 15th (unpublished data), which is another indicator that they estimate the amount of future mast by assessing either pollen densities or seed bud densities [32]. Since pollen densities are highly correlated with the number of juveniles born per year (r = 0.97), both of these mechanisms are possible [33].

Adult edible dormice can gain weight relying on alternatives like leaves or fleshy fruits [23, 34]. A population of edible dormice in Crete has even been shown to survive in *Pinus brutia* forests, but it is not clear if these conifers are their only food source [35]. However, deciduous forests containing beech trees are the preferred habitat resulting in a higher lifetime reproductive success [36].

To date, there has been no systematic investigation of either site-fidelity or of the variables that determine habitat choice in this species. Therefore, we used data from a 9-year capture-recapture study of dormice encountered in a ~15 km² area in the Vienna woods to first determine movements of juveniles, yearling, and adults between nest-boxes. Data on nest-box occupation by a total of ~1100 dormice were also used to see whether all potential territories were used equally, or if certain nest-box locations were preferred. Finally, to assess which type of habitats, if any, are favored and to determine variables that define habitat suitability, we computed an environmental niche factor analysis (ENFA; [37]). This method compares the available niche in a defined space with the area the species is actually using [37, 38]. For this analysis we used capture–recapture data and determined which environmental factor (assessed via a forest inventory of the areas surrounding each nest-box) affected the distribution of individuals at our study site.

Methods

The study site was located close to St. Corona in the Vienna Woods (Lower Austria, 48°05'N/15°54'E; 400–600 m asl). The area (size ~15 km^2) is covered by a mixed forest with most of the site dominated by deciduous beech forest (~60% of the trees) and ~30% coniferous trees.

There were 124 wooden nest-boxes (fixed at 2–3 m height on trees) randomly distributed along the forest trails, which were checked for the presence of edible dormice every second week (May-October; 2006-2014). In the active season, edible dormice use these nest-boxes (in place of natural tree holes in primeval forests) to rest during the day and raise their young. Every newly captured dormouse was marked with a subcutaneously injected PIT-Tag transponder (BackHome BioTec®, 13.8 mm × 2.1 mm or Tierchip Dasmann®, 12.0 mm × 2.0 mm). All dormice were sexed and classified as either juvenile (J, before the first hibernation), yearling (Y, before the second hibernation period) or adult (A, after the second hibernation period) using fur color and size given in Schlund [39].

We recorded environmental variables using wide-ranging forestry based GIS data in a 100 m radius around each nest-box from 2006 with ArcGIS 9.1 ([40]; geographic information system; Table 1). This use of a 100 m radius was based on home ranges determined in three telemetry studies in populations with different densities of edible dormice [28, 41, 42]. Nest-box locations were obtained using a 12-channel GPS receiver (eTrex® Summit, GARMIN Corporation). Small-scale parameters were documented for a 30 m radius around each nest-box (Table 1). Tree species with a very low coverage of the area (mean under 2%) were excluded from the analysis. Herb cover was also excluded from the analysis, because the animals rarely dwell on the ground [23]. A total of seven variables were included in the model to explain the distribution of edible dormice (Table 1).

The density of beech and coniferous trees were included because their seeds and leaves are important food items of edible dormice [34]. Forest age, canopy closure,

Table 1 Environmental variables determined either from a GIS based forest inventory or small scale measurements in the forest

Source	Variable	Abbreviation	Unit
GIS based	Altitude	"alt"	m above sea level
100 m Radius	Forest age	"age"	years
	Fagus sylvatica density	"fag"	%
	Conifer density	"conifers"	%
	Slope	"slope"	degrees
Small scale	Canopy closure	"can"	25%, 50%, 75%, 100%
30 m Radius	Girth of nest-box tree at breast height	"girth"	cm

Canopy closure was measured by partitioning the space around the crown into four quadrants and considering the connection between this crown and neighboring trees. The variable is equal to 100%, if all four quadrants are connected

and slope of the hillside were used because these variables affect the structure and stratification of the forest, which may affect both foraging opportunities and predation risk of the animals. We also included the girths of the trees (as a proxy for both age and height) at which the nest-boxes were mounted on, because dormice may select habitats based on the suitability of the immediate nesting site. Further, we included altitude of the location, as even small differences in elevation can have an influence on the microclimate of the forest.

Statistics

Movements between areas

To analyze dormouse movements between nest-boxes, we calculated distances travelled as well as the time between capture and recapture, the mean and maximum number of captures, mean total distance and the number of nest-boxes used. Animals that had not been recaptured were excluded from the analyses.

To investigate to what extent edible dormice move between nest-boxes in the different age classes, we computed a linear mixed effects model [43] to analyze the mean distance the animals covered using sex, age and timespan between captures as explanatory variables. To adjust for repeated measurements the individual ID was used as a random effect. The response variable distance was log-transformed to achieve a normal distribution of the model residuals, which was confirmed with a Shapiro-Wilks test. To include mast in our analysis we also calculated a linear mixed effects model [43] containing sex, mast and timespan. We could not include age here, because the category juveniles and yearling, were directly associated with full and non-mast year, respectively. We also included individual ID as a random factor in this model and log transformed the response variable mean distance. Subsequently we calculated a type 3 anova in both models. We used a chi^2 test to test if the nest-box occupation was equally distributed over the whole study site.

Environmental niche factor analysis

The environmental preferences of the edible dormice at our study area were calculated using the ENFA approach. The number of captures per nest-box was used as the response variable. This model is based on the concept of the ecological niche and compares the available niche to the niche the species is using. The advantage of this analysis, compared to classical methods, is that it is solely based on presence data. A principal component analysis (PCA) as the first step ensures that the appropriate weights and transformations are provided for the subsequent ENFA. The ENFA summarizes variables into a few uncorrelated factors (as does the PCA), but the so- called marginality and specialization axes have ecological meaning (for details see [38]). We made sure that the environmental variables we included in the ENFA were not highly correlated (all r < 0.5).

In ENFA, the marginality axis is the direction on which the species niche differs at most from the available conditions. After removal of the marginality, a specialization factor can be determined by computing the direction that maximizes the ratio of the variance of the global distribution to that of the species distribution. This determination of the specialization axes is repeated until all the information has been explained [44]. A large specialization corresponds to a narrow niche relative to the habitat conditions available to the species.

For the illustration of the ecological niche in a defined space, we used biplots with one marginality axis and the first specialization axis [38, 44]. The environmental variables were plotted as arrows, where the length of the arrow is a measure of the influence of the variable on the position of the niche in the available habitat. That means the longer an arrow, the more important it is for the explanation of the marginality axis [44]. For the coefficients of marginality a positive value means that the species prefers values higher than the mean, while a negative coefficient indicates a preference for lower values with respect to the study area [37]. The specialization factors have to be handled as absolute values, they represent the variance ratio of the variables. The higher this factor, the higher the degree of specialization with respect to the variable, signs are arbitrary [37]. Since the first specialization axis explains most of the variance (also seen in the histograms in the upper right corner of every ENFA analysis; Figs. 3 and 4) we only used this first axis for the individual biplots. To test if our defined marginality and specialization axes were significant we performed a Monte-Carlo randomization test with 1000 permutations. All analyses were carried out using R version 3.1.1 [45] with the package 'adehabitatHS' [46]. We

calculated the overall situation for all years and conditions (Fig. 3), performed separate ENFA's for female and male dormice (Fig. 4a + b), and additionally for the two extreme food situations in two non-mast and two full mast years (Fig. 4c + d).

Simulations

To investigate the effect of the presence of multiple tree species on the overall year-to-year variability in seed availability, we simulated time series with a given coefficient of variation using the R function rnorm. To test the effect of different degrees of synchrony in seed production between tree species on mean temporal variability, we simulated correlated time series (length 200 years, CV 0.5), with correlation coefficients randomly varying around given desired r-values, ranging from 0 to 1 (Fig. 5b). Each point was determined as the mean from 1000 repeats.

Results

Nest-box occupation and movements

From 2006 to 2014 we caught 1189 individual dormice at our study site. Overall there were 5950 capture events and the mean number of nest-boxes used per animal was 1.58 ± 0.02. One animal was captured 29 times over the course of nine years; the mean number of captures was 3.26 ± 0.06. Overall, a high proportion of animals were captured in the same (52.7%) or neighboring nest-boxes over the years, with especially adult females showing high site fidelity, by staying in the same box. Juvenile dormice and particularly juvenile males had a higher tendency for dispersal and moving longer distances into other areas of the study site, or presumably also into the adjacent forest areas outside of our study area (sex: F = 16.11, P < 0.001; age: F = 36.01, P < 0.001, sex:age: F = 19.33, P < 0.001, timespan: F = 32.57, P < 0.001; Fig. 1).

The mean distance travelled was 49.44 ± 2.62 m for all age and sex categories, also indicating high overall site fidelity. In the second model, were we included mast, sex still had a significant influence (sex: F = 7.71, P < 0.007), but there was no significant result for timespan (F = 2.69, P = 0.134), but a significant difference between non-mast and mast years (F = 12.15, P < 0.005), with adult animals travelling mean distances of 87.7 ± 73.02 m in full mast years and 125.4 ± 91.6 m in non-mast years.

Nest-box occupation was unequally distributed over the study site (χ^2 = 632.26, df = 119, P < 0.001). There were sections of the study site in which nest-boxes were more frequented than in other parts, and some of the nest-boxes were only used rarely during the time of the study (Fig. 2).

Environmental influences

To explain this unequal distribution of animals throughout the forest and to assess which factors influenced

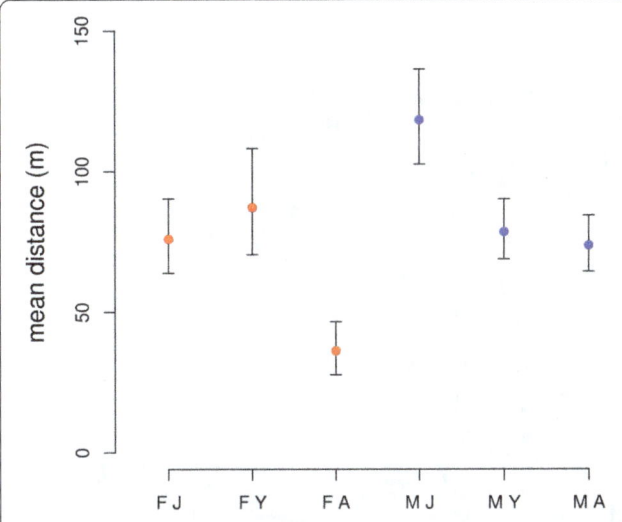

Fig. 1 Mean distance ± SEM of relocations in the different age classes of edible dormice (FJ = juvenile female; FY = yearling female; FA = adult female; MJ = juvenile male; MY = yearling male; MA = adult male)

their site fidelity we performed different ENFA's. These analyses showed that the eigenvalues of the specialization decreased after the first axis for all of the tests performed. Hence, the first axis explained most of the specialization, allowing us to use the first specialization axis only, in all of the computed ENFA's (Figs. 3 and 4).

However, the deviation between available and used niche was moderate in all of the analyses (distance of the available niche to the centroid: overall = 0.1097, non-mast = 0.3801, full-mast = 0.1333, females = 0.1077, males = 0.1234; see also white dots (x-Axis in Figs. 3 and 4)). Especially for the overall analysis, for males and for full-mast years there was an almost complete overlap between used and available

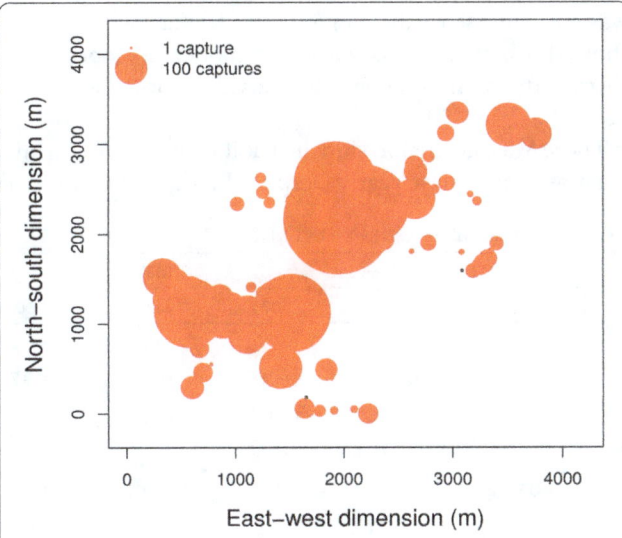

Fig. 2 Distribution of captured animals in the nest-boxes from 2006 to 2014. The sizes of the dots reflect the number of captures

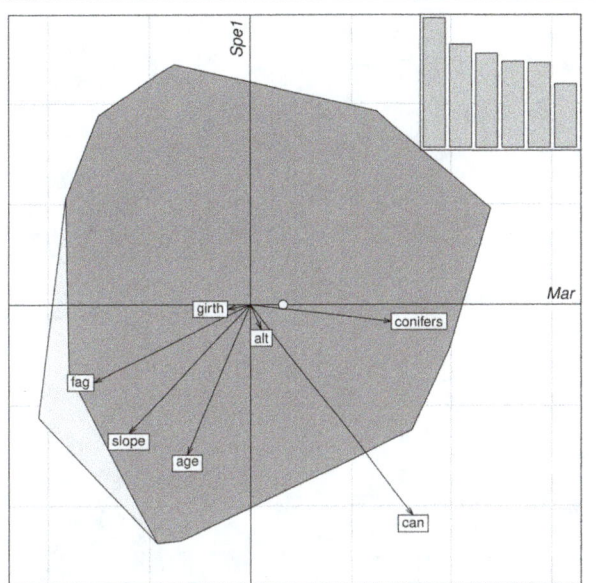

Fig. 3 Result of the overall ENFA carried out to determine the relationship between environmental variables and the distribution of edible dormice in the study area (years 2006–2014). The eigenvalue diagram of the analysis in the *upper right* corner shows the contribution of each specialization axis to the overall specialization, were each barplot represents one specialization axis (*Spe* 1–6; only). The biplot for the analysis is formed by the marginality (x-Axis; *Mar*) and the first of these specialization axes (y-Axis, *Spe1*), which explains most of the variance. The *light grey* area represents the minimum convex polygon enclosing all the projections of the available habitat, whereas the *dark grey* area corresponds to the habitat used by the animals. The *white dot* represents the centroid of the used habitat, while the origin of the plot is the centroid of the available sites. The environmental variables are projected via the *arrows*. The longer an *arrow*, the more important it is for the explanation of the marginality axis. The arrows that have the biggest angle from the marginality axis have the highest specialization, signs are arbitrary in this case. Environmental variable abbreviations: alt = Altitude; age = Forest age; fag = *Fagus sylvatica* density; conifers = Conifer density; slope = Slope; can = Canopy closure; girth = Girth of the nest-box tree. For further explanations of the variables see Table 1

niches (Figs. 3 and 4b and d). When the analysis was restricted to non-mast years or to females only, the overlap was far less pronounced (Fig. 4a and c).

In the overall analysis, canopy closure (preferred), beech tree proportion (high densities avoided), conifer density (high density preferred) and slope of the terrain (large slopes avoided) were the most important variables that defined the used habitat. The highest specialization values in the overall analysis showed that edible dormice were restricted to a limited range of areas with high canopy closure, relatively young trees (at an age of seed production onset) and moderate slopes (i.e., a low tolerance of variation in these three variables, see Table 2 and Fig. 3). The specialization value for conifer density, however, was very low as illustrated by the short distance to

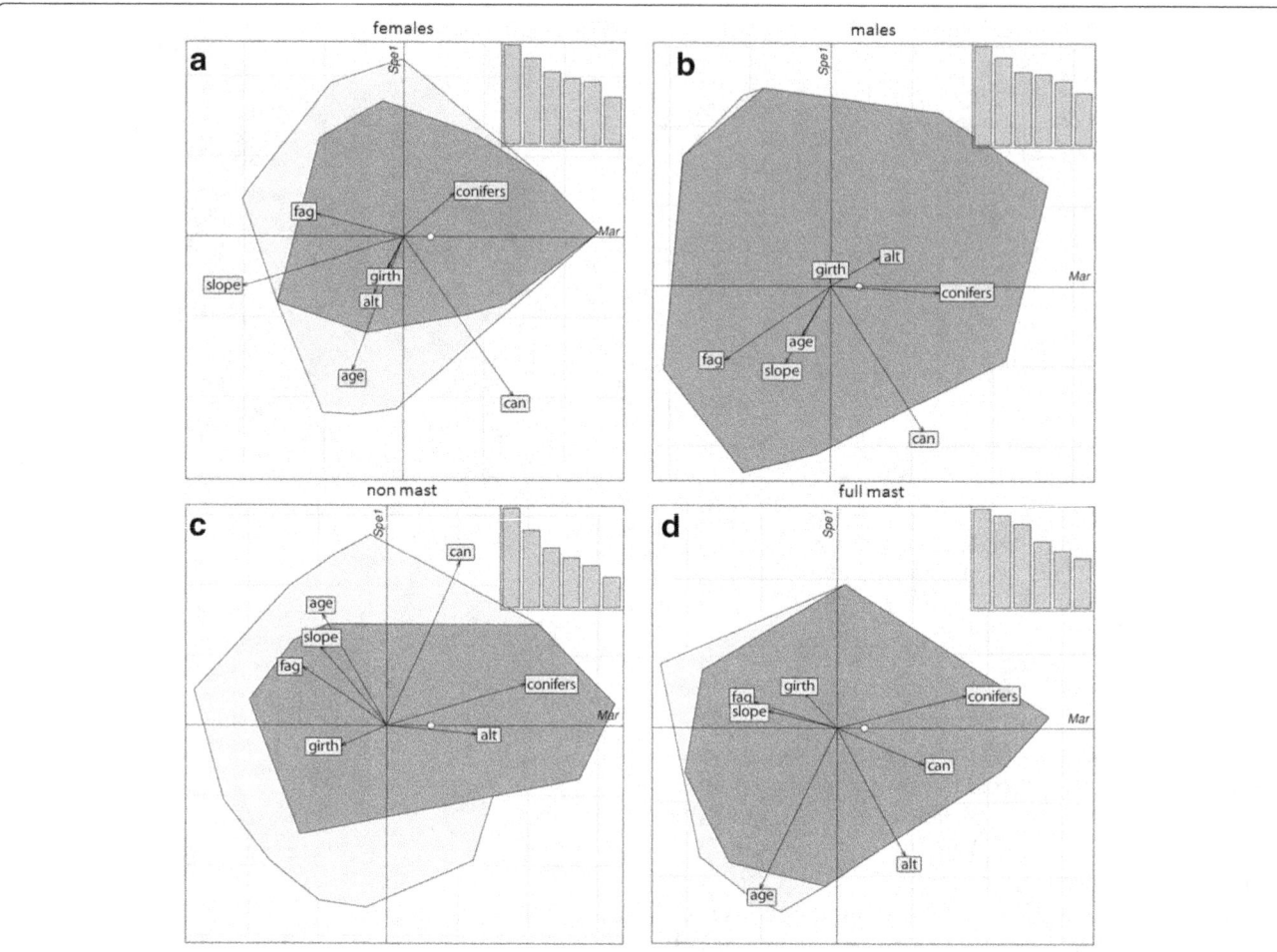

Fig. 4 Results of the ENFA's carried out to determine the relationship between environmental variables and the distribution of edible dormice in **a** two non-mast years 2012 + 2014 **b** two full-mast years 2011 + 2013 **c** only females 2006–2014 and **d** only males 2006–2014. For the detailed explanation of the graphical parameters see Fig. 3

the x-Axis in Fig. 3. In other words, the animals used a wide range of the available conifer cover.

Males avoided areas with high densities of beech trees more than females (Fig. 4 a + b). In addition males preferred areas with high density of coniferous trees, whereas in females the preference for high conifer density was less pronounced. The specialization of females on forest stands

of lower age was more prominent than among males. This was in contrast to the slope of the terrain, where males showed a higher specialization on moderate slopes. Both sexes had a high preference for closed canopy (Fig. 4a + b, Table 2).

There was no major shift in habitat use between full-mast and non-mast years. Despite the difference in food

Table 2 Coefficient values for all calculated ENFA models for the seven environmental variables included

	overall		Non-mast		Full-mast		Females		Males	
	Mar	Spe1	Mar	Spe1	Mar	Spe1	Mar	Spe1	Mar	Spe1
fag	**−0.52**	−0.26	**−0.38**	0.25	**−0.38**	0.12	**−0.39**	0.09	**−0.55**	−0.39
can	**0.54**	−0.70	0.32	0.72	**0.39**	−0.17	**0.48**	−0.69	**0.48**	−0.76
alt	0.04	−0.08	**0.39**	−0.03	0.31	−0.59	−0.13	−0.25	0.25	0.16
age	−0.21	−0.50	−0.29	0.49	−0.34	−0.74	−0.23	−0.58	−0.15	−0.26
girth	−0.07	−0.02	−0.21	−0.09	−0.15	0.16	−0.07	−0.14	−0.01	0.05
slope	−0.40	−0.43	−0.3	0.34	−0.32	0.07	**−0.71**	−0.21	−0.23	−0.40
conifers	**0.47**	−0.06	**0.6**	0.18	**0.59**	0.14	0.22	0.19	**0.56**	−0.03

The three highest values concerning the marginality are printed bold for all models. Abbreviations same as for Table 1 availability in those types of years, dormice remained in habitats with a relatively large density of conifers and a lower density of beech trees (Fig. 4c + d).

Discussion
Site fidelity
Our long-term analysis indicates that edible dormice, especially adult females, show high site fidelity and often stay at the same site over several seasons (Fig. 1). Adult males also showed high overall site fidelity (Fig. 1) and only moved over slightly longer distances in non-mast years, probably due to lower above-ground abundance and competition for good territories in those years (see below). As is typical for rodents in general [47], the only group with an above-average tendency for dispersal was juvenile males, which covered longer distances to explore new territories (Fig. 1). They face a trade-off in the year of their birth between investing in fattening in the mothers territory or exploring new habitats. Dispersal may improve the chances of reproduction for juvenile males in the subsequent year, but may also lead to a higher risk of predation in a foreign territory or during dispersal [47–49]. Our data on the high overall site fidelity of adult dormice confirm previous, shorter studies on this question [18, 27, 35] and further demonstrate that our environmental analysis around the nest-boxes most likely covered the majority of the animals' home ranges.

Habitat choice
The ENFA analysis showed a large general overlap of the used and available niche space (Figs. 3 and 4), indicating that most of the areas around the nest-boxes were a suitable habitat for edible dormice. More importantly, our results indicate that sub-areas of our study site were sufficiently heterogeneous to allow us to identify several habitat characteristics that are clearly preferred or avoided. The most surprising result of our environmental niche analysis was that edible dormice, despite their dependency on beechnut availability for reproduction, avoid forest stands with high beech density and prefer areas with a large proportion of coniferous trees. It has long been known that conifer cones and leaves can be an alternative food source, but there was never an indication for a preference of conifer forest stands [34, 50]. Coniferous trees also have fluctuating masting events, and coefficients of variation (CV) in seed production in individual conifer species are not smaller than in beech (review in [6]). However, in most forests in the distribution range of edible dormice there is only one beech species (*Fagus sylvatica*) but often there are several species of conifers. It can be shown that, when the number of conifer species reaches 4, as was the case at our study site, their collective coefficient of variation in seed production is only half that of

a single tree species, even if all species have the same CV individually (Fig. 5). Since edible dormice can forage on different species of conifers, this results in conifers being a more stable food source. This effect will be reduced if seed production among tree species varies synchronously among conifers, but this degree of synchrony seems only moderate (r < =0.5; [51]). At low to moderate levels of synchrony between tree species, the effect of the presence of multiple species on reducing see variability is still strong (Fig. 5b). Accordingly, Ruf et al. [18] found that dormice in mixed beech and conifer forests survived even better after years of reproduction skipping than dormice in forests dominated by beech. Hence, it seems the optimal habitat for dormice are forest stands that provide a fairly steady food resource such as various conifer seeds, interspersed with a relatively low proportion of trees that

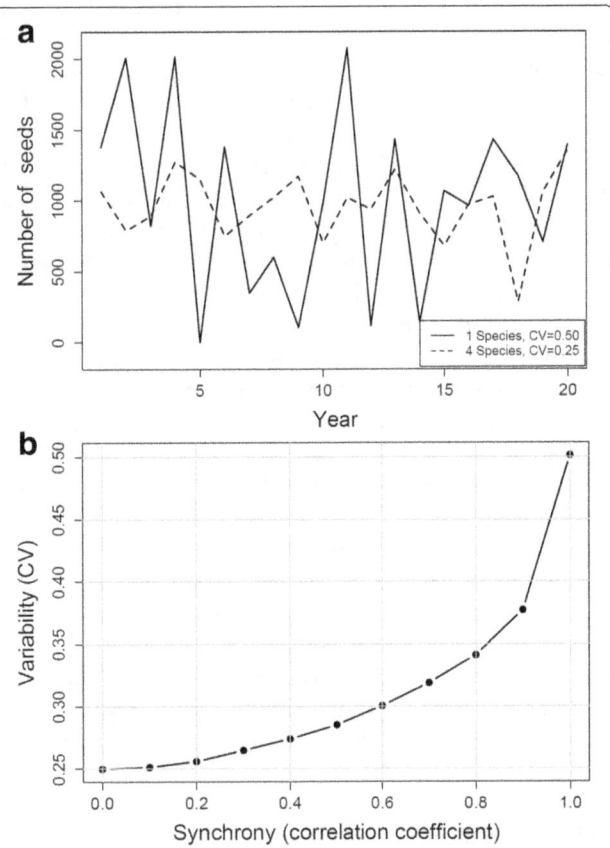

Fig. 5 a Simulated seeding variability over a 20 year period. The *solid line* shows seeding fluctuations in a single tree species (e.g. beech) showing a coefficient of variation (CV) of 0.5. The *dashed line* shows the collective (mean) fluctuation of 4 tree species, e.g. conifers, which individually show the same CV (0.5) and vary independently. The collective CV of 4 species is reduced by 50% (CV = 0.25). **b** Simulated effect of different degrees of synchrony between 4 tree species on mean seed variability. All species show individually the same CV (0.5). Note that even at a correlation coefficient of 0.5, the mean CV (0.28) is strongly reduced below 0.5

show variable masting but produce large seeds, namely beechnuts or acorn, with high energy content [34].

Edible dormice have an alternative option to either stay in their territory and live with lower food availability or move to another site. They can respond to the absence of beech seed not only by foregoing reproduction in certain years, but can also estivate throughout summer, without any food intake [20]. In dormice, states of dormancy in underground burrows are associated with extremely high survival rates, which are thought to reflect low predation risk [19, 20, 52, 53]. Indeed, while remaining largely motionless and odorless in closed hibernacula located ~50 cm below the surface [36], dormice are protected from most predators, particularly from owls, their main predators [35]. Predator avoidance is also a central factor in the high longevity of *Glis glis*, with free-living animals reaching an age of up to 13 years [26, 30].

Because reducing the risk of external mortality is an essential characteristic of the life-history of edible dormice, we suggest that several of the other habitat preferences identified here are also related to minimizing predation risk. The amount of time terrestrial animals spend foraging is mainly influenced by the availability of habitat structures that lower predation risk [54]. Not surprisingly then, one of the most suitable habitats identified by the overall ENFA seem to be closed canopy forests with 75 to 100% closure. Closed canopies should be advantageous in terms of predation avoidance and foraging, as they hamper attacks by birds of prey from above, and allow the animals to move easily between adjacent trees. Further, large birds of prey like owls with wingspans of up to 250 cm (*Strix aluco*) cannot maneuver quickly in forest areas with dense vegetation and therefore prefer open, mature forests as their major hunting grounds [55, 56]. This factor would explain why dormice prefer younger over old-growth forests, as younger stands show higher vertical stratification and mid-canopies [57], which are avoided by owls [55]. Finally, predator avoidance could also explain why dormice avoid stands on steep slopes, because slopes cause layering and vertical opening of the canopy, which may lead to increasing hunting opportunities for birds of prey, especially because owls hunt better under illuminated conditions [58–60].

Males vs females

Females seem to use a slightly smaller proportion of the available habitat (convex polygons; Fig. 4a + b). This higher degree of specialization among females may well be related to their higher costs of reproduction, which requires optimal food resources and foraging conditions. There was a tendency of males to prefer conifer stands and avoid areas with high beech tree proportions even more than females (Table 1). This can be explained by the fact that females share territories with their juveniles, which are highly dependent on energy-rich beechnuts to gain sufficient weight before hibernation [18, 29]. Females with offspring also show a heightened aggressiveness and vigorously defend nest-boxes against intruders in mast years with reproduction (J.S.C. unpublished observation). This observation is in line with the high site-fidelity of adult females (Fig. 1). If females defend their nesting sites this should lead to reduced shelter availability for males (and non-reproductive females) after mating, until weaning of the juveniles. Consequently, males may be forced into territories with higher conifer density, which also provide a good canopy closure, and may have to alter their foraging behavior. This diversification of foraging preferences with different amounts of beech availability was also found by Schlund et al. [31], who detected similar densities of edible dormice in a beech forest with 70% and a coniferous mixed forest with 20% beech trees. The rate of juveniles per female however was far lower in the coniferous-mixed forest than in the beech forest [31], matching our finding of less avoidance of beech among adult females.

Non-mast vs full-mast

In non-mast years the occupied niche was a smaller fraction of the available niche than in full-mast years (Fig. 4c + d), indicating that dormice were apparently more selective in years of mast failure. We attribute this observation to the fact that in non-mast (non-reproductive) years the number of dormice occupying nest-boxes during the active season is ~50% smaller than in mast/reproductive years [19, 20, 26]. This is because in non-mast years a large fraction of the animals, in particular those individuals that have high body fat reserves in spring, retreat to underground burrows for estivation [20]. Hence the abundance of dormice above-ground will significantly differ between years [19, 20]. This provides those animals that remain active and foraging with the opportunity to choose among a larger number of unoccupied nest-boxes in good habitats, which is also reflected by movements over larger distances in non-mast years.

There was, however, no noticeable change in the preference or avoidance of specific habitat characteristics between mast and non-mast years (Table 1), despite lower abundances and a higher mean distance travelled. Together with the overall high site fidelity this finding suggests that, rather than switching between territories with different characteristics, dormice tend to occupy and remain in areas which provide optimal long-term conditions, and which buffer short-term fluctuations in mast seeding.

Conclusions

We showed for the first time that edible dormice avoid forest stands with a high density of beech, likely to evade exposure to large fluctuations in food resources caused by extremely pulsed beech seeding, following the theory of risk sensitive foraging (i.e., risk averse; [61]). This behavior is still fully compatible with the fact that dormice require energy-rich seeds for successful reproduction. It has been estimated that the amount of seeds produced by beech or oak in a full-mast year is ample enough to allow all granivores in a deciduous forest to live ad libitum on beechnuts or acorn alone [62]. Accordingly, a single beech tree in a dormice territory is almost certainly sufficient to provide a female and its offspring with adequate food resources for growth and prehibernation fattening [63]. Interestingly, most other habitat preferences of dormice, such as closed canopies and younger stands with vertical stratification appear to be related to minimizing predation risk, which is a main reason for animals to switch foraging grounds [54]. This points to a potential tradeoff between optimizing resource allocation and predator avoidance, which would be expected from the optimal foraging theory, (e.g. [58, 64, 65]), but deserves further investigation in this species.

Acknowledgements
We thank Karin Lebl for providing us with data from the field. We are grateful to Österreichische Bundesforste AG for their general support for the project and their permission to access the study site and the State of Lower Austria for providing financial support. We also thank Rudolf Litschauer from the Unit of Ecological Genetics and Biodiversity at the Austrian Research Centre for Forests for providing valuable pollen data, Renate Hengsberger for her help with the literature search and Steve Smith for correcting the English.

Funding
This study was financially supported by the Austrian Science Fund (FWF grant no. P25023).

Authors' contributions
TR and CB designed and supervised the study. FH, JSC and BR carried out the fieldwork. TR and JSC analyzed the data and wrote the manuscript. All authors discussed the results and commented on the manuscript. All authors read and approved the final manuscript.

Competing interests
The authors declare that they have no competing interests.

References
1. Yang LH, Bastow JL, Spence KO, Wright AN. What can we learn from resource pulses? Ecology. 2008;89:621–34.
2. Ostfeld RS, Keesing F. Pulsed resources and community dynamics of consumers in terrestrial ecosystems. Trends Ecol Evol. 2000;15:232–7.
3. Bogdziewicz M, Zwolak R, Crone EE. How do vertebrates respond to mast seeding? Oikos. 2016;125:300–7.
4. Holmsgaard E, Olsen HC. Experimental induction of flowering in Beech. Forstliches Forgøgsvaes Danm. 1966;30:1–17.
5. Lindquist B. The ecology of Scandinavian beechwoods. In Volume 29: Svenska skogsvårdsföreningens tidskrift 1931: 486–520
6. Kelly D, Sork VL. Mast seeding in perennial plants: Why, How, Where? Annu Rev Ecol Syst. 2002;33:427–47.
7. Janzen DH. Seed predation by animals. Annu Rev Ecol Syst. 1971;2:465–92.
8. Jansen PA, Bongers F, Hemerik L. Seed mass and mast seeding enhance dispersal by a neotropical scatter-hoarding rodent. Ecol Monogr. 2004;74:569–89.

9. Kelly D. The evolutionary ecology of mast seeding. Trends Ecol Evol. 1994;9:465–70.
10. Crone EE, Rapp JM. Resource depletion, pollen coupling, and the ecology of mast seeding. Ann N Y Acad Sci. 2014;1322:21–34.
11. Pearse IS, Koenig WD, Kelly D. Mechanisms of mast seeding: resources, weather, cues, and selection. New Phytol. 2016;212:546–62.
12. Pucek Z, Jędrzejewski W, Jędrzejewska B, Pucek M. Rodent population dynamics in a primeval deciduous forest (Białowieża National Park) in relation to weather, seed crop, and predation. Acta Theriol. 1993;38:199–232.
13. Zwolak R, Bogdziewicz M, Rychlik L. Beech masting modifies the response of rodents to forest management. For Ecol Manag. 2016;359:268–76.
14. Ostfeld RS, Jones CG, Wolff JO. Of mice and mast. Ecological connections in eastern deciduous forests. Bioscience. 1996;46:323–30.
15. Elias SP, Witham JW, Hunter J. Peromyscus leucopus abundance and acorn mast: population fluctuation patterns over 20 years. J Mammal. 2004;85:743–7.
16. Falls JB, Falls EA, Fryxell JM. Fluctuations of deer mice in Ontario in relation to seed crops. Ecol Monogr. 2007;77:19–32.
17. Boutin S, Wauters LA, McAdam AG, Humphries MM, Tosi G, Dhondt AA. Anticipatory Reproduction and Population Growth in Seed Predators. Science. 2006;314:1928–30.
18. Ruf T, Fietz J, Schlund W, Bieber C. High survival in poor years: life history tactics adapted to mast seeding in the edible dormouse. Ecology. 2006;87:372–81.
19. Lebl K, Bieber C, Adamík P, Fietz J, Morris P, Pilastro A, Ruf T. Survival rates in a small hibernator, the edible dormouse: a comparison across Europe. Ecography. 2011;34:683–92.
20. Hoelzl F, Bieber C, Cornils JS, Gerritsmann H, Stalder GL, Walzer C, Ruf T. How to spend the summer? Free-living dormice (Glis glis) can hibernate for 11 months in non-reproductive years. J Comp Physiol B. 2015;185:931–9.
21. Bergeron P, Réale D, Humphries MM, Garant D. Anticipation and tracking of pulsed resources drive population dynamics in eastern chipmunks. Ecology. 2011;92:2027–34.
22. Bieber C. Population dynamics, sexual activity, and reproduction failure in the fat dormouse (Myoxus glis). J Zool (Lond). 1998;244:223–9.
23. Bieber C, Ruf T. Habitat differences affect life history tactics of a pulsed resource consumer, the edible dormouse (Glis glis). Popul Ecol. 2009;51:481–92.
24. Singer FJ, Otto DK, Tipton AR, Hable CP. Home ranges, movements, and habitat use of European Wild Boar in Tennessee. J Wildl Manag. 1981;45:343–53.
25. Curran LM, Leighton M. Vertebrate responses to spatiotemporal variation in seed production of mast-fruiting Dipterocarpaceae. Ecol Monogr. 2000;70:101–28.
26. Trout RC, Brooks S, Morris P. Nest box usage by old edible dormice (Glis glis) in breeding and non-breeding years. Folia Zool. 2015;64:320–4.
27. Schlund W, Scharfe F, Stauss MJ, Burkhardt JF. Habitat fidelity and habitat utilization of an arboreal mammal (Myoxus glis) in two different forests. Mamm Biol. 1997;62:158–71.
28. Morris PA, Hoodless A. Movements and hibernaculum site in the fat dormouse (Glis glis). J Zool (Lond). 1992;228:685–7.
29. Pilastro A, Tavecchia G, Marin G. Long living and reproduction skipping in the fat dormouse. Ecology. 2003;84:1784–92.
30. Turbill C, Bieber C, Ruf T. Hibernation is associated with increased survival and the evolution of slow life histories among mammals. Proc R Soc B. 2011;278:3355–63.
31. Schlund W, Scharfe F, Ganzhorn JU. Long-term comparison of food availability and reproduction in the edible dormouse (Glis glis). Mamm Biol. 2002;67:219–23.
32. Lebl K, Kürbisch K, Bieber C, Ruf T. Energy or information? The role of seed availability for reproductive decisions in edible dormice. J Comp Physiol B. 2010;180:447–56.
33. Bieber C, Ruf T. Seasonal Timing of Reproduction and Hibernation in the Edible Dormouse (Glis glis). In: Barnes BM, Carey HV, editors. Life in the Cold V: Evolution, Mechanism, Adaptation, and Application, Twelfth International Hibernation Symposium. Fairbanks: Institute of Arctic Biology, University of Alaska Fairbanks; 2004. p. 113–125.
34. Fietz J, Pflug M, Schlund W, Tataruch F. Influences of the feeding ecology on body mass and possible implications for reproduction in the edible dormouse (Glis glis). J Comp Physiol B. 2005;175:45–55.
35. Vietinghoff-Riesch AF. Der Siebenschläfer. Jena: Gustav Fischer Verlag; 1960.
36. Bieber C, Ruf T. Summer dormancy in edible dormice (Glis glis) without energetic constraints. Naturwissenschaften. 2009;96:165–71.

37. Hirzel AH, Hausser J, Chessel D, Perrin N. Ecological-niche factor analysis: how to compute habitat-suitability maps without absence data? Ecology. 2002;83:2027–36.

38. Basille M, Calenge C, Marboutin E, Andersen R, Gaillard JM. Assessing habitat selection using multivariate statistics: some refinements of the ecological-niche factor analysis. Ecol Model. 2008;211:233–40.

39. Schlund W. Die Tibialänge als Maß für Körpergröße und als Hilfsmittel zur Altersbestimmung bei Siebenschläfern (*Myoxus glis* L.). Mamm Biol. 1997;62:187–90.

40. ESRI. ArcGIS desktop: release 10. Redlands: Environmental Systems Research Institute Inc; 2011.

41. Hönel B. Raumnutzung und Sozialsystem freilebender Siebenschläfer (*Glis glis* L.). PhD thesis, University of Karlsruhe (Germany); 1991.

42. Jurczyszyn M. The use of space by translocated edible dormice, *Glis glis* (L.), at the site of their original capture and the site of their release: Radio-tracking method applied in a reintroduction experiment. Pol J Ecol. 2006;54:345–50.

43. Pinheiro J, Bates D, DebRoy S, Sarkar D, and the R Development Core team: nlme: Linear and Nonlinear Mixed Effects Models (R Package Version 3.1-124). 2016.

44. Calenge C, Basille M. A general framework for the statistical exploration of the ecological niche. J Theor Biol. 2008;252:674–85.

45. R Core Team. R: a language and environment for statistical computing. 321st ed. Vienna: R Foundation for Statistical Computing; 2014.

46. Calenge C. The package "adehabitat" for the R software: a tool for the analysis of space and habitat use by animals. Ecol Model. 2006;197:516–9.

47. Wolff JO, Sherman PW. Rodent societies. An ecological & evolutionary perspective. Chicago: University of Chicago Press; 2007.

48. Koppmann-Rumpf B, Scherbaum-Heberer C, Schmidt KH. Influence of mortality and dispersal on sex ratio of the edible dormouse (*Glis glis*). Folia Zool. 2015;64:316–9.

49. Solomon NG. A reexamination of factors influencing philopatry in rodents. J Mammal. 2003;84:1182–97.

50. Santini L. Biology, damage and control of the edible dormouse (*Glis glis* L.) in central Italy. In: Howard WE, Marsh RE editors. Proceedings of the 8th Vertebrate Pest Conference. Sacramento: University of California, Davis, California. 1978. p. 78–84.

51. Koenig WD, Knops JMH. Scale of mast-seeding and tree-ring growth. Nature. 1998;396.6708:225–6.

52. Ruf T, Bieber C, Turbill C. Survival, aging, and life-history tactics in mammalian hibernators. Ruf T, Bieber C, Arnold W, Millesi E editors. Living in a Seasonal World. Berlin Heidelberg: Springer; 2012. p. 123–32.

53. Turbill C, Smith S, Deimel C, Ruf T. Daily torpor is associated with telomere length change over winter in Djungarian hamsters. Biol Lett. 2012;8:304–7.

54. Verdolin JL. Meta-analysis of foraging and predation risk trade-offs in terrestrial systems. Behav Ecol Sociobiol. 2006;60:457–64.

55. Hunter JE, Gutiérrez RJ, Franklin AB. Habitat configuration around Spotted Owl sites in Northwestern California. Condor. 1995;97:684–93.

56. Glutz von Blotzheim UN. Handbuch der Vögel Mitteleuropas. Columbiformes - Piciformes: Tauben, Kuckucke, Eulen, Ziegenmelker, Segler, Racken, Spechte. In: Glutz von Blotzheim UN, Bauer KM editors. Handbuch der Vögel Mitteleuropas, vol. 9, 2nd edition. Wiesbaden: Aula-Verlag GmbH; 1994. pp. 1150.

57. Jones EW. The structure and reproduction of the virgin forest of the north temperate zone. New Phytol. 1945;44:130–48.

58. Kotler BP, Brown J, Mukherjee S, Berger-Tal O, Bouskila A. Moonlight avoidance in gerbils reveals a sophisticated interplay among time allocation, vigilance and state-dependent foraging. Proc R Soc B. 2010;277:1469–74.

59. Clarke JA. Moonlight's influence on predator/prey interactions between short-eared owls (*Asio flammeus*) and deermice (*Peromyscus maniculatus*). Behav Ecol Sociobiol. 1983;13:205–9.

60. Brown JS, Kotler BP, Smith RJ, Wirtz II. WO: the effects of owl predation on the foraging behavior of heteromyid rodents. Oecologia. 1988;76:408–15.

61. Bateson M, Kacelnik A. Risk-sensitive foraging: decision-making in variable environments. In: Dukas R, editor. Cognitive Ecology. Chicago: University of Chicago Press; 1998. p. 297–341.

62. Remmert H. Ökologie - Ein Lehrbuch. 2nd ed. Berlin Heidelberg: Springer; 1980. doi:10.1007/978-3-642-96541-8.

63. Lebl K, Rotter B, Kürbisch K, Bieber C, Ruf T. Local environmental factors affect reproductive investment in female edible dormice. J Mammal. 2011;92:926–33.

64. Brown JS, Laundre JW, Gurung M. The ecology of fear: optimal foraging, game theory, and trophic interactions. J Mammal. 1999;80:385–99.

65. Morris DW, Davidson DL. Optimally foraging mice match patch use with habitat differences in fitness. Ecology. 2000;81:2061–6.

17

Morphogenesis of honeybee hypopharyngeal gland during pupal development

Sascha Peter Klose[1,2], Daniel Rolke[1] and Otto Baumann[1*] 🆔

Abstract

Background: The hypopharyngeal gland of worker bees contributes to the production of the royal jelly fed to queens and larvae. The gland consists of thousands of two-cell units that are composed of a secretory cell and a duct cell and that are arranged in sets of about 12 around a long collecting duct.

Results: By fluorescent staining, we have examined the morphogenesis of the hypopharyngeal gland during pupal life, from a saccule lined by a pseudostratified epithelium to the elaborate organ of adult worker bees. The hypopharyngeal gland develops as follows. (1) Cell proliferation occurs during the first day of pupal life in the hypopharyngeal gland primordium. (2) Subsequently, the epithelium becomes organized into rosette-like units of three cells. Two of these will become the secretory cell and the duct cell of the adult secretory units; the third cell contributes only temporarily to the development of the secretory units and is eliminated by apoptosis in the second half of pupal life. (3) The three-cell units of flask-shaped cells undergo complex changes in cell morphology. Thus, by mid-pupal stage, the gland is structurally similar to the adult hypopharyngeal gland. (4) Concomitantly, the prospective secretory cell attains its characteristic subcellular organization by the invagination of a small patch of apical membrane domain, its extension to a tube of about 100 μm in length (termed a canaliculus), and the expansion of the tube to a diameter of about 3 μm. (6) Finally, the canaliculus-associated F-actin system becomes reorganized into rings of bundled actin filaments that are positioned at regular distances along the membrane tube.

Conclusions: The morphogenesis of the secretory units in the hypopharyngeal gland of the worker bee seems to be based on a developmental program that is conserved, with slight modification, among insects for the production of dermal glands. Elaboration of the secretory cell as a unicellular seamless epithelial tube occurs by invagination of the apical membrane, its extension likely by targeted exocytosis and its expansion, and finally the reorganisation of the membrane-associated F-actin system. Our work is fundamental for future studies of environmental effects on hypopharyngeal gland morphology and development.

Keywords: Exocrine gland, Insect, Epithelial tube, Organogenesis, Cell polarity, Actin cytoskeleton, Apoptosis, Invagination

Background

The European honey bee (*Apis mellifera*) forms highly organized colonies that function as a superorganism [17]. The majority of individuals in a bee colony, the sterile worker bees, support the queen, the drones, and the brood by undertaking various tasks in a temporal sequence [49]. During the first 2 weeks after their emergence, worker bees perform activities within the hive, i.e., cleaning cells, caring for the brood and the queen, ripening nectar, and constructing combs. As the worker bees age to 2-3 weeks, they assume extra-nidal tasks, in particular foraging for pollen, nectar, and water. These behavioral alterations accompany changes in transcriptional and translational activity, physiology, and morphology [9, 11, 18, 23, 30, 44, 46, 48, 49]. In particular, the hypopharyngeal gland in worker bees has a developmental cycle closely related to the division of labor. The paired hypopharyngeal gland is an exocrine gland

* Correspondence: obaumann@uni-potsdam.de
[1]Institute of Biochemistry and Biology, Department of Animal Physiology, University of Potsdam, Karl-Liebknecht-Str. 24/25, 14476 Potsdam, Germany
Full list of author information is available at the end of the article

specific to hymenopterans, is located in the front of the head capsule, and delivers its proteinaceous secretory product via a large collecting duct to the hypopharynx [7]. In nursing bees, this gland is voluminous, has a high secretory activity, and contributes to the production of royal jelly, which is fed to future queens and, to a lesser extent, to worker larvae [33]. As the worker bees start foraging, their hypopharyngeal glands decrease in size, secrete at a lower rate, and produce a different protein blend including enzymes involved in carbohydrate metabolism [9, 30, 44].

The hypopharyngeal gland in worker bees has a characteristic morphology (Fig. 1). It is composed of thousands of two-cell units, a secretory cell and a duct cell [7, 22]. The secretory cell discharges its products into the canaliculus, a blind-ending membrane-bound tubule that meanders within the cell and that is covered on its lumenal side by a thin fenestrated cuticular lining termed the end apparatus [22, 35]. At the open end of the canaliculus, the secretory cell forms a tube-joint-like connection to the duct cell, a long thin ductule lined by a cuticular layer. Based on these morphological characteristics, hypopharyngeal glands thus belong to class III of the insect dermal glands [28, 29]. Groups of 6–20 two-cell units are clustered to form acini, with the duct cells extending in a bundle toward the collecting duct. In each hypopharyngeal gland, about 800 such acini are arranged around and along the 60-μm-wide collecting duct that delivers the secretion to the hypopharynx [7].

From a cell-biological perspective, the canaliculus of the secretory cells is peculiar. This structure has been suggested to represent the apical domain of the plasma membrane involuted into the cell [5]. Only recently, however, has molecular evidence been provided in support of this notion. Richter et al. [35] have demonstrated that phosphorylated (=activated) moesin, an apical membrane marker that links actin filaments to integral membrane proteins, is associated with and confined to the canalicular membrane. The cytoskeletal system affiliated to the canaliculus is also special. Rings of actin filaments encircle the membrane tubule at regular distances, whereas a sparse web of actin filaments is associated with the inter-ring portions of the canaliculus [21, 22, 35]. These actin rings are thought to provide a stabilizing framework to the canalicular membrane system during phases of high exocytic activity [22].

The anatomy and the cellular and subcellular organization of the hypopharyngeal glands in the adult worker bee have been studied extensively by use of various techniques, i.e., histology, electron microscopy, and fluorescence microscopy (e.g., [9, 16, 21–23, 31, 35]). In particular, several studies have addressed the structural changes that occur in hypopharyngeal gland as worker bees age and/or adopt other tasks [9, 23, 31, 35]. Of special interest is also the influence of pesticides on the morphology and physiology of adult hypopharyngeal glands and, thus, of the adverse side effects of these substances on honeybee vigor [16, 45]. In contrast, the organogenesis of the hypopharyngeal glands and the morphogenesis of the various gland cells have not been characterized as yet in detail. By use of histological techniques, Painter [32] has examined pupal gland cells, without noting the exact developmental stage, and has provided evidence for the transient presence of an

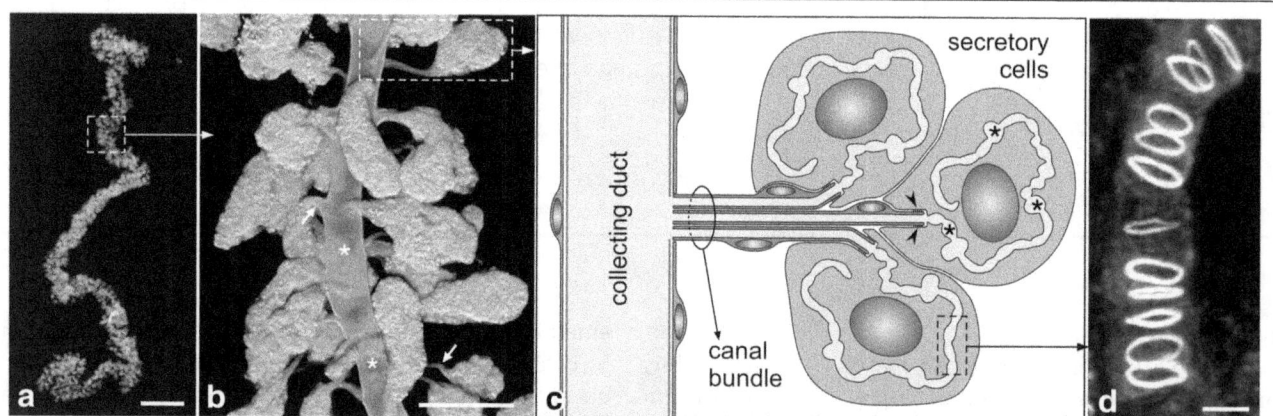

Fig. 1 Hypopharyngeal gland in adult worker bee. **a** and **b** Macroscopic and microscopic views of hypopharyngeal glands. The gland consists of several hundred acini that are arranged around a collecting duct (*white asterisks*) and connected to the latter via a bundle of microcanals (*white arrows*). **c** Schematic presentation of the organization of an acinus. Each acinus is composed of several secretory cells that have their apical membrane involuted to form a long canaliculus (*black asterisks*). Each secretory cell is attached via a junctional complex (*arrowheads*) to a canal cell that forms a microcanal for the delivery of secretory products to the collecting duct. **d** The actin cytoskeleton along the membrane of the canaliculus. Maximum intensity projection of a confocal image stack through a secretory cell labeled with fluorophore-tagged phalloidin. Rings of actin filaments girdle the canaliculus at regular intervals. Faint staining for F-actin is associated with the inter-ring sections of the canaliculus. Bars, **a** 1 mm, **b** 100 μm, **d** 2.5 μm

additional cell type besides the secretory cell and duct cell during the morphogenesis of the gland units. Subsequently, da Cruz-Landim and Mello [8] have analyzed hypopharyngeal gland morphogenesis during pupal life by using histological techniques, but in the stingless bee *Melipona quadrifasciata anthidioides*. Since none of these studies has examined the morphogenesis of secretory cells at the subcellular level, the time and the manner in which the secretory cells form their distinctive apical membrane system, the canaliculus, remain mysterious. This topic is of genuine interest in view of the recent finding that insecticides impair brood development [45].

In the present study, we have attempted to track the origin of the secretory and duct cells during pupal hypopharyngeal gland development by using the DNA-binding dye DAPI and the F-actin-binding phalloidin to visualize nuclei and the cell outline, respectively. We confirm the transient existence of an additional cell type between the secretory cell and duct cell, and that this cell is lost at the mid-pupal stage by apoptosis. Moreover, we examine the development of the canalicular system and the establishment of the associated rings of actin filaments in secretory cells.

Results

Gross morphology of hypopharyngeal glands during pupal development

Morphogenetic events during hypopharyngeal gland development were studied by staining with fluorophore-conjugated phalloidin in conjunction with serial confocal sectioning and three-dimensional (3D) image reconstruction. Since F-actin is enriched on the plasma membrane [4], the morphology of entire hypopharyngeal glands can be depicted by using appropriate parameters for image acquisition and modes for 3D presentation (Fig. 2).

Pupae were staged from P1 to P9, equivalent to days of pupal development [12, 14]. At developmental stage P1, two saccule-like evaginations, representing hypopharyngeal gland primordia, extended from the ventral side of the pharynx (Fig. 2a and c). The saccules consisted of a transparent epithelium enclosing a large lumen, had a smooth outer surface, and measured about 0.5 mm in length and 0.2 mm in width. During subsequent days of pupal developmental up to stage P4, the hypopharyngeal gland primordia increased in length to about 5 mm, their width declined to about 0.1 mm, and their outer surface became undulating (Fig. 2d and e). By stage P5, a collecting duct of approximately 40 μm in width extended on the medial axis along the entire hypopharyngeal glands. Numerous cauliflower-like structures, representing future acini, were arranged around the duct, being linked to it by stalks that were approximately 20 μm long and 20 μm thick (Fig. 2f). Because of the large number of acini and their proximity to the duct and to each other, the collecting duct was almost completely masked from sight. By stage P6, the hypopharyngeal

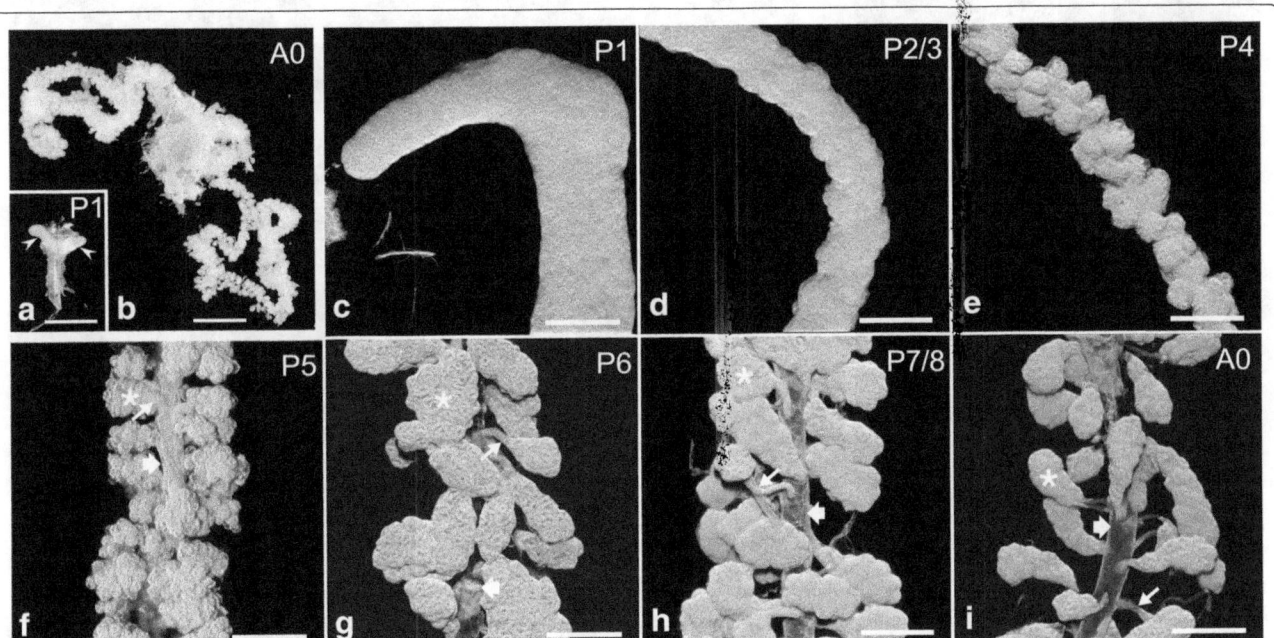

Fig. 2 Hypopharyngeal gland morphogenesis. **a** and **b** Macroscopic images of hypopharyngeal gland primordia (arrowheads) at pupal stage P1 (**a**) and of hypopharyngeal glands in a newly emerged worker bee (**b**; A0). Note the size difference. **c-i** Microscopic views of hypopharyngeal glands during pupal development (P1-P7/P8) and of a newly emerged worker bee (A0). Confocal image stacks of phalloidin-labeled glands are presented in 3D shadow mode. Asterisks, acini; thin arrows, bundle of ductules; broad arrows, collecting duct. Bars, **a** and **b** 1 mm; **c-i** 100 μm

glands had adopted a gross morphology similar to that of adult glands (Fig. 2g-i), with numerous acini of ovoid shape linked via bundles of ductules of approximately 50 μm in length to the collecting duct that extended over the entire length of the gland.

Mitotic events

To identify mitotic events during hypopharyngeal gland development, entire glands were labeled with the DNA-binding dye 4′,6-diamidino-2-phenylindole (DAPI). At pupal stage P1, mitotic nuclei were detected in the apical portion of the epithelium, with the division plane in most but not all cases being oriented horizontally in the epithelial layer (Fig. 3a-f; Additional file 1). The middle and basal regions of the epithelium contained numerous interphase nuclei. In addition, the basal region had nuclei that contained condensed chromatin and that were sometimes fragmented, probably representing apoptotic

cells. In order to validate the above results on the mitotic events in P1 gland primordia, organs were labeled with an antibody against histone H3 phosphorylated at Ser10 (H3-P; Fig. 3g and h). Anti-H3-P is known to be a reliable marker for mitosis in insect tissues [26, 27]. H3-P-positive nuclei were present in the apical region of the epithelium. Moreover, a few H3-P-positive nuclei were detected in the basal zone of the epithelium, suggesting that mitotic events also occurred in this region, although at low frequency. From stage P2/P3 on, no mitotic cells were detected by DAPI staining or anti-H3-P labeling. These results suggest that mitotic events are completed during the P1 phase.

Cellular morphology of developing hypopharyngeal glands

In hypopharyngeal gland primordia at stage P1, the bounding epithelium was pseudostratified and about

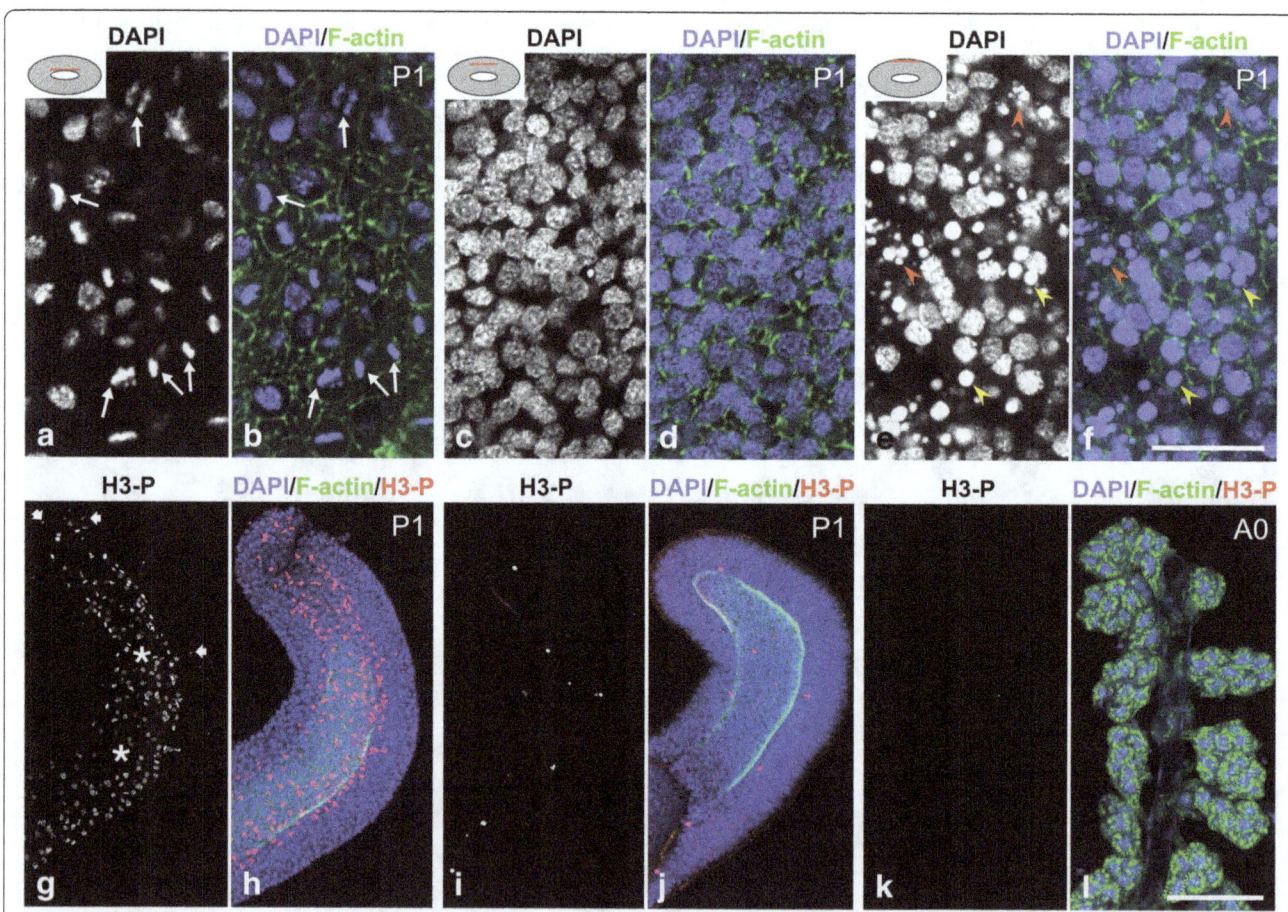

Fig. 3 Mitotic cells in hypopharyngeal gland primordia at stage P1. **a-f** Optical sections at various levels (indicated by the *red* line in the inset) through a hypopharyngeal gland at stage P1 labeled with DAPI and fluorophore-tagged phalloidin (F-actin). **a** and **b** Few nuclei are located in the apical region of the epithelium, but most of these are in mitotic stages (*arrows*). **c** and **d** Mid-epithelial sections are crowded with interphase nuclei. **e** and **f** The basal region of the epithelium contains nuclei with condensed chromatin (*yellow* arrowheads), some of them seem fragmented (*red arrowheads*). **g** and **h** Hypopharyngeal gland of P1 pupa labeled with an antibody against phosphorylated histone 3 (H3-P), DAPI and, phalloidin (F-actin) and presented as maximum intensity projections. H3-P stained numerous nuclei close to the lumen (*asterisks*) and a few nuclei in the basal region of the epithelium (*arrows*). Bars, **a-f** 20 μm; **g** and **h** 100 μm

40 µm thick (Fig. 4a, e and i). In addition to mitotic cells in the apical region, the epithelium consisted of flask-like interphase cells with their nuclei positioned at various levels in the mid and basal region of the epithelium. Interphase nuclei were oval and measured about 5 µm by 3 µm, with the long axis oriented in an apicobasal direction in the epithelial layer. A cellular process that was 1–2 µm thick and 10–20 µm long extended from the cell body to the luminal surface. Intense staining with phalloidin of the apicolateral sides of these processes indicated that F-actin occurred at adherens junctions (Fig. 4a inset). In addition, weaker staining over the entire apical surface of the cell processes suggested the presence of microvilli-like structures.

At developmental stage P2/3, the epithelium retained a uniform thickness of about 40 µm over its entire expanse (Fig. 4b). However, several cell types could be distinguished by their differences in morphology and position (Fig. 5a-d; Additional file 2). Of these, three different cells appeared to be organized into units, with each unit being characterized by a short F-actin-bounded tubule, probably representing the prototype of a ductule. Based on the layout of these three cells and their further morphogenesis, two of them were identified as the future duct cell and the secretory cell, respectively. The third cell type within the unit was interposed between the two above-mentioned cells and had no equivalence in the adult hypopharyngeal gland (subsequently, this cell is termed accessory cell). The body of the future secretory cell was located basally within the epithelium and contained a nucleus of about 6 µm in diameter (Figs. 4j and 5a,b). A cell process of 10–15 µm in length and 1–2 µm in width extended from the cell body in an apical direction to close off the distal end of the ductule. This secretory cell process was wrapped by two sheaths, the interior being formed by an accessory cell and the exterior by a future duct cell (Figs. 4f and 5a-d). The accessory cell had a nucleus of size and position similar to the future secretory cell. At the end of the secretory cell process, an extension of the accessory cell formed the distal portion of the ductule precursor. The nucleus of the prospective duct cell was smaller than the nuclei of the other cell types, being horizontally flattened and located in the mid-epithelial region (Fig. 5a and b). The future duct cell formed, in the basal portion of the three-cell unit, the outer sheath but reached with a narrow process above the end of the accessory cell process all the way to the luminal surface of the epithelium to build the proximal portion of the ductule precursor. Five to 15 of these three-cell units were arranged in clusters next to each other, with the ductule precursors extending in a radial fashion from the luminal surface basally for several micrometers (Fig. 4b). A few cells with a round nucleus in the apical portion of the

epithelium were localized between the clusters and covered the remaining area of the luminal surface (Fig. 4b). We suggest that the clusters of the three-cell units represent future acini and their associated ductules, and that the intermediary cells will configure the collecting duct.

At developmental stage P4, acini primorida bulged in a basal direction from the epithelial layer (Fig. 4c). Ductule precursors had increased in length to about 20 µm and were composed of a prospective duct cell almost over their entire length, except for a short segment lying next to the secretory cell and produced by an accessory cell (Fig. 5e-h; Additional file 3). The secretory cell process was shortened, retracted in a basal direction, and contained an onion-shaped F-actin-rich structure next to the distal end of the ductule. We consider that this structure corresponds to an array of microvilli and represents the origin for the development of the secretory cell canaliculus (see below).

At developmental stage P5, the acini had moved basally out of the epithelial layer, remaining connected to it by short bundles of ductules (Fig. 4d). The remaining epithelial layer, the future collecting duct, was a monolayer of isoprismatic cells with an apical seam of F-actin, indicative of short microvilli (Fig. 4h). Rounded secretory cells with large nuclei were positioned on the periphery of the acini, whereas duct cell bodies with flattened nuclei were located in the interior of the acini or in the stalk between the ductules (Figs. 4d,l and 5i-l). The last-mentioned were formed over their entire length by duct cells, whereas accessory cells were restricted to a collar around the distal portion of the ductules, abutting the secretory cell (Fig. 5i-m; Additional file 4). Thus, except for the presence of accessory cells, P5 acini had adopted an organization similar to that of the hypopharyngeal glands of adult worker bees. Accessory cells seem to undergo apoptosis during pupal stages P6 to P8, since DAPI staining visualized fragmented nuclei in the interior of the acini and in the stalk (Fig. 6). Moreover, only secretory cells and duct cells were detected in gland units at later developmental stages.

Differentiation of the canaliculus

Since the timing and manner of formation of the canaliculus of the secretory cells are unknown, we wished to analyze this morphogenetic process by the use of probes specific for this membrane domain. We have shown previously that anti-phosphorylated ERM (anti-pERM) and anti-phosphotyrosine selectively stain the canalicular system of adult secretory cells. Whereas anti-pERM outlines membrane segments between adjacent actin rings, anti-phosphotyrosine identifies dot-like structures that are associated with the canaliculus and that may represent microvillar tips [35]. Unfortunately, however, neither anti-pERM nor anti-phosphotyrosine stained any structures in prospective secretory cells of the pupal

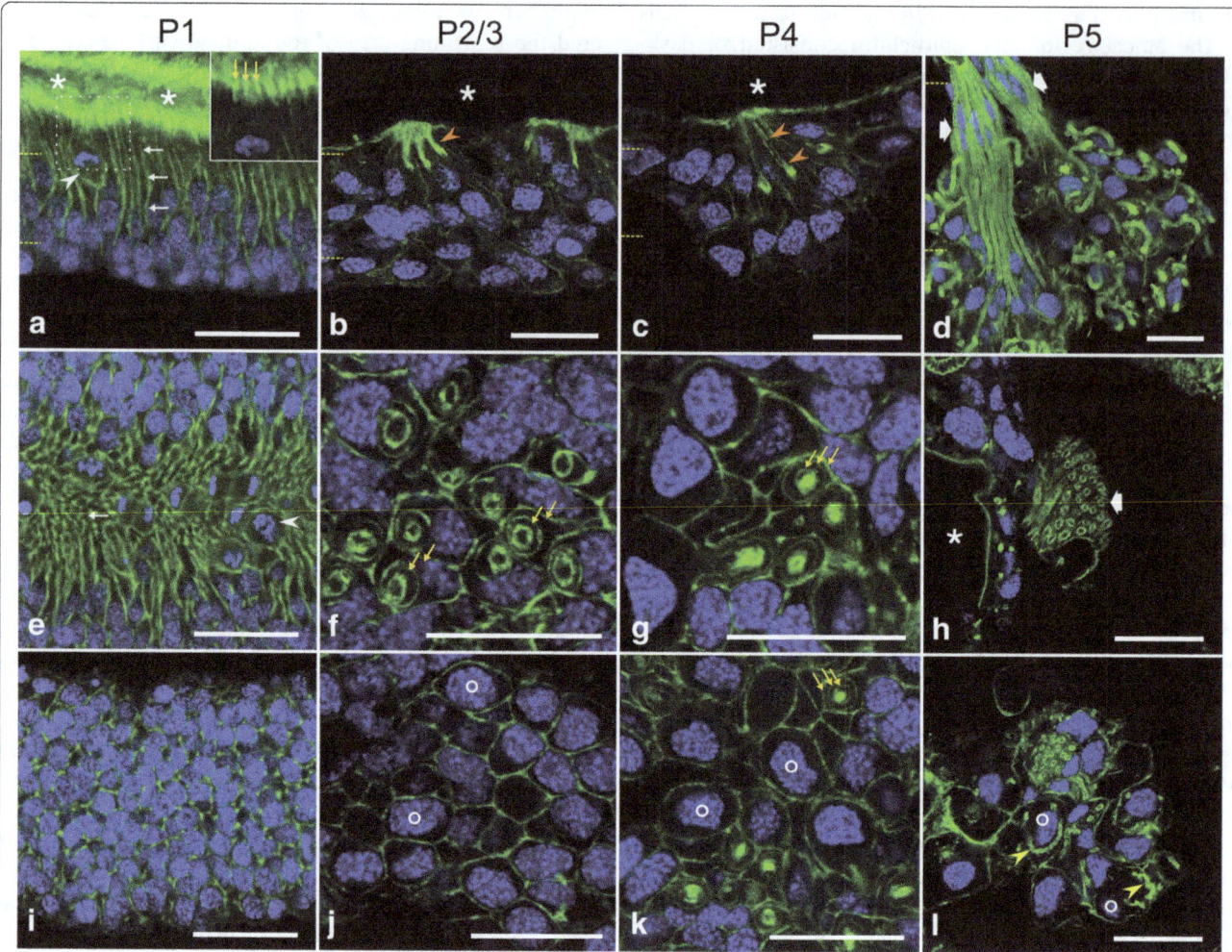

Fig. 4 Differentiation of hypopharyngeal gland during the first half of pupal life. Glands were isolated, fixed, labeled with phalloidin (*green*) and DAPI (*blue*), and imaged by confocal serial sectioning. In the case of phalloidin images, gamma correction was set to 0.5 to visualize areas of faint staining. **a-d** Sagittal sections through the gland epithelium or acini at the developmental stages as indicated. Lumen of the gland primordium or the collecting duct is indicated by asterisks. **e-l** Horizontal optical sections through the epithelial layer or acini. Dashed lines in **a-d** indicate relative positions of section planes. At P1 (**a,e,i**), the hypopharyngeal gland primoridium is composed of a pseudostratified columnar epithelium with mitotic cells (arrowheads) in the apical region. Flask-like cells have their nucleus in the basal half and a narrow process (*white* arrows) extending toward the gland lumen. The area outlined by **a** dashed line is presented at higher magnification in the inset (*green*, no gamma correction). Intense phalloidin staining at the apicolateral side (*orange* arrows) indicates junctional complexes; the fainter staining between and above the apicolateral sides suggests the presence of microvilli. From P2-P4 on, F-actin-rich tubulous structures (red arrowheads) extend in bundles from the gland lumen basally and are wrapped by two to three concentric rings of cell processes (*yellow* arrowheads). Future secretory cells (circles) have a large nucleus in the basal region and are as yet devoid of canaliculus-like structures. At P5, the gland is organized into acini that are connected by bundles of ductules (broad arrows) to the collecting duct (asterisk). Future secretory cells (circles) contain an F-actin-rich tubulous structure, the future canaliculus (*yellow* arrowheads in l). Bars, 20 μm

hypopharyngeal glands, suggesting that either these cells lack a canaliculus throughout pupal development, or that the expression of these proteins and/or their localization to the canalicular membrane occurs after the formation of the canalicular membrane system. Thus, we could only rely on the subcellular distribution of F-actin to probe the formation of the canalicular membrane, assuming that the F-actin assemblies in the interior of the secretory cells are associated with this membrane domain and/or its precursors.

At developmental stages P2-P4, prospective secretory cells contained, at the contact site of their cell process with the ductule, an F-actin-rich structure (Fig. 5c, d, g and h). We interpret this structure as being an array of microvilli, as noted previously for developing secretory cells in female accessory glands in *Rhodnius prolixus* [24]. No other F-actin-rich assemblies were identified within the prospective secretory cells during these developmental stages, suggesting that the canaliculus had not yet formed. In P5 secretory cells, a single continuous

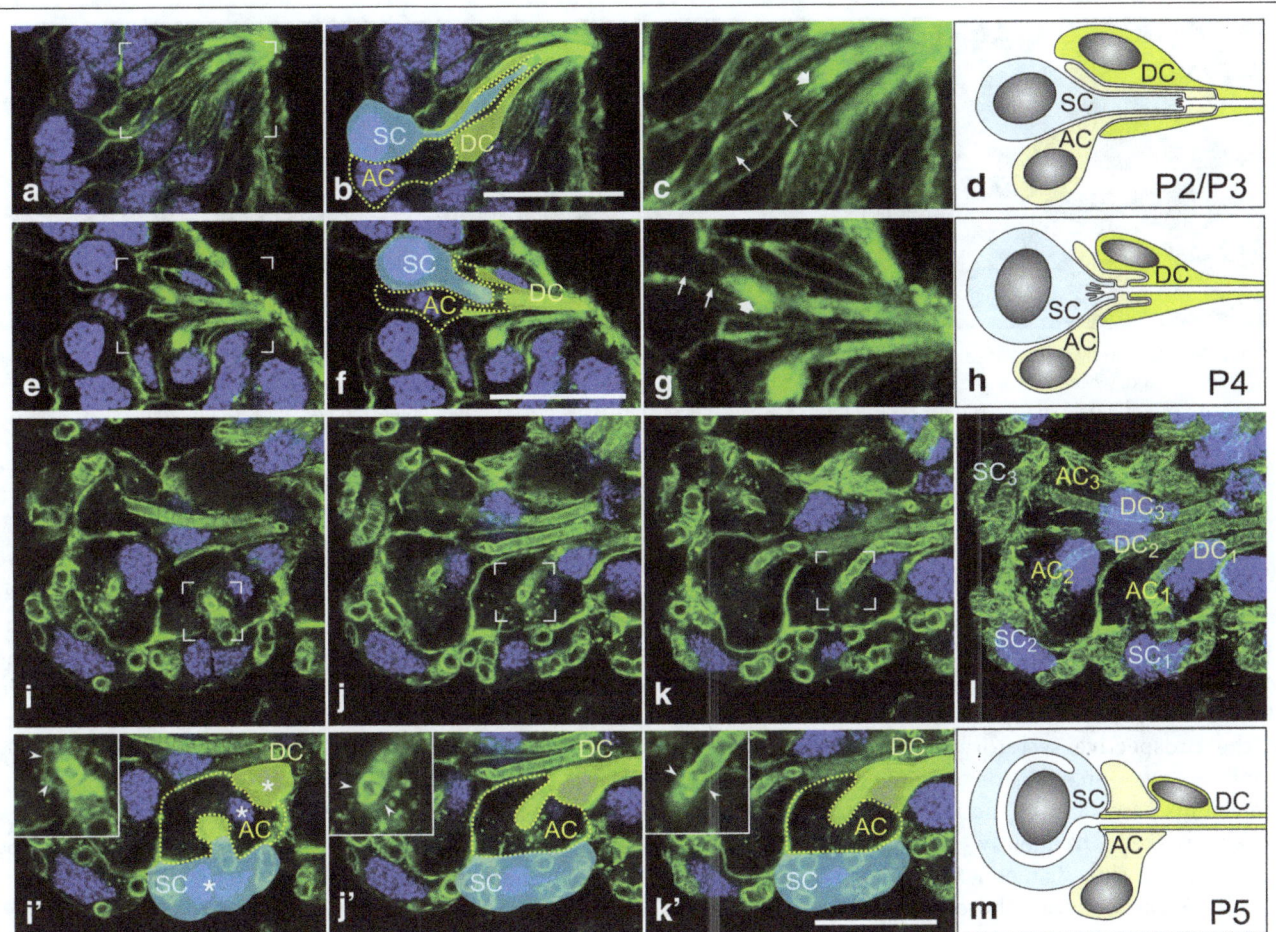

Fig. 5 Gland units in pupal hypopharyngeals glands are composed of three cell types. Cryo-sections of hypopharyngeal glands were labeled with fluorophore-tagged phalloidin (*green*) and DAPI (*blue*) and imaged by confocal serial sectioning. Gamma correction was set to 0.5 for phalloidin images. Gland units were examined at developmental stages P2/P3 (**a-c**), P4 (**e-g**), and P5 (**i-l**). **d**, **h** and **m** Schematic presentations of the spatial arrangement of the secretory cell (SC), duct cell (DC), and accessory cell (AC) at developmental stages P2/P3, P4, and P5, respectively. **a,b** An optical plane visualizing a P2/P3 gland unit in longitudinal section, with the three cell types highlighted in (**b**). **c** The area outlined in **a** at higher magnification, showing the SC process (arrowheads) contacting (arrow) the distal end of the future ductule. **e** and **f** An optical plane visualizing a P4 gland unit in longitudinal section, with the three cell types indicated in (**f**). **g** The area outlined in **e** at higher magnification, showing the F-actin-rich terminus (arrow) of the short secretory cell process (arrowheads) that extends toward the future ductule. **i-k** Three image planes with an inter-plane distance of ~1.2 µm through an acinus at developmental stage P5. The indicated area is shown at higher magnification in the insets below. **i'-k'** Lower regions of images i-k, with secretory cell (SC), accessory cell (AC), and duct cell (DC) outlined. Nuclei of the three cell types are indicated by asterisks. Insets in **i'-k'** The distal part of the duct cell (arrowheads) is enclosed by the accessory cell. Arrowheads indicate the plasma membrane of the duct cell. **l** Maximum intensity projection of the entire image stack, representing a thickness of ~5 µm and showing three gland units. The presence of an acessory cell (AC$_n$) is consistent for all secretory cell / duct cell assemblies. Bars, 20 µm

tube-like F-actin structure extended from the basal terminal of the ductule into the secretory cell for various distances. In some specimens, the tube ended after a few micrometers (Fig. 7a; Additional file 5:). However, in the most extended version, the tube-like F-actin array had a length of about 100 µm and adopted a meandering path around the nucleus (Fig. 7b-f; Additional files 6 and 7), similar to the canaliculus in adult secretory cells [35]. We suggest that these F-actin tubes are associated with membrane on their inside, and tubes of increasing length represent sequential developmental stages of canaliculus formation. This length increase was accompanied by fine-

structural changes (Fig. 8). Relatively short F-actin tubes had an external diameter of about 0.8 to 1.0 µm and an internal diameter of about 0.3 µm (Fig. 8a and g). Long F-actin arrays had a tube wall of uniform (apparent) thickness of 0.2 µm, although their outer diameter varied between ~1.2 and ~2.5 µm in a periodic manner. Hence, these long F-actin tubes had the appearance of a long series of conjoined oblate spheroids (Fig. 8b and h). Some individual F-actin spheres with a diameter of 1.0 to 2.5 µm were observed either in contact with or apart of the long F-actin tube (Fig. 7c-e). We consider that these structures represent material for the growth of the canaliculus.

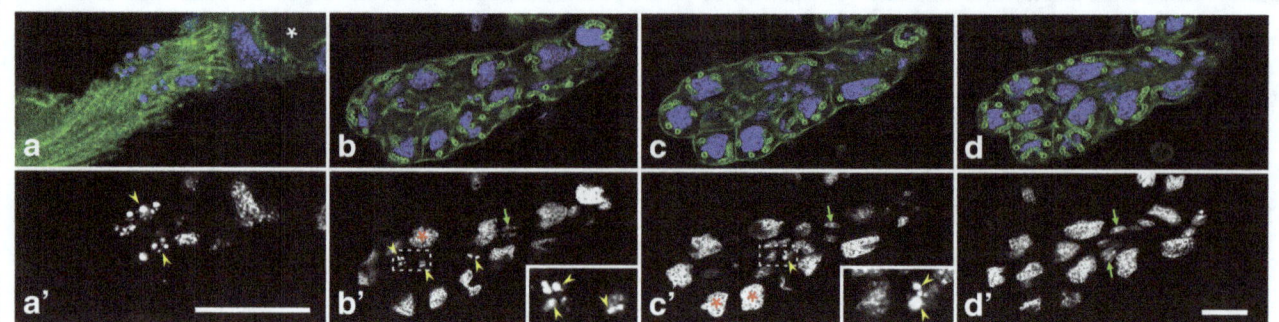

Fig. 6 Apoptosis in hypopharyngeal gland acini at pupal stage P6. Glands were fixed, labeled with fluorophore-tagged phalloidin (*green*) and DAPI (*blue* in **a-d**, *white* in a'-d'), and imaged by confocal serial sectioning. **a** and **a'** A bundle of ductules that connects the acinus (outside the field of view at the bottom left) with the collecting duct (*white* asterisk). **b-d** Three optical planes (inter-plane distance 3.8 µm) through an acinus. *Red* asterisks, secretory cell nuclei; *green* arrows, nuclei of duct cells or accessory cells; *yellow* arrowheads, nuclei with condensed and fragmented chromatin (shown at a higher magnification in insets). Bars, 25 µm

At stage P6, the canaliculus was radially expanded to a diameter of 2.8 to 3.5 µm over its entire length, without any periodic constrictions. Caniculus-associated F-actin was organized in a planar irregular web (Fig. 8c and i). Other than the canaliculus, no other F-actin assemblies were detected in the interior of the prospective secretory cells from this developmental stage on. By developmental stage P7/P8, canaliculus-associated F-actin was concentrated in closely spaced, frequently interconnected or fused rings, with a ring diameter of 2.5 to 3.0 µm (Fig. 8d and j). Areas between these F-actin assemblies were covered by a sparse matrix of actin filaments. During subsequent pupal development, the actin rings became more pronounced, the extent of the interconnections decreased, and the distance between the adjacent actin rings increased (Fig. 8e and k). In addition, the amount of F-actin associated with inter-ring portions seemed to decrease during the last few days of pupal development. Hence, canaliculi-associated actin rings were prominent and were regularly spaced with a few residual interconnections by the eclosion of the worker bees (Fig. 8f and l), as described previously [35].

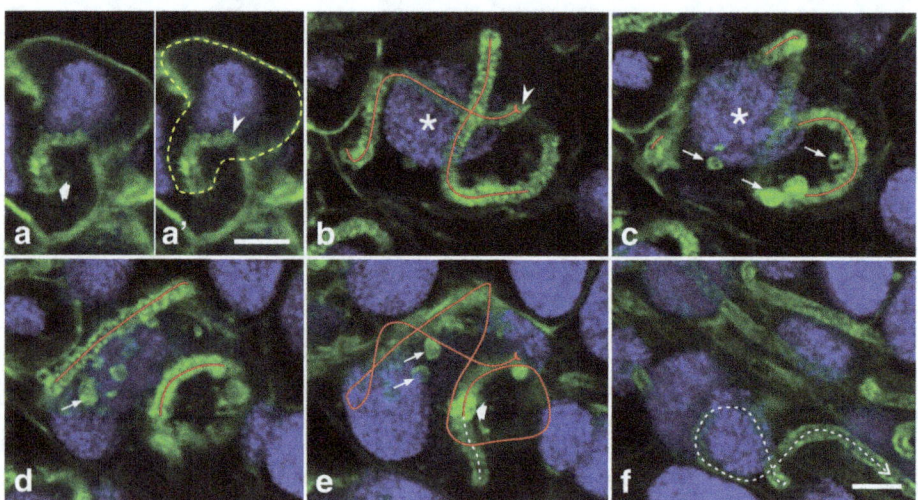

Fig. 7 Canaliculi formation at pupal stage P5 proceeds from the ductule end. Entire glands were fixed, labeled with phalloidin (*green*) and DAPI (*blue*), and imaged by confocal serial sectioning. **a** and **a'** In the secretory cell outlined by a dashed *yellow* line (**a'**), an array of F-actin defines a short tube that originates at the distal ductule end (broad arrow) and terminates blindly in the cytoplasm (arrowhead). Continuity between the F-actin tube and ductule was backtraced in confocal serial sections, although only one optical plane is shown here. **b-f** Five optical planes (inter-plane distance 1.15 µm) through a P5 acinus. A continuous blind-ending (arrowhead) tube is outlined by phalloidin staining and takes a convoluted path around the nucleus (asterisks) of the prospective secretory cell. Tube segments in each section plane are indicated by *red* lines and summarized in **e**. Spheroidal structures (thin arrows) close to the tube are also delineated by phalloidin staining. The F-actin tube is connected to the ductule (broad arrow); the latter is indicated by a dashed *white* line (**e,f**). Bars, 5 µm

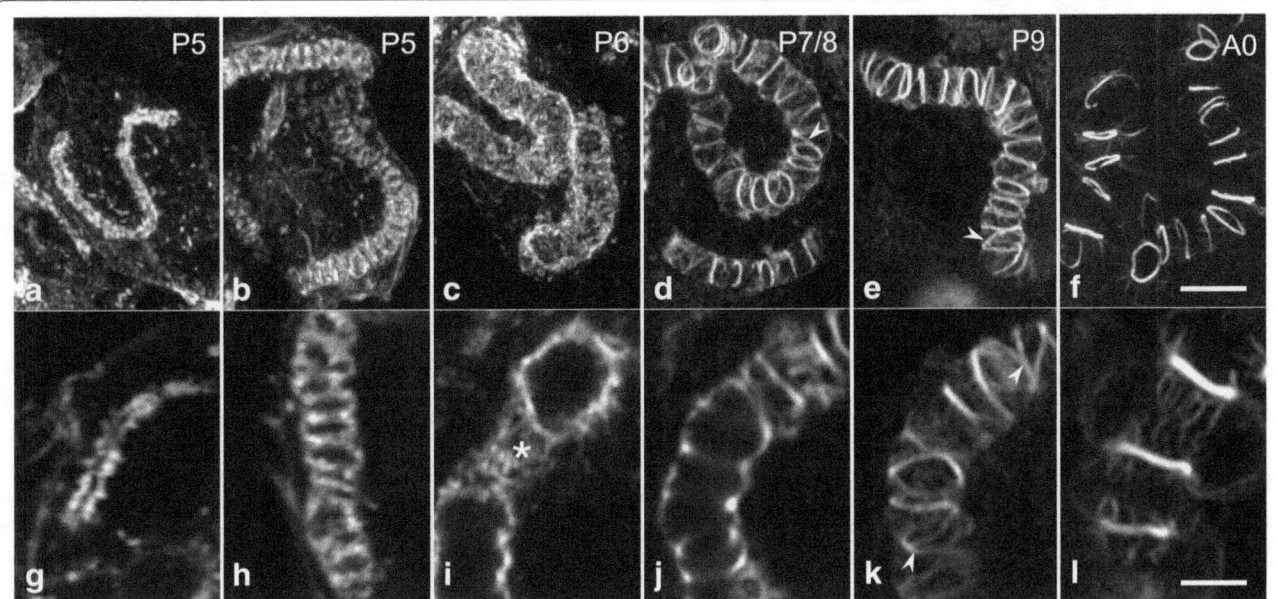

Fig. 8 Morphogenesis of the canaliculus in secretory cells during the second half of pupal development. Cryosections through hypopharyngeal glands at various developmental stages were labeled with phalloidin and imaged by confocal serial sectioning. **a-f** Maximum intensity projections of image stacks. **g-l** Individual optical sections at a higher magnification. **a** and **g** At P5, the F-actin tube, representing F-actin associated with the developing canaliculus, is thin and short with a narrow lumen. **b** and **h** In developmentally more advanced P5 specimens, the F-actin tube has increased in length and diameter. As a result of periodic constrictions, the tube resembles a series of oblate spheroids on a string. **c** and **i** At P6, a dense web of F-actin (asterisk) forms an expanded tube of uniform diameter. **d-f** and **j-l** Between P7 and eclosion of the worker bees, F-actin in the tube becomes reorganized and concentrated into rings. Ring distance increases during developmental progression, whereas the number of interconnections (arrowheads) and the amount of F-actin in association with the inter-ring segments decreases. Bars, **a-f** 5 μm; **g-l** 2.5 μm

Discussion

During the pupal development of worker bees, the hypopharyngeal gland primordium, which is a simple saccule enclosed by a pseudostratified epithelium, develops into an elaborate organ composed of hundreds of acini that are connected to and arranged around a collecting duct. Our results demonstrate that this developmental process can be subdivided into several key events (Fig. 9a): (1) During the first day of pupal life, at pupal stage P1, mitoses in the epithelial layer produce precursors for all the cells that compose the adult hypopharyngeal gland. (2) During pupal phase P2/P3, the epithelium becomes organized into rosettes of three cells, i.e., a prospective duct cell, an accessory cell, and a prospective secretory cell. Sets of five to 15 of these three-cell units become arranged in clusters within the epithelium. Eventually, such a patch of cells will differentiate into an acinus and its associated canal bundle, whereas cells between these clusters will produce the collecting duct epithelium. (3) By P5, morphogenesis of the three-cell units has produced a gland that is organized into hundreds of acini linked by canal bundles to a collecting duct, as in the adult hypopharyngeal gland. (4) Between pupal stages P6 to P8, the three-cell units are converted to two-cell units by the apoptotic elimination of the accessory cells. (5) During pupal stage P5, the canalicular system in the prospective secretory cells is formed by

the invagination of a small apical domain, its extension to a tube of approximately 100 μm in length, and its expansion to a diameter of about 3 μm. (6) After the establishment of the canaliculus to its full extent, the membrane-associated F-actin system becomes reorganized and concentrated at regular distances along the canaliculus to form actin rings.

Hypopharyngeal glands, being specific to hymenopterans, are quite diverse in morphological aspects [7]. Whereas the two-cell unit of secretory cell and canal cell seems to be common, the modes in which these units are organized to hypopharyngeal glands vary between species. In particular, the units may be attached directly to the hypopharyngeal plate or deliver their secretory product via a more or less elaborate collecting duct to the hypopharynx. Moreover, two-cell units may be individually attached to the collecting duct, like in the stingless bee *Melipona quadrifasciata anthidioides*, or several two-cell units may be assembled to acini with bundled ductules, as in *Apis mellifera*. Unfortunately, hypopharyngeal gland development has been examined as yet only in *M. quadrifasciata anthidioides*, but by histological techniques [8]. Nevertheless, since all hypopharyngeal glands are built of two-cell units, it may be supposed that the developmental program leading to these units is also shared among hymenopterans.

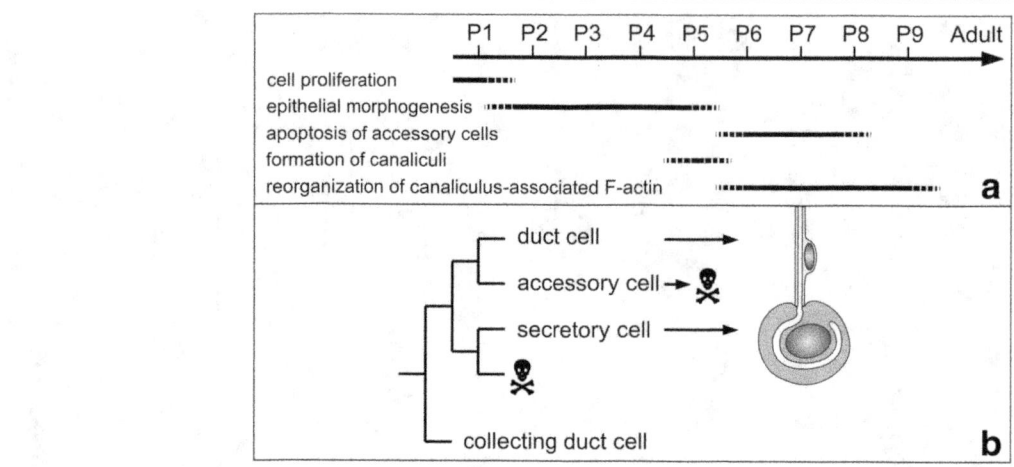

Fig. 9 Schematic outline of key events during pupal development of hypopharyngeal gland. **a** Timeline for major developmental events. P1-P9, pupal stages, termed according to the literature [12, 14]. **b** Putative cell lineage in worker bee hypopharyngeal gland, based on the cell lineage of other insect dermal glands [38, 40] and modified by taking into account our findings

Development of secretory units

Dermal glands of insects exhibit an enormous diversity. On morphological criteria, they have been classified into three types, with class III being organized in basic units of secretory and support cells [28, 29]. The development of class III glands has been proposed to be based on a common mechanism. Each glandular unit is an isogenic group of four cells that are derived from one progenitor cell by sequential mitoses [29]. Initially, these cells are concentrically wrapped one around another with, first, the innermost cell termed either the ciliary or basal cell, then the future secretory cell, and finally, the two outermost duct-forming cells [34, 38]. Subsequent apoptosis leads to mature gland units composed of fewer than four cells; at minimum, just one cell remains as a secretory cell.

In the present study, we visualized mitotic cells in the hypopharyngeal gland primordium, but we did not track the developmental fate of daughter cells. Nevertheless, our data are congruent with the above model of class-III-gland development in insects. Since secretory units in the hypopharyngeal gland of adult worker bees are composed merely of two cells, namely a secretory cell and a duct cell, two cells seem to be lacking in the case of the four-cell isogenic group. One of the cells missing in the adult state is the accessory cell that resides between the future duct and secretory cells during the first half of pupal life. The wrapping of these three cells around each other, with the future duct cell being the outermost and the prospective secretory cell lying on the inside, is reminiscent of the situation in the developing mandibular glands of the death's head cockroach *Blaberus craniifer* and tergal glands of the male German cockroach *Blattella germanica* [34, 38]. The presence of an accessory cell in pupal hypopharyngeal glands has previously been described [32]. Painter suggested that

this cell, being transiently present, becomes "absorbed" by the secretory cell. Our data indicate, instead, an apoptotic fate for this cell after the formation and elaboration of the gland units. The other cell missing in the secretory units of the hypopharyngeal gland is the basal/ciliary cell that is located basal of the secretory cell in other insect glands. Conceivably, this cell is not produced in honeybee hypopharyngeal glands because of a change in cell lineage during gland evolution, as demonstrated for *Drosophila* spermathecae [40]. Alternatively, the basal cell may undergo apoptosis soon after its production and thus does not contribute to the further development of the gland units. In support of the second option are the numerous apoptotic nuclei in the basal region of the P1 epithelium, concurrently with mitotic events.

In summary, we suggest the following model of cell lineage for the secretory units in the worker bee hypopharyngeal gland (Fig. 9b): mitosis of a progenitor cell produces four cells; one of them, localized basally, undergoes apoptosis shortly after genesis, whereas another one, the accessory cell, contributes to secretory unit morphogenesis but suffers cell death during late pupal development. The remaining two cells form the gland units (see below).

Development of the canaliculus

The two-cell secretory units of adult hypopharyngeal glands can be considered as epithelial tubes, with the tube lumen being circumscribed by the apical surface of the cells [35]. The canal cell forms a continuous conduit that opens into the collecting duct, whereas the adjoining secretory cell contains the blind ending of the tube. Electron-microscopic imaging has demonstrated that both cells are linked by an intercellular junctional complex, but that the cells do not form autocellular junctions [2, 5, 35]. These fine-structural details enable the secretory units of the hypopharyngeal gland to be classified as unicellular seamless tubes.

The morphogenesis of seamless epithelial tubes has been analyzed in three different model systems, i.e., the vertebrate vascular system, the tracheal system of *Drosophila*, and the excretory system of *Caenorhabditis elegans*. These studies have demonstrated three ways in which unicellular epithelial tubes can be created [36, 37, 41]. First, vesicular structures in the cytoplasm of the cell merge and form a lumen that extends over the entire length of the cell and fuses finally at its two ends with the plasma membrane. This mechanism, termed cell hollowing, is involved in the formation of vertebrate blood vessels and of the excretory canal cell in *C. elegans* [6, 19]. Second, a patch of apical membrane circumscribed by intercellular junctions enlarges by exocytosis and extends internally into the cell. Such apical invagination in combination with apically directed exocytosis has been detected in the case of the development of the tracheal terminal cells of *Drosophila* [13] and of the blood vessels of the zebrafish [15]. Finally, as in the case of the excretory duct cell of *C. elegans*, an epithelial cell wraps itself up with its apical surface towards the inside, forms an autocellular junction to close the tube, and subsequently removes the junction [39].

At P2/P3, the earliest time-point at which a tubular structure could be identified in the hypopharyngeal gland primordium, the short ductule was connected to the apical surface of the epithelium, was continuous, and was apparently formed by three cells arranged in a row, with the future duct cell and the accessory cell molding the tube and with the prospective secretory cell closing off the distal end of the tube. Unfortunately, we were unable to visualize the initial step of tube formation. However, since all cells in the pseudostratified epithelium reach to the luminal surface and hold an apical domain, cell wrapping or invagination may account for the creation of the ductule precursor. Discrimination between these possibilities requires the imaging of the junctional complexes. Since antibodies against *Drosophila* junctional proteins did not cross-react with their honeybee homologues [35], we are currently unable to characterize the initial process of ductule formation.

In the case of secretory cells and their canaliculus, our data are in agreement with an invagination and apically targeted exocytosis. At developmental interval P2 to P4, the prospective secretory cell contacts the distal end of the ductule with a small surface area that is enriched with F-actin. We suggest that the F-actin at this site reflects the presence of microvilli, as demonstrated in glands by electron microscopy of developing secretory cells in the female accessory of *Rhodnius prolixus* [24]. During the P5 stage, a continuous tube grows from this site inwards into the future secretory cell to reach a final length of about 100 µm. Spheroidal F-actin structures in the vicinity of the developing tube are indicative of membrane material for tube extension.

In various insect dermal glands, yet another, completely different mechanism has been reported for the formation of the canaliculus of secretory cells [3, 34, 38]. A basal cell, located basally to the future secretory cell, extends a ciliary process that pierces the future secretory and duct cells to form a mold for the canaliculus and the duct. Subsequently, the ciliary process retracts and the basal cell degenerates. Although we did not attempt to localize cilia, such a mechanism can be rejected in the case of hypopharyngeal gland morphogenesis since basal cells are absent, at least during the period of canaliculus formation. Similarly, ductule morphogenesis in the female accessory glands in *Rhodnius prolixus* and in *Drosophila* spermathecae has been reported to occur without the contribution of a ciliary process [24, 40].

After the generation of the canaliculus in full length in the secretory cell, the tube becomes elaborated. Accordion-like folding of the tube wall at this developmental stage indicates an increase in surface area, apparently as a stockpile for later expansion. The presence of F-actin-bounded vesicular structures in the cytoplasm alongside the canaliculus during this developmental stage indicates that the surface increase is fed by the lateral fusion of vesicles. Likewise, new membrane material is added by way of vesicles along the tube of the apical membrane in the terminal tracheal cells of *Drosophila* [13]. Subsequently, the canaliculus expands to a uniform diameter of about 3 µm. This process may be driven by ion and water transport into the canaliculus, like the aquaporin-dependent increase in lumen size in the excretory canal cell of *C. elegans* [20]. Alternatively or concomitantly, the canaliculus might become inflated and stabilized by the deposition of chitin in the lumen [10, 37, 43].

Differentiation of the F-actin system associated with the canalicular membrane

A network of F-actin is generally associated with the luminal membrane of epithelial tubes [13, 37]. This feature enabled the imaging of the developing canaliculus in the hypopharyngeal gland secretory cells, despite antibodies against marker proteins for the apical membrane not working in the present study. We have demonstrated that the F-actin system that is attached to the canalicular membrane is reorganized during the second half of pupal life. At developmental stage P6, when the canalicular system is formed to its full extent, a web of actin filaments with a seemingly random orientation surrounds the canalicular tube. F-actin then becomes gradually concentrated in rings, with the amount of interconnections decreasing and the inter-ring distance increasing. This change in actin cytoskeletal organization probably is concomitant or is attributable to a switch in the actin-binding proteins (ABPs) associated with the actin filaments on the canalicular membrane. Whereas cross-linking ABPs such as spectrin produce orthogonal arrays of actin filaments, bundling ABPs such as the Kelch protein of *Drosophila* can produce tight bundles of parallel actin filaments [47]. In agreement with this hypothesis is the finding that spectrin is not detectable with canaliculus-associated F-actin in adult secretory cells [35]. Moreover, the formation of actin rings in the ovarian ring canals of *Drosophila* depends on Kelch [42]. The expression of hDKIR, a human homologue of the *Drosophila* Kelch protein, produces ring-like actin structures in cultured mammalian cells [25].

Actin rings seem to be characteristic of secretory cells in hymenopteran hypopharyngeal glands. However, species-specific differences occur with respect to the diameter, the level of interconnections, and the distance between the actin rings [1, 22]. In particular, in the stingless bee *Tetragonula carbonaria*, actin rings on the canaliculus have frequent connections and are often not closed [22], thus resembling the P7/P8 intermediate stage in worker bees. We suggest that the differences in the relative expression of cross-linking and bundling ABPs account for these differences in actin ring organization between species.

Conclusions

We have described the various steps of hypopharyngeal gland development from the pupal primordium to the intricate organ that adult worker bees possess at emergence. The gland develops as follows: cell proliferation in a pseudostratified epithelium, formation and morphogenesis of three-cell units within the epithelial layer, removal of accessory cells from the three cell units to obtain the final units of a duct cell and a secretory cell, elaboration of the canaliculus in the latter cell by invagination, extension and expansion of apical membrane, and finally reorganization of the canaliculus-associated actin cytoskeleton to form distinctive actin rings. Based on these findings, the effects of environmental factors, such as insecticides, on gland development can be explored. Moreover, since species-specific differences in the organization of the canaliculus-associated F-actin system have been reported, an analysis of hypopharyngeal gland development in other hymenopteran species might be informative.

Methods
Animals and preparation

Pupae of worker bees (*Apis mellifera*) were taken from combs with sealed broods and were kept in a humidified incubator at 34 °C. Pupae were staged from P1 to P9 by using morphological criteria, i.e., pigmentation of the eyes, bodies, and legs [12, 14]. Pupal stages P2 and P3 and stages P7 and P8 could not be discriminated unambiguously; hence, we pooled these stages into P2/P3 and P7/P8, respectively. Newly emerged worker bees (stage A0) were collected off the comb just after emergence. Animals were decapitated, the head capsule was opened on the frontal side with a microscalpel, and the hypopharyngeal glands were removed in Ringer solution (270 mM NaCl, 3.2 mM KCl, 1.2 mM CaCl$_2$, 10 mM MgCl$_2$, 10 mM morpholinopropansulfonic acid, pH 7.3) and immediately transferred to fixative (3% paraformaldehyde, 1 mM dithiobis(succinimidyl proprionate), 0.1 M phosphate buffer, pH 7.0).

Antibodies

The following antibodies were used: monoclonal rabbit antibody against phosphorylated ezrin/radixin/moesin (pERM; product # 3149; Cell Signaling, Danvers, MA, USA), monoclonal mouse antibody clone PY99 against phosphotyrosine (pY; Santa Cruz Biotechnology Inc., Santa Cruz, CA, USA), and polyclonal rabbit antibody against histone H3 phosphorylated at Ser10 (H3-P; product # 06–570; Merck Millipore, Billerica, MA, USA). Cross-reactivity of anti-pERM with honeybee moesin has been demonstrated previously [35].

Fluorescence staining and imaging

After fixation for 1 h at room temperature, specimens were washed, cryofixed in melting isopentane (ca. -150 °C), cryosectioned at a thickness of about 10 μm, and stained with antibodies, 4′,6-diamidino-2-phenylindole (DAPI), and AlexaFluor 488 phalloidin (Life Technologies GmbH; Darmstadt, Germany) or CF488A phalloidin (Biotium Inc., Hayward, CA, USA) as described in detail previously [50]. To label entire glands, fixed glands were (1) washed 3 × 10 min in phosphate-buffered saline (PBS), (2) permeabilized with 0.01% Tween20 in PBS for 10 min, (3) treated with 50 mM NH$_4$Cl in PBS for 10 min, and (4) washed 1 × 10 min in PBS. After (5) treatment with blocking solution (1% normal goat serum, 0.8% bovine serum albumine, 0.5% Triton X-100 in PBS) for 15 min, specimens were (6) incubated overnight at 4 °C with anti-H3-P diluted in blocking solution, (7) washed 3 × 15 min in PBS, (8) incubated for 3 h at room temperature with Cy3-conjugated goat anti-rabbit IgG, fluorophore-tagged phalloidin, and DAPI in PBS, (9) washed again 3 × 15 min in PBS, and (10) embedded in Mowiol 4–88 mounting medium supplemented with 2% propyl gallate as an antifade reagent. In the case of the labeling of entire glands with phalloidin and DAPI only, steps 2–7 were omitted, and the specimens were incubated for 3 h or over-night with fluorophore-tagged phalloidin and DAPI, diluted in blocking solution. Fluorescence images were recorded with LSM 510, LSM 710, or LSM 880-Airyscan confocal microscopes and processed (3D presentation, gamma correction as noted in figure legends) with ZEN software (Carl Zeiss Microscopy GmbH, Jena, Germany).

Additional files

Additional file 1: Animation of an image stack through a hypopharyngeal gland primordium, stained with phalloidin (green) and DAPI (blue). Whole-mount specimen; inter-plane distance, 0.34 μm; objective lens, Zeiss C-Apochromat 40x/1.2 W. See Fig. 3a-f for details. (AVI 10486 kb)

Additional file 2: Animation of an image stack through a hypopharyngeal gland at pupal stage P2/3, stained with phalloidin (green) and DAPI (blue). Cryosection; inter-plane distance, 0.37 μm; objective lens, Zeiss Plan-Apochromat 63x/1.4 Oil. See Fig. 5a-c for details. (AVI 514 kb)

Additional file 3: Animation of an image stack through a hypopharyngeal gland at pupal stage P4, stained with phalloidin (green) and DAPI (blue). Cryosection; inter-plane distance, 0.24 μm; objective lens, Zeiss Plan-Apochromat 63x/1.4 Oil. See Fig. 5e-g for details. (AVI 2590 kb)

Additional file 4: Animation of an image stack through a hypopharyngeal gland at pupal stage P5, stained with phalloidin (green) and DAPI (blue). Cryosection; inter-plane distance, 0.23 μm; objective lens, Zeiss Plan-Apochromat 63x/1.4 Oil. See Fig. 5i-l for details. (AVI 1144 kb)

Additional file 5: Animation of an image stack through a hypopharyngeal gland at pupal stage P5, stained with phalloidin (green) and DAPI (blue). Whole-mount specimen; inter-plane distance, 0.29 μm; objective lens, Zeiss C-Apochromat 40x/1.2 W. See Fig. 7a for details. (AVI 670 kb)

Additional file 6: Animation of an image stack through a hypopharyngeal gland at pupal stage P5, stained with phalloidin (green) and DAPI (blue). Whole-mount specimen; inter-plane distance, 0.29 μm; objective lens, Zeiss C-Apochromat 40x/1.2 W. See Fig. 4d for details. (AVI 8006 kb)

Additional file 7: Animation of an image stack through a hypopharyngeal gland at pupal stage P5, stained with phalloidin (green) and DAPI (blue). Whole-mount specimen; inter-plane distance, 0.29 μm; objective lens, Zeiss C-Apochromat 40x/1.2 W. This movie shows an area of the acinus presented in movie 6. See Fig. 7b-f for details. (AVI 411 kb)

Abbreviations
ABPs: Actin-binding proteins; DAPI: 4′,6-diamidino-2-phenylindole; H3-P: Histone H3 phosphorylated at Ser10; PBS: Phosphate-buffered saline; pERM: Phosphorylated ezrin/radixin/moesin

Acknowledgements
We are grateful to Ricarda Scheiner and to Markus Thamm for providing honeybees, to Carl Zeiss Microscopy GmbH for providing access to a LSM880-Airyscan, to Bärbel Wuntke for technical assistance, and to Dr. Theresa Jones for language editing. We acknowledge the support of the Deutsche Forschungsgemeinschaft and Open Access Publishing Fund of University of Potsdam.
Funding
Not applicable.

Authors' contributions
SPK, DR, and OB conceived and designed the experiments; SPK and DR performed animal dissection and immunochemistry; SPK and OB collected and analyzed the data; OB wrote the paper. All authors read and approved the final manuscript.

Competing interests
The authors declare that they have no competing interests.

Author details
[1]Institute of Biochemistry and Biology, Department of Animal Physiology, University of Potsdam, Karl-Liebknecht-Str. 24/25, 14476 Potsdam, Germany. [2]Present Address: Institute of Biology, Department of Molecular Parasitology, Humboldt University, Philippstrasse 13, 10115 Berlin, Germany.

References
1. Albert S, Spaethe J, Grübel K, Rössler W. Royal jelly-like protein localization reveals differences in hypopharyngeal glands buildup and conserved expression pattern in brains of bumblebees and honeybees. Biol Open. 2014;3:281–8.
2. Beams HW, Tahmisian TN, Anderson E, Devine RL. An electron microscope study on the pharyngeal glands of the honeybee. J Ultrastr Res. 1959;3:155–70.
3. Berry SJ, Johnson E. Formation of temporary flagellar structures during insect organogenesis. J Cell Biol. 1975;65:489–92.
4. Bretscher A. Microfilament structure and function in the cortical cytoskeleton. Annu Rev Cell Biol. 1991;7:337–74.
5. Britto FB, Caetano FH. Ultrastructural features of the hypopharyngeal glands in the social wasp Polistes versicolor (Hymenoptera: Vespidae). Insect Sci. 2008;15:277–84.
6. Buechner M. Tubes and the single C. elegans excretory cell. Trends Cell Biol. 2002;12:479–84.
7. da Cruz-Landim C, Costa RAC. Structure and function of the hypopharyngeal glands of Hymenoptera: a comparative approach. J Comp Biol. 1998;3:151–63.
8. da Cruz-Landim C, Mello ML. The post-embryonic changes in Melipona quadrifasciata anthidioides Lep. (Hym. Apoidea). II. Development of the salivary glands system. J Morphol. 1967;123:481–502.
9. Deseyn J, Billen J. Age-dependent morphology and ultrastructure of the hypopharyngeal gland of Apis mellifera workers (Hymenoptera, Apidae). Apidologie. 2005;36:49–57.
10. Devine WP, Lubarsky B, Shaw K, Luschnig S, Messina L, Krasnow MA. Requirement for chitin biosynthesis in epithelial tube morphogenesis. Proc Natl Acad Sci U S A. 2005;102:17014–9.
11. Feng M, Fang Y, Li J. Proteomic analysis of honeybee worker (Apis mellifera) hypopharyngeal gland development. BMC Genomics. 2009;10:645.
12. Ganeshina O, Schäfer S, Malun D. Proliferation and programmed cell death of neuronal precursors in the mushroom bodies of the honeybee. J Comp Neurol. 2000;417:349–65.
13. Gervais L, Casanova J. In vivo coupling of cell elongation and lumen formation in a single cell. Curr Biol. 2010;20:359–66.
14. Groh C, Rössler W. Caste-specific postembryonic development of primary and secondary olfactory centers in the female honeybee brain. Arthropod Struct Dev. 2008;37:459–68.
15. Herwig L, Blum Y, Krudewig A, Ellertsdottir E, Lenard A, Belting HG, Affolter M. Distinct cellular mechanisms of blood vessel fusion in the zebrafish embryo. Curr Biol. 2011;21:1942–8.
16. Heylen K, Gobin B, Arckens L, Huybrechts R, Billen J. The effects of four crop protection products on the morphology and ultrastructure of the hypopharyngeal gland of the European honeybee, Apis mellifera. Apidologie. 2011;42:103–16.
17. Hölldobler B, Wilson EO. The superorganism: the beauty, elegance, and strangeness of insect societies. 1st ed. New York: W. W. Norton & Company; 2009.
18. Hrassnigg N, Crailsheim K. Adaptation of hypopharyngeal gland development to the brood status of honeybee (Apis mellifera L.) colonies. J Insect Physiol. 1998;44:929–39.
19. Kamei M, Saunders WB, Bayless KJ, Dye L, Davis GE, Weinstein BM. Endothelial tubes assemble from intracellular vacuoles in vivo. Nature. 2006; 442(7101):453–6.

20. Khan LA, Zhang H, Abraham N, Sun L, Fleming JT, Buechner M, Hall DH, Gobel V. Intracellular lumen extension requires ERM-1-dependent apical membrane expansion and AQP-8-mediated flux. Nat Cell Biol. 2013;15:143–56.

21. Kheyri H, Cribb BW, Merritt DJ. Comparing the secretory pathway in honeybee venom and hypopharyngeal glands. Arthropod Struct Dev. 2013; 42:107–14.

22. Kheyri H, Cribb BW, Reinhard J, Claudianos C, Merritt DJ. Novel actin rings within the secretory cells of honeybee royal jelly glands. Cytoskeleton. 2012; 69:1032–9.

23. Knecht D, Kaatz HH. Patterns of larval food production by hypopharyngeal glands in adult worker honey bees. Apidologie. 1990;21:457–68.

24. Lococo D, Huebner E. The development of the female accessory gland in the insect Rhodnius prolixus. Tissue Cell. 1980;12:795–813.

25. Mai A, Jung SK, Yonehara S. hDKIR, a human homologue of the Drosophila kelch protein, involved in a ring-like structure. Exp Cell Res. 2004;300:72–83.

26. Malun D, Moseleit AD. Grünewald. B 20-Hydroxyecdysone inhibits the mitotic activity of neuronal precursors in the developing mushroom bodies of the honeybee, Apis mellifera. J Neurobiol. 2003;57:1–14.

27. Micchelli CA, Perrimon N. Evidence that stem cells reside in the adult Drosophila midgut epithelium. Nature. 2006;439:475–9.

28. Noirot C, Quennedey A. Fine-structure of insect epidermal glands. Annu Rev Entomol. 1974;19:61–80.

29. Noirot C, Quennedey A. Glands, gland cells, glandular units: some comments on terminology and classification. Ann la Soc Entomol Fr. 1991;27:123–8.

30. Ohashi K, Natori S, Kubo T. Change in the mode of gene expression of the hypopharyngeal gland cells with an age-dependent role change of the worker honeybee Apis mellifera L. Eur J Biochem. 1997;249:797–802.

31. Painter TS, Biesele JJ. The fine structure of the hypopharyngeal gland cell of the honey bee during development and secretion. Proc Natl Acad Sci U S A. 1966;55:1414–9.

32. Painter TS. Nuclear phenomena associated with secretion in certain gland cells with especial reference to the origin of cytoplasmic nucleic acid. J Exp Zool. 1945;100:523–47.

33. von Planta A. Über den Futtersaft der Bienen. Hoppe-Seyler's Z Physiol Chem. 1888;12:327–54.

34. Quennedey A. The moulting process of perennial class 3 gland cells during the postembryonic development of two heterometabolous insects: Blaberus (Dictyoptera) and Dysdercus (Heteroptera). Ann la Soc Entomol Fr. 1991;27:143–61.

35. Richter KN, Rolke D, Blenau W, Baumann O. Secretory cells in honeybee hypopharyngeal gland: polarized organization and age-dependent dynamics of plasma membrane. Cell Tissue Res. 2016;366:163–74.

36. Schottenfeld-Roames J, Ghabrial AS. Osmotic regulation of seamless tube growth. Nat Cell Biol. 2013;15:137–9.

37. Sigurbjörnsdóttir S, Mathew R, Leptin M. Molecular mechanisms of de novo lumen formation. Nat Rev Mol Cell Biol. 2014;15:665–76.

38. Sreng L, Quennedey A. Role of a temporary ciliary structure in the morphogenesis of insect glands. An electron microscope study of the tergal glands of male Blattella germanica L. (Dictyoptera, Blattellidae). J Ultrastruct Res. 1976;56:78–95.

39. Stone CE, Hall DH, Sundaram MV. Lipocalin signaling controls unicellular tube development in the Caenorhabditis elegans excretory system. Dev Biol. 2009;329:201–11.

40. Sun J, Spradling AC. NR5A nuclear receptor Hr39 controls three-cell secretory unit formation in Drosophila female reproductive glands. Curr Biol. 2012;22:862–71.

41. Sundaram MV, Cohen JD. Time to make the doughnuts: Building and shaping seamless tubes. Semin Cell Dev Biol. 2016. doi:10.1016/j.semcdb. 2016.05.006.

42. Tilney LG, Tilney MS, Guild GM. Formation of actin filament bundles in the ring canals of developing Drosophila follicles. J Cell Biol. 1996;133:61–74.

43. Tonning A, Hemphälä J, Tång E, Nannmark U, Samakovlis C, Uv A. A transient luminal chitinous matrix is required to model epithelial tube diameter in the Drosophila trachea. Dev Cell. 2005;9:423–30.

44. Ueno T, Takeuchi H, Kawasaki K, Kubo T. Changes in the gene expression profiles of the hypopharyngeal gland of worker honeybees in association with worker behavior and hormonal factors. PLoS One. 2015;10(6):e0130206.

45. Wessler I, Gärtner HA, Michel-Schmidt R, Brochhausen C, Schmitz L, Anspach L, Grünewald B, Kirkpatrick CJ. Honeybees produce millimolar concentrations of non-neuronal acetylcholine for breeding: possible adverse effects of neonicotinoids. PLoS One. 2016;11(6):e0156886.

46. Whitfield CW, Cziko AM, Robinson GE. Gene expression profiles in the brain predict behavior in individual honey bees. Science. 2003;302:296–9.

47. Winder SJ, Ayscough KR. Actin-binding proteins. J Cell Sci. 2005;118:651–4.

48. Winnington AP, Napper RM, Mercer AR. Structural plasticity of identified glomeruli in the antennal lobes of the adult worker honey bee. J Comp Neurol. 1996;365:479–90.

49. Winston ML. The biology of the honey bee. 1st ed. Cambridge: Harvard University Press; 1987.

50. Zimmermann B, Dames P, Walz B, Baumann O. Distribution and serotonin-induced activation of vacuolar-type H⁺-ATPase in the salivary glands of the blowfly Calliphora vicina. J Exp Biol. 2003;206:1867–76.

18

Group or ungroup – moose behavioural response to recolonization of wolves

Johan Månsson[1*], Marie-Caroline Prima[2], Kerry L. Nicholson[1], Camilla Wikenros[1] and Håkan Sand[1]

Abstract

Background: Predation risk is a primary motivator for prey to congregate in larger groups. A large group can be beneficial to detect predators, share predation risk among individuals and cause confusion for an attacking predator. However, forming large groups also has disadvantages like higher detection and attack rates of predators or interspecific competition. With the current recolonization of wolves (*Canis lupus*) in Scandinavia, we studied whether moose (*Alces alces*) respond by changing grouping behaviour as an anti-predatory strategy and that this change should be related to the duration of wolf presence within the local moose population. In particular, as females with calves are most vulnerable to predation risk, they should be more likely to alter behaviour.

Methods: To study grouping behaviour, we used aerial observations of moose (*n* = 1335, where each observation included one or several moose) inside and outside wolf territories.

Results: Moose mostly stayed solitary or in small groups (82% of the observations consisted of less than three adult moose), and this behavior was independent of wolf presence. The results did not provide unequivocal support for our main hypothesis of an overall change in grouping behaviour in the moose population in response to wolf presence. Other variables such as moose density, snow depth and adult sex ratio of the group were overall more influential on grouping behaviour. However, the results showed a sex specific difference in social grouping in relation to wolf presence where males tended to form larger groups inside as compared to outside wolf territories. For male moose, population- and environmentally related variables were also important for the pattern of grouping.

Conclusions: The results did not give support for that wolf recolonization has resulted in an overall change in moose grouping behaviour. If indeed wolf-induced effects do exist, they may be difficult to discern because the effects from moose population and environmental factors may be stronger than any change in anti-predator behaviour. Our results thereby suggest that caution should be taken as to generalize about the effects of returning predators on the grouping behaviour of their prey.

Keywords: *Alces alces*, anti-predator, Behaviour, *Canis lupus*, Group size, Predator, Prey, Ungulate

Background

Predation risk influences the adoption of potentially costly anti-predatory behaviour by prey. To avoid predation, prey may modify their vigilance, habitat selection, movement patterns, spatial and temporal distribution, or sexual segregation [1–4]. However, prey cannot be solely devoted to predator avoidance behaviours as they are obligated to obtain necessary resources for growth, reproduction, and survival. Therefore anti-predator adaptations need to be balanced against the present risk

level [1]. For instance, a spatio-temporal variation in predation risk may lead to prey adopting different behavioural strategies including more pronounced anti-predator behaviour in certain areas or during periods of higher relative risk [5].

Predation risk is thought to be one of the primary motivations for animals to change their grouping strategy e.g., form larger groups or avoid conspecifics and thereby decrease group size [6–10]. The choice of strategy is context- and species dependent and a general pattern seems to be that prey in open terrain aggregate more while prey in closed environments are more solitary [9]. As group size increases, there are more eyes

* Correspondence: johan.mansson@slu.se
[1]Department of Ecology, Grimsö Wildlife Research Station, Swedish University of Agricultural Sciences, SE-730 91 Riddarhyttan, Sweden
Full list of author information is available at the end of the article

scanning the environment, more individuals will share the predation risk among several group members and more individuals may also cause confusion to the predator when being attacked [1, 11, 12]. With increased group size any one individual forager can devote less time to vigilance and more time to feeding [13–15]. However, there are also negative consequences of forming groups as larger groups may imply increased risk of detection and attack rate by predators [16, 17]. Forming larger groups may also increase competition over critical resources [18–20]. As a consequence, an alternative strategy to minimize predation risk can also be to form small groups that rarely are encountered by predators [17]. Moreover, the net gain of forming groups is dependent on the individual's condition and placement in the group; for example low status and marginal individuals in the group may have a higher predation risk [10, 21]. Individuals may therefore respond differently to variation in predation risk according to the costs and benefits associated with grouping behaviour. It is therefore expected that prey individuals should show different grouping behaviour in response to the presence of predators but also that this may be linked to the individual's vulnerability to predation.

Prey vulnerability to predation can be directly linked to the characteristics of the predator and predator population (such as search image, and population density) but also to individual traits and physical condition of the prey and to environmental conditions [22–24]. Several studies have reported higher risk for low ranked (e.g., immature and small-sized) individuals [24–26] and increased risk during times of harsh climatic conditions as for example deep snow cover [22, 25, 27, 28]. Heard [23] showed that musk-ox (*Ovibos moschatus*) group size increased with an increase of wolf (*Canis lupus*) density but also that group size was dependent on the prey type targeted by wolves (i.e., larger groups during seasons when wolves preyed more on musk-ox). Moreover, an individual's status or potential ranking within a group can necessitate formation or avoidance of grouping behaviour [18]. For instance, groups formed by individuals with the same physical condition may benefit from dilution and distribute the probability of capture amongst group members. In contrast, a low ranked individual may benefit from the many eyes and overall increased vigilance in larger groups. Conversely, a low ranked individual could be at disadvantage as it is considered a marginal individual in the group and may be exposed to competitive disadvantage for resources [29, 30]. It is therefore not always easy to predict which individual that will obtain most benefits by group formation.

Among ungulates, the tendency to form groups varies between species, with variation in population density, food distribution, and between habitats [31, 32]. Moose

(*Alces alces*) have been described as a "quasi-solitary" species since they show both solitary and group living behaviour, possibly as a response to increased predation risk in open habitats [2, 18, 33]. For instance, in Alaska, moose were found to form larger groups at greater distance from cover, which suggests that social grouping in moose, in addition to other factors e.g., rutting and mobility in deep snow, is an adaptation to increased predation risk [18].

Starting in the early 1980s moose have been re-exposed to predation risk from wolves, in south-central Scandinavia, as wolves re-colonized this region after being extinct for more than 100 years [34]. Here, we study whether moose grouping behaviour change as a response to the recolonization of wolves. Moose in Scandinavia are related to forested areas i.e., closed environments. According to earlier studies on browsing deer (concentrate selectors) and ungulates living in terrain with cover, these species are less prone to form larger groups due to both foraging and/or antipredator behaviour [9, 10, 35]. It's therefore not obvious if and how moose should change their grouping behaviour when predation risk increase. However, if the solitary pattern of moose has been relaxed due to the period without wolves a decrease rather than an increase in group size may be expected. Moose can deploy both flight and fight strategies to avoid predation [36] but do not form defensive formations as groups of musk-oxen and mule deer [10, 37] which also support the prediction that moose will not form larger groups when re-exposed to predation. Here, we test whether moose grouping behaviour changes as a response to the recolonization of wolves. If a change in grouping behaviour exist, we predict that females with calves would be most prone to change their behaviour because calves of the year are the main prey by wolves in Scandinavia [38, 39]. In addition, we predict that if the presence of wolves is important for the grouping behaviour of moose, the strength of this behaviour would be linked to the duration of wolf presence (i.e., time since territory establishment). Moose within recently established territories may therefore express a less pronounced change in grouping behaviour than moose that has been exposed to wolves for a longer time period. In addition to the potential effects of wolf predation risk, we also considered population and environmental variables that may influence grouping behaviour such as moose density, adult sex-ratio and snow depth.

Methods

Study area

The study area encompasses approximately 50 000 km^2 within the boreal zone of south-central Sweden (Fig. 1, 58.58°-62.16 °N, 13.45°-16.64 °E). The area mainly consists of forests mixed with agriculture fields, bogs and

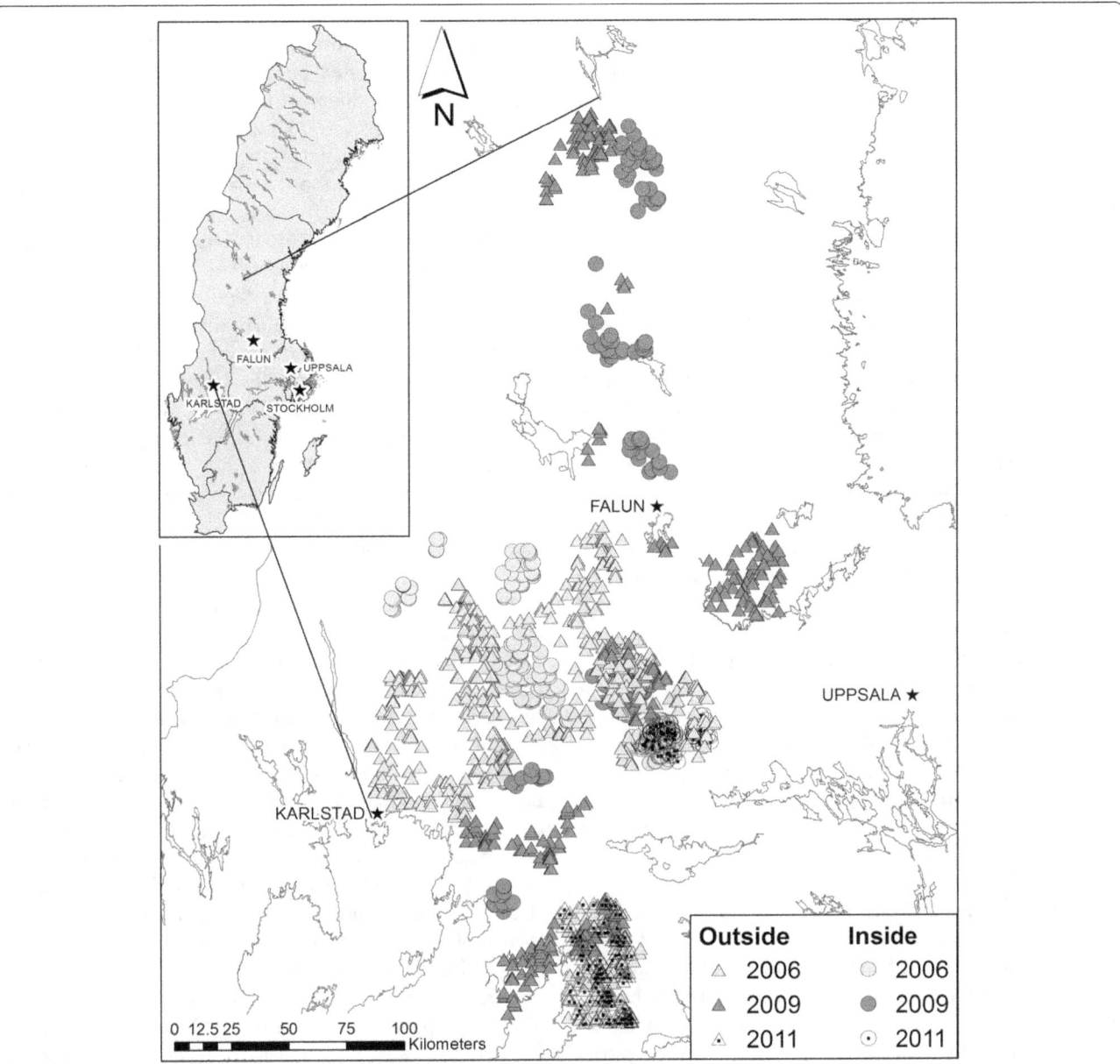

Fig. 1 The study area in south-central Sweden with the distribution of aerial moose observations and whether they were classified as inside or outside wolf territories. Note, that some areas can be considered as both inside and outside due to the successive establishment of new wolf territories

lakes. Forests are dominated by Norway spruce (*Picea abies*) and Scots pine (*Pinus sylvestris*) mixed with deciduous trees, such as birch (*Betula spp.*), aspen (*Populus tremula*), alder (*Alnus incana*) and willow (*Salix spp*), and is intensively managed for timber and pulp. Mature stands are harvested by clear-cutting and reforested by planting or natural regeneration, resulting in an even-aged forest stand mosaic. Average monthly temperature range between +15 °C and -5 °C with the coldest month in January and the warmest month in July [40]. The ground is usually snow covered between late November and early April with a mean snow depth of 20 cm in mid-January [41].

Studied populations

Moose populations are managed in Sweden and approximately 25-30% of the population is harvested annually [42]. Management in many areas is female biased to promote high productivity [42]. Moose density during winter commonly ranged between 0.6 and 2.5 moose/km^2 [43, 44]. Roe deer (*Capreolus capreolus*) are distributed over the whole study area at variable densities, whereas red deer (*Cervus elaphus*) and wild boar (*Sus scrofa*) only occur in scattered populations in the south [45].

In 1983, wolves started a re-colonization of Scandinavia and has increased in numbers and distribution since the early 1990s [34, 46]. The wolf distribution area in Scandinavia currently covers approximately 100 000 km^2 [47]. In the winter of 2010/2011 the population consisted of 31 family groups and 30 scent-marking pairs [48]. Brown bear (*Ursus arctos*) also occur at variable densities within our study area and prey mainly on natal moose calves during early summer [49].

Moose aerial counts

We used aerial surveys made by helicopter to obtain data on the number and spatial distribution of moose in the landscape. These data were further used for estimating the size of moose groups, the relative density and adult sex ratio of moose. Aerial surveys were conducted in 2006, 2009 and 2011 by Svensk Naturförvaltning (www.naturforvaltning.se) in order to estimate moose densities that could be used for estimating appropriate hunting quotas. The moose is counted by transect survey and two different methods were used to subsample surveyed area; "square sampling" (782 observations) and "distance sampling" (553 observations), including a total of 1335 observations where each observation included one or several moose. For both survey methods, line transects were used and the sampling method used unlikely influence the observed grouping behaviour by moose. Moose surveys were conducted from mid-December to mid-February during short time periods with snow cover in order to increase detectability (for detailed description of aerial counts and distance sampling; [50, 51]. For each moose observation, the location was recorded by GPS (±10 m accuracy) and all moose classified according to age class (calves or adults) and sex.

Presence of wolves and territory range

Presence/absence data of wolves for each area of aerial sampling was obtained by the ongoing annual monitoring of the Scandinavian wolf population throughout the study period (Liberg et al. 2012). Snow tracking in combination with DNA-analysis of scats and oestrus bleedings are used to monitor wolf family groups and scent-marking pairs which also gives a minimum size of territorial wolves [52]. In addition to snow tracking, several wolves were radio-collared (VHF and GPS) during the study period [53]. Seventeen wolf territories were included in the study and were defined by applying a 100% minimum convex polygon (MCP; [54]) on either locations from collared wolves or information collected from repeated snow tracking events in an area. Only scent-marking pairs and family groups were included to ensure that moose had regularly experienced wolf presence. Because the establishment of all wolf territories in Scandinavia has been annually registered since the founding of the population in 1983, the duration, i.e., the number of years, that moose had been exposed to territorial wolves could be estimated for each wolf

territory and ranged from 1 to 13 years. Because non-collared and some of the collared wolves were monitored during a limited portion of the year, we buffered the estimated boundary of these wolf territories using a conservative approach to ensure that moose observations actually were outside any wolf territory (see below for classification of moose observations). We generated buffer zones by using two different methods because accuracy differs between MCPs estimated by snow tracking versus radio-tracking data. For MCPs ($n = 10$), based on snow-tracking data, we added a buffer zone so that the buffered territory equalled the maximum potential wolf territory observed in Scandinavia [1700 km2 based on GPS-data; ,53]. For MCPs based on radio-tracking data ($n_{total} = 7$; 2VHF and 5GPS), for which we have less than 12 months of data ($n = 3$; data range 1-7 months), we generated buffer zones so that the total area corresponded to one year of radio-tracking (see [53] for the proportion of annual home range covered in relation to studied time period).

Classification of moose observations and estimation of snow depth

Moose observations were classified as inside wolf territories (presence of wolves), outside wolf territories (absence of wolves), or in a buffer zone (uncertain wolf presence). Observations inside the buffer zone were excluded from the analysis as were any moose observation with missing data on sex and age. All adult moose observed within 100 m from each other were considered as a group, i.e., one group can consist of one or several moose [55, 56]. Calves were not included in the group size estimate because of their propensity to move with their mother (not independent). Moose observations were classified in 5 categories depending on the composition of the group: 1) females + calves; 2) mixed group + calves; 3) mixed group no calves; 4) females; 5) males. Adult sex-ratio of moose was estimated both inside and outside wolf territories as the proportion of females in the adult (≥ 1 year) population. The average snow depth was measured to the nearest centimetre (range 0-75) at randomly selected sites where the helicopter could land during the aerial count (i.e., open areas in the terrain) using a metre stick.

Statistical analysis

We used winter aerial survey data of moose spatial distribution and group size to investigate if moose within wolf territories employed different grouping strategies as compared to moose in areas not yet re-colonized by wolves. We used a generalized linear model with a Poisson distribution to model group size as a function of the variables of interest [57]. We tested the spatial auto-correlation of the response variable with a permutation

Table 1 Total number of moose groups observed, total number of moose and group size range for each group category of moose in Sweden from aerial counts 2006, 2009, and 2011 inside (In) and outside (Out) wolf territories

Category	Females + calves		Mixed group no calves		Mixed group + calves		Females		Males	
	In	Out	In	Out	In	Out	In	Out	In	Out
Groups (n)	148	425	64	111	27	87	75	173	69	156
Adult moose (n)	170	471	228	378	80	250	109	239	145	235
Group size range	1-5	1-5	2-9	2-8	2-7	2-9	1-4	1-4	1-8	1-5

Moran's test [58]. Two observations (group size) were considered as spatially auto-correlated if they were within 16 km which should be considered as conservative as it equals a distance ten times the radius of winter home range of moose [59]. The test showed a significant spatial autocorrelation in adult moose group size (Moran's I = 0.04, p-value = 0.001). Therefore to take into account and correct for spatial autocorrelation, we computed the local Moran's index [60] for each observation using the same neighbourhood weights matrix and included this as an explanatory variable in all the models. We defined a set of models including our five explanatory variables (i.e., either wolf absence/presence or time since wolf territory establishment, moose density, snow depth, adult sex ratio, spatial autocorrelation) and identified the most parsimonious model based on Akaike Information Criteria (AIC) and AIC weight (ωi; [61]). Then model fit was assessed using the pseudo R^2 [57].

We conducted two separate analyses using the same model set. First, the response variable was the group size of adult moose (all categories together). Second, we considered each moose group category (listed above) separately and evaluated the explanatory variables. When considering each category separately, moose group size was no longer spatially auto-correlated, therefore we did not include this variable in this part of the analysis. Finally, we used multi-model inference from the top-ranked models (ΔAIC <2) to estimate coefficients and 85% confidence intervals. Because over dispersion parameters ($\hat{c}$) ranged 0.17 – 0.88 we did not have to correct for this [57]. Adding the variable year of survey as a random effect did not increase model fit (i.e., variation in AIC between the model with random effect and the model without random effect <2) and was therefore excluded from further analysis.

Results

Of the total 1335 moose observations, 383 were inside wolf territories and 952 were outside (Table 1). Moose density inside and outside wolf territories averaged 1.08 (±3.43 S.D.) moose/km^2 and 1.00 (±3.89) moose/km^2, respectively. The proportion of females among adult moose ranged 0.58 – 0.64 inside and 0.62 – 0.69 outside wolf territories (over the three years included in the study).

The median group size inside and outside wolf territories were both equal to 1. Overall, smaller groups were more common than larger groups and 82% of the observations consisted of groups with less than three adult moose (Fig. 2). Moreover, groups composed of a mix of females and males were the most gregarious category with 70% of the groups including three or more adult moose (Fig. 3). Females with calves were the most common group category both inside and outside wolf territories (Table 1) but also the least gregarious category (Fig. 3). This category had a minimum group size of 1 adult and a maximum group size of 5 adults (Table 1) but more than 90% of the observations contained groups of only 1 adult female (Fig. 3c) and was not dependent on wolf presence. In both unisex and mixed groups, the presence of calves was linked to smaller groups (Fig. 3).

We evaluated 12 *a priori* models in order to explain moose grouping behaviour (Additional file 1: Table S1). The best model from our candidate models list included moose density, snow depth, and adult sex-ratio (Table 2). This model explained 13% of the variation in adult moose group size (pseudo R^2 = 0.13). Wolf presence was not included in the best model. Including a variable for the time since wolf territory establishment did not increase model fit (i.e., difference in AIC between the model with wolf presence and the model with number

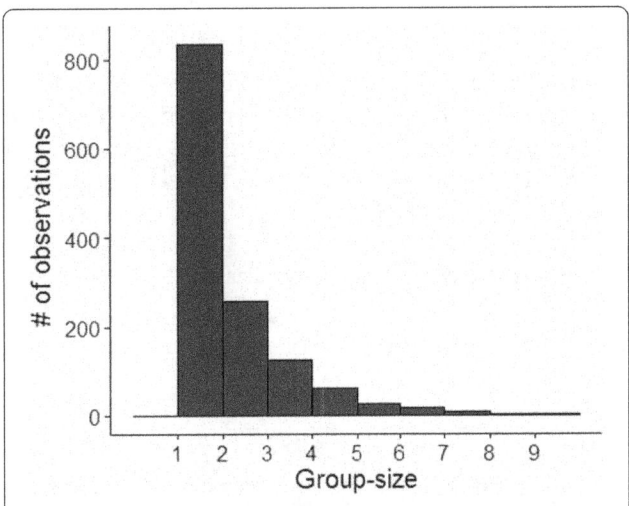

Fig. 2 Distribution of adult moose group size in Sweden from aerial survey data collected in 2006, 2009, and 2011

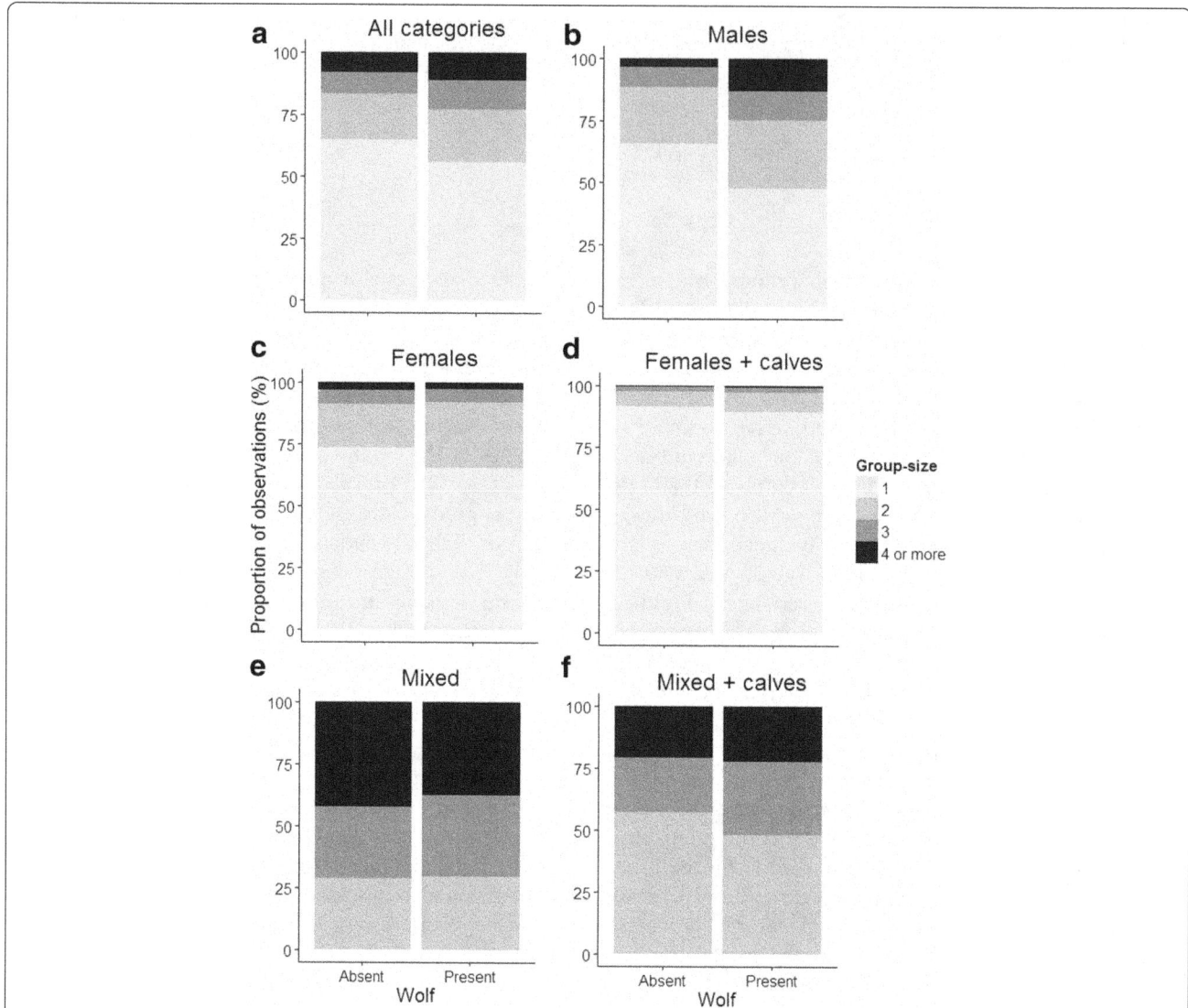

Fig. 3 Proportion of group size observations of moose inside (wolf present) and outside (wolf absent) wolf territories by moose group category (**a**-all categories, **b**-males, **c**-females, **d**-females+calves, **e**-mixed, **f**-mixed with calves) from aerial survey data collected in Sweden in 2006, 2009, and 2011

of years since wolf territory establishment <2). Moose density and snow depth both correlated positively to adult moose group size. In contrast, the proportion of females (sex-ratio) was negatively correlated to adult moose group size (Table 3).

We then used the same set of 12 models to evaluate the performance of explanatory variables for each moose

group category separately. For four of the five different categories (females + calves, mixed group + calves, mixed group no calves, and females), the NULL model (i.e., intercept only) was the most parsimonious to explain variation in moose group size. In contrast, the best model to explain variation in male group size included both moose density and the presence of wolves and

Table 2 Model selection to predict adult moose group size in Sweden applied to 1335 observations of moose groups from survey data collected in 2006, 2009, and 2011

Variable	-logLik	AIC	N parameters	Δ_i	ω_i
Moose density + Snow depth + Sex-ratio + *I*	1954.00	3917.99	5	0	0.65
Wolf presence + Moose density + Snow depth + Sex-ratio + *I*	1953.77	3919.55	6	1.60	0.30

Models are shown in order of decreasing rank with model log-likelihood (logLik), number of model parameters (N parameters), Akaike's information criterion (AIC), AIC differences (Δ_i) and AIC weights (ω_i)
I = spatial autocorrelation

Table 3 Coefficients estimates (β) and 85% confidence intervals (CI) of moose density, snow depth, sex-ratio and spatial autocorrelation to predict adult moose group size (n = 1335) in Sweden from aerial counts 2006, 2009, and 2011

Variable	B	85% CI Lower - Upper
Moose density	0.017	0.0082 - 0.025
Snow depth	0.0037	0.00011 - 0.0063
Sex-ratio	-1.64	-2.52 - 0.77
Spatial autocorrelation (I)	0.44	0.38 - 0.51
Wolf presence (inside/outside territories)	-0.013	-0.13 - 0.049

these variables together explained 13% of the variation in male group size (Table 4). In these models, wolf presence was positively correlated to male group size. However, two other models including a combination of wolf presence, moose density, snow depth and sex-ratio also received high empirical support (Table 4). Moose density and snow depth were both positively correlated to group size of males whereas sex-ratio had the opposite effect (Table 5). The most important variable to predict male group size was moose density, followed by wolf presence, snow depth and sex-ratio (Table 6).

Discussion

In this study, moose did not form large groups and mostly stayed solitary or in small groups consisting of two to three animals. The results show, that wolf re-establishment in Sweden has not resulted in an overall change in social grouping behaviour among moose. Nor did we find any support that females with calves, i.e., the category most vulnerable to predation [38, 39], change their grouping behaviour. In fact, other variables such as moose density, snow depth and adult sex ratio of the group seemed to be more important for the overall pattern of grouping behaviour. However, when analyzing moose categories separately, males were found to form larger groups inside wolf territories.

In the light of other studies of ungulates presenting evidence of changed grouping behaviour as a response to predation risk, especially for females with calves [3, 24], our results may at first glance seem unexpected. Young of the year are more susceptible to direct predation

[24, 38, 39] and female with calves should thereby be expected to respond more strongly to an increase in predation risk [62–64]. Although increased grouping behaviour in general is viewed as an effective anti-predator response, the plasticity of this behaviour is likely to be both species- and context dependent. For instance, elk (*Cervus elaphus*) formed smaller herds when wolves were present, which is thought to reduce the likelihood of being detected by wolves [24]. Another study suggests that elk may adopt different strategies to minimize predation risk whereby they either choose to live in small herds that are rarely encountered by wolves, or they choose to live in large herds that reduce their predation risk by dilution and many eyes scanning [17]. Solitary or small groups are strategies that have been suggested to be best suited for concentrate selectors and animals in forested terrain and close to cover [18, 65] and specifically for animals less likely to benefit of group living because of the high probability of being attacked by a selective predator once encountered by a predator [18]. However, as the grouping pattern of females with calves was not at all related to the presence of wolves in our study this observation provides support for that the grouping behaviour of female moose in Scandinavia is more a result of other factors affecting the benefit of grouping, e.g., foraging and competition for access to food [65]. Further, as food competition is likely to affect low ranked individuals more [29, 30] it is also possible that these costs may differ among categories of moose, i.e., a higher cost for females with calves than for males, which may explain why females with calves did not exhibit a change in grouping behaviour while males did. Similarly, Creel [24] showed that elk male groups increased in size when wolves were present contrary to mixed herds that decreased in size. Given our data, we cannot address the underlying mechanism for this divergent pattern among moose categories in our study. However, Creel [24] suggested that poor condition of males post rut forced elk males to spend more time foraging and less time being vigilant and thereby they should benefit more than other animals from forming larger groups. Early detection of approaching wolves has been shown to increase survival of moose targeted by wolves [27, 66]. Moreover, a large number of males in the Swedish moose population are young (e.g., yearlings) mainly because of high turnover rate in the population due to intensive

Table 4 Top models to predict male group size in Sweden applied to 225 observations of male moose groups from aerial survey data collected in 2006, 2009, and 2011

Model	-logLik	AIC	N parameters	Δᵢ	ωᵢ
Wolf presence + Moose density	321.04	648.08	3	0	0.41
Moose density + Snow depth + Sex-ratio	320.46	648.91	4	0.83	0.27
Wolf presence + Moose density + Snow depth + Sex-ratio	319.75	649.51	5	1.40	0.20

Models are shown in order of decreasing rank with model log-likelihood (logLik), number of model parameters (N parameters), Akaike's information criterion (AIC), AIC differences (Δᵢ) and AIC weights (ωᵢ)

Table 5 Coefficients estimates (β) and 85% confidence intervals (CI) of moose density, snow depth, sex-ratio and wolf presence to predict male group size in Sweden from aerial survey data collected in 2006, 2009, and 2011

Variable	B	85% CI	
		Lower	Upper
Moose density	0.039	0.018	0.060
Snow depth	0.0028	-0.0014	0.012
Sex-ratio	-1.92	-6.41	-0.98
Wolf presence	0.20	0.085	0.50

harvest strategies (and especially so for males) [42, 67] and therefore these groups may mainly consists of young males similar to "bachelor herds" in other species of ungulates [9]. Yearlings are the second most common age class among moose killed by wolves in Scandinavia but still less vulnerable than calves (Sand et al. unpubl. data).

Our results are partly in line with several recent studies that have investigated changes in moose behaviour as a response to wolf re-colonization in Scandinavia. In general, there is no or only weak support for that moose behaviour has changed with increased wolf predation risk, and if these effects exist they are small relative to the effect of population structure of prey and environmental factors [36, 43, 68–72]. Compared to the other studies investigating behaviorally mediated effects of wolf return in Scandinavia this study may provide some support for a behavioural response though it was not consistent across the entire population.

However, the response by male moose was opposite to the one predicted, i.e., male groups was larger within than outside wolf territories. Similar to Sand et al. [43] and Gervasi et al. [70], we did not find any relationship between the degree of behavioural change and time since establishment of wolf territories. However, that social population structure may be important for the behaviour of ungulates as well as other factors than predation risk such as habitat type, population density, snow conditions, and the distribution and availability of

food is well known [22, 73, 74]. For example, a positive relationship between group size and population density is documented [17, 65, 73] as is the relation with snow [74]. Increased snow cover decreases food availability by creating limited and irregular food patches but also restricts movements of moose [74]. Thus, moose individuals tend to group more in areas that provide high energy intake and where the costs of mobility may be reduced by taking advantage of the tracks made by conspecifics [4, 23, 74]. Moreover, habitat type and composition can affect grouping behaviour and for example in open habitats, ungulates tend to form larger groups [18, 31, 32]. However, moose do not change habitat selection in relation to predation risk by wolves in Sweden [69]. Also in the current study, moose used the habitats similarly inside and outside wolf territories (forested areas (inside 95% of observed groups, outside 96%), wetlands and lakes (inside 4%, outside 3%) and other (urban and agriculture; inside 1%, outside 1%)). Furthermore, the habitat composition inside and outside wolf territories were similar (dominated by forested areas (inside 74% of total area, outside 80%) wetlands and lakes (inside 17%, outside 18%) and other (urban and agriculture; inside 9%, outside 2%)).

Conclusions

The results did not give support for that wolf recolonization in Scandinavia has resulted in an overall change in moose grouping behaviour. Nor could we confirm our two predictions that females with calves should be the category most prone to change their grouping behaviour and that moose grouping behaviour was related to the time since wolf territory establishment. Rather, the results showed a sex specific difference in social grouping in relation to wolf presence where males group more in areas with wolves. However, even for this moose category population and environmentally related variables was also important for the pattern of grouping. If indeed wolf-induced effects on behavior do exist in the moose population, they may be difficult to discern because the effects from population and environmental factors may be much stronger and thus conceal any subtle change in anti-predator behaviour. These results suggest that caution should be taken as to generalize about the effects of returning predators on the grouping behaviour of their prey.

Table 6 Relative importance of moose density, wolf presence, snow depth and sex-ratio to predict male group size in Sweden from aerial survey data collected in 2006, 2009, and 2011

Variable	Importance	N
Moose density	0.89	5
Wolf presence	0.70	5
Snow depth	0.52	5
Sex ratio	0.48	4

N indicates the number of models in which the variable was used

Acknowledgements
We thank P. Ahlqvist, J.M. Arnemo, U. Grinde, P. Grängstedt, K. Sköld, and T.H. Strømseth for capture and handling of wolves. The aerial counts of moose were provided by the company Svensk Naturförvaltning.

Funding
This study was financed by the Swedish Environmental Protection Agency, FORMAS, Norwegian Environment Agency, Swedish Association for Hunting and Wildlife Management, World Wildlife Fund for Nature (Sweden), Olle and Signhild Engkvists Foundations, Carl Tryggers Foundation, Oscar and Lili Lamms Foundation, and Swedish Carnivore Association.

Authors' contributions
JM initiated the study and acquired moose data. HS and CW handled and contributed with the wolf data. JM, M-CP and KN made substantial contributions to conception and design of the study. M-CP and KN analyzed data. JM, M-CP, KN, CW and HS were involved in interpretation of data and writing. All authors read and approved the final manuscript.

Competing interests
The authors declare that they have no competing interests.

Author details
[1]Department of Ecology, Grimsö Wildlife Research Station, Swedish University of Agricultural Sciences, SE-730 91 Riddarhyttan, Sweden. [2]Département de Biologie, 1045, av de la Médecine, Université Laval, Québec G1V 0A6, Canada.

References

1. Lima SL, Dill LM. Behavioral decisions made under the risk of predation: a review and prospectus. Can. J. Zool. NRC Research Press Ottawa, Canada; 1990;68:619–40
2. Miquelle DG, Peek JM, Van Ballenberghe V. Sexual segregation in Alaskan moose. Wildl Monogr. 1992;122:3–57.
3. Creel S, Winnie J, Maxwell B, Hamlin K, Creel M. Elk alter habitat selection as an antipredator response to wolves. Ecology. 2005;86:3387–97.
4. Gude JA, Garrott RA, Borkowski JJ, King F. Prey Risk Allocation In A Grazing Ecosystem. Ecol Appl. 2008;16:285–98.
5. Ripple WJ, Beschta RL. Wolves and the ecology of fear: can predation risk structure ecosystems? Bioscience. Oxford University Press; 2004;54:755.
6. Berger J. Group size, foraging, and antipredator ploys: An analysis of bighorn sheep decisions. Behav. Ecol. Sociobiol. SPRINGER VERLAG, 175 FIFTH AVE, NEW YORK, NY 10010; 1978;4:91–9.
7. Li C, Jiang Z, Li L, Li Z, Fang H, Li C, et al. Effects of reproductive status, social rank, sex and group size on vigilance patterns in Przewalski's gazelle. PLoS One. Public Library of Science; 2012;7:e32607.
8. Buuveibaatar B, Fuller TK, Fine AE, Chimeddorj B, Young JK, Berger J. Changes in grouping patterns of saiga antelope in relation to intrinsic and environmental factors in Mongolia. J Zool. 2013;291:51–8.
9. Jarman PJ. The Social Organisation of Antelope in Relation To Their Ecology. Behaviour. E J Brill, PO Box 9000, 2300 PA Leiden, Netherlands; 1974;48:215–67
10. Lingle S. Anti-Predator Strategies and Grouping Patterns in White-Tailed Deer and Mule Deer. Ethology. Blackwell Science Ltd.; 2001;107:295–314.
11. Landeau L, Terborgh J. Oddity and the "confusion effect" in predation. Anim Behav. 1986;34:1372–80.
12. Nelson ME, Mech LD. Deer social organization and wolf depredation in northeastern Minnesota. Wildl Monogr. 1981;77:3–53.
13. Lipetz VE, Bekoff M. Group Size and Vigilance in Pronghorns. Z Tierpsychol. 2010;58:203–16.
14. Lima SL. Back to the basics of anti-predatory vigilance: the group-size effect. Anim Behav. 1995;49:11–20.
15. Shi J, Li D, Xiao W. Influences of sex, group size, and spatial position on vigilance behavior of Przewalski's gazelles. Acta Theriol (Warsz). 2010;56:73–9.
16. Vine I. Risk of visual detection and pursuit by a predator and the selective advantage of flocking behaviour. J Theor Biol. 1971;30:405–22.
17. Hebblewhite M, Pletscher DH. Effects of elk group size on predation by wolves. Can. J. Zool. NRC Research Press Ottawa, Canada; 2002;80:800–9
18. Molvar EM, Bowyer RT. Costs and Benefits of Group Living in a Recently Social Ungulate: The Alaskan Moose. J. Mammal. The Oxford University Press; 1994;75:621–30
19. Janson CH, Goldsmith ML. Predicting group size in primates: foraging costs and predation risks. Behav Ecol. 1995;6:326–36.
20. Creel S, Winnie JA. Responses of elk herd size to fine-scale spatial and temporal variation in the risk of predation by wolves. Anim Behav. 2005;69:1181–9.
21. Hamilton WD. Geometry for the selfish herd. J Theor Biol. 1971;31:295–311.
22. Nelson ME, Mech LD. Relationship between snow depth and gray wolf predation on white- tailed deer. J Wildl Manage. 1986;50:471–4.
23. Heard DC. The effect of wolf predation and snow cover on musk-ox group size. Am. Nat. Univ Chicago Press, 5720 S Woodlawn Ave, Chicago, IL 60637; 1992;139:190.
24. Winnie J, Creel S. Sex-specific behavioural responses of elk to spatial and temporal variation in the threat of wolf predation. Anim Behav. 2007;73:215–25.
25. Huggard DJ. Effect of snow depth on predation and scavenging by gray wolves. J Wildl Manage. 1993;57(2)382–388.
26. Carbyn LN. Wolf Predation on Elk in Riding Mountain National Park, Manitoba. J. Wildl. Manage. Wildlife Soc, 5410 Grosvenor Lane, Bethesda, MD 20814-2197; 1983;47:963.
27. Peterson RO. Wolf ecology and prey relationships on Isle Royale. U. S. Natl. Park Serv. Sci. Monogr. Ser. Washington; 1977;11:1–210.
28. Jędrzejewski W, Schmidt K, Theuerkauf J, Jędrzejewska B, Selva N, Zub K, et al. Kill rates and predation by wolves on ungulate populations in Bialowieza Primeval Forest (Poland). Ecology. 2002;83:1341–56.
29. Barrette C, Vandal D. Social rank, dominance, antler size, and access to food in snow- bound wild woodland caribou. Behaviour. 1986;97:118–45.
30. Thouless CR. Feeding competition between grazing red deer hinds. Anim Behav. 1990;40:105–11.
31. Hirth DH. Social behavior of white-tailed deer in relation to habitat. Wildl Monogr. 1977;53:3–55.
32. Gerard JF, Bideau E, Maublanc ML, Loisel P, Marchal C. Herd size in large herbivores: Encoded in the individual or emergent? Biol Bull. 2002;202:275–82.
33. Baskin LM. Behaviour of moose in the USSR. Swedish Wildlife Research. 1989;(suppl 1):377–387.
34. Wabakken P, Sand H, Liberg O, Bjärvall A. The recovery, distribution, and population dynamics of wolves on the Scandinavian peninsula, 1978-1998. Can J Zool. 2001;79:710–25.
35. Hofmann RR. Evolutionary steps of ecophysiological adaptation and diversification of ruminants: a comparative view of their digestive system. Oecologia. Springer-Verlag; 1989;78:443–57.
36. Wikenros C, Sand H, Wabakken P, Liberg O, Pedersen HC. Wolf predation on moose and roe deer: chase distances and outcome of encounters. Acta Theriol (Warsz). 2009;54:207–18.
37. Lent PC. Ovibos moschatus. Mamm Species. 1988;1–9.
38. Sand H, Zimmermann B, Wabakken P, Andrén H, Pedersen HC. Using GPS technology and GIS cluster analyses to estimate kill rates in wolf—ungulate ecosystems. Wildl Soc Bull. 2005;33:914–25.
39. Sand H, Wabakken P, Zimmermann B, Johansson O, Pedersen HC, Liberg O. Summer kill rates and predation pattern in a wolf-moose system: can we rely on winter estimates? Oecologia. 2008;156:53–64.
40. Vedin H. Lufttemperatur. In: Raab B, Vedin H, editors. Klimat, sjöar och vattendrag. Gävle: Kartförlaget; 1995. p. 44–57.
41. Dahlström B. Snötäcke. In: Raab, B, H V, editors. Klimat, sjöar och vattendrag. Gävle: Kartförlaget; 1995. p. 91–7
42. Lavsund S, Nygrén T, Solberg EJ. Status of moose populations and challenges to moose management in Fennoscandia. Alces. 2003;39:109–30.
43. Sand H, Wikenros C, Wabakken P, Liberg O. Cross-continental differences in patterns of predation: will naive moose in Scandinavia ever learn? Proc R Soc B. 2006;273:1421–7.
44. Sand H, Vucetich JA, Zimmermann B, Wabakken P, Wikenros C, Pedersen HC, et al. Assessing the influence of prey-predator ratio, prey age structure and packs size on wolf kill rates. Oikos. 2012;121:1454–63.
45. Liberg O, Bergström R, Kindberg J, von Essen H. Ungulates and their management in Sweden. In: Apollonio M, Andersen R, Putman R, editors. Eur. ungulates their Manag. 21st century. Cambridge: Cambridge University Press; 2010. p. 37–70.
46. Svensson L, Wabakken P, Kojola I, Maartmann E, Strømseth TH, Åkesson M, et al. Elverum: Varg i Skandinavien och Finland - Slutrapport från inventering av varg vintern 2013-2014. 2014.
47. Chapron G, Kaczensky P, Linnell JDC, von Arx M, Huber D, Andrén H, et al. Recovery of large carnivores in Europe's modern human-dominated landscapes. Science (80-.). American Association for the Advancement of Science; 2014;346:1517–9.

48. Wabakken P, Aronsson Å, Strømseth TH, Sand H, Maartmann E, Svensson L, et al. Ulv i Skandinavia - statusrapport for vinteren 2010-2011. 2011. Elverum.

49. Kindberg J. Monitoring and management of the Swedish brown bear (Ursus arctos) population. Elverum: Swedish University of Agricultural Sciences; 2010.

50. Rönnegård L, Sand H, Andrén H, Månsson J, Pehrson Å. Evaluation of four methods used to estimate population density of moose Alces alces. Wildlife Biol. 2008;14:358–71.

51. Thomas L, Buckland ST, Rexstad EA, Laake JL, Strindberg S, Hedley SL, et al. Distance software: design and analysis of distance sampling surveys for estimating population size. J Appl Ecol. 2010;47:5–14.

52. Liberg O, Aronson Å, Sand H, Wabakken P, Maartmann E, Svensson L, et al. Monitoring of wolves in Scandinavia. Hystrix, Ital J Mammal. 2011;23:29–34.

53. Mattisson J, Sand H, Wabakken P, Gervasi V, Liberg O, Linnell JDC, et al. Home range size variation in a recovering wolf population: evaluating the effect of environmental, demographic, and social factors. Oecologia. 2013;173:813–25.

54. Mohr CO. Table of equivalent populations of North American small mammals. Am Midl Nat. 1947;37:223–49.

55. Sweanor PY, Sandegren F. Winter behavior of moose in central Sweden. Can. J. Zool. NRC Research Press Ottawa, Canada; 1986;64:163–7.

56. Proffitt KM, Grigg JL, Hamlin KL, Garrott RA. Contrasting Effects of Wolves and Human Hunters on Elk Behavioral Responses to Predation Risk. J Wildl Manage. 2009;73:345–56.

57. Zuur AF, Ieno EN, Walker N, Saveliev AA, Smith GM. Mixed effects models and extensions in ecology with R. New York: Springer; 2009.

58. Cliff AD, Ord JK. Spatial Processes: Models and Applications. London: Pion; 1981.

59. Cederlund G, Sand H. Home-Range Size in Relation to Age and Sex in Moose. J. Mammal. The Oxford University Press; 1994;75:1005–12.

60. Anselin L. Local indicators of spatial association - LISA. Geogr. Anal. Ohio State Univ Press, 1050 Carmack Rd, Columbus, OH 43210; 1995;27:93–115.

61. Burnham KP, Anderson DR. Model Selection and Inference: A Practical Information-theoretic Approach. New York: Springer; 1998.

62. Edwards J. Diet shifts in moose due to predator avoidance. Oecologia. 1983;60:185–9.

63. Dussault C, Ouellet J-P, Courtois R, Huot J, Breton L, Jolicoeur H. Linking moose habitat selection to limiting factors. Ecography (Cop). 2005;28:619–28.

64. Bjørneraas K, Herfindal I, Solberg EJ, Sæther B-E, van Moorter B, Rolandsen CM. Habitat quality influences population distribution, individual space use and functional responses in habitat selection by a large herbivore. Oecologia. 2012;168:231–43.

65. Peek JM, LeResche RE, Stevens DR. Dynamics of Moose Aggregations in Alaska, Minnesota, and Montana. J. Mammal. The Oxford University Press; 1974;55:126–37

66. Mech LD. Species, The wolf: the ecology and behavior of an endangered. New Jersey: Natural History Press; 1970.

67. Jonzén N, Sand H, Wabakken P, Swenson JE, Kindberg J, Liberg O, et al. Sharing the bounty—Adjusting harvest to predator return in the Scandinavian human–wolf–bear–moose system. Ecol Modell. 2013;265:140–8.

68. Eriksen A, Wabakken P, Zimmermann B, Andreassen HP, Arnemo JM, Gundersen H, et al. Activity patterns of predator and prey: a simultaneous study of GPS-collared wolves and moose. Anim Behav. 2011;81:423–31.

69. Eriksen A, Wabakken P, Zimmermann B, Andreassen HP, Arnemo JM, Gundersen H, et al. Encounter frequencies between GPS-collared wolves (Canis lupus) and moose (Alces alces) in a Scandinavian wolf territory. Ecol Res. 2008;24:547–57.

70. Gervasi V, Sand H, Zimmermann B, Mattisson J, Wabakken P, Linnell JDC. Decomposing risk: Landscape structure and wolf behavior generate different predation patterns in two sympatric ungulates. Ecol Appl Ecological Society of America. 2013;23:1722–34.

71. Nicholson KL, Milleret C, Månsson J, Sand H. Testing the risk of predation hypothesis: the influence of recolonizing wolves on habitat use by moose. Oecologia. 2014;176:69–80.

72. Wikenros C, Balogh G, Sand H, Nicholson KL, Månsson J. Mobility of moose – comparing the effects of wolf predation risk, reproductive status and seasonality. Ecol. Evol. 2016;6:8870–8880.

73. Borkowski J. Influence of the density of a sika deer population on activity, habitat use, and group size. Can. J. Zool. NRC Research Press Ottawa, Canada; 2011;78:1369–74

74. Renecker LA, Schwartz CC. Food habits and feeding behavior. In: Franzmann AW, Schwartz CC, editors. Ecol. Manag. North Am. moose. Washington: Smithsonian institution press; 1998. p. 403–40.

PERMISSIONS

LIST OF CONTRIBUTORS

Elodie F. Briefer and Edna Hillmann
Institute of Agricultural Sciences, ETH Zürich, Universitätstrasse 2, 8092 Zürich, Switzerland

Roi Mandel
Institute of Agricultural Sciences, ETH Zürich, Universitätstrasse 2, 8092 Zürich, Switzerland
Koret School of Veterinary Medicine, Robert H. Smith Faculty of Agriculture, Food and Environment, the Hebrew University, Rehovot 76100, Israel

Anne-Laure Maigrot
Institute of Agricultural Sciences, ETH Zürich, Universitätstrasse 2, 8092 Zürich, Switzerland
Division of Animal Welfare, Veterinary Public Health Institute, Vetsuisse Faculty, University of Bern, Länggassstrasse 120, 3012 Bern, Switzerland

Sabrina Briefer Freymond and Iris Bachmann
Agroscope, Swiss National Stud Farm, Les Longs Prés, 1580 Avenches, Switzerland

Sebastian Büsse and Stanislav N. Gorb
Department of Functional Morphology and Biomechanics, Institute of Zoology, Christian-Albrechts-Universität zu Kiel, Am Botanischen Garten 9, 24118 Kiel, Germany

Thomas Hörnschemeyer
Senckenberg Gesellschaft für Naturforschung, Senckenberganlage 25, 60325 Frankfurt, Germany

Chung-Chi Lin
Department of Biology, National Changhua University of Education, No. 1, Jin-De Rd., Changhua 50007, Taiwan

Ching-Chen Lee
Department of Biology, National Changhua University of Education, No. 1, Jin-De Rd., Changhua 50007, Taiwan
Master Program for Plant Medicine, National Taiwan University, No.1, Sec. 4, Roosevelt Rd., Taipei, Taiwan

Hirotaka Nakao and Fuminori Ito
Faculty of Agriculture, Kagawa University, Ikenobe, Miki 761–0795, Japan

Hung-Wei Hsu and Gwo-Li Lin
Department of Entomology, National Taiwan University, No.1, Sec. 4, Roosevelt Rd., Taipei, Taiwan

Chin-Cheng (Scotty) Yang
Research Institute for Sustainable Humanosphere, Kyoto University, Gokasho, Uji, Kyoto 611-0011, Japan

Shu-Ping Tseng
Department of Entomology, National Taiwan University, No.1, Sec. 4, Roosevelt Rd., Taipei, Taiwan
Research Institute for Sustainable Humanosphere, Kyoto University, Gokasho, Uji, Kyoto 611-0011, Japan

Jia-Wei Tay
Department of Entomology, University of California, Riverside, CA 92521, USA

Johan Billen
K.U. Leuven, Zoological Institute, Naamsestraat 59, box 2466, B-3000 Leuven, Belgium

Chow-Yang Lee
Urban Entomology Laboratory, Vector Control Research Unit, School of Biological Sciences, Universiti Sains Malaysia, 11800 Penang, Malaysia

Sandra A. Heldstab, Carel P. van Schaik and Karin Isler
Department of Anthropology, University of Zurich, Winterthurerstrasse 190, 8057 Zurich, Switzerland

Giacomo Tavecchia
Population Ecology Group, IMEDEA (CSIC-UIB), c. Miquel Marqués 21, 07190 Esporles, Spain

Miguel-Angel Miranda, David Borrás, Carlos Barceló and Claudia Paredes-Esquivel
Laboratory of Zoology, Department of Biology, University of the Balearic Island, c. Valldemossa s/n, Palma de Mallorca, Spain

Mikel Bengoa
Consultoria Moscard Tigre, c. Gremi Passamaners 24, Local 15, 07009 Palma de Mallorca, Spain

Carl Schwarz
Department of Statistics and Acutarian Science, Simon Fraser University, Burnaby, BC, Canada

Aidan O'Hanlon, Kristina Feeney and Michael J. Gormally
Applied Ecology Unit, School of Natural Sciences, National University of Ireland Galway, Galway, Ireland

Peter Dockery
Centre for Microscopy and Imaging, National University of Ireland Galway, Galway, Ireland

Adrian Brückner, Nico Blüthgen and Michael Heethoff
Ecological Networks, Technische Universität Darmstadt, Schnittspahnstr. 3, 64287 Darmstadt, Germany

Félix B. Rosumek
Ecological Networks, Technische Universität Darmstadt, Schnittspahnstr. 3, 64287 Darmstadt, Germany
Department of Ecology and Zoology, Federal University of Santa Catarina, Campus Trindade, Florianópolis 88040-900, Brazil

Florian Menzel
Institute of Organismic and Molecular Evolution, Johannes Gutenberg-Universität Mainz, Johannes-von-Müller-Weg 6, 55128 Mainz, Germany

Molly A. Albecker and Michael W. McCoy
Department of Biology, Howell Science Complex, East Carolina University, Greenville, NC, USA

Miriam Linnenbrink and Sophie von Merten
Max- Planck Institute for Evolutionary Biology, Plön, Germany

Ute Radespiel and Elke Zimmermann
Institute of Zoology, University of Veterinary Medicine Hannover, Buenteweg 17, 30559 Hannover, Germany

Christina Strube
Institute for Parasitology, Centre for Infection Medicine, University of Veterinary Medicine Hannover, Buenteweg 17, 30559 Hannover, Germany

May Hokan
Institute of Zoology, University of Veterinary Medicine Hannover, Buenteweg 17, 30559 Hannover, Germany

Institute for Parasitology, Centre for Infection Medicine, University of Veterinary Medicine Hannover, Buenteweg 17, 30559 Hannover, Germany

Charlotte Katharina Maria Schielke, Yoshiyuki Henning and Sabine Begall
Faculty of Biology, University of Duisburg-Essen, Essen, Germany

Hynek Burda
Faculty of Biology, University of Duisburg-Essen, Essen, Germany
Faculty of Forestry and Wood Sciences, Czech University of Life Sciences, Praha, Czech Republic

Jan Okrouhlík
Faculty of Science, University of South Bohemia, České Budějovice, Czech Republic

Tom Weihmann
Department of Animal Physiology, Institute of Zoology, University of Cologne, Zülpicher Strasse 47b, 50674 Cologne, Germany

Pierre-Guillaume Brun
Ecole Normale Supérieure de Lyon Département de Biologie, Lyon, France

Emily Pycroft
Department of Zoology, University of Cambridge, Downing Street, Cambridge CB2 3EJ, UK

Pawel Fedurek
Department of Primatology, Max Planck Institute for Evolutionary Anthropology, Leipzig, Germany

Klaus Zuberbühler
Institute of Biology, University of Neuchâtel, Neuchâtel, Switzerland
School of Psychology and Neuroscience, University of St Andrews, St Andrews, Scotland, UK

Stuart Semple
Centre for Research in Evolutionary, Social and Interdisciplinary Anthropology, University of Roehampton, London, UK

Clare C. Rittschof
Department of Entomology, University of Kentucky, S-225 Ag. Science Center North, Lexington, KY 40546, USA

Jin-Hua Zhang, Jian-Xu Zhang and Yao-Hua Zhang
State Key Laboratory of Integrated Management of Pest Insects and Rodents in Agriculture, Institute of Zoology, Chinese Academy of Sciences, No.1-5 Beichen West Road, Chaoyang District, Beijing 100101, China

Hong-Ling Guo and Hua-Jing Teng
State Key Laboratory of Integrated Management of Pest Insects and Rodents in Agriculture, Institute of Zoology, Chinese Academy of Sciences, No.1-5 Beichen West Road, Chaoyang District, Beijing 100101, China
College of Life Sciences, University of Chinese Academy of Sciences, Beijing 100049, China

Jessica S. Cornils, Franz Hoelzl, Birgit Rotter, Claudia Bieber and Thomas Ruf
Department of Integrative Biology and Evolution, University of Veterinary Medicine, Savoyenstraße 1, 1160 Vienna, Austria

Daniel Rolke and Otto Baumann
Institute of Biochemistry and Biology, Department of Animal Physiology, University of Potsdam, Karl-Liebknecht-Str. 24/25, 14476 Potsdam, Germany

Sascha Peter Klose
Institute of Biochemistry and Biology, Department of Animal Physiology, University of Potsdam, Karl-Liebknecht-Str. 24/25, 14476 Potsdam, Germany
Institute of Biology, Department of Molecular Parasitology, Humboldt University, Philippstrasse 13, 10115 Berlin, Germany

Johan Månsson, Kerry L. Nicholson, Camilla Wikenros and Håkan Sand
Department of Ecology, Grimsö Wildlife Research Station, Swedish University of Agricultural Sciences, SE-730 91 Riddarhyttan, Sweden

Marie-Caroline Prima
Département de Biologie, 1045, av de la Médecine, Université Laval, Québec G1V 0A6, Canada

Index